PRINCIPLES OF ELECTRON DEVICES

PRINCIPLES

OF ELECTRON

DEVICES

Angelo C. Gillie

Niagara County Community College
Formerly Ward School of Electronics
A Division of the University of Hartford

McGraw-Hill Book Company, Inc.

NEW YORK   TORONTO   LONDON

To Dorothea, Glenda, and Leslie

PREFACE

This book is designed to assist in the developing of technicians to the full capabilities industry requires today. It has been written for use in technical institutes, junior and community colleges, service schools, and industrial in-plant programs. Modern techniques of presentation that enable the student to advance on his own with a minimum of supervision have been used in its preparation.

The material presented here continues the logical step-by-step method employed in "Electrical Principles of Electronics." The latest techniques for the design of new circuits as well as troubleshooting analysis of electronic problems are applied to design circuitry. In this text more than thirty-five control device circuits have been designed for analysis and development—all having been validated by extensive laboratory tests and actual class application.

To make full use of the technician's mathematical background and to offer opportunity for further application and development in advanced mathematics, the design computations are analyzed mathematically by algebraic processes. In addition, each circuit design includes visual evaluation techniques and procedures through the use of the voltmeter and oscilloscope. Several of the most significant bias networks are designed in detail to provide the student with sufficient background to design his own circuits. Since the practical approach method is used throughout, the student is able to translate the theoretical circuits to working designs, thus bridging the gap between preparatory theory and final practice.

An entire chapter is devoted to basic transistor and vacuum-tube logic circuits because of their increasing importance to modern industry. Design procedures for the majority of these circuits are included. Basic transistor and vacuum-tube modulation and detection circuits are introduced, with design techniques developed for the majority of these circuits.

To facilitate the use of the text in the classroom and provide background for efficient home study, hundreds of problems have been provided throughout the text. In working out these problems, the student will be calculating the values of the components for the circuits analyzed in each chapter.

The author wishes to express his appreciation to the industrial concerns that generously offered the photographs and characteristic curves appearing in this book as well as to those industries which permitted the author to illustrate some of their circuits and to utilize their latest information on several special devices.　Special thanks are extended to two coworkers: Chester A. Gehman, Assistant Director of Training, Ward School of Electronics, for his excellent ideas and suggestions, and Edward Cushing, Laboratory Instructor, Ward School of Electronics, for the many hours spent with the author over a period of several years in the construction and evaluation of the circuit designs incorporated in the text.

Angelo C. Gillie

CONTENTS

CHAPTER 1

FUNDAMENTALS OF NONLINEAR RESISTIVE CONTROL DEVICES

Vacuum tubes and transistors are nonlinear resistive control devices. Although they are physically different, they possess striking similarities in their over-all circuit behavior.   Both devices can be analyzed in terms of their volt-ampere characteristics, which show they are nonlinear resistances.   One of the chief characteristics of the transistor and vacuum tube is that their resistance can be varied by an input current (in the case of the transistor) or input voltage (in the case of the vacuum tube).   Because of their fundamental similarities, this chapter is devoted to the study of those properties which are common to both.   The advantages and disadvantages of each are considered in the appropriate chapters.

1.1   Methods of Computing Nonlinear Resistances.   Figures 1.1*a* and *b* illustrate a portion of the volt-ampere characteristics of a nonlinear

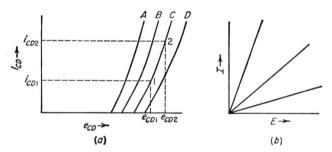

FIG. 1.1   (*a*) Partial *E-I* characteristics of a nonlinear resistive control device, (*b*) *E-I* characteristics of linear resistance.

resistive control device.   Notice the difference between the characteristics of nonlinear and linear resistance.   With a fixed input voltage (or current), the relationship of the current through the nonlinear resistive device i_{CD} is plotted as a function of the voltage across the device e_{CD}. In the case of transistors, the fixed E_I for each curve is replaced with a

1

constant I_I, since the resistance of the vacuum tube is a function of the input voltage while the input current serves the same purpose in the transistor. With the input parameter fixed, the relationship between the control device voltage and current is readily seen. Since the characteristics of nonlinear resistive devices are obtained with constant input conditions, they are known as *static characteristic curves*. The behavior of the control device under actual load conditions can be partially predicted by an analysis of the static curves.

One of the important factors which often must be known about a control device in a particular application is its resistance. Examination of Fig. 1.1 reveals that obtaining a resistance value at various points on a characteristic curve would result in a different value for each point considered.

Example 1: Refer to Fig. 1.1. $i_{CD1} = 2$ ma; $i_{CD2} = 8$ ma; $e_{CD1} = 100$ volts; $e_{CD2} = 140$ volts. Find the control device resistance at points 1 and 2 on curve C.

Solution:

$$R_{CD1} = \frac{e_{CD1}}{i_{CD1}} = \frac{100}{2 \times 10^{-3}} = 50 \text{ kilohms}$$

$$R_{CD2} = \frac{e_{CD2}}{i_{CD2}} = \frac{140}{8 \times 10^{-3}} = 17.5 \text{ kilohms}$$

The values obtained in example 1 are called *static resistance values* and are generally not useful in control device work.

Several techniques may be utilized in arriving at a good approximation of the resistance of the control device. The following method results in determining *the dynamic resistance r_{CD}, defined as the opposition of the control device to changes in current*. In equation form,

$$r_{CD} = \frac{\Delta e_{CD}}{\Delta i_{CD}} \qquad \text{with } E_I \text{ or } I_I \text{ fixed}$$

where r_{CD} = dynamic resistance, ohms
Δe_{CD} = change in voltage across the control device
Δi_{CD} = change in current through the control device

NOTE: E_I is fixed for vacuum tubes, while I_I is constant for transistors.

Example 2: Refer to Fig. 1.1. Using the parameters stated in example 1, calculate the dynamic resistance of the control device while operating between points 1 and 2 of curve C.

Solution:

$$r_{CD} = \frac{e_{CD2} - e_{CD1}}{i_{CD2} - i_{CD1}} = \frac{140 - 100}{(8 - 2) \times 10^{-3}} = 6.6 \text{ kilohms}$$

The dynamic resistance value is considerably different from the static values obtained in example 1.   The accuracy of the dynamic resistance technique hinges on the magnitude of the current and voltage changes considered.   Generally, greater accuracy is obtained when using only small changes of voltage and current.

1.2 Series Control Device Circuit: General Considerations. Figure 1.2*a* illustrates a series control device circuit where the device could be a transistor or vacuum tube, which is basically a three-terminal device. The manner in which the resistance of the control device can be varied is discussed in the following section.   Let us proceed with a waveform analysis of the circuit.   Assume E_{bb} is a constant-voltage source with negligible internal resistance.

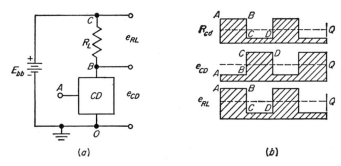

FIG. 1.2   Series control device circuit: (*a*) circuit, (*b*) waveforms.

R_{cd} at Rest Condition.   The original setting of R_{cd}, which is called the "rest" or "no signal" condition, is designated by line Q in the R_{cd} waveform of Fig. 1.2*b*.   Since R_{cd} and R_L together possess the entire circuit resistance, then

$$e_{RL} + e_{CD} = E_{bb} \qquad \text{at all times}$$
and
$$e_{CD} = E_{bb} - e_{RL} \qquad \text{at all times}$$
and
$$e_{RL} = E_{bb} - e_{CD} \qquad \text{at all times}$$

Also, the voltage drop across R_L or CD may be determined from the following relationships:

$$e_{CD} = \frac{R_{cd}}{R_{cd} + R_L} E_{bb}$$
and
$$e_{RL} = \frac{R_L}{R_{cd} + R_L} E_{bb}$$

Thus, the voltage drop across the two series resistive components hinges directly upon the ratio of the resistance of the component to the total series-circuit resistance, and the sum of the two voltage drops is always equal to E_{bb}.

R_{cd} at Maximum Resistance Condition. Assume the resistance of CD is increased to a maximum value in some manner. At this time, e_{CD} will increase from its Q value to point A of the e_{CD} waveform. As dictated by the fundamental series-circuit voltage relationships, e_{RL} must decrease by the same amount that e_{CD} increases by in order to maintain a sum equal to the magnitude of E_{bb}. In equation form,

$$\Delta e_{CD} = -\Delta e_{RL}$$

R_{cd} at Minimum Resistance Condition. Let us now examine the circuit waveforms at the time R_{cd} is at a minimum resistance value. Since R_{cd} is now a smaller portion of the total series-circuit resistance, e_{CD} undergoes a decrease in accordance with the reduction in the R_{cd}/R_t ratio, as shown in the e_{CD} waveform as point C. Since the R_{cd}/R_t ratio is decreased, then the R_L/R_t ratio must be greater, and e_{RL} undergoes a corresponding increase (see point C of the e_{RL} waveform) in order to maintain a sum equal to the magnitude of E_{bb}.

The preceding waveform analysis reveals several significant features of the series nonlinear resistive control device circuit. The output voltage of this circuit may by choice be either e_{CD} or e_{RL}, which are similar in magnitude but have inverted voltage changes with respect to one another. For example, when e_{CD} increases by 10 volts, e_{RL} must decrease by 10 volts. For a given E_{bb}, the magnitude of these two voltage excursions is directly dependent upon the R_{cd} changes.

Example 1: Assume the circuit of Fig. 1.2 has the following parameters: $E_{bb} = 100$ volts; $R_L = 5$ kilohms; R_{cd} varies from 1 to 10 kilohms. Find (a) Δe_{CD} from rest to maximum; (b) Δe_{RL} at this same time.

Solution:

(a) Find Δe_{CD} from rest to maximum:

When $R_{cd} = 1$ kilohm:

$$e_{CD} = \frac{1K}{1K + 5K} \times 100 = 16.7 \text{ volts}$$

When $R_{cd} = 10$ kilohms:

$$e_{CD} = \frac{10K}{10K + 5K} \times 100 = 66.7 \text{ volts}$$
$$\Delta e_{CD} = e_{CD,\max} - e_{CD,\min} = 66.7 - 16.7 = 50 \text{ volts}$$

(b) Find Δe_{RL} at the same time:

Δe_{RL} is the same as Δe_{CD} but in the opposite direction; hence,

$$\Delta e_{RL} = -50 \text{ volts}$$

NOTE: A negative change is graphically downward, and a positive change is graphically upward and not related to the actual polarity of the circuit.

Example 2: Assume the circuit of Fig. 1.2 has the following parameters: $E_{bb} = 100$ volts; $R_L = 5$ kilohms; R_{cd} varies from 3 to 8 kilohms. Find (a) Δe_{CD} from rest to maximum; (b) Δe_{RL} at this same time.

Solution:

(a) Find Δe_{CD} from rest to maximum:

When $R_{cd} = 3$ kilohms:

$$e_{CD} = \frac{3K}{3K + 5K} \times 100 = 37.5 \text{ volts}$$

When $R_{cd} = 8$ kilohms:

$$e_{CD} = \frac{8K}{8K + 5K} \times 100 = 61.5 \text{ volts}$$

$$\Delta e_{CD} = e_{CD,\max} - e_{CD,\min} = 61.5 - 37.5 = 24 \text{ volts}$$

and

$$\Delta e_{RL} = -24 \text{ volts}$$

Notice that the magnitude of the voltage change is greatest with the larger R_{cd} changes.

In many of the common resistive control device applications where a symmetrical input signal is applied, the output voltage (e_{RL} or e_{CD}) must be symmetrical with respect to its rest value. For example, if e_{CD} increases by 20 volts when R_{cd} is increased, then R_{cd} must be decreased from its rest value such that e_{CD} decreases by 20 volts. This is a requirement for linear applications of the control device.

One method by which this type of circuit behavior can be ensured is to match r_{CD} to R_L by making $R_L = r_{CD}$. It should be pointed out that matching the value of R_L to r_{CD} is not utilized in all linear control device applications for practical reasons. R_L, in some cases, is necessarily larger than r_{CD} of the control device. In such instances, a second method of ensuring a symmetrical output voltage is used, which is discussed in the following chapter in conjunction with load lines.

1.3 Several Methods of Varying the Resistance of the Control Device.

The resistance of a control device may be varied as a function of:

1. e_{CD}
2. e_i or i_i

The effect of e_{CD} upon R_{CD} is limited to values determined by the slope of the volt-ampere characteristics.

Figure 1.3 illustrates the generalized volt-ampere characteristics of several nonlinear resistive control devices. Refer to Fig. 1.3a, which is the volt-ampere characteristic of a certain type of vacuum tube. Let us calculate the static resistance values at points A, B, and C:

$$R_A = \frac{e_{CD,a}}{i_{CD,a}} = \frac{50}{1 \times 10^{-3}} = 50 \text{ kilohms}$$

$$R_B = \frac{e_{CD,b}}{i_{CD,b}} = \frac{125}{3 \times 10^{-3}} = 41.7 \text{ kilohms}$$

$$R_C = \frac{e_{CD,c}}{i_{CD,c}} = \frac{200}{5 \times 10^{-3}} = 40 \text{ kilohms}$$

The calculations reveal that the resistance of the control device decreases with increases in e_{CD}. Changing the resistance in this manner is limited by the maximum voltage and current of the device. If the characteristic possessed a more vertical slope, the resistance would have varied between lower values.

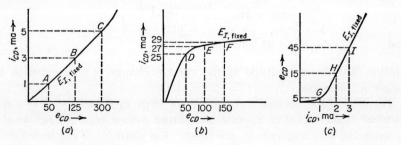

FIG. 1.3 E-I characteristics of several nonlinear resistive control devices.

Figure 1.3b is the generalized volt-ampere characteristic of another type of vacuum tube. Let us calculate the static resistance values at points D, E, and F.

$$R_D = \frac{e_{CD,d}}{i_{CD,d}} = \frac{50}{25 \times 10^{-3}} = 2 \text{ kilohms}$$

$$R_E = \frac{e_{CD,e}}{i_{CD,e}} = \frac{100}{27 \times 10^{-3}} = 3.7 \text{ kilohms}$$

$$R_F = \frac{e_{CD,f}}{i_{CD,f}} = \frac{150}{29 \times 10^{-3}} = 5.14 \text{ kilohms}$$

Notice that the effect of e_{CD} upon the resistance of the device, whose volt-ampere characteristics are illustrated in Fig. 1.3b, is not the same as the preceding example. From the origin up to point D, the effect of e_{CD} upon R_{CD} is similar to that in Fig. 1.3a. From point D, called the "knee" of the characteristic, increases in e_{CD} result in an increase in R_{CD}. This is due to the decreasing effect of e_{CD} upon further i_{CD} increases beyond point D. A more vertical slope would, as in the previous case, indicate resistances between lower values.

Figures 1.3a and b are volt-ampere characteristics of nonlinear resistive control devices which are voltage-controlled (vacuum tubes). The volt-ampere characteristics of current-controlled devices (transistors) are often plotted in the manner illustrated in Fig. 1.3c. Since the current is the factor which varies the resistance, i_{CD} is plotted on the x axis and e_{CD} on the y axis. Some manufacturers illustrate transistor characteristics in the same manner as those for vacuum tubes, however.

Let us calculate the static resistance values of points G, H, and I of Fig. 1.3c.

$$R_G = \frac{e_{CD,g}}{i_{CD,g}} = \frac{5}{1 \times 10^{-3}} = 5 \text{ kilohms}$$

$$R_H = \frac{e_{CD,h}}{i_{CD,h}} = \frac{15}{2 \times 10^{-3}} = 7.5 \text{ kilohms}$$

$$R_I = \frac{e_{CD,i}}{i_{CD,i}} = \frac{45}{3 \times 10^{-3}} = 15 \text{ kilohms}$$

Increases of i_{CD} result in increases in R_{CD} for the control device whose characteristic is illustrated in Fig. 1.3c. As in the two preceding cases, the extent to which the resistance can be varied by this method is limited by the maximum voltage and current ratings of the device. A more vertical slope would indicate a variable resistance between higher values. It should be pointed out that the resistance of the control device is not varied in this manner in all applications. The voltage across the device and the current passing through it are usually fixed in the most linear region of the characteristic with zero signal applied to the device in linear applications. In this way the effect of e_{CD} and/or i_{CD} upon R_{CD} is minimized, thereby enabling the input signal to develop the changes in R_{CD}.

R_{CD} is most often varied by the input signal in linear applications. In the case of the vacuum tube, positive excursions of input voltage reduce R_{CD}, and negative input voltage changes produce the opposite effect. *A positive excursion is meant as a change which is graphically upward.* Following are several examples of positive voltage changes: 0 to 5 volts; -10 to -3 volts. The resistance of the transistor, on the other hand, is varied by the input current in a manner determined by the type of transistor. The specific manner in which the input signal alters the resistance of each type of control device is analyzed in the appropriate chapters. Our only concern at this point is that the resistance of a control device can be varied in the two general ways described in this section:

1. Variations of input current (transistors)
2. Variations of input voltage (vacuum tubes)

1.4 The Three Basic Configurations. In terms of the input and output signals, the common types of nonlinear resistive control devices possess three external terminals, as shown in Fig. 1.4. The input and output signals (which may be voltages or currents) require the use of two external terminals each. The input and output terminal possibilities are illustrated in Fig. 1.5.

Since the input and output signals each require two terminals and only three external terminals are available, one of the terminals must be common to both the input and output signals. Three types of signal applications are possible, which are called *configurations*. In Fig. 1.5*a*, terminal *A* is common to both the input and output signals, and this connection may be called the common terminal *A* configuration. Since terminal *B* is common to both signals in 1.5*b*, this connection may be called the

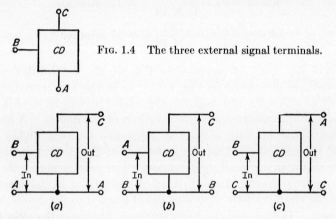

FIG. 1.4   The three external signal terminals.

FIG. 1.5   The basic configurations: (*a*) common terminal *A*, (*b*) common terminal *B*, (*c*) common terminal *C*.

common terminal *B* configuration. Using the same type of analysis, the control device of Fig. 1.5*c* is called the common terminal *C* configuration.

Each of the external terminals is given a specific name in accordance with the type of the control device being considered, and the configuration is named after the common terminal. The terminals used for the input and output signals have much to do with the over-all behavior of the device, as is studied in the appropriate chapters. The great majority of transistor and vacuum-tube circuits incorporate one of these three basic configurations.

1.5   Reproduction and Amplification.   Recall that the resistance of a control device can be varied by changes in the input signal. *When the input voltage or current excursions change the resistance of the control device in such a manner that e_{CD} and i_{CD} possess the same variations as the input signal, the control device has duplicated the input variations in its output;*

this is called reproduction. Reproduction is a requirement only in those circuits which are considered to be linear applications of the control device. There are many control device circuits where complete or even partial reproduction is not required or desired.

Amplification may be defined as the process by which the output current and/or voltage variations are made larger than the input signal variations which produced them. In linear applications, faithful reproduction and amplification are generally both desired, whereas only amplification is required in most nonlinear applications.

In conclusion, to reproduce means to duplicate, and to amplify means to enlarge. It is important that the distinction between these two terms be clear in the mind of the reader.

1.6 Control Ratio and Amplification Factor. The merits of a nonlinear resistive control device for a stated application are best evaluated when the effect of the input signal and e_{CD} upon i_{CD} can be predicted. One of the relationships useful in this type of analysis is the control ratio.

The control ratio is defined as the ratio of the change in e_{CD} required to produce the same change in i_{CD} as produced by a 1-volt change in e_i. In equation form,

$$A_v = \frac{\Delta e_{CD}}{\Delta e_I}$$

where A_v = control ratio or voltage gain

Δe_{CD} = change in output voltage across the device to change i_{CD} by the same amount as caused by a Δe_i of 1 volt

e_I = input voltage

In analyzing the circuit of Fig. 1.2 in Sec. 1.2, it was found that $\Delta e_{RL} = \Delta e_{CD}$, but the change was in the opposite direction, which can be indicated by the negative sign. Therefore, the preceding A_v equation may be expanded to

$$A_v = \frac{\Delta e_{CD}}{\Delta e_I} = -\frac{\Delta e_{RL}}{\Delta e_I}$$

where the negative sign indicates that the voltage change is in the opposite direction.

Example: Assume the circuit of Fig. 1.2 has the following parameters: R_L = 10 kilohms; when e_I is changed from 2 to 3 volts, i_{CD} increases from 10 to 15 ma. Find A_v.

Solution: Since R_L and CD are in series, the current is the same. Knowing the ohmic value of R_L and the two current values, the two e_{RL} values can be computed.

When $e_I = 2$ volts, $i_{CD} = 10$ ma, and

$$e_{RL} = i_{CD}R_L = 10 \times 10^{-3} \times 10 \times 10^3 = 100 \text{ volts}$$

When $e_I = 3$ volts, $i_{CD} = 15$ ma, and

$$e_{RL} = i_{CD}R_L = 15 \times 10^{-3} \times 10 \times 10^3 = 150 \text{ volts}$$

Hence
$$\Delta e_{RL} = 150 - 100 = 50 \text{ volts}$$

and
$$A_v = -\frac{\Delta e_{RL}}{\Delta e_i} = -\frac{50}{1} = -50$$

A 1-volt input signal excursion resulted in a 50-volt change in the output voltage; the control device amplified the input voltage by a factor of 50. Examining this in terms of the definition of A_v, a 50-volt change in e_{CD} is required to produce the same effect upon i_{CD} as a 1-volt change in e_i.

1.7 Voltage Amplifiers.

A control device, when incorporated into a circuit which is specifically designed to develop an output voltage greater than the input voltage, is called a voltage amplifier. Low values of current through the device are usual because of the high R_L values (as compared to R_{CD}) customarily used. It is not uncommon for R_L to be anywhere from two to ten times greater than r_{cd} of the control device, since this would increase the voltage amplification. Voltage amplifiers generally develop low output power because of the low current. In some applications, R_L may be small and the output current large in order to develop a large voltage across the "ultimate load," which is fed from R_L.

1.8 Control Current Amplification Ratio.

The ratio of the output current change to a given input current excursion which occurs at the same time is referred to as the control current amplification factor. In equation form,

$$A_i = \frac{\Delta i_{CD}}{\Delta i_I} = \frac{\Delta i_o}{\Delta i_i}$$

where all current values are in amperes.

This ratio has only limited use in vacuum-tube work since the input current of a vacuum tube is frequently small enough to be neglected. But in transistor work, where the input current is of importance, the preceding relationship is an invaluable aid in predicting the current amplification performance of the circuit. When used in transistor circuits, the ratio is further specified in terms of the configuration with a fixed e_{CD} and reveals the number of times a variation of i_i is magnified in the output of the transistor.

Example: Assume a series control device circuit has the following parameters. When i_i changes from 50 to 100 μa, i_{CD} increases from 0.5 to 2.5 ma. Find A_i.

Solution:

$$A_i = \frac{2 \times 10^{-3}}{50 \times 10^{-6}} = 40$$

In the preceding example, a given change of input signal current resulted in an output current variation which is 40 times greater. This ratio may also be used with rms current values.

1.9 Current Amplifiers. *A control device which is designed into a circuit such that relatively high values of output current flow may be defined as a current amplifier.* Current amplifiers are often called power amplifiers since the large output currents result in a substantial output power. They are used where medium to large magnitudes of load currents are required. Control devices used as current amplifiers are specially designed in their manufacture so that they can withstand larger current values without undergoing physical damage.

In order to permit rather large values of output current in the ultimate load, R_L is frequently small in ohmic value, resulting in a small voltage amplification. Such a situation is permissible, since the magnitude of the load current is of primary interest in a current amplifier while the voltage amplification is of lesser significance.

1.10 Power Ratio. The power ratio is another relationship between the input and output signals which is frequently useful in evaluating the merits of a control device in a particular application.

The signal power ratio is the ratio of the signal power output to the input signal power:

$$G_{\text{sig}} = \frac{P_{o,\text{sig}}}{P_{i,\text{sig}}}$$

where G_{sig} = signal power ratio

$P_{o,\text{sig}}$ = power developed by the output signal

$P_{i,\text{sig}}$ = power developed by the input signal

Another power relationship worthy of consideration is the ratio of the total output power to the total input power, which may be called the control power ratio.

$$G_{\text{control}} = \frac{P_{o,t}}{P_{i,t}}$$

The output current of a nonlinear control device is usually a composite current made up of the signal variations superimposed upon a constant unidirectional current. Figure 1.6b illustrates such an output current.

Consider a voltage-controlled CD circuit such as Fig. 1.6a. In the absence of an input signal, i_{CD} is at its *quiescent* or *rest value*, indicated by line Q of Fig. 1.6b. Assuming the control device and R_L are properly matched for faithful reproduction, the positive excursion of the input signal results in a corresponding increase in i_{CD} from its Q value to point A. When the input signal changes from its maximum positive value to maximum negative value, i_{CD} swings to its minimum value (point B). This follows the relationship stated in an earlier discussion between the input voltage and control device resistance (positive input voltage changes decrease R_{CD} and vice versa).

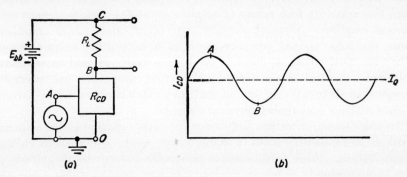

FIG. 1.6   Voltage-controlled CD circuit: (a) circuit, (b) waveform.

The varying portion of i_{CD} is the output signal current (those variations above and below the Q point), and the power dissipation which results from this current is the signal power output. The constant unidirectional portion of i_{CD} is the current required to operate the control device at its rest resistance value.

In analyzing the power dissipated by a varying unidirectional current, the current may be considered as the composite of a constant unidirectional current upon which the bidirectional component is superimposed. Using the superposition technique, the sum of the power dissipation due to each current component is equal to the total power dissipation. Recall that rms values must be used when determining average power dissipation where bidirectional current is involved. The average power dissipated by R_L in the circuit of Fig. 1.6a may be determined from

$$P_{RL,\text{av}} = P_{o,\text{sig}} + P_{o,\text{unidir}}$$

which is $\qquad P_{RL,\text{av}} = (I_{\text{bidir,rms}})^2 R_L + (I_{\text{unidir}})^2 R_L$

Factoring out R_L,

$$P_{RL,\text{av}} = [(I_{\text{bidir,rms}})^2 + (I_{\text{unidir}})^2]R_L$$

The preceding equation may also be expressed in the following manner:

$$P_{RL,av} = [(I_{o,sig,rms})^2 + (I_Q)^2]R_L$$

Example 1: Assume the circuit of Fig. 1.6a has the following parameters, where R_L is the output and is 10 kilohms: $E_i = 0.4$ volt; $I_i = 50$ μa (both are rms values); $i_{CD,max} = 6.2$ ma; $i_{CD,min} = 3.8$ ma. The variations above and below rest are symmetrical. Find (a) G_{sig}; (b) $P_{RL,av}$; (c) $G_{control}$.

Solution:

(a) Find G_{sig}:

We must first find the input and output signal powers.

$$P_{i,sig} = E_iI_i = 4 \times 10^{-1} \times 50 \times 10^{-6} = 20 \times 10^{-6} \text{ watt}$$
$$P_{o,sig} = (I_{o,sig,rms})^2 R_L$$

Since the bidirectional component of the output signal waveform is sinusoidal (see Fig. 1.6b), the rms value is equal to 0.707 of the peak bidirectional current.

$$I_{o,sig} \text{ (peak to peak)} = i_{CD,max} - i_{CD,min}$$

and

$$I_{o,sig \text{ (peak)}} = I_{o,sig,m} = \frac{i_{CD,max} - i_{CD,min}}{2}$$

hence

$$I_{o,sig,rms} = \frac{0.707(i_{CD,max} - i_{CD,min})}{2}$$

Substituting values,

$$I_{o,sig,rms} = \frac{0.707(6.2 \text{ ma} - 3.8 \text{ ma})}{2} = 0.833 \text{ ma}$$

The output signal power may now be determined:

$$P_{o,sig} = (8.33 \times 10^{-4})^2 \times 10 \times 10^3 = 69.4 \times 10^{-4} \text{ watt}$$

Knowing the input and output signal powers, we may calculate the signal power ratio:

$$G_{sig} = \frac{P_{o,sig}}{P_{i,sig}} = \frac{69.4 \times 10^{-4}\text{w}}{20 \times 10^{-6}\text{w}} = 347$$

That is, the output signal power is 347 times greater than the input signal power.

(b) Find $P_{RL,av}$:

The constant unidirectional component leads to an additional power dissipation through R_L. Since the output signal power has been determined, we need only compute the power dissipation due to the unidirectional current, since the sum of these two powers is $P_{RL,av}$.

$$P_{o,unidir} = (I_Q)^2 R_L$$

Since the bidirectional variations above and below rest are symmetrical, I_Q is exactly between $i_{CD,max}$ and $i_{CD,min}$. I_Q in such cases may be found from either of the two following relationships:

$$I_Q = i_{CD,\max} - \frac{i_{CD,\max} - i_{CD,\min}}{2}$$

or

$$I_Q = i_{CD,\min} + \frac{i_{CD,\max} - i_{CD,\min}}{2}$$

Using the first equation, let us find I_Q:

$$I_Q = 6.2 \text{ ma} - \frac{6.2 \text{ ma} - 3.8 \text{ ma}}{2} = 5 \text{ ma}$$

and

$$P_{o,\text{unidir}} = (5 \times 10^{-3})^2 \times 10 \times 10^3 = 2.5 \times 10^{-1} \text{ watt}$$

We may now find $P_{RL,\text{av}}$:

$$P_{RL,\text{av}} = P_{o,\text{sig}} + P_{o,\text{unidir}} = 6.94 \times 10^{-3} + 2.5 \times 10^{-1}$$
$$= 2.57 \times 10^{-1} \text{ watt}$$

(c) Find G_{control}:

$$G_{\text{control}} = \frac{P_{o,t}}{P_i} = \frac{P_{RL,\text{av}}}{P_i} = \frac{2.57 \times 10^{-1}}{20 \times 10^{-6}} = 12,850$$

That is, when the power dissipated across R_L by the unidirectional current component is included in P_o, the circuit under consideration develops an output power which is 12,850 times greater than the input power.

Example 2: Assume the circuit of Fig. 1.6a has a symmetrical output current and has the following parameters: $R_L = 25$ kilohms; $i_{CD,\max} = 10$ ma; $i_{CD,\min} = 5$ ma. Find $P_{RL,\text{av}}$.
Solution: We must first find I_Q:

$$I_Q = i_{CD,\min} + \frac{i_{CD,\max} - i_{CD,\min}}{2}$$

Substituting values,

$$I_Q = 5 \text{ ma} + \frac{10 \text{ ma} - 5 \text{ ma}}{2} = 7.5 \text{ ma}$$

Since $I_{o,\text{sig,max}} = i_{CD,\max} - I_Q$
then $I_{o,\text{sig,max}} = 10 \text{ ma} - 7.5 \text{ ma} = 2.5 \text{ ma}$

Since a sinusoidal variation is assumed, then

$$I_{o,\text{sig,rms}} = 2.5 \times 0.707 \times 1.77 \text{ ma}$$

We may now find $P_{RL,\text{av}}$:

$$P_{RL,\text{av}} = [(1.77 \times 10^{-3})^2 + (7.5 \times 10^{-3})^2]25 \times 10^3$$
$$= 1.485 \text{ watts}$$

Example 3: Assume the circuit of Fig. 1.6a has a symmetrical output current and has the following parameters: $E_i = 0.5$ volt; $I_i = 0.07$ ma (both input parameters are rms values); $i_{CD,\text{max}} = 8$ ma; $i_{CD,\text{min}} = 4$ ma; $R_L = 5$ kilohms. Find (a) I_Q; (b) A_v; (c) A_i.

Solution:

(a) Find I_Q:

$$I_Q = i_{CD,\text{max}} - \frac{i_{CD,\text{max}} - i_{CD,\text{min}}}{2}$$

$$= 8 \text{ ma} - \frac{8 \text{ ma} - 4 \text{ ma}}{2} = 6 \text{ ma}$$

(b) Find A_v:

Since the input voltage rms value is stated, the rms output signal voltage must be found, since the ratio of these two voltages is A_v. Let us first find $I_{o,\text{sig,rms}}$.

$$I_{o,\text{sig,rms}} = 0.707(i_{CD,\text{max}} - I_Q) = 0.707(8 \text{ ma} - 6 \text{ ma})$$
$$= 1.414 \text{ ma}$$

Knowing the value of R_L and the rms value of the output current variations, we can determine the rms variation of the voltage across R_L:

$$E_{rl} = I_{rl}R_L = 1.414 \times 10^{-3} \times 5 \times 10^3 = 7.07 \text{ volts}$$

We may now compute A_v:

$$A_v = \frac{E_{o,\text{rms}}}{E_{i,\text{rms}}} = \frac{7.07}{5 \times 10^{-1}} = 14.14$$

(c) Find A_i:

$$A_i = \frac{I_{o,\text{sig,rms}}}{I_{i,\text{sig,rms}}} = \frac{1.414 \times 10^{-3}}{7 \times 10^{-5}} = 20.2$$

Example 4: Find G_{sig} for example 3:

Solution:

$$G_{\text{sig}} = A_v A_i$$
$$= 14.14 \times 20.2 = 284.8$$

Let us verify this relationship by computing the actual input and output signal powers and determine the value of their ratio (which is G_{sig}):

$$P_{i,\text{sig}} = E_i I_i = 5 \times 10^{-1} \times 7 \times 10^{-5} = 35 \times 10^{-6} \text{ watt}$$
$$P_{o,\text{sig}} = E_{o,\text{sig,rms}} I_{o,\text{sig,rms}} = 7.07 \times 1.414 \times 10^{-3}$$
$$= 10 \times 10^{-3} \text{ watt}$$

and
$$G_{\text{sig}} = \frac{P_{o,\text{sig}}}{P_{i,\text{sig}}} = \frac{10 \times 10^{-3}}{35 \times 10^{-6}} = 285$$

Thus we see that the product of A_v and A_i does equal G_{sig}. The small discrepancy results from rounding off $P_{o,\text{sig}}$ in the last ratio.

Output Terminal Efficiency.   The output terminal efficiency is considered as the ratio of the signal power output to the product of the rest current $i_{CD,Q}$ and rest control device voltage $e_{CD,Q}$. Multiplying the ratio by 100 results in the output terminal efficiency expressed as a percentage:

$$\text{Output terminal efficiency} = \frac{P_{o,\text{sig}}}{i_{CD,Q}e_{CD,Q}} \times 100$$

Example 5: A control device, whose output signal variations are symmetrical, has the following parameters: $R_L = 1$ kilohm; $i_{CD,Q} = 4$ ma; $e_{CD,Q} = 6$ volts; $i_{CD,\text{max}} = 5$ ma. Find the output terminal efficiency.
Solution: We must first determine $P_{o,\text{sig}}$:

$$P_{o,\text{sig}} = [0.707 \times (i_{CD,\text{max}} - i_{CD,Q})]^2 R_L$$
$$= [0.707 \times 1 \times 10^{-3}]^2 \times 1 \times 10^3 = 5 \times 10^{-4} \text{ watt}$$
$$\text{Output terminal efficiency} = \frac{5 \times 10^{-4}}{4 \times 10^{-3} \times 6} \times 100 = 2.08\%$$

1.11 Power Amplifiers. *A control device which is designed such that a relatively large power output value is obtained may be defined as a power amplifier.* Like the current amplifiers discussed in a preceding section, they are usually of special construction so as to permit the development of moderate or large values of power without exceeding the wattage dissipation rating of the control device.

Power amplifiers are used in a great variety of applications where faithful reproduction of the input signal variations may or may not be a requirement. In some cases, a periodic development of power output is the sole requirement imposed upon the device. Where faithful reproduction is important, R_L must be selected such as to satisfy the requirements of linear operation and the proper level of power output. The best results in both reproduction and power output at the same time are difficult to achieve in some instances, requiring the incorporation of special power amplifier techniques and usually a compromise choice of R_L. There are a number of special power amplifier circuits where both faithful reproduction and high power output are achieved with the use of two power amplifiers instead of the single unit. These techniques are studied in the appropriate chapters.

1.12 Selection of Control Device Resistance. *The minimum, rest, and maximum control device resistances to be used are determined in great part by the particular circuit application.* In switching circuits, for example, a minimum resistance approaching zero and a very large maximum resistance value are the ideal condition (since a "perfect" switch would display zero ohms when closed and infinite ohms while open). In some other applications, the difference between maximum and minimum R_{CD} is comparatively small. The rest value of R_{CD} also hinges on the application (linear or nonlinear operation). The relationship of the R_{CD} variations is discussed with each type of application.

PROBLEMS

1.1 A series control device has the following parameters: When $e_{CD} = 50$ volts, $i_{CD} = 4$ ma; when $e_{CD} = 80$ volts, $i_{CD} = 2.7$ ma. Find r_{CD}.

1.2 A series control device has the following parameters: When $e_{CD} = 120$ volts, $i_{CD} = 40$ ma; when $e_{CD} = 150$ volts, $i_{CD} = 28$ ma. Find r_{CD}.

1.3 Assume that faithful reproduction is a requirement of the circuit of Prob. 1.1. What relationship between r_{CD} and R_L could be utilized in determining the R_L to be used?

1.4 Faithful reproduction is a requirement of the circuit in Prob. 1.2. Find R_L.

1.5 Refer to the parameters of Prob. 1.1. Find the e_{CD} value when R_{CD} is (a) maximum; (b) minimum.

1.6 Refer to the parameters of Prob. 1.2. Find the e_{CD} value when R_{CD} is (a) maximum; (b) minimum.

1.7 Refer to Fig. 1.2. $e_{RL} = 115$ volts; $e_{CD} = 85$ volts. Find E_{bb}.

1.8 Refer to Fig. 1.2. $E_{bb} = 450$ volts; $e_{RL} = 160$ volts. Find e_{CD}.

1.9 Refer to Fig. 1.2. $E_{bb} = 300$ volts; $e_{CD,Q} = 150$ volts; $R_L = 4.7$ kilohms. Find (a) $e_{RL,Q}$; (b) $R_{CD,Q}$; (c) I_Q.

1.10 Repeat Prob. 1.9 for a circuit with the following parameters: $E_{bb} = 120$ volts; $e_{CD,Q} = 40$ volts; $R_L = 150$ kilohms.

1.11 Repeat Prob. 1.9 for a circuit with the following parameters: $E_{bb} = 22$ volts; $e_{CD,Q} = 5$ volts; $R_L = 50$ ohms.

1.12 The circuit of Fig. 1.2 has the following parameters: $E_{bb} = 12$ volts; $R_L = 1.2$ kilohms; R_{CD} varies from 120 ohms to 12 kilohms; $R_{CD,Q} = 1,200$ ohms. Find (a) e_{CD} (maximum, rest, minimum); (b) E_{RL} (minimum, rest, maximum); (c) Δe_{CD} (peak to peak).

1.13 In Prob. 1.12, find (a) i_{CD} (minimum, rest, maximum); (b) r_{CD} (using peak-to-peak variations); (c) r_{CD} (using peak-to-rest variations).

1.14 Assume the circuit of Fig. 1.2 has the following parameters: $R_L = 20$ kilohms; e_{RL} variations are 90 volts (maximum), 60 volts (rest), and 30 volts (minimum); $E_{bb} = 200$ volts. Find:

(a) i_{CD} (minimum, rest, maximum)   (b) e_{CD} (maximum, rest, minimum)

(c) r_{CD} (using peak-to-peak variations)   (d) r_{CD} (using peak-to-rest variations)

1.15 State the relationship between the slope of a volt-ampere characteristic and R_{CD}.

1.16 How many basic configurations are possible with the three-external-terminal control device? Explain.

1.17 State the difference between the terms "reproduction" and "amplification."

1.18 Refer to the circuit of Fig. 1.2. $R_L = 150$ kilohms; when e_i is changed from -2 volts to -1 volt, i_{CD} increases from 1 to 1.4 ma. Find A_v.

1.19 Repeat Prob. 1.18 with $R_L = 20$ kilohms.

1.20 Explain why A_v of Prob. 1.18 and 1.19 are not the same.

1.21 A series control device circuit has the following parameters: When i_i changes from 100 to 200 μa, i_{CD} increases from 5 to 30 ma. Find A_i.

1.22 i_i of a series control device circuit changes from 25 to 50 μa, causing i_{CD} to increase from 0.5 to 3 ma. Find A_i.

1.23 In Prob. 1.21, $R_L = 200$ ohms; $\Delta e_i = 0.1$ volt. Find A_v.

1.24 In Prob. 1.22, $R_L = 2.4$ kilohms; $\Delta e_i = 0.02$ volt. Find A_v.

1.25   Refer to the circuit and waveform of Fig. 1.6. $R_L = 5$ kilohms; $E_{i,\text{rms}} = 0.5$ volt; $I_{i,\text{rms}} = 60$ μa; $i_{CD,\text{max}} = 6$ ma; $i_{CD,\text{min}} = 1$ ma.   The variations above and below rest are sinusoidal and symmetrical.   Find:

     (a) A_v (b) A_i

     (c) G_{sig} (d) $P_{RL,\text{av}}$

     (e) G_{control} (f) output terminal efficiency

1.26   Assume the circuit of Prob. 1.25 has the following input parameters: $E_{i,\text{rms}} = 0.2$ volt; $I_{i,\text{rms}} = 35$ μa; i_{CD} values and R_L are the same as Prob. 1.25.   Find:

     (a) A_v (b) A_i

     (c) G_{sig} (d) $P_{RL,\text{av}}$

     (e) G_{control} (f) output terminal efficiency

1.27   In Prob. 1.25, find I_Q if R_L were changed to 18 kilohms.   Explain why I_Q changed in this manner.

1.28   Refer to the circuit and waveform of Fig. 1.6. $R_L = 220$ kilohms; $E_{i,\text{rms}} = 1$ volt; $I_{i,\text{rms}} = 50$ μa; $i_{CD,\text{max}} = 1.8$ ma; $i_{CD,\text{min}} = 0.3$ ma.   Symmetrical and sinusoidal variations are assumed.   Find:

     (a) A_v (b) A_i

     (c) G_{sig} (d) $P_{RL,\text{av}}$

     (e) G_{control} (f) output terminal efficiency

1.29   Repeat Prob. 1.28 with an R_L of 25 kilohms.

1.30   In Probs. 1.28 and 1.29: State the effect produced by a reduction of R_L in:

     (a) A_v (b) A_i

     (c) G_{sig} (d) $P_{RL,\text{av}}$

     (e) G_{control} (f) output terminal efficiency

CHAPTER 2

VOLT-AMPERE CHARACTERISTICS OF NONLINEAR RESISTIVE CONTROL DEVICES

The volt-ampere characteristics of vacuum tubes (voltage-controlled devices) and transistors (current-controlled devices) are analyzed in this chapter. Both the input and output characteristics of each type of device are studied. Upon becoming familiar with the volt-ampere relationship, several of the common techniques employed in the design of amplifiers are investigated. One of the most effective techniques in designing an amplifier circuit, the load-line analysis, is studied in considerable detail with illustrated examples.

2.1 Output Characteristics of a Voltage-controlled Device. The graphical relationship of e_{CD} versus i_{CD} with a fixed input signal is called the volt-ampere characteristic of the device. *Since e_{CD} is the output voltage and i_{CD} is the output current of the device, the graphical relationship is in effect the output characteristic of the device. Such a graph may be plotted for many fixed input voltage values. A group of these curves plotted on one set of axis for the same control device is called a family of output characteristics.* Figure 2.1a and b illustrate the family of output characteristics for two of the most important types of voltage-controlled devices:

1. Triode-type vacuum tube (having three important internal elements)
2. Pentode-type vacuum tube (having five important internal elements)

Refer to Fig. 2.1a. The curve representing the relationship between e_{CD} and i_{CD} is most nonlinear in the region closest to the point of origin. The slope of a curve may be determined by

$$\text{Slope} = \frac{\Delta y}{\Delta x} = \frac{\Delta i_{CD}}{\Delta e_{CD}}$$

Example 1: Find the slope of the $E_I = -2$ volt curve between points A and B.

Solution:

$$\text{Slope} = \frac{10 - 5}{120 - 75} = \frac{5 \times 10^{-3}}{45} = 1.11 \times 10^{-4}$$

Example 2: Find the slope of the $E_I = -22$ volt curve between points C and D.

Solution:

$$\text{Slope} = \frac{5 - 2}{400 - 350} = \frac{3 \times 10^{-3}}{50} = 6 \times 10^{-5}$$

Notice that the E_I curve is closer to the horizontal x axis than the $E_I = -2$ volt curve. Therefore the volt-ampere characteristics which are closer to the horizontal axis have the smallest slope values.

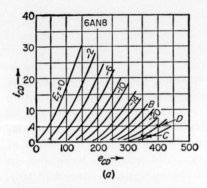 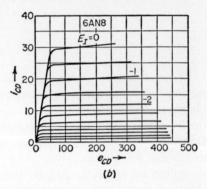

FIG. 2.1   Output characteristics of two types of voltage-controlled devices: (*a*) triode VT, (*b*) pentode VT.   (*RCA*)

Examination of the slope relationship reveals that it is the reciprocal of the dynamic resistance r_{CD} equation; that is,

$$r_{CD} = \frac{1}{\text{slope}} = \frac{\Delta e_{CD}}{\Delta i_{CD}} = r_p$$

where r_{CD} of a vacuum tube is termed the dynamic plate resistance r_p.

Example 3: Calculate the dynamic resistance between points AB of the $E_I = -2$ volt curve.

Solution:

$$r_p = \frac{\Delta e_{CD}}{\Delta i_{CD}} = \frac{120 - 75}{10 - 5} = \frac{45}{5 \times 10^{-3}} = 9.01 \text{ kilohms}$$

As a check,

$$r_p = \frac{1}{\text{slope}} = \frac{1}{1.11 \times 10^{-4}} = 9.01 \text{ kilohms}$$

Example 4: Calculate the dynamic resistance between points CD of the $E_I = -22$ volt curve.

Solution:

$$r_p = \frac{\Delta e_{CD}}{\Delta i_{CD}} = \frac{400 - 350}{5 - 2} = \frac{50}{3 \times 10^{-3}} = 16.67 \text{ kilohms}$$

As a check,

$$r_p = \frac{1}{\text{slope}} = \frac{1}{6 \times 10^{-5}} = 16.67 \text{ kilohms}$$

Examples 1 to 4 prove that a reduced slope is indicative of a larger dynamic resistance value. The dynamic resistance of the voltage-controlled device whose output characteristics are shown in Fig. 2.1a tends to be greater for more negative values of input voltages. This is a general trend in voltage-controlled devices.

The family of output characteristics of the pentode type of vacuum tube is shown in Fig. 2.1b. It will be noted that each curve begins at the origin. When e_{CD} is small, comparatively small changes of e_{CD} result in large changes in i_{CD}. Refer to the curve where $E_I = 0$ volt, for example. When e_{CD} increases from 0 to 50 volts, i_{CD} increases from 0 to 28 ma. The slope of the characteristic undergoes an abrupt change in this vicinity—from a near vertical direction to almost horizontal. The area where the abrupt slope change takes place is often called the "knee" of the curve. To the right of the knee, i_{CD} is relatively independent of e_{CD}, which is one of the most important properties of the pentode-type vacuum tube. In other words, relatively large e_{CD} changes result in comparatively small i_{CD} changes in this area where r_p is greatest.

Example 5: Refer to Fig. 2.1b. Calculate r_p of the $E_I = 0$ volt curve between the following e_{CD} values: (a) 0 and 50 volts; (b) 100 and 150 volts.

Solution:

(a) Calculate r_p between 0 and 50 volts:

$$r_p = \frac{\Delta e_{CD}}{\Delta i_{CD}} = \frac{50 - 0}{28 - 0} = \frac{50}{28 \times 10^{-3}} = 1.786 \text{ kilohms}$$

(b) Calculate r_p between 100 and 150 volts:

$$r_p = \frac{150 - 100}{30 - 29.5} = \frac{50}{5 \times 10^{-4}} = 100 \text{ kilohms}$$

As indicated by the reduced slope to the right of the characteristic's knee, the pentode-type vacuum tube is capable of displaying larger values of dynamic resistance than the triode type. The chapter devoted to vacuum tubes (Chap. 11) investigates both tube types and their internal actions and explains the reason for this difference.

2.2 Input Characteristics of a Voltage-controlled Device. *The transfer characteristics of a vacuum tube are a plot of the relationship e_I versus i_{CD} for various values of fixed e_{CD}.* The transfer characteristic curves for both the triode and pentode types take essentially the same shape, as illustrated in Fig. 2.2.

The transfer characteristics illustrate the effect of the input voltage changes upon the output current. Notice that the lower negative e_I changes produce the largest excursions in i_{CD}. With a fixed e_{CD}, a more negative e_I increases R_{CD}, which results in a decrease of i_{CD}. A sufficiently negative input voltage, for a given fixed e_{CD}, can reduce i_{CD} to zero, which is the "cutoff" point of the control device.

The family of transfer characteristics in Fig. 2.2 does not consider positive values of input voltage, which is a customary practice with voltage-controlled devices. The manner in which the input voltage is

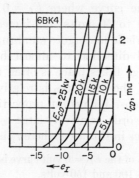

Fig. 2.2 Family of transfer characteristic curves. (*RCA*)

maintained at negative values for all signal conditions is studied in Sec. 2.4, which is devoted to bias.

2.3 Transconductance. *Transconductance g_m is a factor which describes the effect of a small input voltage change upon the output current with a constant voltage across the device.* The "mho" is the basic unit used to express g_m, although micromho (μmho) is more often used because of the small values encountered. In equation form,

$$g_m = \frac{\Delta i_{CD}}{\Delta e_I} \quad \text{with } e_{CD} \text{ constant}$$

The transconductance is also a factor which is the ratio of the amplification factor (or control ratio) to the dynamic resistance. That is,

$$g_m = \frac{A_v}{r_p} = \frac{\mu}{r_p}$$

where μ = amplification factor. Any one of the three factors can be determined if the remaining two are known.

The transconductance during a given input voltage change can be computed from the family of transfer characteristic curves.

Example 1: Refer to the transfer characteristics of Fig. 2.2. With e_{CD} fixed at 25,000 volts, find g_m when e_I varies from -7.5 to -10 volts.
Solution: From the 25,000-volt curve, the following values are obtained:

When $e_I = -7.5$ volts, $i_{CD} = 1.25$ ma.
When $e_I = -10$ volts, $i_{CD} = 0.38$ ma.
We may now find g_m:

$$g_m = \frac{\Delta i_{CD}}{\Delta e_I} \qquad \text{with } e_{CD} \text{ fixed}$$

Substituting values,

$$g_m = \frac{1.25 - 0.38}{10 - 7.5} = \frac{8.7 \times 10^{-4}}{2.5}$$
$$= 3.48 \times 10^{-4} \text{ mho} = 348 \ \mu\text{mho}$$

The output conductance g_o can be determined from the output characteristics of the control device. $g_o = \dfrac{\Delta i_{CD}}{\Delta e_{CD}}$. g_m is utilized in vacuum-tube work and g_o in transistor analysis (studied in Chap. 4).

Example 2: Refer to the output characteristics of Fig. 2.1a. With e_{CD} fixed at 200 volts, find g_m when E_i is varied from -4 to -6 volts.
Solution: Since e_{CD} is fixed at 200 volts, the two points under consideration are those where the -4- and -6-volt curves intersect the e_{CD}-200-volt vertical line.
When $E_i = -4$ volts, $i_{CD} = 20.5$ ma.
When $E_i = -6$ volts, $i_{CD} = 15$ ma.
g_m may now be computed:

$$g_m = \frac{20.5 - 13}{6 - 4} = \frac{7.5 \times 10^{-3}}{2}$$
$$= 3.75 \times 10^{-3} \text{ mho} = 3,750 \ \mu\text{mho}$$

Example 3: Refer to the output characteristics of Fig. 2.1b. Find g_m when e_I is varied from -0.5 to -1.0 volt; e_{CD} is fixed at 200 volts.
Solution: The i_{CD} value for each e_I is obtained from the intersection of the vertical 200-volt e_{CD} line and the two stated input voltage values.
When $e_I = -0.5$ volt, $i_{CD} = 25$ ma.
When $e_I = -1.0$ volt, $i_{CD} = 20$ ma.

$$g_m = \frac{25 - 20}{1 - 0.5} = \frac{5 \times 10^{-3}}{5 \times 10^{-1}}$$
$$= 1 \times 10^{-2} = \text{mho} = 10,000 \ \mu\text{mho}$$

2.4 Bias. *Bias may be defined as a potential applied in conjunction with the input signal so as to place the control device at the desired rest resistance value or "operating point" on the characteristics.* Bias voltages for vacuum tubes usually place the grid at a negative potential with respect to the cathode.   The zero signal output current of the device $i_{CD,Q}$ can be controlled by selection of the proper bias potential.   The input signal is superimposed upon the bias potential, as illustrated in Fig. 2.3.

Refer to Fig. 2.3.   At point A, the resistance of the vacuum tube is at its rest value, since the signal voltage is zero.   The rest resistance can be made greater by increasing the magnitude of the negative bias potential and by the same token can be reduced by decreasing the magnitude of the negative bias potential.

As the input signal voltage swings in the positive direction, the input terminal is made less negative with respect to common by the magnitude of the input signal excursion.   This results in a corresponding decrease in R_{CD}, and i_{CD} accordingly increases.   The negative excursion of the input signal (B to D of Fig. 2.3) results in a correspondingly more negative input terminal with respect to common, which increases R_{CD} to its

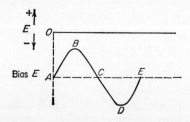

FIG. 2.3   Input voltage with bias.

rest value (point C) and then to its maximum value at the negative peak of the input voltage (point D). $i_{CD,Q}$ flows at the condition which exists at point C, and $i_{CD,\min}$ occurs at the point D condition.   As the input voltage changes from D to E, R_{CD} again decreases to its rest value, resulting in the output current increasing to $i_{CD,Q}$.

The bias potential offers a high degree of control over the magnitude of i_{CD} by its ability to vary R_{CD}.   In some nonlinear applications, the selection of proper bias values are chosen so as to prohibit the flow of i_{CD} during portions of the input cycle.   In this way, the application possibilities are greatly enhanced.

The bias potential used in transistor work may be positive or negative with respect to common, depending upon the transistor type.   The purpose of the bias voltage is fundamentally the same, however: to set the rest resistance of the transistor at some desired value.

There are a number of techniques that are used to obtain the bias voltages for vacuum tubes and transistors, which are analyzed in the appropriate chapters.

2.5   Selection of the Operating Point.   *The e_{CD}, i_{CD}, and bias parameters at zero input signal condition are the rest values of the control device, and this* *point on the output characteristics is known as the operating point.* The selection of the operating point hinges on a number of factors, including the following:

1. Maximum current and voltage ratings of the device
2. Desired excursions in output current and voltage
3. Magnitude of input signal excursions
4. Type of amplification desired (symmetrical or nonsymmetrical)
5. Power supply potential E_{bb} available
6. Voltage, current, and power ratio considerations

A convenient method by which the operating point and other parameters may be selected involves the *load line*, which is analyzed in the following section.

2.6 Principles of Load-line Construction and Uses. *A load line represents the ohmic value of the resistance connected in series with the control*

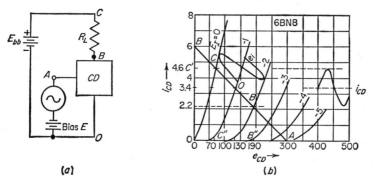

Fig. 2.4   Load-line construction: (a) circuit, (b) output characteristics. [(b) *Characteristics from RCA.*]

device (R_L *in Fig.* 2.4a).   This may be an actual resistor or the reflected primary resistance of a transformer.   Since the resistor is of fixed value, its slope is constant and the load line is a straight line.

The load-line construction upon the output characteristics of the control device enables us to determine a number of factors, including:

1. Useful output signal voltage
2. Maximum input signal voltage permitted for linear operation
3. Waveform of the output signal voltage and current
4. Voltage, current, and power ratios
5. Power output
6. Selection of the best load resistance value

With a given R_L and E_{bb}, the following procedure is useful when constructing the load line:

Step 1. Find the output current which would flow if $R_{CD} = 0$ ohms. At this condition, the entire circuit resistance is R_L; hence by Ohm's law,

$$I \text{ (when } R_{CD} = 0 \text{ ohms)} = \frac{E_{bb}}{R_L}$$

This is point *B*, one of the extreme points on the load line (see Fig. 2.4*b*).

Step 2. When R_{CD} is infinite ohms, the output current is zero, the entire supply potential is dropped across the control device, and $e_{CD} = E_{bb}$ at this condition. This is point *A*, the second extreme point of the load line.

Step 3. Draw a straight line between points A and B, which is the load line for the given conditions.

Upon completion of the load-line construction, the next task is to determine the operating point on the load line, which in effect is to determine the bias voltage to be used. Since this technique is best explained with a practical example, let us consider the common case where faithful reproduction is desired.

Example: The control device circuit of Fig. 2.4*a* has the following parameters: $E_{bb} = 300$ volts; $R_L = 50$ kilohms. The input signal is a sinusoidal voltage whose maximum value is 1 volt. Assume the input terminal must be zero or negative in potential with respect to the common terminal at all signal conditions. (*a*) Construct the load line. (*b*) Determine a suitable operating point.

Solution:

(*a*) Construct the load line:

Following the steps for constructing a load line as outlined in a preceding paragraph, we obtain the two extreme points (refer to Fig. 2.4*b*):

$$I \text{ at point } B = \frac{E_{bb}}{R_L} = \frac{300}{50\text{K}} = 6 \times 10^{-3} \text{ amp}$$

Point *A* is at $e_{CD} = 300$ volts, and points *A* and *B* are joined by a straight line.
(*b*) Determine a suitable operating point:

Since each curve is at a fixed input voltage, the input signal will cause i_{CD} and e_{CD} to vary along the load line up to the maximum excursions of the input signal potential. Point *C* is the highest point allowed on the load line, since the input terminal is not to become positive at any time in this application. That is, point *C* is the maximum i_{CD} and minimum e_{CD} point. Since the input signal has a maximum value of 1 volt, the bias must be at least -1 volt, in order to avoid a positive input voltage. One of the requirements of this circuit is faithful reproduction of the input signal; therefore, e_{RL} above and below rest (the operating point) and i_{RL} above and below rest must be symmetrical and sinusoidal, since these are the characteristics of the input signal voltage waveform. We must select an operating point such that this symmetry is achieved. An approximate indication of meeting this requirement is to have equal lengths of the load line to the left and right of the operating point. In relation to Fig. 2.4*b*, there is a good possibility of achieving a symmetrical output if $CO = OB$. This is most conveniently achieved when R_L is so selected that the load line traverses the characteristic curves where they are most linear.

Let us tentatively place the operating point O at the intersection of the load line and the $E_I = -1$ volt characteristic. In order to determine whether this point is satisfactory, we must check the current and voltage excursions above and below point O. From the load line of Fig. 2.4b, we can read the following values:
Zero signal condition:

$$e_{CD} = 130 \text{ volts} \qquad i_{CD} = 3.4 \text{ ma}$$

Maximum positive signal condition:

$$e_{CD} = 70 \text{ volts} \qquad i_{CD} = 4.6 \text{ ma}$$

Maximum negative signal condition:

$$e_{CD} = 190 \text{ volts} \qquad i_{CD} = 2.2 \text{ ma}$$

The e_{CD} variations above and below the operating point:

$$e_{CD,\text{max}} - e_{CD,Q} = 190 - 130 = 60 \text{ volts}$$
$$e_{CD,Q} - e_{CD,\text{min}} = 130 - 70 = 60 \text{ volts}$$

Since $\Delta e_{RL} = -\Delta e_{CD}$, the e_{RL} variations above and below the operating point are also symmetrical.
The i_{CD} variations above and below the operating point:

$$i_{CD,\text{max}} - i_{CD,Q} = 4.6 - 3.4 = 1.2 \text{ ma}$$
$$i_{CD,Q} - i_{CD,\text{min}} = 3.4 - 2.2 = 1.2 \text{ ma}$$

Hence the i_{CD} variations are symmetrical, and the selected point O is a suitable operating point for the stated conditions. In many cases, there are a number of suitable operating points, thereby allowing some flexibility in the final selection.

The manner in which the input signal undergoes its variations is shown upon the characteristics above the load line.

It should be pointed out that load lines can be constructed with given conditions other than those stated here. These techniques, along with the specific uses of the load line with various vacuum-tube amplifiers, are analyzed in a later chapter.

PROBLEMS

2.1 The following points are obtained from a volt-ampere characteristic of a voltage-controlled device (triode type): When $e_{CD} = 150$ volts, $i_{CD} = 40$ ma; when $e_{CD} = 200$ volts, $i_{CD} = 60$ ma. Find the slope between the indicated points.

2.2 In Prob. 2.1, calculate the dynamic resistance between the two stated points.

2.3 Another curve from the family of volt-ampere characteristics of the device considered in Prob. 2.1 has the following point values: When $e_{CD} = 350$ volts, $i_{CD} = 10$ ma; when $e_{CD} = 400$ volts, $i_{CD} = 20$ ma. Find the slope between the two stated points.

2.4   In Prob. 2.3, calculate the dynamic resistance between the two stated points.

2.5   Which of the two conditions, based on the computed r_p values of Probs. 2.2 and 2.4, must be the characteristic with the greatest bias value?

2.6   The following points are obtained from a volt-ampere characteristic of a voltage-controlled device (pentode type): When $e_{CD} = 160$ volts, $i_{CD} = 12$ ma; when $e_{CD} = 280$ volts, $i_{CD} = 12.5$ ma.   Find the slope between the two stated points.

2.7   In Prob. 2.6, calculate the dynamic resistance between the two stated points.

2.8   Another curve from the family of volt-ampere characteristics of the device considered in Prob. 2.6 has the following point values: When $e_{CD} = 160$ volts, $i_{CD} = 8.5$ ma; when $e_{CD} = 280$ volts, $i_{CD} = 8.55$ ma.   Find the slope between the two stated points.

2.9   In Prob. 2.8, calculate the dynamic resistance between the two points.

2.10 Which curve must have been at the greatest bias value, that of Prob. 2.6 or 2.8? Explain.

2.11 What is the "knee" of a pentode output characteristic?

2.12 What is the "plateau region" of a pentode output characteristic?

2.13 State the relationship illustrated by the transfer characteristics of a voltage-controlled device.

2.14 Refer to the input characteristics of Fig. 2.2.   Consider the curve where e_{CD} is fixed at 20,000 volts.   Find g_m between the i_{CD} values of (a) 1 and 2 ma; (b) 0.5 and 1 ma.

2.15 Refer to the input characteristics of Fig. 2.2.   Consider the curve where e_{CD} is fixed at 15,000 volts.   Find g_m between i_{CD} values of (a) 1.5 and 1 ma; (b) 1 and 0.5 ma.

2.16 Refer to the output characteristics of Fig. 2.1a. e_{CD} is fixed at 100 volts; e_I is varied from 0 to -2 volts.   Find g_m.

2.17 Refer to the output characteristics of Fig. 2.1a. e_{CD} is fixed at 250 volts; e_I is varied from -6 to -8 volts.   Find g_m.

2.18 Refer to the output characteristics of Fig. 2.1b. e_{CD} is fixed at 150 volts; e_I is varied from -0.5 to -1.0 volt.   Find g_m.

2.19 Refer to the output characteristics of Fig. 2.1b. e_{CD} is fixed at 300 volts; e_I is varied from -2.5 to -3 volts.   Find g_m.

2.20 An input signal voltage has a maximum value of 2.5 volts.   What is the minimum bias voltage required to prevent the input terminal from becoming positive during any portion of the input cycle?

2.21 Refer to the circuit and output characteristics of Fig. 2.5a and b. $E_{bb} = 400$ volts; $R_L = 100$ kilohms.   Construct the load line.

2.22 In Prob. 2.21, find the most suitable operating point for faithful reproduction when the input signal voltage maximum value is (a) 0.5 volt; (b) 1 volt; (c) 1.5 volts.

2.23 Refer to the circuit and output characteristics of Fig. 2.5a and b. $E_{bb} = 400$ volts; $R_L = 200$ kilohms.   Construct the load line.

2.24 In Prob. 2.23, find the most suitable operating point for faithful reproduction when the input signal voltage maximum value is (a) 0.5 volt; (b) 1 volt; (c) 1.5 volts.

2.25 Refer to the circuit of Fig. 2.5a and output characteristics of Fig. 2.5c. $E_{bb} = 300$ volts; $R_L = 6$ kilohms.   Construct the load line.

2.26 In Prob. 2.25, find the most suitable operating point for faithful reproduction when the input signal voltage maximum value is (a) 1 volt; (b) 2 volts.

2.27 Refer to the circuit of Fig. 2.5a and the output characteristics of Fig. 2.5c. $E_{bb} = 250$ volts; $R_L = 5$ kilohms.   Construct the load line.

2.28   In Prob. 2.27, find the most suitable operating point for faithful reproduction when the input signal voltage maximum value is (a) 1 volt, (b) 2 volts.

2.7   Output Characteristics of a Current-controlled Device.

The output characteristics of a transistor (current-controlled device) are illustrated in Fig. 2.6.   It should be noted that these curves are somewhat similar to the pentode characteristics.   As pointed out in an earlier section, the control device voltage is often plotted on the x axis for the

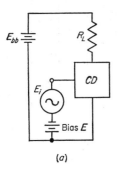

(a)

FIG. 2.5   (a) Series voltage-controlled device circuit, (b) output characteristics of a triode unit, (c) output characteristics of another triode unit. [(b) and (c) from RCA.]

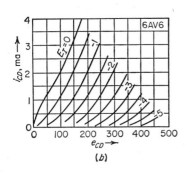

(b)

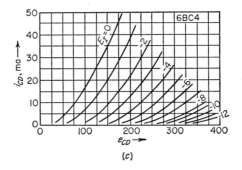

(c)

transistor output characteristics, in spite of the fact that it is a current-controlled device.   The input current is held at a constant value for each curve.   Let us calculate the dynamic resistance r_o of the transistor whose output characteristics are illustrated in Fig. 2.6.

Example 1: Calculate the dynamic resistance between the e_{CD} values of -3 and -4 volts.   Input current is fixed at 80 μa.

Solution:

$$r_o = \frac{\Delta e_{CD}}{\Delta i_{CD}} = \frac{4 - 3}{19.5 - 19} = 2 \text{ kilohms}$$

Example 2: Calculate the dynamic resistance between the e_{CD} values of -6 and -7 volts.   Input current is fixed at 80 μa.

Solution:

$$r_o = \frac{7 - 6}{20.3 - 20} = 3.33 \text{ kilohms}$$

The preceding two examples point out that *larger values of e_{CD}, with a fixed input current, result in the transistor's displaying larger dynamic resistance values.*

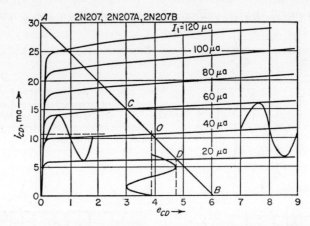

Fig. 2.6   Output characteristics of a transistor.   (*Lansdale Tube Co.*)

Example 3: Calculate the dynamic resistance between e_{CD} values of -3 and -4 volts, and -6 and -7 volts.   Input current is fixed at 40 μa.

Solution:

(a) Find r_o between -3 and -4 volts:

$$r_o = \frac{4 - 3}{10.8 - 10.6} = 5 \text{ kilohms}$$

(b) Find r_o between -6 and -7 volts:

$$r_o = \frac{7 - 6}{11.2 - 11.1} = 10 \text{ kilohms}$$

Example 3 shows the effect of the input current upon the resistance of the transistor: *Reducing the magnitude of the input current results in a increase in the dynamic resistance of the transistor.*   As in the case of the vacuum tube, the customary application of the transistor is to first establish a predetermined voltage across the device and R_L in series, and then to vary its resistance by means of the input signal (current in the case of the transistor).

As in vacuum tubes, an input bias would serve to set the rest resistance of the transistor at some predetermined value.   The base voltage is chosen to establish the desired zero signal input current, since this determines the rest resistance.

In order to determine the control ratio of the transistor in a particular circuit, the input voltage variation associated with the input signal current must be known, since the control ratio is the ratio of $\Delta e_o/\Delta e_I$ with e_{CD} fixed.

2.8 Input Characteristics of a Current-controlled Device.

Figure 2.7 illustrates the input characteristics of a current-controlled device. e_{CD} is a constant value for each curve within the family of characteristics.

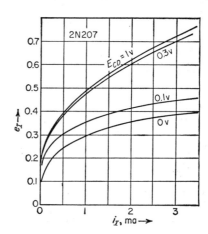

Fig. 2.7 i_I versus e_I characteristics.
(*Lansdale Tube Co.*)

The input current is plotted on the x axis, since it is the controlling factor, and the input voltage is placed on the y axis. It should be noted that current changes near the origin of a given curve are associated with a larger change in input voltage than is the case with current changes further along the same curve, which indicates a reduction of input resistance.

Example 1: Refer to Fig. 2.7. Find the dynamic input resistance r_i between 0 and 0.5 ma of the curve where e_{CD} is fixed at 0 volts.

$$r_i = \frac{\Delta e_I}{\Delta i_I} \quad \text{with } e_{CD} \text{ constant}$$

$$= \frac{0.24 - 0.1}{(0.5 - 0) \times 10^{-3}} = 280 \text{ ohms}$$

Example 2: Refer to Fig. 2.7. Find the dynamic input resistance between -2 and -2.5 ma of the curve where e_{CD} is 0 volts.

$$r_i = \frac{0.37 - 0.35}{(2.5 - 2) \times 10^{-3}} = 40 \text{ ohms}$$

The two preceding examples indicate the effect of the input current upon the input resistance of the transistor, showing that the resistance is reduced by larger magnitudes of input current.

The input characteristics of the transistor differ from those of the vacuum tube. It will be recalled that e_I versus i_{CD} (with e_{CD} fixed) is the relationship for the vacuum-tube input characteristics whereas i_I versus e_I (with a fixed e_{CD}) is utilized in describing the transistor's input characteristics.

2.9   Use of the Load Line with Current-controlled Devices.   The techniques for construction of a load line were discussed in Sec. 2.6 and apply for the transistor as well.   Let us consider a typical example where R_L and V_{cc} are given and a suitable operating point for a given input signal must be determined.   Note that the power supply notation for transistor circuitry is V_{cc} rather than E_{bb}.

Example 1: Refer to Fig. 2.6 for the output characteristics of a transistor which is connected in series with an R_L of 200 ohms and a voltage supply of 6 volts.   The maximum i_{CD} must be less than 20 ma: $i_{I,m} = 20$ μa.
(a) Construct the load line.
(b) Select an operating point such that faithful reproduction is achieved.

Solution:

(a) Construct the load line:

Following the steps outlined in Sec. 2.6:
1. We find i_{CD} when R_{CD} is zero ohms:

$$i_{CD} = \frac{V_{cc}}{R_L} \text{ (when } R_{CD} = 0 \text{ ohms)} = \frac{6}{200} = 30 \text{ ma}$$

This is point A of the load line.
2. e_{CD} (when R_{CD} is infinite ohms) $= V_{cc} = 6$ volts, which is point B of the load line.
3. Points A and B are joined by a straight line, which is the load line for the given conditions.
(b) Select a suitable operating point for faithful reproduction:

Since $i_{CD,\max}$ cannot exceed 20 ma, I_I values larger than 80 μa cannot be used. Let us consider placing the zero signal operating point O at the intersection of the $I_I = 40$-μa curve and the load line.   That is, the input bias is 40 μa.   The input signal without bias has a maximum value of 20 μa; therefore the composite input current varies from 60 μa (point C) to 20 μa (point D).   The input signal is sinusoidal in waveform, as illustrated to the right of the load line.   We may read the e_{CD} and i_{CD} values at rest, maximum, and minimum conditions from the characteristics:

Rest values:

$$i_{CD} = 10.7 \text{ ma} \qquad e_{CD} = 3.85 \text{ volts}$$

Point C values:
$$i_{CD} = 15 \text{ ma} \qquad e_{CD} = 3 \text{ volts}$$
Point D values:
$$i_{CD} = 6.4 \text{ ma} \qquad e_{CD} = 4.70 \text{ volts}$$

The i_{CD} variations above rest:
$$15 - 10.7 = 4.3 \text{ ma}$$
and below rest:
$$10.7 - 6.4 = 4.3 \text{ ma}$$

Hence i_{CD} is symmetrical above and below rest.
The e_{CD} variations above rest:

$$3.85 - 3 = 0.85 \text{ volt}$$
and below rest:
$$4.70 - 3.85 = 0.85 \text{ volt}$$

e_{CD} and therefore e_{RL} are symmetrical above and below the operating point.

A_v, A_i, and G, along with other information, can be easily determined with the aid of the load line.

Example 2: Refer to the load line and operating point of example 1 (see Fig. 2.6). $e_{I,\text{max}} = 0.04$ volt. Find the following information: (a) A_v; (b) A_i; (c) $P_{i,\text{sig}}$; (d) $P_{i,t}$; (e) $P_{RL,\text{sig}}$; (f) $P_{RL,\text{av}}$; (g) G_{sig}; (h) G_{control}.

Solution:

(a) Find A_v:
$$A_v = \frac{\Delta e_O}{\Delta e_I} = \frac{8.5 \times 10^{-1}}{4 \times 10^{-2}} = 21.25$$

(b) Find A_i:
$$A_i = \frac{\Delta i_O}{\Delta i_I} = \frac{4.3 \times 10^{-3}}{20 \times 10^{-6}} = 215$$

(c) Find $P_{i,\text{sig}}$:

$$P_{i,\text{sig}} = i_{i,\text{rms}} e_{i,\text{rms}}$$
$$= 0.707 \times 20 \times 10^{-6} \times 0.707 \times 4 \times 10^{-2} = 40 \times 10^{-8} \text{ watt}$$

(d) Find $P_{i,t}$:
Since 0.04 volt is associated with 20 μa of signal current, then the 40 μa of bias current must be associated with 0.08 volt of bias voltage. Hence

$$P_{i,\text{rest}} = E_{I,\text{rest}} I_{I,\text{rest}} = 8 \times 10^{-2} \times 40 \times 10^{-6}$$
$$= 320 \times 10^{-8} \text{ watt}$$

We may now determine $P_{i,t}$:

$$P_{i,t} = P_{i,\text{sig}} + P_{i,\text{rest}}$$
$$= 40 \times 10^{-8} \text{ watt} + 320 \times 10^{-8} \text{ watt} = 3.60 \times 10^{-6} \text{ watt}$$

(e) Find $P_{RL,\text{sig}}$:

$$P_{RL,\text{sig}} = (I_{o,\text{sig,rms}})^2 R_L$$
$$= (4.3 \times 0.707 \times 10^{-3})^2 200 = 18.48 \times 10^{-4} \text{ watt}$$

(f) Find $P_{RL,\text{av}}$ (which is $P_{o,t}$):

$$P_{RL,\text{av}} = R_L[(I_{o,\text{sig,rms}})^2 + (I_{o,\text{rest}})^2]$$
$$= 200[(3.04 \times 10^{-3})^2 + (10.7 \times 10^{-3})^2] = 2.47 \times 10^{-2} \text{ watt}$$

(g) Find G_{sig}:

$$G_{\text{sig}} = \frac{P_{o,\text{sig}}}{P_{i,\text{sig}}} = \frac{18.48 \times 10^{-4}}{40 \times 10^{-8}} = 4{,}620$$

(h) Find G_{control}:

$$G_{\text{control}} = \frac{P_{o,t}}{P_{i,t}} = \frac{2.47 \times 10^{-2}}{3.6 \times 10^{-6}} = 6{,}861$$

The voltage, current, and power output of the device may be expressed in decibels, based on the following relationships:
For voltage:

$$\text{db} = 20 \log \frac{E_o(R_i)^{\frac{1}{2}}}{E_i(R_o)^{\frac{1}{2}}}$$

where R_i, R_o = input and output impedances
E_o, E_i = input and output voltage variations

NOTE: When the input and output impedances are equal, R_i and R_o cancel in the preceding equation, simplifying the relationship to

$$\text{db} = 20 \log \frac{E_o}{E_i} = 20 \log A_v$$

For current:

$$\text{db} = 20 \log \frac{I_o(R_o)^{\frac{1}{2}}}{I_i(R_i)^{\frac{1}{2}}}$$

where I_o, I_i are the output and input current variations.
NOTE: When the input and output impedance are equal, R_i and R_o cancel out, simplifying the relationship to

$$\text{db} = 20 \log \frac{I_o}{I_i} = 20 \log A_i$$

For signal power:

$$\text{db} = 10 \log \frac{P_{o,\text{sig}}}{P_{i,\text{sig}}} = 10 \log G_{\text{sig}}$$

For power average:

$$\text{db} = 10 \log \frac{P_{o,t}}{P_{i,t}} = 10 \log G_{\text{control}}$$

Example 3: Express the signal power ratio and control power ratio of example 2 in decibels.

Solution:

(*a*) Signal power ratio in decibels:

$$db = 10 \log G_{\text{sig}} = 10 \log 4{,}620$$
$$= 10 \times 3.665 = 36.65 \text{ db}$$

(*b*) Control power ratio in decibels:

$$db = 10 \log G_{\text{control}} = 10 \log 6{,}861$$
$$= 10 \times 3.836 = 38.36 \text{ db}$$

PROBLEMS

2.29 Refer to the input characteristics of Fig. 2.8*a*. Calculate the input dynamic resistance between the following i_I values on the $e_{CD} = 0$ volt curve:
(*a*) 0 and 100 μa
(*b*) 500 and 600 μa

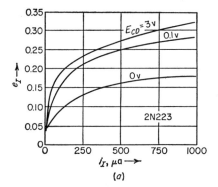
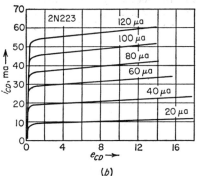

Fig. 2.8 Input and output characteristics of a transistor: (*a*) i_I versus e_I characteristics, (*b*) e_{CD} versus i_{CD} characteristics. (*Lansdale Tube Co.*)

2.30 Repeat Prob. 2.29 for the $e_{CD} = 0.1$ volt curve.
2.31 Repeat Prob. 2.29 for the $e_{CD} = 3$ volt curve.
2.32 State the relationship between the magnitude of the input current and transistor resistance.
2.33 Refer to the output characteristics of Fig. 2.8*b*. Calculate the output dynamic resistance between the following e_{CD} values when i_I is fixed at 100 μa:
(*a*) 2 and 4 volts
(*b*) 6 and 8 volts
(*c*) 12 and 14 volts
2.34 Repeat Prob. 2.33 with i_I fixed at 60 μa.
2.35 Repeat Prob. 2.33 with i_I fixed at 120 μa.
2.36 Refer to the output characteristics of Fig. 2.8*b*. $R_L = 200$ ohms; $E_{bb} = 12$ volts; $i_{I,\text{max}} = 100$ μa; $e_{I,\text{max}} = 0.03$ volt.
(*a*) Construct the load line.
(*b*) Select a suitable operating point.

2.37 From the load line of Prob. 2.36, find:
    (a) A_V (b) A_i
    (c) $P_{i,\text{sig}}$ (d) $P_{i,t}$
    (e) $P_{RL,\text{sig}}$ (f) $P_{RL,\text{av}}$
    (g) G_{sig} (h) G_{control}

2.38 In Prob. 2.9, express the following in decibels:
    (a) G_{sig}
    (b) G_{control}

2.39 Assume $R_i = 35$ ohms; $R_o = 2$ kilohms in the transistor of Prob. 2.37. Express the voltage and current ratios in decibels.

2.40 Refer to the output characteristics of Fig. 2.8b. $R_L = 360$ ohms; $V_{cc} = 18$ volts; $i_I = 100$ μa; assume $e_i = 0.04$ volt.
    (a) Construct the load line.
    (b) Select a suitable operating point.

2.41 From the load line of Prob. 2.40, find
    (a) A_V (b) A_i
    (c) $P_{i,\text{sig}}$ (d) $P_{i,t}$
    (e) $P_{RL,\text{sig}}$ (f) $P_{RL,\text{av}}$
    (g) G_{sig} (h) G_{control}

2.42 In Prob. 2.41, express G_{sig} and G_{control} in decibels.

2.43 In Prob. 2.41, assume $R_i = 100$ ohms; $R_o = 3$ kilohms. Express the voltage and current ratios in decibels.

CHAPTER 3

SEMICONDUCTOR PHYSICS

This chapter is devoted to analyzing the electrical characteristics of semiconductors along with their physical make-up. The properties of germanium which are important in the electrical behavior of this material are considered first. The principles of electron energy levels and bands are introduced and then applied to electron behavior in crystalline structures. The electrical effects of various crystal structure imperfections and impurities are considered. The formation of the junction and its electrical characteristics, which is one of the chief features of the junction transistor, is analyzed under several externally imposed conditions. Junction diode circuit action is investigated along with its volt-ampere characteristics. Many of the foundations for the understanding of transistor action are laid in this chapter.

3.1 The Germanium Atom. The Bohr model of an atom is that of a positively charged nucleus around which electrons revolve within concentric

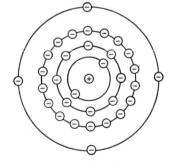

Fig. 3.1 Bohr model of a germanium atom.

shells, and Fig. 3.1 illustrates the model of the germanium atom. Notice that the three innermost shells are complete (2, 8, and 18 electrons, respectively). An unexcited germanium atom possesses 32 electrons, 28 of which are located within the three completed shells, while the remaining 4 electrons occupy positions within the fourth shell. *Since the four outer electrons of the germanium atom exist in an incomplete shell, they are called valence electrons. Electrical action within a material is restricted to the valence electrons, and our analysis of semiconductor behavior is primarily confined to studying the actions of these electrons.*

3.2   Theory of Energy Levels and Bands.   Prior to delving further into the nature of germanium and other semiconductor materials, it is best that we consider the theory of electron energy levels and bands.   *An electron revolving about its mother nucleus possesses both kinetic and potential energy.*   The kinetic energy is due to its orbital motions within the confines of its shell and is equal to $mv^2/2$ (where m is its mass, and v is its velocity).   The electron, being of a negative charge, is attracted to the nucleus with a force that is directly proportional to the core charge and inversely proportional to its distance (radius) from the core.   It will be noted that this is Coulomb's law:

$$f = \frac{(+Q_c)(-Q_e)}{s^2}$$

where Q_c = core charge
$\quad Q_e$ = electron charge
$\quad\quad s$ = distance of electron from core (radius)
$\quad\quad f$ = force of attraction

The potential energy of the electron under consideration is determined from the following relationship:

$$\text{pe} = \frac{(+Q_c)(-Q_e)}{s}$$

The total energy of the orbital electron is equal to the sum of the kinetic and potential energy, and is known to be equal to

$$\text{te} = \text{pe} + \text{ke} = \frac{-Q_cQ_e}{s} + \frac{mv^2}{2} = \frac{Q_cQ_e}{2s}$$

where te = total energy possessed by the electron within its shell
$\quad$ pe = potential energy of the electron
$\quad$ ke = kinetic energy of the electron

The kinetic energy of the orbiting electron is equal to about half of the potential energy but is positive in sign.   The dimension s is the radius of the path of the orbital electron, which becomes smaller for distances closer to the nucleus and results in the total energy possessed by the electron in orbit being negative (primarily potential energy).   On the other hand, if the electron is further away from the nucleus, s becomes larger and the electron's potential energy becomes small in value.   The electron, unless additional energy were supplied, would possess zero energy at an infinite distance from the nucleus. *The implication is that an electron must be furnished additional energy if it is to be removed from its normal shell to regions which are further away from the nucleus.* The manner in which this requirement is achieved is investigated in a subsequent section.

In an unexcited atom, every electron in each shell occupies a definite position with respect to distance from the nucleus. Again noting the total electron energy equation, we see that each electron must possess a definite amount of energy, which can be called an energy level. In an isolated, unexcited atom, the number of possible electron energy levels in a shell is determined by the number of electrons permitted in that shell. This is equally true for those atoms whose outer shell is not filled.

The permissible energy levels of electrons within a specified atom hinges directly upon the positions occupied by the electrons within the atomic structure. Figure 3.2 illustrates the energy level closest to the zero energy level, which is called the valence energy level. This is the

Fig. 3.2 Upper energy level of an isolated atom.

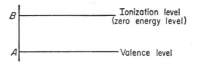

energy level of the electron in the outermost shell at the greatest radius from the nucleus and therefore has the smallest amount of potential energy. The valence electron is closest to the zero energy level, which is the energy level at which the electron would be free. In accordance with the quantum energy theory, the valence electron must receive the correct increment of energy in one package if it is to be excited into mobility.

The behavior of an aggregate of atoms is important to this discussion. As pointed out, each electron in the atom occupies a specific energy level which is determined by its position in the shell arrangement. In a solid, where there are many atoms, there are identical electrons in terms of the shell arrangement in each atom. Consider the outermost valence electron of each atom, which is the electron furthest removed from the nucleus. Assume the energy level of this electron in the isolated atom is the valence energy level illustrated in Fig. 3.2. *Upon placing the various atoms sufficiently close to one another, so as to form a crystal, this valence electron of each atom undergoes a shift in its orbit, thereby altering its energy level. This is based on a physics law, the "Pauli exclusion principle," which states that two electrons in a crystal cannot occupy exactly the same energy level. The result is that the valence electrons of the crystal have a cluster of energy levels which are similar in value, termed the valence energy band, as illustrated in Fig. 3.3.*

The number of energy levels within the valence band hinges upon the number of atoms within the crystal. The common-level electrons of each atom acquire new levels, some above and some below the common level, thereby forming an energy band. Thus it is possible to have as many discrete energy bands as there are electrons in the individual atom.

The ionization level, point C of Fig. 3.3, is the minimum electron energy level required of an electron to be free and hence available for conduction. At this point, the electron's potential energy content is considered zero, since there is no force of attraction between the electron and its mother nucleus. This band, whose lowest level is the ionization level, is the *conduction band*. In order for an electron to be excited into this level, it must receive energy from an external source. Its chances of accomplishing this are determined by the applied energy and the height of the forbidden energy gap between the valence and conduction bands,

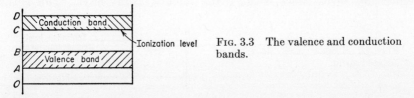

Fig. 3.3   The valence and conduction bands.

which cannot be occupied by electrons of the particular material. *The valence electrons must acquire, at one time, all of the necessary increment of energy to change from a valence level to a conduction level. The energy height of this forbidden gap is one method of making a distinction between conductors, semiconductors, and insulators.*

3.3   Crystalline Structures: Insulators, Conductors, and Semiconductors.

Figure 3.4 is a simplified two-dimensional illustration of the atomic arrangement of a crystal structure. The chemically inert portion of each atom is considered to be its nucleus and inner-shell electrons, while the valence electrons are greatly responsible for the binding force developed between the atoms. Each atom, without its valence electrons, has an over-all positive charge and is represented by the encircled positive charges. The valence electrons are shown as the negative charges in Fig. 3.4.

Let us consider some of the actions of one valence electron of two neighboring atoms, such as electrons A and B of Fig. 3.4. Assume electron A is the outermost electron of its atom, and electron B is similarly the outermost valence electron of an adjacent atom. As pointed out in the preceding section, electrons A and B occupy similar energy levels if each atom is completely isolated. Because of the proximity of the atoms in a solid structure, the orbits of electrons A and B overlap, as shown in Fig. 3.5.

Recall that electrons A and B have similar charges, resulting in a repulsion effect between them. Because of their mutual repulsion, they attempt to maintain as large a separation as possible. Electrons A and B possessed identical energy levels when each atom was in an isolated state. Upon merging into a crystal structure, these valence electrons occupy valence levels which are not identical (the energy level of electron A is slightly greater or smaller than the energy level of electron B). The difference between the energy levels of the overlapping valence electrons is directly related to the extent of overlapping. In crystals, where the valence electrons engage in greater overlapping orbits than in other structures, the difference in energy levels of the involved valence electrons is greater (and vice versa of course). The spreading of valence

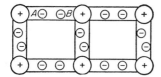

Fig. 3.4 Crystal atomic arrangement.

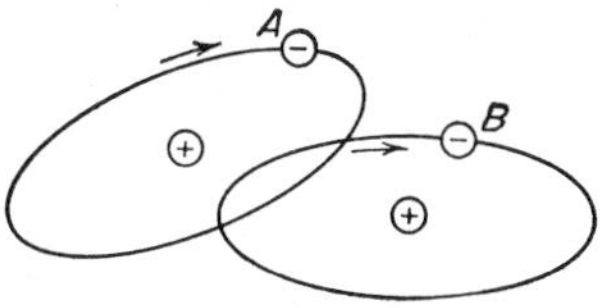

Fig. 3.5 Overlapping orbits of valence electrons.

levels, as here described, occurs throughout the crystal structure, resulting in the development of a cluster of valence energy levels, called the *valence energy band*. The number of energy levels within this band is about equal to the number of atoms within the crystal structure.

The inert portion of each atom has an over-all positive charge and is locked into the structure in an orderly manner. Each ion is repelled by the similarly charged surrounding ions from various directions, thereby stabilizing the position of the ion in the structure. The relatively large mass of the ion also contributes to this stability. The overlapping orbits of the valence electron contribute to the binding action, making the position of each ion even more stable by forming what are called covalent bonds. In a perfect crystal, the orderliness is maintained throughout the structure with no discontinuities in the rows of ions, from which the term lattice structure is derived.

Figure 3.6 illustrates a simplified lattice structure. It should be pointed out that perfect lattice structures do not exist in actual practice owing to several types of lattice imperfections, such as:

1. Interstitial atoms
2. Atomic vacancies
3. Crystal growth dislocations

Figure 3.7 illustrates these lattice defects. Atom A is an interstitial atom, which has no position in the orderly lattice arrangement. An interstitial atom, in effect, is an extra atom wedged into the crystal structure. When one of the rows lacks an atom in a particular area, such as area B in Fig. 3.7b, the structure has an atom vacancy. The third defect is an incomplete row of atoms, as shown in Fig. 3.7c. These defects, along with others not discussed here, have a significant effect on

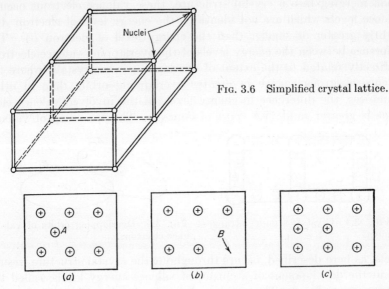

Fig. 3.6 Simplified crystal lattice.

Fig. 3.7 Several lattice structure defects: (a) interstitial atom, (b) atomic vacancy, (c) crystal growth dislocation.

the electrical characteristics of the crystal, as is discussed in a subsequent section.

Also important to the electrical characteristics of a crystal is its chemical purity. The first step in the preparation of a crystal is the removal of as many impurities as possible. Germanium and silicon can be purified to such a great degree that the remaining impurities have a negligible effect in most crystal applications. A later section reveals that certain chemically compatible impurities are then added to the crystal after it has been purified to a high degree.

A preceding section stated that insulators, semiconductors, and conductors can be conveniently described in terms of the valence band, conduction band, and the forbidden energy gap between these two bands.

Insulator. *An insulator may be defined as a material which offers high resistance to the flow of electrons.* The fundamental reason for offering such a high opposition to the flow of electrons is that free electrons are relatively scarce in such materials. Let us determine why mobile electrons are not in great abundance, by referring to Fig. 3.8a, which illustrates the energy diagram of an insulator's valence and conduction band. The energy levels within the insulator's valence band are filled, and the energy levels in the conduction band are empty. This means the valence electrons have very little opportunity to undergo changes from one valence energy level to another. Furthermore, the increment of energy required of a valence electron to be excited into the ionization level (bottom of the conduction band) is very large when compared to the thermal packages of energy (called phonons) which are available at room temperature. *The probability of a valence electron becoming free is quite remote because of the large forbidden energy gap.* Within the valence band, the electron motions are restricted to their orbital motions, and virtually no mobile electrons are available for participation in the conduction process.

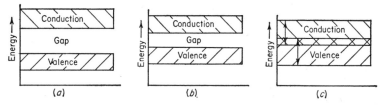

FIG. 3.8 Energy band diagrams: (a) insulator, (b) semiconductor, (c) conductor.

Semiconductors. Figure 3.8b is the energy diagram of a semiconductor. The energy levels within the valence band are filled, as in the case of insulators. The valence electrons have very little opportunity to change levels in the valence band, since there are no vacant levels. The forbidden gap between the valence and conduction band is not as great as in the typical insulator. *The probability of valence electrons acquiring the necessary increment of thermal energy is better, since smaller packages of thermal energy are required. This energy gap, at room temperature, is about 0.7 ev for germanium and 1.1 ev for silicon. Because of the lower height of the forbidden energy gap, semiconductors have resistivity lower than insulators, but considerably higher than conductors.* An electron volt (ev) is the kinetic energy acquired by an electron when accelerated through a one-volt potential difference.

Conductors. The energy diagram of a conductor is illustrated in Fig. 3.8c. Notice the overlap between the upper valence levels and lower conduction levels. *Valence electrons, in such a material, require very small increments of thermal energy to be excited into a conduction level. The probability of this action occurring at room temperature is very high, and the material displays low resistivity (high conductivity).* Such is the condition which exists in most metallic conductors. In certain metals (particularly those which are alkaline in nature), the valence band is not filled. In such cases, conduction can occur by the excitation of valence electrons into vacant valence energy levels.

3.4 Temperature Effects upon Intrinsic Semiconductors. *The temperature coefficient of resistance of a material is defined as the small change in resistance per degree rise in temperature.* Most metals have a positive temperature coefficient (resistance increases with temperature rises). On the other hand, semiconductors possess negative temperature coefficients, indicating a rise in conductivity with temperature rises. Let us examine the effect of thermal energy upon the crystal structure of conductors and semiconductors in order to determine a reason for the opposite temperature coefficient of resistance.

At temperatures above absolute zero, phonons (which are packages of thermal energy) are present throughout the material. Temperature increases result in a more pronounced vibration and movement of the atoms within the lattice structure. Increasing the atomic agitation results in an increase in the number of collisions between the mobile electrons and the atoms. With the presence of an electric field, which imposes a directional characteristic to the mobile electron movements, the mean free path of the mobile electrons (distance between collisions) is reduced, tending to scatter the directions of the electrons. This effect is produced in both semiconductors and conductors.

It will be recalled that a copious supply of mobile electrons are already present at room temperature within a conductor. An increase of temperature does not increase the number of free electrons to any great extent. *In a conductor, the scattering effect due to the increased atomic agitation is more detrimental to the electron drift velocity than the small increase in mobile electrons is beneficial.* The over-all result is that the drift velocity is reduced with temperature increases.

In a semiconductor, the number of electrons liberated by temperature rises is more significant when compared with the number of electrons that were free at the initial or reference temperature. It will be recalled that in a pure semiconductor (also called intrinsic semiconductor), the number of mobile electrons is indeed small. A temperature rise results in a number of valence electrons acquiring the correct package of energy required to leave the valence level and enter the conduction band. To be more specific, the number of mobile electrons increases with temperature in an exponential manner. *The increase in mobile electrons affects the conductivity more than the scattering action, resulting in the conductivity of the semiconductor increasing in an exponential fashion (and the resistivity decreasing exponentially).*

In a perfect germanium lattice at a temperature of absolute zero, there would be no mobile electrons in the conduction band and all the valence energy levels would be filled. At room temperature, thermal energy is present throughout the lattice structure. Valence electrons in pure germanium at room temperature require only about 0.7 ev of energy to break the covalent band and achieve liberation. (Recall that an electron volt is the kinetic energy acquired by an electron when accelerated through a 1 volt potential difference.) The thermal energy within the lattice is in the form of phonons, which are packages of thermal energy, and are of various amounts. The average phonon in intrinsic germanium at room temperature contains considerably less thermal energy than this requirement, and only a relatively few valence electrons succeed in achieving liberation, however. *These electrons, upon making contact with the sufficiently large phonon, break out of their covalent bonds, thereby creating a vacancy in the lattice structure, which is called a hole. In terms of energy, a hole may be defined as a vacant valence energy level.*

The action results in the generation of a free electron in the conduction region and a hole in the valence band, which is the intrinsic generation of an electron-hole pair. The reverse action very likely occurs in a relatively short time: The free electrons give up their recently acquired energy and return to valence energy levels. A free electron and a hole have been taken out of circulation; this process is called *recombination*. The lifetime of the mobile electron, also called the negative carrier, is equal to the time lapse between the instant it was generated and the instant it underwent recombination. The lifetime of the positive carrier, or hole, is determined in the same manner. The average lifetime of an electron-hole pair is 1×10^{-4} sec. At a fixed temperature and constant electric field in an intrinsic semiconductor, generation and recombination of electron-hole pairs take place at the same rate, resulting in a constant average number of mobile electrons and holes in the crystal. A rise in temperature increases the number of electron-hole pairs in the crystal in an exponential manner, up to some maximum temperature.

3.5 The Effect of Light Energy upon Intrinsic Semiconductors.

As previously stated, a valence electron may be excited into the conduction band if it acquires a quantum of energy sufficiently large to enable the electron to acquire the necessary increment of energy. This is called intrinsic conduction. The minimum increment of energy is equal to the energy of the forbidden energy gap between the valence and conductance bands. A limited amount of this action occurs at room temperature because of the thermal packages of energy (phonons).

The conductivity of the semiconductor may be enhanced by the increase in electron-hole pairs made possible by the absorption of light energy, and the result is termed photoconductivity. The packages of energy associated with light are called photons, and they are capable of exciting valence electrons into the conduction band under certain conditions. The energy content of a photon is determined by

$$e = hf$$

where e = light energy
h = Planck's constant
f = light frequency

The wavelength and frequency of light are related in accordance with the following equation:

$$\lambda = \frac{c}{f}$$

where λ = wavelength
c = velocity of light
f = light frequency

The preceding two equations reveal that:
1. Photon energy content increases with light frequency.
2. Wavelength decreases with frequency.

Therefore, light of shorter wavelength possesses photons of larger energy content. At first thought it would seem that light of shorter wavelength, having photons of larger energy content, results in an increased number of electron-hole pairs. This statement is true within certain limits only. *The maximum wavelength which has sufficient energy to contribute to conduction, defined as the long wavelength threshold,* is determined from the following:

$$\lambda_m = \frac{c}{f_{\min}} = \frac{hc}{\Delta e}$$

where λm = long wavelength threshold, μ
hc = 1.237
Δe = electron volts

NOTE: 1 micron = 1×10^{-6} meter.

The long wavelength threshold may further be defined as the wavelength which has a photon of minimum energy content Δe required for electron excitation. The minimum useful wavelength in the photoconducting process is that wavelength which has photons whose energy content is equal to the distance Δe in Fig. 3.9 (energy height of the forbidden energy gap between the valence and conduction bands).

It should be pointed out that the momentum of the electron must be maintained constant when it is excited into the conduction band. *Thus the excitation of a valence electron results only upon fulfillment of two requirements:*

1. *The correct increment of energy must be absorbed.*

2. *The momentum of the electron must be the same before and after excitation.*

The momentum of an electron in the valence or conduction band is maximum at the center of the band and approaches zero near the edges of the band. This imposes a limitation upon the number of electrons which can be liberated by photons. For example, a bound electron at the top of the valence band cannot be excited into the center of the conduction band because of the difference of momentum of these states (zero at the edge of the valence band and maximum at the conduction-band center). Also, an electron near the center of the valence band can only absorb a photon whose energy content will place it in the center of the conduction band (where the momentum state is the same). This is the condition at which the probability of electron excitation is greatest (shown as A to B of Fig. 3.9).

At the long wavelength threshold, the minimum energy required for excitation is available, and the probability of photoelectron excitation

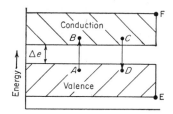

Fig. 3.9 Electron generation and recombination.

is greater than zero. As the wavelength is reduced, the photon energy content increases. At some wavelength, the available energy is equal to the energy required to excite an electron from the center of the valence band into the center of the conduction band. This is the wavelength where the probability of electron-hole generation is greatest. As the wavelength is reduced still further, until the photon energy content is equal to EF of Fig. 3.9 (bottom of the valence band to the top of the conduction band), the direct relationship between wavelength and excitations no longer exists (the probability of photoelectron excitation is zero). The photoconductive response of a semiconductor is a function of wavelength between the long wavelength threshold and the wavelength whose energy content spans the distance between the bottom of the valence band and the top of the conduction band. The chapter on photoelectric devices (Chap. 15) delves further into this topic.

3.6   The Effect of Controlled Impurities.   Practical semiconductors contain traces of impurities deliberately added to the material after the natural impurity level has been reduced to a negligible degree. The types of added impurities fall under two categories:

1. Elements containing five outer-shell electrons
2. Elements containing three outer-shell electrons

Let us examine the effect of each type of impurity upon the energy and electrical characteristics of germanium.

The Donor or N-type Crystal. There are a number of elements possessing five electrons in their outermost shells which are chemically compatible with germanium. Arsenic is a typical example and is used in our discussion. The arsenic atoms occupy positions in the lattice structure, as illustrated in Fig. 3.10a. Four of the outer-shell electrons engage in the formation of covalent bonds with the electrons of the neighboring germanium atoms, thereby completing these bonds. The fifth outer electron of the arsenic atom, finding no vacancy in the lattice, is left out of the locked arrangement. This results in the

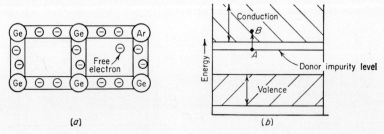

Fig. 3.10 Donor-type germanium: (a) crystal structure, (b) energy distribution.

fifth electron being loosely held to its nucleus even at very low temperatures; it is, in effect, at an energy level above the valence band but below the conduction band (see point A of Fig. 3.10b). There is one electron in such a state for each arsenic atom in the lattice. The energy gap between the energy level of the fifth electron and that of the conduction band is very small—so small that the probability of the fifth electron's being free at room temperature is very great. The energy required to elevate this fifth electron into conduction is considerably lower than that possessed by the phonons at room temperature, and this electron is probably in the conduction band at room temperature, shown as point B in Fig. 3.10b.

A crystal containing an impurity whose atoms possess five outer-shell electrons is called a donor crystal, since it readily donates electrons to the conduction process. The conduction which occurs because of these mobile electrons, when an external electric field is applied, is called *extrinsic conduction,* as opposed to *intrinsic conduction,* which is due to electrons that are elevated to the conduction region from the valence energy band. Because of the great difference in energy requirement between the two types of conduction, extrinsic conduction occurs at a much lower temperature than intrinsic conduction (where the increment of energy must be at least as high as the forbidden energy gap). At a given temperature the conductivity of the donor crystal is higher than that of a pure crystal to an extent determined by the number of impurity atoms added. The major carriers of charge in this type of material are negative, hence the term *N-type* crystal.

The Acceptor or P-type Crystal. Figure 3.11*a* illustrates the lattice structure when a small per cent of the element gallium is added to the purified crystal. Gallium has three valence electrons, leaving one of its covalent bonds incomplete at very low temperatures. At room temperature, the gallium atom readily captures an electron from another atom to complete its covalent bonding arrangement, and any such three-valence elements are called "acceptor" atoms. When the gallium atom has acquired one more electron than it should normally have, it becomes a

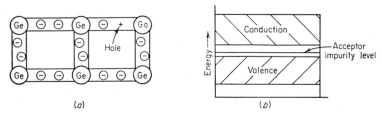

(a)   (b)

FIG. 3.11   Acceptor-type germanium: (a) crystal structure, (b) energy distribution.

negative ion. This extra electron must be at a greater distance from the gallium nucleus than the normal three-valence electrons and is therefore closer to the conduction band in terms of energy. The energy level of the acquired electron is immediately above the valence band, as shown in Fig. 3.11*b*. Notice that only a small increment of energy is required to excite a valence electron into the acceptor impurity energy level, since this level is closer to the valence band than the conduction region. This action usually occurs even at low temperatures, where relatively few valence electrons are thermally excited across the entire forbidden energy gap. Each valence electron excited into the acceptor impurity level leaves a vacancy (hole) in the valence energy band, so it is possible for valence electrons to leave one covalent band and move to another with the help of an externally applied electric field.

The holes may be considered as possessing the mass of an electron and an equal but opposite (positive) charge. With the application of an external electric potential, the valence electrons are encouraged to move from vacancy to vacancy in a direction toward the positive side of the external potential. It should be noted that a valence electron creates a hole when it moves on to fill an adjacent hole, so that the number of holes is constant and determined by the number of acceptor impurity atoms present in the lattice. *As valence electrons are drifting toward the positive terminal of the external potential, the holes are drifting toward the negative terminal of the external potential. Within the extrinsic range of temperatures, at which intrinsic conduction is negligible, the significant portion of conduction occurs by this hole current in the P-type semiconductor.* Actually, the major carriers of change in this material are positive, hence the term *P-type* semiconductor.

3.7   Theory of Deathnium Center Action.   *A deathnium center is defined as a structural defect in the crystal not purposely caused by the addition of controlled impurities.*   Three of the major defects were discussed in an earlier section, namely:

1. Interstitial atoms
2. Atomic vacancies
3. Crystal growth dislocations

Because of these defects, a number of actions may occur within the crystal which cannot be predicted from its normal behavior pattern. The preceding section pointed out the manner in which the number of mobile carriers can be increased by the addition of chemically compatible impurities.   We found that free electrons are more prevalent in the N-type crystal, while holes in the valence band are most common in the P-type crystal.   *Mobile electrons are the majority carriers in the N-type, and holes are the majority carriers in the P-type.   The minority carriers are holes for the N-type and electrons for the P-type crystal.*   In a crystal having no defects other than those created by the addition of the chemical impurity, there would be virtually no minority carriers in either type of crystal.   As it turns out, however, there are significant quantities of minority carriers in even the highest-quality semiconductors.   To explain their presence, the theory of deathnium center action has been adopted.

Deathnium centers are presumably located in those areas of the crystal where structural defects are present, where some of the normally forbidden energy levels are permissible states.   At room temperature, the probability of a particular valence electron acquiring sufficient energy to become mobile is not too great.   But the vacant energy level within the deathnium center may lie within the forbidden energy gap.   A valence electron has a much better chance of obtaining sufficient thermal energy to leave its normal energy level, occupy this normally forbidden level, and create a hole.   Shortly thereafter, the electron may be successful in obtaining the necessary energy to render it free.   Hence both a hole and a mobile electron have been generated by this process.   *Notice that the deathnium center serves as a "stepping stone" in the liberation process.*

Of even greater significance in semiconductor action is that deathnium center behavior generates minority carriers in the crystal.

Again, because of structural defects, a hole may exist in a normally forbidden energy level. Since the energy gap between this level and that of a valence electron is comparatively small, a valence electron can succeed in acquiring the required energy to make this small change in energy level, where it then has a better chance of liberation. The hole may return to its original energy level after the electron has been liberated, or possibly assist another valence electron in the stepping-stone process. A minority carrier free *electron* in a P-type crystal has been generated each time the process is completed. Deathnium center action is relatively independent of external potential but is sensitive to crystal temperatures. Higher temperatures increase the lattice structure vibrations, which produces a larger amount of deathnium center action. Deathnium center action, both recombination and generation, occurs at a comparatively rapid rate, thereby not affecting the over-all carrier charge density of the crystal. The importance of minority carriers within a crystal is treated in several of the following chapters.

3.8 Some Electrical Properties of Common Semiconductor Materials.

Carrier mobility, conductivity, and resistivity are electrical properties of semiconductors which warrant special attention. Let us analyze them in more detail.

Carrier Mobility. It will be recalled that mobile electrons in all materials undergo random motion owing to heat when not subjected to an external electric field. The over-all net displacement of the particle over a period of time is zero. In a semiconductor, either holes or electrons may undergo random motion within the valence band.

The direction in which the charge moves is altered by collisions within the lattice structure, creating a scattering effect. The result is a zigzag, haphazard pattern with no over-all directional characteristic on the part of the carrier. *Drift is the result of an external electric field exerting an additional influence upon the carriers that were undergoing random motion. If the carriers are electrons, they will be urged toward the positive terminal of the potential, whereas holes are made to drift toward the negative terminal. The average distance between the points of collision is called the mean free path of the carrier.* The over-all displacement of the carriers in a direction parallel to the external field within a certain period of time may be considered the uniform velocity of the carriers. *Carrier mobility (μ) is defined as the carrier velocity per unit electric field:*

$$\mu = \frac{v}{E} \qquad cm^2/volt\text{-}sec$$

where

v = average carrier velocity, cm/sec

E = volts/centimeter

Reexpressing this for electron and hole mobility,

$$\mu_n = \frac{v_n}{E} = 3{,}800 \ \text{cm}^2/\text{volt-sec} \qquad \text{pure germanium at 20°C}$$

and $\quad \mu_p = \frac{v_p}{E} = 1{,}800 \ \text{cm}^2/\text{volt-sec} \qquad \text{pure germanium at 20°C}$

where μ_n = electron mobility

μ_p = hole mobility

In intrinsic silicon at 20°C,

$$\mu_n = 1{,}300 \ \text{cm}^2/\text{volt-sec}$$
$$\mu_p = 500 \ \text{cm}^2/\text{volt-sec}$$

Conductivity. Mobile carrier density is defined as the number of free carriers per unit volume of the material. A greater mobile density indicates that a larger number of free carriers exist in the semiconductor. It will be recalled that current is determined by

$$I = \frac{Q}{t}$$

where Q = number of charges moved past a point

t = time

With an external potential applied across the semiconductor, the current is found by

$$I = nqv$$

where n = number of carriers

q = charge per carrier

v = average velocity per carrier

With only electrons present, the preceding equation becomes

$$I = nqv_n$$

where v_n is the average velocity per electron.

With only holes present, the equation becomes

$$I = nqv_p$$

where v_p is the average velocity per hole.

With both types of carriers present, the current equation becomes

$$I = nqv_n + nqv_p = nq(v_n + v_p)$$

From the carrier-mobility equation, the average velocity v is equal to the product of the carrier mobility μ and the unit electric field E:

$$v = \mu E$$

Thus
$$v_n = \mu_n E$$
and
$$v_p = \mu_p E$$

We may now restate the current equations with the proper substitution:

$$I = nqv_n = nq\mu_n E \qquad \text{electrons only}$$
$$I = nqv_p = nq\mu_p E \qquad \text{holes only}$$
$$I = nq(v_n + v_p) = nq(\mu_n E + \mu_p E) = nqE(\mu_n + \mu_p)$$

Thus the current is directly proportional to (1) the number of mobile carriers, (2) the carrier mobility, and (3) the external potential. With a larger number of free carriers in the semiconductor, more become involved in the drift action. With higher carrier mobility, the carriers transverse a given distance parallel to the external potential in less time. The velocity of the mobile carriers is directly proportional to the magnitude of the applied potential.

Conductivity σ is the reciprocal of resistance; therefore

$$\sigma = \frac{1}{R} = \frac{I}{E} = \frac{nqE(\mu_n + \mu_p)}{E}$$

Canceling E from the numerator and denominator, intrinsic conductivity becomes

$$\sigma = nq(\mu_n + \mu_p)$$

For the N-type crystal:

$$\sigma_n = nq\mu_n$$

For the P-type crystal:

$$\sigma_p = nq\mu_p$$

Resistivity. *Resistivity is the reciprocal of conductivity;* hence:

For the N-type crystal:

$$p_n = \frac{1}{\sigma_n} = \frac{1}{nq\mu_n}$$

For the P-type crystal:

$$p_p = \frac{1}{\sigma_p} = \frac{1}{nq\mu_p}$$

For the intrinsic crystal:

$$p = \frac{1}{\sigma} = \frac{1}{nq(\mu_n + \mu_p)}$$

The intrinsic resistivity of germanium at room temperature (20°C) is 60 ohms-cm, and 230 kilohms-cm for silicon.

Figure 3.12 is a graph of the resistivity of a typical germanium crystal (either type) as a function of temperature. The Kelvin scale reads like the centigrade scale, except that it starts at absolute zero, which is $-273°C$ (there are no moving charges at $0°K$, according to some theories). Room temperature is about $20°C$ or $293°K$. At the low temperatures, the majority carriers are primarily the result of the crystal's impurities. In effect then, the mobile carriers are primarily the result of extrinsic excitation at low temperatures. *Between points A and B, which are at very low temperatures, the mobile carriers made possible by the impurities begin to become available at just above $0°K$. The resistivity decreases between A and B because of the increased number of carriers produced by extrinsic excitation.* At point B, the number of mobile carriers does not materially increase, since the impurity atoms have already contributed their mobile carriers and yet the temperature is not high enough to

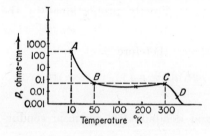

FIG. 3.12 Temperature versus resistivity for a typical germanium crystal.

promote intrinsic excitation (elevation of valence electrons into the conduction band). Therefore, there is relatively little change in resistivity between points B and C. *Point C, which is slightly above room temperature, is the point where intrinsic conduction becomes appreciable, resulting in a significant decrease in resistivity.* The region between points C and D is the intrinsic excitation region, since the number of valence electrons excited into the conduction band is large as compared to the number of impurity states possessed by the crystal. The highest safe operating temperature is somewhere near point D, where the intrinsic excitations dominate the resistivity characteristic of the crystal. Temperatures greater than D result in lattice structure damage. The conductivity of the crystal is the inverse value of its resistivity.

3.9 Formation and Behavior of a PN Junction (Diode).

Figure 3.13 illustrates the conditions inside a PN junction immediately after its fabrication and prior to the time when diffusion and recombination take place. The actual manner in which such a junction is manufactured is not discussed in this text, as the fabrication techniques are not of primary concern here. Let us assume that the P- and N-type crystals are made to resemble the illustration of Fig. 3.13 and examine its electrical characteristics at that time.   Also assumed is that the junction is at room temperature.

Being at room temperature, the holes in the P-type material have been occupied by nearby valence electrons.   This results in the acceptor impurity atoms becoming negative ions, which are represented by the encircled negative charges.   The holes are located in the valence band.   The over-all charge of the P-type crystal is zero.

In the N-type material, which is located on the right side of the junction in Fig. 3.13, the donor impurity atoms lose their fifth outer-shell electron at room temperature.   This results in the donor impurity atoms becoming positive ions, represented by the encircled positive charges.   The

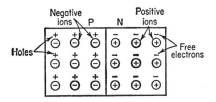

Fig. 3.13   A PN junction prior to diffusion and recombination.

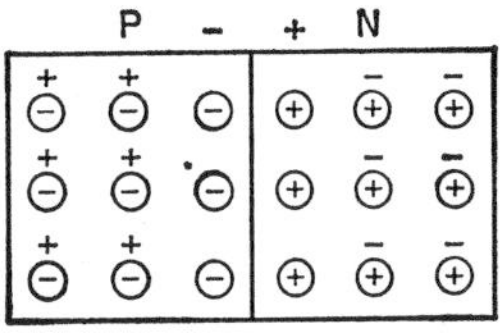

Fig. 3.14   A PN junction after diffusion and recombination.

fifth outer-shell electron of the donor impurity atoms is free.   The over-all charge of the N crystal is zero.

With no further action taking place, each acceptor ion $(-)$ is neutralized by a hole, and each donor ion $(+)$ is neutralized by a free electron. Upon formation of the junction, diffusion and recombination take place across the junction area.   Diffusion is brought about by the mutual repulsion which similarly charged mobile carriers exert upon each other. Some of the free electrons within the N side of the junction wander into the P-type material at the opposite side of the junction because of diffusion.   It will be recalled that a number of holes exist in the P region. The free electrons, upon crossing the junction, undergo recombination with the holes.   As this process goes on, the N side of the junction becomes positive, which is the charge polarity of the unbalanced donor impurity ions, and the P side becomes negative.   These charges develop as the outcome of free electrons leaving the N region and recombining with holes in the P region.   A junction potential is developed, as illustrated in Fig. 3.14, even though the total net charge of the crystal is zero.

The magnitude of the junction potential is dependent upon the number of free electrons and holes involved in the process just described.   Notice that the polarity of this potential is such that free electrons in the N-type region require additional energy to overcome the repulsion (negative) effect at the P side of the junction.   The same conditions exist for holes which may otherwise tend to wander from the P-type to the N-type region.   *The junction potential serves as a potential barrier for majority carriers attempting to migrate across the junction.*   It should be noted that the junction potential exists almost immediately after the junction is formed.

The junction potential develops because of the initial diffusion and recombination activity, which depletes the free carriers within the junction region.   Because of the scarcity of mobile carriers (both holes and electrons), each side of the junction takes on the charge of the unbalanced impurity ions found in each region.   *The junction region where the mobile carriers have been recombined is often called the depletion region.*   The resistance of the depletion region is similar to that of the intrinsic material, since virtually no free carriers are present, whereas the resistance of the P and N regions outside the junction area is considerably lower.

Increasing the width of the depletion region increases the junction potential.   This must be so, since removal of holes in the P region and free electrons in the N region for greater distances from the junction result in a greater number of involved impurity ions.   Hence the P side becomes more negative, and the N side becomes more positive.

The N side of the junction has its greatest charge density in the area closest to the junction, where the probability of a free electron leaving the N region to cross the junction is greatest.   Farther away from the junction, the positive charge density is reduced.   At some distance, virtually none of the free electrons are affected by the formation of the junction potential, and the charge density from that point to the end of the N region is zero.   Similarly, the charge density on the P-region side of the junction is predictable.   Immediately adjacent to the junction, the probability of the holes undergoing recombination with electrons from the N region is greatest, and the negative ion density is greatest.   The probability of a hole not recombining with an electron increases with distance from the junction, until at some point, virtually zero charge density exists.

A logical question at this point is: To what extent does junction ionation go on?   When a free electron from the N region diffuses across the junction and recombines with a hole, the P side of the junction becomes more negative by an amount equal to the electron charge.   This presents a repulsion effect upon the next free electron migrating across the junction.   Eventually, after enough recombinations, the P side of the junction is sufficiently negative to prevent any further migration of free electrons across the junction, and the junction potential is formed.   *The junction potential is considered a potential hill or potential barrier, since majority carriers from either region must acquire additional energy to pass over the junction.*

Figure 3.15 illustrates the potential hill as seen by free electrons and holes. Let us first consider the effect of this junction barrier upon free electrons. Free electrons are the majority carriers in the N region and minority carriers in the P region. In order for an electron to leave the N region, cross the junction, and arrive in the P region, it must acquire sufficient energy to enable it to overcome the repulsion force exerted by the negative ions residing on the P side of the junction. Once the junction is formed, the probability of a free N-region electron acquiring this needed increment of energy by thermal means is small. Free electrons in the P region, on the other hand, would be attracted across the junction, if they were sufficiently close to the junction, because of the positive ions residing on the N side of the depletion region. The probability of this type of electron action is also remote because there are only minority carriers in the P region. The following sections go into more detail

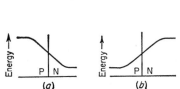

FIG. 3.15   Potential hill of a PN junction: (a) electrons, (b) holes.

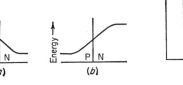

FIG. 3.16   PN junction with no external potential.

with regard to mobile carrier behavior due to the junction potential. At this point, it is important to note that once the depletion region is formed, no further carrier migrations are experienced for practical purposes unless an external potential is applied across the junction.

3.10   Junction Action with No External Potential.   Figure 3.16 illustrates a PN junction with no external potential applied to the junction and the switch in its open position. A junction such as this one is also called a junction crystal diode. Let us examine its behavior in the absence of an external voltage.

Recall that the polarity of the junction potential is such that majority carriers must acquire additional energy if they are to be diffused across the depletion region. At room temperature, a free electron from the N region or a hole from the P region has little probability of obtaining the required increment of energy, so the flow of majority carriers across the junction barrier can be only a very small current. Also present in each region are a number of minority carriers (free electrons in the

P region and holes in the N region) as generated by the deathnium center action described in a preceding section. The lifetime of the minority carriers is considerably shorter than that of the majority carriers. A hole generated in the N region, for example, would rapidly undergo recombination because of the comparatively many free electrons in the N region, while a free electron in the P region would encounter relatively many holes in the area. But some of the minority carriers, upon generation, are close enough to the junction to diffuse into the depletion region, where they are no longer so likely to undergo recombination. Those minority carriers will very likely cross the junction and arrive in the opposite-type region. The probability of this minority carrier action occurring is equal to the probability of majority-carrier flow across the junction, with no external potential present. Equal currents of the same type carriers flowing in opposite directions result in the total junction current being zero. Figure 3.17 illustrates the four individual currents discussed in this paragraph (the majority- and minority-carrier flow of each region).

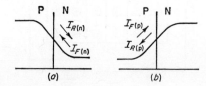

FIG. 3.17   Carrier flow in a diode with no external potential: (a) electron flow, (b) hole flow.

If we assign positive polarity to majority current and negative to minority current, *the majority carrier flow is the forward current I_F and the minority carrier flow is the reverse current I_R.* Refer to Fig. 3.17a. In equation form,

$$I = I_F - I_R$$

Since

$$I_F \cong I_R$$

then

$$I \cong 0$$

where I = total electron current across the junction

I_F = forward electron current

I_R = reverse electron current

The same analysis is applicable to the hole currents (shown in Fig. 3.17b). *The total junction current, with no external potential applied, is zero.*

With the switch in Fig. 3.16 closed, a readable forward current (called the diode-contact current) will flow. When the P and N regions are externally connected with zero external voltage, I_F is greater than I_R by a small value, owing to the "battery effect" of the junction potential (the negative side of the junction repels electrons out of the P crystal, and the positive side of the junction drains electrons from the N crystal).

3.11 Junction Action with Forward and Reverse Potential

Forward Potential. Figure 3.18a illustrates the PN junction with an applied external "forward" potential. Notice that the positive terminal of the external potential is connected to the P-type region (which is the negative side of the junction) and the negative terminal to the N-type region of the diode (which is the positive side of the junction). Such a connection results in the donor electrons in the N crystal being repelled toward the junction while the acceptor holes in the P crystal also are repelled toward the junction. Therefore, some of the impurity atoms in the original junction region are no longer ionized, the width of the depletion region is compressed because of the reduction in negative ions on the P side and positive ions on the N side of the junction, and there is a corresponding reduction in charge density because of the decreased number of unbalanced ions. The potential barrier is reduced in accordance with the decrease in junction charge density. Figure 3.18b and c

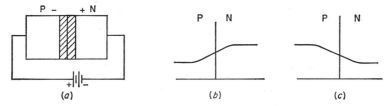

Fig. 3.18 (a) PN junction with forward potential, (b) potential hill (holes), (c) potential hill (electrons).

depicts the potential hill for holes and electrons. *Forward potential or bias is of such a polarity that the height of the junction barrier is reduced.* The effect of this upon carrier flow is important. Majority carriers must acquire additional energy to cross the depletion region because of the potential barrier, and the increment of this required energy is equal to the height of the potential. *A reduction in the potential hill reduces the energy increment required by the majority carriers to cross the junction barrier.* The probability of majority carriers crossing the depletion region is directly proportional to the forward potential which reduces the junction barrier.

Minority-carrier flow, as stated in the preceding section, is possible because of the deathnium center activity in each crystal region. With a fixed temperature, the deathnium center activity goes on at a constant rate. Being unaffected by external potential, the minority carrier flow is the same, with or without voltage applied to the diode. *With no potential across the diode (and the P and N regions externally not connected), the forward and reverse currents are equal, yielding a total current of zero.*

With the application of forward potential, the forward current becomes many times greater than the unchanging reverse current. Hence for all practical purposes, the diode current is equal to the forward (majority) current under conditions of forward bias. The relationship between the forward potential and the current is nonlinear up to the potential which cancels the junction potential. With sufficiently large forward bias, the barrier potential is effectively nonexistent since all the free electrons can cross the junction, and the forward current becomes a linear function of any series resistance in the external circuit alone.

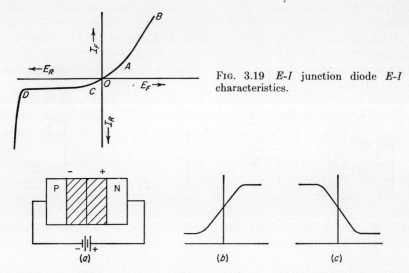

Fig. 3.19 *E-I* junction diode *E-I* characteristics.

Fig. 3.20 (*a*) PN junction with reverse potential, (*b*) potential hill (holes), (*c*) potential hill (electrons).

Figure 3.19 illustrates the volt-ampere characteristic of the junction diode. Forward external voltage is plotted on the positive x axis, and the forward current is plotted on the positive y axis. The characteristic between O and A represents the nonlinear $E-I$ relationship for those forward voltages which are smaller in magnitude than the barrier potential. From points A to B, the $E-I$ characteristic is essentially a linear function of the circuit resistance alone because the barrier potential is effectively overwhelmed at point A. Point B is determined by the maximum forward current allowed in terms of *not* overheating the diode unit. *The portion of the characteristics from the origin to point B is the forward volt-ampere characteristic of the PN junction diode.*

Reverse Potential (Bias). The PN junction with the application of reverse bias is illustrated in Fig. 3.20*a*. *Reverse potential or bias is a voltage the effect of which is to increase the height of the junction potential* *hill* (see Fig. 3.20*b* and *c*). The negative terminal of the external voltage is connected to the P region and the positive terminal to the N crystal. Holes are majority carriers in the P layer and are attracted toward the negative terminal of the external potential. The hole density in the junction area is reduced, thereby increasing the negative ion charge density in the P side of the junction area. Also, the influence of this external potential is such that unbalanced negative ions exist at greater distances from the junction in the P region. This widens the depletion region (the region where there is an absence of mobile carriers) within the P-type crystal.

The positive terminal of the external potential, being connected to the N-type material, creates the same effect with the mobile electrons in the N-type crystal: The positive unbalanced donor ions exist for a greater distance from the junction in the N region, with a corresponding increase in the depletion region width on this side of the junction.

Since more unbalanced negative acceptor ions (on the P side) and unbalanced positive donor ions (on the N side) are created by an external potential of this polarity, the junction potential is greater (the potential hill is higher).

The original potential hill height (which exists with zero external potential) need be increased by only a small amount to completely prohibit majority current through the junction. The minority-carrier flow is unaffected by the potential hill height because it flows "down" the hill. Since a constant diode temperature is assumed, *the minority-carrier flow remains at a constant value, being insensitive to potential hill changes.* Considering the electron currents, we have the relationship

$$I = I_F - I_R$$

At small values of reverse bias, I_F becomes smaller, since fewer electrons from the N region are capable of acquiring the larger increment of energy required to cross the heightened junction barrier. But I_R remains the same, since the deathnium center action remains constant at a fixed temperature. Therefore the total diode electron current is a minority electron current equal to the difference between the majority and minority electron currents. *As the reverse bias is increased, the total diode current approaches a value equal to the reverse current. This occurs at the potential where all forward current is prohibited by the increased barrier height.* This relationship, from zero diode current to total reverse current, occurs in the region between points *O* and *C* in the volt-ampere characteristic of Fig. 3.19. Further increases in the magnitude of the reverse bias have virtually no effect upon the reverse current for a relatively large voltage range (points *C* to *D* of Fig. 3.19). Since the total reverse current flows at the point *C* condition, this reverse current remains constant, assuming the diode temperature is steady. *The maximum reverse current which remains relatively unaffected by an external potential is called the saturation current I_S of the diode.* In Fig. 3.19, the saturation current region exists between points C and D. This phenomenon is used to advantage in "constant-current" regulators.

3.12 Boltzmann's Law. A semiconductor diode's current hinges upon the temperature of the material. The relationship between carrier energy and temperature is conveniently evaluated with the use of Boltzmann's law.

A carrier in any solid may be assumed to possess three degrees of freedom, such as (1) to the right and left (x axis); (2) up and down (y axis); (3) toward and away from the reader (z axis). These three degrees of freedom are possible for vibrational motion. It is assumed that the vibrational energy of the carrier is equally distributed between these three degrees of freedom. *Boltzmann's law states that the magnitude of vibrational energy of a carrier is directly proportional to the temperature of the solid:*

$$E_{\text{vib}} = \frac{kT}{2}$$

where E_{vib} = vibrational energy of the carrier
k = Boltzmann's constant
= 1.38×10^{-16} erg/deg of freedom
= 8.62×10^{-5} ev/deg of freedom
T = absolute temperature

From Boltzmann's law, the relationship between carrier energy level and semiconductor temperature can be more readily seen. Boltzmann's constant is incorporated in the diode-current equation in the following section.

3.13 Semiconductor Diode Equations. Refer to Fig. 3.19, which is a typical volt-ampere characteristic of a semiconductor diode. The following equation is a reasonable representation of this volt-ampere characteristic:

$$I \cong I_S(\epsilon^{qV/kT} - 1)$$

where I = diode current in amperes
I_S = reverse saturation current
ϵ = natural log base (about 2.72)
V = diode applied voltage
k = Boltzmann's constant
q = charge per carrier
T = absolute temperature of the diode

For room temperature conditions, the exponent of epsilon ϵ may be simplified:

$$\frac{qV}{kT} \cong \frac{qV}{8.62 \times 10^{-5} \times 2.93 \times 10^2}$$

Setting $q = 1$, then

$$\frac{q}{kT} = \frac{1}{2.5 \times 10^{-2}} = 40$$

hence

$$\frac{qV}{kT} = 40V \text{ at room temperature}$$

The preceding diode equation, for room temperature conditions, may be simplified to

$$I \cong I_S(2.72^{40V} - 1)$$

The equation is a general form for applied voltages which are less than the junction barrier height. With forward voltage values larger than the junction potential hill, the rectifier equation may be further simplified to the following:

$$I \cong I_S(2.72^{qV/kT})$$

With large values of reverse bias, $I \cong I_S$ for all practical purposes (since the saturation current is relatively independent of the magnitude of the reverse voltage).

Example: A semiconductor rectifier has a reverse saturation current of 30 μa and is operated at room temperature. Find the forward current when the diode forward bias is 0.2 volt.

Solution: Using the simplified equation:

$$I \cong I_S(2.72^{40V} - 1) = 30 \times 10^{-6}(2.72^8 - 1)$$
$$\cong 30 \times 10^{-6}(\text{antilog}_e\ 8 - 1) = 30 \times 10^{-6} \times 2.999 \times 10^3$$
$$\cong 89.97 \times 10^{-3} \text{ amp}$$

3.14 Electrical Breakdown.

Every PN junction will undergo "electrical breakdown" at some magnitude of reverse voltage, as characterized by a sharp increase in the reverse current through the junction. *In the volt-ampere characteristic illustrated in Fig. 3.19, electrical breakdown occurs at point D. The reverse voltage at point D is called the breakdown voltage of the diode.* The magnitude of this breakdown voltage imposes a limitation upon the voltage which may be applied to the rectifier.

Junction breakdown occurs by the Zener effect in very thin junctions. In such cases, the Zener effect occurs at lower reverse voltages when the junction is at a higher temperature. *The Zener effect is the passing of valence electrons through the forbidden energy gap into the conduction band. In junctions of average and above average thickness, breakdown occurs because of the avalanche effect.*   This is a process by which mobile electrons collide with valence electrons, imparting sufficient energy upon collision to liberate an electron-hole pair.   *The actions of one free electron result in the creation of three mobile charges in such cases (the original free electron and electron-hole pair).   When this action takes place in a wholesale fashion, avalanche breakdown occurs.*

The electrical breakdown leads to destruction of the junction if it is accompanied by excessive reverse current and high temperature.   Placement of the proper series resistance and "heat sinks" in the diode circuit will limit the breakdown current to well within safe values in those cases where the breakdown region is utilized (such as certain switching applications).

The diode may also electrically break down with excessive forward current, which overheats the junction.   The maximum forward current as well as the breakdown voltage values must be observed when using the semiconductor diode.

3.15 Metal-semiconductor Diodes.   A metal-semiconductor diode is formed by making contact between the semiconductor and metal.

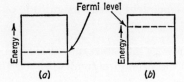

(a)   (b)

Fig. 3.21   Energy levels before contact: (a) metal, (b) N-type semiconductor.

Assume the semiconductor is the N type, which is the usual case.   Prior to making the contact, the mobile electrons in the N-type material are at higher energy levels than the free electrons in the metal.   Recall that a semiconductor free electron must acquire a larger increment of energy to achieve freedom than would be the case for a free electron in a metal conductor, as represented in Fig. 3.21.

The Fermi level is the average energy level occupied by an electron.   Hence, the Fermi level of the metal is lower than the Fermi level of the N-type semiconductor.   *Upon electrical contact, many of the mobile electrons in the N region will flow into the metal in order to seek a lower energy level. Hence, the donor impurities in the vicinity of the junction become unbalanced positive ions (see Fig. 3.22).   The electrons which leave the N region remain near the contact, thereby making that side of the contact negative, which forms a potential barrier.   Electrons continue to flow from the N region into the metal until the Fermi levels of the two materials are equalized.*   Electrons leaving the N region reduce the semiconductor Fermi level, and these same electrons raise the Fermi level of the metal.   When the Fermi levels are in balance, the contact is in equilibrium, and the electron migration stops.

Forward Bias. In Fig. 3.23, the positive side of an external potential is connected to the metal, and the negative terminal is connected to the N-type semiconductor. This potential reduces the barrier height, which enables more electrons to pass from the N region through the barrier into the metal. At some value of forward potential, the barrier potential is completely overwhelmed, making it possible for all the free electrons in the N region to cross the contact. The electron drift throughout the

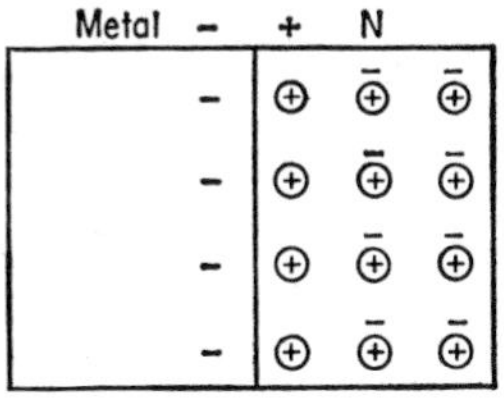

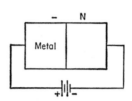

Fig. 3.22 Metal N-type semiconductor contact.

Fig. 3.23 A metal-semiconductor diode with forward bias.

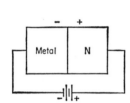

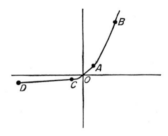

Fig. 3.24 A metal-semiconductor diode with reverse bias.

Fig. 3.25 *E-I* characteristics of the metal-semiconductor diode.

external circuit replenishes the N region with the same number of electrons. Greater current values can be achieved by increasing the velocity of these electrons. *E* and *I* have a nonlinear relationship in the forward position of the characteristic up to the point where the barrier voltage is overwhelmed (*O* to *A* of Fig. 3.25). From that point on to greater forward voltages (*A* to *B* of Fig. 3.25), *E* and *I* have a linear relationship since the current is limited only by the external circuit resistance.

Reverse Bias. An external potential with the polarity shown in Fig. 3.24 causes an increase in the contact barrier height, resulting in relatively little reverse current. After a small magnitude of reverse bias, the reverse current reaches its saturation level (see Fig. 3.25). The general behavior of the "point-contact" or metal-semiconductor diode is similar to that of the PN junction type. The rectifier equations stated in the previous section hold approximately true for the metal-semiconductor diode as well.

PROBLEMS

3.1 State the number of orbital electrons in the unexcited germanium atom.

3.2 Which electrons in the germanium atom play the most significant part in the lattice structure? Explain.

3.3 What is an energy level?

3.4 Briefly describe the principles behind the formation of an energy band.

3.5 Why is the ionization level referred to as the zero energy level?

3.6 State the relationship between the distance of an orbital electron and its potential energy content. Explain.

3.7 What is Pauli's exclusion principle?

3.8 Explain the development of the valence energy band in terms of Pauli's exclusion principle.

3.9 In terms of the forbidden energy gap, describe the difference between insulators, semiconductors, and conductors.

3.10 Briefly describe the development of a covalent bond.

3.11 State three types of lattice structure imperfections.

3.12 What is a deathnium center? Briefly describe its possible actions.

3.13 In what energy band does electron conduction occur in an insulator? Explain.

3.14 In what energy band does electron conduction occur in most metal conductors? Explain.

3.15 In what energy band does hole conduction occur in semiconductors? Explain.

3.16 In what energy band does electron conduction occur in a semiconductor? Explain.

3.17 What are thermal packages of energy called?

3.18 What are radiation (light) packages of energy called?

3.19 What is an intrinsic semiconductor?

3.20 Explain the reason for semiconductors possessing a negative temperature coefficient of resistance and metals possessing a positive temperature coefficient of resistance.

3.21 What is the absolute temperature at (a) 0°C; (b) 20°C?

3.22 Intrinsic conduction (in a semiconductor) occurs only with the generation of electron-hole pairs. Explain.

3.23 What is photoconductivity?

3.24 Define the long wavelength threshold.

3.25 Assume the forbidden energy gap of intrinsic germanium to be 0.72 ev. Calculate the long wavelength threshold in microns.

3.26 Assume the forbidden energy gap of intrinsic silicon is 1.1 ev. Calculate the long wavelength threshold in microns.

3.27 Assume the fifth outer-shell electrons of the N-type impurity in a germanium crystal require 0.03 ev to achieve the ionization potential. Calculate the long wavelength threshold in microns.

3.28 State the two requirements for exciting a valence electron into mobility. Explain.

3.29 What is the relationship between the wavelength of light and photon energy?

3.30 State the two general types of chemical impurities added to germanium.

3.31 Describe the energy distribution within the donor-type crystal at (a) absolute zero; (b) room temperature.

3.32 Describe the energy distribution with the acceptor-type crystal at (a) absolute zero; (b) room temperature.

3.33 Describe intrinsic conduction.

3.34 Describe extrinsic conduction.

3.35 Extrinsic conduction occurs at a much lower temperature than intrinsic conduction. Explain.

3.36 What is the donor impurity energy level? Explain its effect upon conductivity.

3.37 What is the acceptor impurity energy level? Explain its effect upon conductivity.

3.38 Assume a P-type and an N-type crystal contain the same impurity density and are at the same temperature. Which crystal has the highest conductivity? Explain.

3.39 State the minority carriers for the (a) N-type crystal, (b) P-type crystal.

3.40 State the majority carriers for the (a) N-type crystal; (b) P-type crystal.

3.41 Briefly describe the manner in which deathnium center action generates minority carriers.

3.42 Deathnium center action is directly proportional to temperature. Explain.

3.43 Deathnium center action is independent of external potential. Explain.

3.44 What is the difference between random motion and drift?

3.45 What produces the scattering effect in drift action?

3.46 Define carrier mobility.

3.47 Electron mobility in intrinsic germanium at room temperature is about 3,800 cm² per volt-sec. A potential of 10 volts per cm is applied to the crystal. Find the average electron velocity in centimeters per second.

3.48 Repeat Prob. 3.47 with an external potential of 30 volts per cm.

3.49 Hole mobility in intrinsic germanium at room temperature is about 1,800 cm² per volt-sec. A potential of 10 volts per cm is applied to the crystal. Find the average hole velocity in centimeters per second.

3.50 Repeat Prob. 3.49 with an external potential of 30 volts per cm.

3.51 Electron drift mobility in silicon at room temperature is about 1,300 cm² per volt-sec. An external potential of 10 volts per cm is applied to the crystal. Find the average electron velocity.

3.52 Repeat Prob. 3.51 with an external potential of 30 volts per cm.

3.53 Hole drift mobility in silicon at room temperature is 500 cm² per volt-sec. An external potential of 10 volts per cm is applied to the crystal. Find the average hole velocity.

3.54 Repeat Prob. 3.53 with an external potential of 30 volts per cm.

3.55 Assume an N-type crystal contains 1.9×10^7 impurity atoms. An electron charge is 1.60×10^{-19} coulomb. Assume the average velocity per electron is 5,000 cm per sec when an external potential is applied to the crystal. Find (a) the number of free electrons at room temperature; (b) the crystal electron current.

3.56 Find the intrinsic conductivity of germanium at room temperature.

3.57 Find the intrinsic conductivity of silicon at room temperature.

3.58 What is the approximate absolute temperature range of extrinsic conduction for a typical germanium crystal?

3.59 At what absolute temperature does intrinsic conduction for a typical germanium crystal begin?

3.60 Why is there a temperature gap between the upper end of the extrinsic conduction range and lower end of the intrinsic conduction region?

3.61 Explain the manner in which the barrier potential of a PN junction is developed.

3.62 Why is the charge distribution greatest in the region immediately adjacent to the junction?

3.63 What is a potential hill?

3.64 Why is there no resultant junction current with zero external potential?

3.65 Define forward current and reverse current.

3.66 What is forward potential?

3.67 What is reverse potential?

3.68 What effect does forward potential have on the potential hill?   Explain.

3.69 What effect does reverse potential have on the potential hill?   Explain.

3.70 For what region of the forward characteristics is the relationship between E and I nonlinear?   Explain.

3.71 In what portion of the forward characteristic is the $E\text{-}I$ relationship linear? Explain.

3.72 Why doesn't the full reverse saturation current flow for the first increment of reverse bias?

3.73 Why does the reverse characteristic of the diode have a comparatively large "constant-current" region?

3.74 What is zener action?

3.75 What is avalanche action?

3.76 Explain the diode's "constant-voltage" region.

3.77 What precautions must be taken if the constant-voltage region is to be utilized?

3.78 A diode has the following parameters: $I_S = 15$ μa; V varies from 0 to 8 volts, both forward and reverse.   Using the simplified diode equation, plot the forward and reverse characteristics of the diode in steps of 1 volt.

3.79 Repeat Prob. 3.78 for a diode with $I_S = 30$ μa.

3.80 Repeat Prob. 3.78 for a diode with $I_S = 5$ μa.

3.81 A diode has the following parameters: $I_S = 10$ μa; V(forward) = 2 volts. Using the simplified diode equation, find the ratio of forward to reverse current at room temperature.

3.82 In Prob. 3.81, assume $T = 230°$K.   Find the ratio of forward to reverse current at this temperature.   Explain the difference between this ratio and the ratio found in Prob. 3.81.

3.83 What is the concept of Fermi level?

3.84 Explain the development of the metal-semiconductor barrier in terms of the Fermi level concept.

CHAPTER 4

JUNCTION TRANSISTOR CHARACTERISTICS

The junction diode and the formation of the junction barrier potential were considered in the preceding chapter. The junction transistor is one logical step beyond the diode analysis, since it may be viewed as two diodes connected "back to back," sharing a common element—the *base*. The first diode is customarily in the forward bias condition, and the second is under the influence of reverse bias. The reasons for this arrangement are developed in the sections that follow.

Since the junction transistor is the most promising type of transistor, this entire chapter is devoted to its analysis. Its basic features are noted in detail, along with a thorough discussion of its behavior. Although the device does not know how it is connected, the configuration has much to do with the amplification and gain possibilities, as well as the input and output resistances. The theory of operation of both junction types (NPN and PNP) are examined when connected in each configuration.

4.1 Formation and Behavior of the NPN Type. Figure 4.1 illustrates the NPN junction transistor, which has three crystal regions, each with

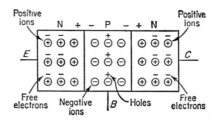

Fig. 4.1 The NPN transistor.

an associated terminal. The middle region is the P-type base B, the N-type emitter E is to the left of the base, and the N-type collector C is to the right. The emitter region customarily contains a higher impurity concentration than the base. In the NPN junction transistor, this means the emitter region possesses a greater number of donor atoms per unit volume than the base region does acceptor atoms per unit volume.

The base-region width is generally much smaller than that of the emitter. Higher impurity concentration in the emitter region and a narrow base region lead to more efficient transistor operation, as will be revealed.

The junction transistor can be made by one of several fabrication processes which are not discussed in this text since the ability to understand and utilize the device does not depend on knowledge of its manufacture. The junction between the emitter and base regions is called the emitter-base junction; and by the same reasoning, the junction formed between the collector and the base is called the collector-base junction. Let us briefly review the manner in which the two junction potentials are developed.

The Emitter-base Junction (with No External Circuit Connections). Upon formation of the junction, a number of the mobile electrons in the

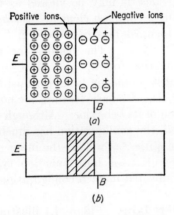

FIG. 4.2 The effect of higher impurity concentration in the emitter region on the depletion area: (*a*) impurity charge distribution, (*b*) depletion width.

N region diffuse into the P region and recombine with holes. The free electrons closest to the junction on the N side and free holes nearest the junction on the P side are the first participants in the process of diffusion and recombination. As pointed out in the preceding chapter, this action continues until the potential hill is sufficiently high to prohibit further electron diffusion without the assistance of an external field.

It may be assumed that the number of holes filled at the base side of the junction is equal to the number of electrons which diffused across the junction from the emitter region. Since the emitter region possesses a higher concentration of impurities than the base, the depletion region extends deeper into the base than into the emitter side of the depletion region.

Figure 4.2 illustrates this effect. Assume six electrons diffused from the emitter across the emitter-base junction and recombined with six holes in the base region. Also assume that the impurity concentration of the emitter region is twice that of the base layer, and that the impurity concentration of each is uniformly distributed in its respective region. The depletion on the base side would be twice the width of that on the emitter side of the junction (see Fig. 4.2b).

The Collector-base Junction. Upon contact, this junction potential is also developed in the manner just described. The result of the two junction barriers (with no external circuit connections) is the potential hill diagram of Fig. 4.3. Notice that the potential as seen by electrons passing from the emitter into the base is similar to the potential viewed by electrons in the collector region passing into the base. This bidirectional characteristic is undesirable in the majority of transistor applications. *In order to change the conductivity characteristics to a unidirectional*

Fig. 4.3   NPN potential hill diagram (electrons).

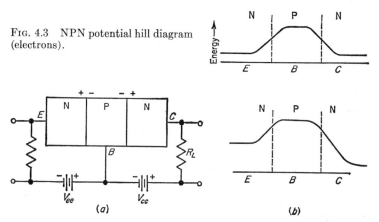

Fig. 4.4   (a) NPN with bias voltages, (b) potential hill diagram (electrons) with bias voltages.

nature, the emitter-base junction is usually biased in the forward direction and the collector-base junction is biased in the reverse direction, as shown in Fig. 4.4.

Refer to Fig. 4.4. Because of the forward bias applied to the emitter-base junction V_{ee}, electron flow from the emitter into the base region is enhanced. The magnitude of this current is directly proportional to the magnitude of the forward bias, since forward bias reduces the junction potential.

Electrons in the collector region, on the other hand, are prohibited from passing into the base because of the heightened junction potential produced by the reverse bias across that junction V_{cc}.

Notice that the emitter-base junction actually functions as a diode under the influence of forward potential, while the collector-base junction functions as a diode in the reverse bias condition. Let us next consider the development of current in the NPN junction transistor.

4.2 Transistor Currents. In Fig. 4.5, notice that the emitter-base junction is in the forward bias condition, and the collector-base junction is under the influence of reverse bias. *Free electrons are the majority carriers in the emitter region, and a number of them will flow across the emitter-base junction into the base region because of the application of forward bias; this is called the emitter current.* The base region is the P-type material, with holes being the majority carriers. A small percentage of the electrons which arrive in the base region from the emitter undergo recombination with the base region holes, resulting in recombination current. The remainder of these emitter electrons, however, diffuse across the base region through the collector-base junction into the collector region.

With a narrow base region, the diffusion distance of the emitter majority carriers is reduced, enabling them to travel through the base in a short period of time, called the *transit time*. With a base region which

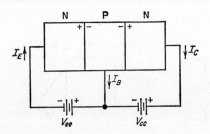

Fig. 4.5 NPN terminal currents.

has lower conductivity than the emitter, the greater portion of the depletion region exists in the base area. Recall that there are no mobile electrons or holes in the depletion region. Therefore a wider depletion region in the base area means the mobile electrons originating in the emitter have a better chance of diffusing across the base and into the collector without undergoing recombination. For efficient transistor operation, a large percentage of the emitter majority carriers must avoid recombination in the base region and succeed in arriving at the collector. This is the prime reason for utilizing a base region whose conductivity is lower than that of the emitter and whose width is relatively narrow. *A reduction of emitter majority-carrier transit time and base-region conductivity leads to increased transistor efficiency, since the objective is to pass charges across the base-collector junction.*

Examination of the collector-base junction potential reveals that mobile electrons in the P-type base region view this as a "downhill" potential. Like the reverse current of a PN junction, these electrons will readily pass from the P-type base to the N-type collector once they arrive at the base side of the depletion region. This current is directly dependent upon the number of emitter majority carriers that arrive at the collector-base junction and is relatively independent of the collector-base reverse voltage. *The emitter majority-carrier flow across the collector-base junction is called the collector current.*

It will be recalled that minority carriers are generated by deathnium center action. Electrons are minority carriers in the base region of the NPN junction transistor. Because of the predominance of holes in the P base, most of the deathnium-center-generated electrons in this region undergo recombination. A number of holes are filled by the emitter electrons that do not succeed in arriving at the collector, which enhances the probability of deathnium-center-generated electrons reaching the base terminal. *The number of deathnium-center-generated electrons which avoid recombination and succeed in reaching the base terminal is equal to the number of emitter electrons that underwent recombination in the base region and forms what is called the base current.*

The relationship between the three terminal currents in the junction transistor is

$$I_E = I_B + I_C$$

where I_E = emitter current
I_B = base current
I_C = collector current

4.3 Junction Transistor Circuit Properties. Several transistor circuit properties are analyzed in this section in order to facilitate their use in later portions of this text.

Dynamic Resistance. Three dynamic resistance properties are considered here:

1. Dynamic resistance of the junction r_j
2. Dynamic resistance of the emitter-base circuit $r_{e\text{-}b}$
3. Dynamic resistance of the collector-base circuit $r_{c\text{-}b}$

In equation form, the dynamic resistance of a junction in germanium is

$$r_j \cong \frac{1}{39I_f}$$

where I_f = forward junction current in amperes.

Example 1: Find the dynamic resistance of a germanium diode when its forward current is 2 ma.

Solution:

$$r_j = \frac{1}{39 \times 2 \times 10^{-3}} = 12.8 \text{ ohms}$$

Example 2: Find the dynamic resistance of the same germanium diode when its forward current is 5 ma.

Solution:

$$r_j = \frac{1}{39 \times 5 \times 10^{-3}} = 5.1 \text{ ohms}$$

The examples indicate that the dynamic resistance of the junction decreases with larger forward currents, which result from the application of larger forward bias values.

The dynamic resistance of the emitter-base circuit is

$$r_{e\text{-}b} = \frac{\Delta e_E}{\Delta i_E} \qquad \text{with } E_C \text{ constant}$$

where e_E = instantaneous total emitter voltage
i_E = instantaneous total emitter current
E_C = constant-collector voltage

NOTE: $r_{e\text{-}b}$ is equivalent to the hybrid parameter h_{11} which is analyzed in Sec. 4.16.
Example 3: A junction transistor has the following parameters: With a constant-collector voltage, an emitter-voltage change of 0.2 volt is accompanied by a 2-ma change in emitter current. Find the dynamic resistance of the emitter-base circuit.

Solution:

$$r_{e\text{-}b} = \frac{2 \times 10^{-1}}{2 \times 10^{-3}} = 100 \text{ ohms}$$

NOTE: $r_{e\text{-}b}$ is usually 100 ohms or less in typical applications.

The dynamic resistance of the collector-base circuit is

$$r_{c\text{-}b} = \frac{\Delta e_C}{\Delta i_C} \qquad \text{with } I_E \text{ constant}$$

where e_C = instantaneous total collector voltage
i_C = instantaneous total collector current
I_E = fixed emitter current

Example 4: A junction transistor has the following parameters: With a fixed emitter current, a 4-volt change in the collector voltage is accompanied by a 5-μa collector current change. Find the dynamic resistance of the collector-base circuit.

Solution:

$$r_{c\text{-}b} = \frac{4}{5 \times 10^{-6}} = 800 \text{ kilohms}$$

NOTE: $r_{c\text{-}b}$ is usually between 200 kilohms and 2 megohms.

Dynamic Conductance. Dynamic conductance is the reciprocal of dynamic resistance, and the dynamic conductance of the collector-base circuit is the most significant of those that could be considered. This is also called the output conductance h_{22} and is further analyzed in Sec. 4.16 as a hybrid parameter.

$$g_{c\text{-}b} = \frac{\Delta i_C}{\Delta e_C} \qquad \text{with } I_B \text{ constant}$$

and

$$g_{c\text{-}b} = \frac{1}{r_{c\text{-}b}}$$

Example 5: Find the collector-base circuit dynamic conductance of example 4. *Solution:* The collector-base circuit dynamic conductance may be found by using either of the two preceding equations:

$$g_{c\text{-}b} = \frac{\Delta i_C}{\Delta e_C} = \frac{5 \times 10^{-6}}{4} = 1.25 \ \mu\text{mho}$$

or, using the reciprocal of $r_{c\text{-}b}$,

$$g_{c\text{-}b} = \frac{1}{r_{c\text{-}b}} = \frac{1}{8 \times 10^5} = 1.25 \ \mu\text{mho}$$

NOTE: $g_{c\text{-}b}$ is usually 5 to 0.5 μmho in typical applications.

Current Amplification. *The current amplification may be defined as the ratio of a change in output terminal current to a change in input terminal current.* Since three configurations are possible, there is a specific ratio for each. In the common-base configuration, the emitter is the input and the collector is the output. This current amplification is given the general symbol α, but it may be further specified as α_{ce} where the first subscript is the output terminal and the second subscript denotes the input terminal. In equation form, the alpha current amplification is

$$\alpha = \alpha_{ce} = \frac{\Delta i_C}{\Delta i_E} \qquad \text{with } E_C \text{ constant}$$

The change in collector current cannot exceed the emitter current change, since the collector current is less than the emitter current by an amount equal to the magnitude of the base current at that time. In equation form, since

$$i_E = i_B + i_C$$

then

$$i_C = i_E - i_B$$

A typical value of α or α_{ce} is 0.98. This ratio is an indication of the emitter efficiency, since it reveals the fraction of the emitter majority carriers that succeed in diffusing across the base to arrive at the collector.

With the common-emitter configuration, the base is the input terminal and the collector is the output; hence the current amplification factor is given the symbol α_{cb}, or β.

$$\beta = \alpha_{cb} = \frac{\Delta i_C}{\Delta i_B}$$

where Δi_C is the instantaneous total change in the collector current caused by the instantaneous total change in the base current Δi_B.

NOTE: The manner in which the variations in base current cause collector current changes is discussed in the following section. Beta is analyzed as a hybrid parameter h_{21} in Sec. 4.16.

Let us examine the relationship between α and β:

Since
$$\alpha = \frac{\Delta i_C}{\Delta i_E}$$

an instantaneous α value can be determined by

$$\alpha = \frac{i_c}{i_e}$$

Transposing,
$$i_c = \alpha i_e$$

The base current, it will be recalled, is equal to the difference between the emitter and collector currents at any instant:

$$i_b = i_e - i_c$$

Substituting into the beta amplification factor, we obtain

$$\beta = \frac{i_c}{i_b} = \frac{\alpha i_e}{i_e - i_c} = \frac{\alpha i_e}{i_e - \alpha i_e}$$
$$= \frac{i_e(\alpha)}{i_e(1 - \alpha)}$$

Canceling i_e in the numerator and denominator,

$$\beta = \frac{\alpha}{1 - \alpha}$$

It was earlier stated that a typical value of alpha is about 0.98. From this, the typical value of beta can be found:

$$\beta = \frac{0.98}{1 - 0.98} = 49$$

Notice that beta increases as alpha becomes closer to unity. Beta values of 100 and greater are possible.

The third current amplification to be considered is used when the transistor is connected in the common-collector configuration, where the base is the input terminal and the emitter is the output. Using α to denote current amplification factor, the first subscript denotes the output terminal, and the input terminal is represented by the second subscript. The current amplification factor for the common-collector configuration is

$$\alpha_{eb} = \frac{\Delta i_E}{\Delta i_B}$$

where Δi_E = instantaneous total change in emitter current

Δi_B = instantaneous total change in base current

It will be recalled that the base current is equal to a small fraction of the emitter current, and the emitter current is slightly greater than the collector current. This results in the ratio of $\Delta i_E/\Delta i_B$ (which is α_{eb}) being slightly greater than the $\Delta i_C/\Delta i_B$ ratio (which is α_{cb} or β). The common-collector current amplification factor is slightly greater than the common-emitter current amplification factor.

Examination of the three current amplification factors indicate that they are at their highest values when the ratio of collector to emitter current (α or α_{ce}) approaches unity. Let us examine the relationship between transistor current and the following:

1. Emitter efficiency
2. Transport factor
3. Collector-base voltage
4. Collector reverse saturation current I_{CO}

The *emitter efficiency* may be defined as the ratio of the emitter majority-carrier flow across the emitter-base junction to the total emitter current. The total emitter current includes those emitter majority carriers which undergo recombination in the emitter region before they are able to pass across the emitter-base junction. This ratio is made very high (0.99 is typical) by utilizing an emitter region whose impurity concentration is much higher than that of the base.

The *transport factor* may be defined as the ratio of the emitter majority carriers which arrive at the collector to the emitter majority carriers that pass from the emitter into the base. In other words, the transport factor is that portion of the emitter current which does not undergo recombination in the base but succeeds in arriving at the collector. From this it becomes obvious that anything which reduces the probability of recombination occurring in the base will increase the transport factor. Recall that the base impurities offer the recombination possibilities in that region. A low base impurity concentration increases the chances of the emitter majority carriers' diffusing across the base region into the collector without undergoing recombination. The width of the base has much to do with the transport factor as well, since the emitter majority carriers rely on diffusion to pass through the base into the collector. When the diffusion path is reduced in length, as is the case with a more narrow base width, the time spent in the base by the emitter majority carriers is reduced; and their chances of arriving at the collector are improved. A transport factor of 0.99 or greater is achieved in common-junction transistors by utilizing a base of comparatively low impurity concentration and narrow width.

The emitter and collector currents are affected by the *collector-base voltage* to some extent. Recall that the collector-base voltage is a reverse potential across that junction, which widens the depletion region. This effectively reduces the base width in that the region in which recombination can occur is made more narrow by widening the depletion region in the base. The diffusion distance of the emitter majority carriers in the base is correspondingly reduced, thereby improving their chances of successfully arriving at the collector. The ratio of collector to emitter current becomes closer to unity with higher values of collector-base voltage.

The *collector reverse saturation current I_{CO}* is similar to the saturation current of a reverse-biased diode, which is described in the preceding chapter. This current is sufficiently small in value to be considered negligible when the transistor is operated at normal temperatures.

4.4 The Common-emitter Configuration (NPN). In Fig. 4.6, notice the symbol used for NPN junction transistor. The emitter is always designated by an arrow, whose direction specifies the transistor type (NPN or PNP). *The direction of the emitter arrow is outward for the NPN and inward for the PNP. A second simple technique for determining the transistor type is to observe the bias potential polarities, remembering that the collector is biased in reverse and the emitter utilizes forward bias.* Several transistor units are shown in Fig. 4.6 *c, d,* and *e.*

The forward bias for the emitter-base junction is customarily not obtained by a separate supply potential. Several emitter-base junction bias techniques are analyzed later. In order to investigate the action of the common-emitter circuit in its most simple aspects, separate bias supplies for the emitter and collector are used at this time. V_{ee} is the emitter-base bias supply, and the collector-base bias potential is V_{cc}.

The input signal is applied across the base-emitter junction in this configuration. Depending upon the specific application of the circuit, the output may be taken across R_L (see Fig. 4.6a) or across the transistor (as shown in Fig. 4.6b). In actual applications Fig. 4.6a applies where the output is coupled to a load resistance by means of a transformer, in which case R_L is the reflected primary impedance of the transformer. Figure 4.6b illustrates the case where RC coupling is incorporated. The aspects of both coupling techniques are treated in later sections.

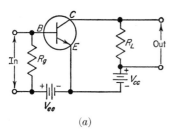

(a)

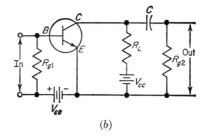

(b)

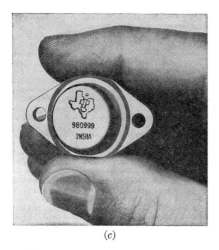

(c)

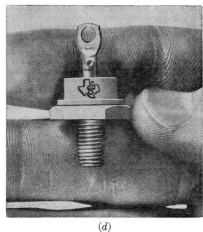

(d)

Fig. 4.6 The basic NPN common-emitter (CE) circuits: (a) output across R_L, (b) output across transistor, (c) close-up view of a transistor, (d) side view of a power transistor, (e) top view of a transistor. [(c), (d), and (e) *from Texas Instruments, Inc.*]

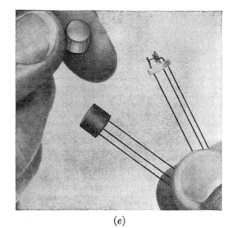

(e)

Let us define current polarity in the following manner: Electron current leaving a given terminal is a positive current, and electron current entering a terminal is a negative current. In the NPN transistor, electrons enter the emitter terminal and leave via the base and collector terminals; hence, the emitter current is negative while base and collector currents are positive, by this definition (which is applicable to a voltage source).

4.5 Waveform Analysis of the NPN (CE) Circuit. The behavior of the junction transistor is easily understood by a waveform analysis of Fig. 4.7. An alternating square wave has been selected for the input signal

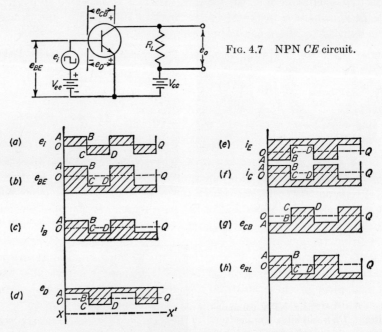

FIG. 4.7 NPN *CE* circuit.

FIG. 4.8 Waveforms for the circuit of Fig. 4.7.

waveform e_i, whose changes from maximum positive to maximum negative (and vice versa) are assumed to occur instantaneously. The external instantaneous total voltage from the base to the emitter is here called e_{B-E}, while the base to emitter junction potential is e_D. Since the collector-base junction is reverse-biased, the external potential from collector to base is the same as the internal junction potential and is e_{C-B}. The output is taken across R_L.

Figure 4.8a (e_i). This is the alternating square wave used as the voltage to vary the forward bias of the base-emitter junction. We will assume that it is symmetrical and that all circuit parameters are so selected that faithful reproduction is assured.

Figure 4.8b ($e_{B\text{-}E}$). The sum of e_i and V_{ee} is called $e_{B\text{-}E}$ in this discussion. Notice that e_i and V_{ee} are series-aiding during the half cycle when the base side of e_i is positive, resulting in a large $e_{B\text{-}E}$ value. Conversely, these voltages are series-opposing when the base side of e_i is negative, thereby reducing $e_{B\text{-}E}$ by the magnitude of e_i. The forward bias actually applied to the base-emitter junction at any instant is the algebraic sum of e_i and V_{ee}. It is important to recognize that the magnitude of the forward bias contains the variations of e_i. When e_i is zero, $e_{B\text{-}E}$ is at the Q value (point O), which is the magnitude of V_{ee} alone. During the positive excursion of e_i, the forward bias $e_{B\text{-}E}$ is increased to point A. At point B, where e_i instantaneously reverses its polarity, the forward bias is reduced by the magnitude of $-e_{i,\max}$.

Example: $V_{ee} = 0.5$ volt; $e_{i,\max} = 0.3$ volt (alternating square wave). Determine the $e_{B\text{-}E}$ variations.

Solution:

At point O: $\qquad\qquad$ $e_{B\text{-}E} = 0.5 + 0 = 0.5$ volt
At points A-B: $\qquad$ $e_{B\text{-}E} = 0.5 + 0.3 = 0.8$ volt
At point C: $\qquad\qquad$ $e_{B\text{-}E} = 0.5 - 0.3 = 0.2$ volt

The input signal voltage serves to vary the base-emitter junction forward bias.

Figure 4.8c (i_B). Since the junction transistor under consideration is the NPN type, electron flow out of the base is a minority-carrier flow. That is, the electrons drawn out of the base region via the base terminal are primarily deathnium-center-generated minority carriers. It will be recalled that the generation of these carriers is independent of potential. Assuming the temperature of the transistor is constant, the generation of electrons in the base region of the NPN transistor occurs at a fixed rate, and most of these electrons rapidly participate in recombination action, since the base region is the P-type crystal.

With a forward bias applied across the emitter-base junction, the positive terminal of the bias supply is connected nearest to the base terminal. Depending upon the magnitude of this forward bias, a number of the deathnium-center-generated electrons are drawn out of the base region (electrons are attracted to the positive side of a potential). This leaves an additional number of holes in the base region, since a number of the electrons moved out of the base before they became "lost" in the recombination process.

Let us determine the relationship between base current and depletion width. With no forward bias applied to the emitter-base junction, its potential barrier is that value developed at the time the junction was formed. The application of a forward bias, such as V_{ee} of Fig. 4.7, results in the flow of minority carriers out of the base region (electrons in the NPN junction transistor). The number of holes in the base increase because some of the deathnium-center-generated electrons leave the base via the base terminal. Holes now exist closer to the edges of the emitter and collector regions, thereby reducing the width of the two depletion regions. It will be recalled that reducing the width of the barrier is a reduction in the potential hill (junction potential), resulting in an increase of emitter and collector currents. Figure 4.9a illustrates this effect: The junction width prior to the application of forward bias is shown by the dotted lines while the solid lines indicate the junction width after the application of V_{ee} alone (point O of waveform in Fig. 4.8b).

When e_i swings in the positive direction, the forward bias of the base-emitter junction is increased (e_{B-E} increases from point O to A in Fig. 4.8b). More minority electrons are drawn out of the base, which increases the number of base region holes, resulting in a reduction in the widths of both depletion regions. This effect is illustrated in Fig. 4.9b.

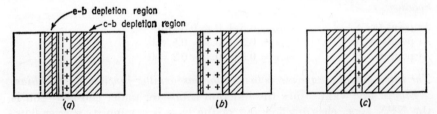

FIG. 4.9 Base current effect on the depletion regions width: (a) point O, (b) points A and B, (c) point C.

At point B, e_i instantaneously reverses polarity, and the over-all effect is a reduction in the base-emitter forward bias (e_{B-E} decreases from B to C of Fig. 4.8b). Since the base side of e_{B-E} is less positive, a smaller number of minority electrons are drawn out of the base. This results in an increase in minority electron-hole recombinations within the base region, thereby reducing the number of holes there, which enlarges the depletion areas within the base, as shown in Fig. 4.9c.

The following conclusion may be drawn: *In the NPN junction transistor, an increase in the base electron current reduces the junction potentials, and vice versa.*

Figure 4.8d (e_D). e_D is the base-emitter junction potential. Checking the polarity of the base side of this junction with respect to the emitter (see Fig. 4.7) reveals that it is a negative potential in the common-emitter configuration. The dotted line XX' indicates the magnitude of this voltage in the absence of a bias potential. Since V_{ee} is a forward bias, its placement in the base-emitter circuit in the manner illustrated results in a reduction of e_D from point X to point O. The forward bias is increased by the positive e_i excursion, and e_D undergoes a corresponding reduction (O to A). The negative e_i swing decreases the forward bias causing e_D to increase from point B to C.

Note the effect of e_i upon e_D: *The base-emitter potential hill is modulated (varied) in accordance with the input signal e_i excursions.*

Figure 4.8e (i_E). Note that i_E is negative by the previous definition. *The emitter current variations follow the changes in the base-emitter potential barrier height e_D in an inverse fashion.* A reduction in this junction potential permits a greater number of mobile emitter electrons to acquire the necessary energy increment to pass through the junction. Recall that a potential hill reduction means that the energy difference between the energy level of an emitter-free electron and that of the base region is reduced. The probability of a mobile emitter electron acquiring the necessary thermal energy increment becomes improved as the potential hill is reduced. This is the situation when e_i undergoes its positive excursion; e_D is reduced and i_E becomes greater (O to A). Since e_D is enlarged by the negative e_i swing, i_E undergoes a corresponding reduction in amplitude (B to C).

Figure 4.8f (i_C). i_C is a positive current by the preceding definition. The collector current is relatively independent of the collector-base potential hill, since the majority carriers passing into the collector go down this hill (energy decrease). *The collector is a direct function of the emitter current; i.e., when i_E increases in magnitude i_C does likewise, and vice versa. In terms of the NPN junction transistor, an increase of electron flow into the emitter is accompanied by an increase of an electron flow out of the collector and base (since $i_E = i_C + i_B$).*

Figure 4.8g ($e_{C\text{-}B}$). The behavior of this junction is described in the preceding paragraphs which investigate the effect of the base current. An increase in electrons leaving the base of the NPN transistor results in more holes in that region, which reduces the width of both depletion areas in the base portion of the transistor. Both junction potentials, e_D and $e_{C\text{-}B}$, vary in step with each other. *The voltage across the collector-base junction varies inversely as the potential applied across the base-emitter terminals.* Since the input signal creates the change in $e_{B\text{-}E}$ and $e_{C\text{-}B}$ may be considered as the output voltage, the common-emitter circuit is said to create a signal inversion. In other words, as e_i swings in the positive direction, $e_{C\text{-}B}$ changes in the negative direction.

Figure 4.8h (e_{RL}). Since a positive electron current flows through R_L, its voltage is assigned the positive polarity. R_L is a fixed resistor connected between the collector and the positive terminal of V_{cc}. Using Kirchhoff's series loop technique, let us start from the emitter and work through to the collector, R_L, V_{cc}, and back to the emitter in order to determine the voltage relationships:

$$e_D + e_{C-B} + e_{RL} + V_{cc} = 0$$

e_D is generally small enough to be considered negligible in many practical cases, hence

$$e_{C-B} + e_{RL} + V_{cc} = 0$$

Since V_{cc} is a constant potential, e_{C-B} and e_{RL} must vary inversely so as to maintain a fixed sum equal to V_{cc}. For example, when e_{C-B} decreases by 5 volts, e_{RL} must increase by 5 volts. e_{C-B} and e_{RL} possess the same magnitude of voltage variations but inverted with respect to each other. This relationship is further verified by Ohm's law: Since R_L is of fixed ohmic value, an increase in the current passing through $R_L(i_C)$ must be accompanied by an increase in e_{RL}. The increase in e_{RL} is equal to the decrease in e_{C-B}, which occurs with i_C increases, such that

$$e_{RL} + e_{C-B} = V_{cc} \qquad \text{at all times}$$

4.6 Volt-Ampere Characteristics (*CE* Configuration). Figure 4.10 illustrates the volt-ampere characteristics of an NPN transistor in the

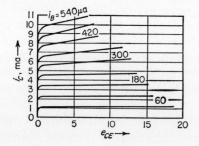

FIG 4.10　　e_{C-E} versus i_C characteristics.

common-emitter configuration. The collector-to-emitter voltage is plotted on the x axis and the collector current on the y axis in this case (but they may be reversed). Each characteristic is plotted with a fixed base current, whose magnitude is stated on the curve. For lower values of fixed base current, the collector current is relatively independent of the collector-emitter voltage, while with higher base current values, the collector voltage has a greater effect. Let us consider the influence of several fixed base currents upon the dynamic resistance of the collector-base circuit, which may also be called the output resistance of the transistor in the common-emitter configuration.

Example 1: Refer to Fig. 4.10: Find the dynamic resistance of the collector-base circuit between the collector-voltage values of 1 and 5 volts for the following base currents: (*a*) 60 μa; (*b*) 120 μa; (*c*) 240 μa; (*d*) 360 μa.

Solution:

(a) With $i_B = 60$ μa:

$$r_o = r_{c\text{-}b} = \frac{\Delta e_C}{\Delta i_C} = \frac{5 - 1}{1.1 \text{ ma} - 1.07 \text{ ma}} = \frac{4}{3 \times 10^{-5}} = .133 \text{ megohms}$$

(b) With $i_B = 120$ μa:

$$r_o = r_{c\text{-}b} = \frac{5 - 1}{1.2 \text{ ma} - 1.15 \text{ ma}} = 80 \text{ kilohms}$$

(c) With $i_B = 240$ μa:

$$r_o = r_{c\text{-}b} = \frac{5 - 1}{4.74 \text{ ma} - 4.65 \text{ ma}} = 40 \text{ kilohms}$$

(d) With $i_B = 360$ μa:

$$r_o = r_{cb} = \frac{5 - 1}{6.95 \text{ ma} - 6.55 \text{ ma}} = 10 \text{ kilohms}$$

NOTE: $e_{C\text{-}E} \cong e_{C\text{-}B}$ in most cases; therefore using $e_{C\text{-}E}$ does not materially affect the calculations to any great extent.

The dynamic resistance of the collector-base circuit is greatest for smallest values of base current.

Example 2: Using the dynamic-resistance values of example 1, find the dynamic conductance of the collector-base circuit for the four stated conditions:

Solution:

(a) With $i_B = 60$ μa:

$$g_{c\text{-}b} = \frac{1}{r_{c\text{-}b}} = \frac{1}{1.33 \times 10^6} = 0.75 \text{ }\mu\text{mho}$$

(b) With $i_B = 120$ μa:

$$g_{c\text{-}b} = \frac{1}{8 \times 10^4} = 12.5 \text{ }\mu\text{mho}$$

(c) With $i_B = 240$ μa:

$$g_{c\text{-}b} = \frac{1}{4 \times 10^4} = 25 \text{ }\mu\text{mho}$$

(d) With $i_B = 360$ μa:

$$g_{c\text{-}b} = \frac{1}{1 \times 10^4} = 100 \text{ }\mu\text{mho}$$

The dynamic conductance of the collector-base circuit is lowest with the smallest magnitude of base current.

Example 3: Assume the NPN transistor considered in examples 1 and 2 has an alpha value of 0.96.   Find beta:

Solution:

$$\beta = \frac{\alpha}{1 - \alpha} = \frac{0.96}{1 - 0.96} = 24$$

Example 4: Assume the alpha value of the transistor under consideration is 0.99. Find beta.

Solution:

$$\beta = \frac{0.99}{1 - 0.99} = 99$$

The current amplification factor in the common-emitter configuration β or α_{cb} is directly dependent upon the alpha current amplification factor.

4.7 Formation and Behavior of the PNP Type.

Figure 4.11a depicts the charge distribution within the PNP junction transistor upon formation, and the symbol for the PNP type is illustrated in Fig. 4.11b. Notice

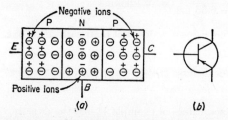

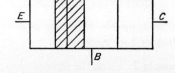

FIG. 4.11 The PNP transistor: (a) charge distribution, (b) symbol.

FIG. 4.12 Wider depletion region in the base region because of its lower impurity concentration.

that the emitter terminal has the arrow directed inward for the PNP transistor.

The encircled charges are those of the balanced impurity ions. Recall that positive impurity ions reside in the *N*-type crystal, and negative impurity ions exist in the P regions. In the P regions, a hole and a negative ion balance to an over-all charge of zero, while a free electron and a positive ion have an over-all zero charge in the N region. When the P-type and N-type regions are made to form a junction, a process of diffusion occurs across the junction, as described in the formation of the PN junction in Chap. 3. This action occurs across both the emitter-base and collector-base junctions. *Each free electron, prior to crossing the junction, balanced off a positive ion in the base region. By leaving the base side of each junction, these mobile electrons no longer balance off the positive ions, and the base side of both junctions becomes positive in charge. The free electrons which diffused across each junction undergo recombination in the P region, thereby creating that number of negative ions at the emitter and collector sides of the two junctions. Mobile electron diffusion across the junctions stops when the potential hills are sufficiently great, at which time the junctions are said to be formed.*

As in the case of the NPN, the emitter region customarily has a much greater impurity concentration than the base (i.e., the emitter region has a much greater number of free holes than the base has free electrons). Because of this inequality, the conductivity of the emitter crystal is much higher than that of the base; and the depletion region extends further into the base region than it does in the emitter (as shown in Fig. 4.12).

Extending the depletion region further into the base effectively reduces the base width. Since the barrier area has the characteristics of the intrinsic semiconductor, an emitter majority carrier encounters very little possibility of undergoing recombination when in the depletion region. By extending the depletion region further into the base, the diffusion path in the base region, where recombination can occur, is correspondingly reduced. All this results in a large percentage of the emitter majority carriers arriving at the collector, which renders an alpha current amplification factor closer to unity (0.98 is typical). *The emitter efficiency (ratio of emitter majority-carrier flow across the emitter-base junction to the total emitter current) and the transport factor (ratio of emitter majority carriers which arrive at the collector to the emitter majority carriers which pass into the base) are increased by utilizing a base region whose impurity concentration is lower than that of the emitter region.*

Current Inside the PNP Transistor. The emitter and collector regions possess a number of free holes while the base majority carriers are electrons. As in the NPN, the emitter-base junction is placed under forward bias while the collector-base junction is reverse-biased. Emitter region electron current flow occurs in the valence energy band; the holes are urged toward the emitter-base junction, and valence electrons drift toward the emitter terminal as a result of the emitter forward bias. It is relatively easy for a valence electron to occupy another valence energy level, since the change between energy levels is small. The valence electron is urged toward a hole closer to the emitter terminal, thereby leaving a hole nearer the emitter-base junction. This is hole flow. Because of the forward bias, valence electrons in the base region are also able to gain the relatively small energy increase required to occupy a hole on the emitter side of the junction, thereby creating holes in the base. Valence electrons in the collector region likewise find it an easy matter to occupy these holes in the base region, creating holes in the collector region. *In the PNP transistor, the hole current from the emitter through the base to the collector is called the collector current. Emitter current is that hole flow from the emitter to the base. The base current is the difference between the emitter and collector currents, since*

$$I_E = I_C + I_B$$

Hole current flows in the transistor only; the external terminal currents are electron currents. Hence, the operation of the PNP transistor may be ana-lyzed in terms of electron currents.

4.8 The Common-emitter Configuration (PNP).

Figure 4.13 illustrates the PNP junction transistor in the common-emitter configuration. The emitter-base junction is placed under the influence of forward bias by V_{ee}, and the collector-base junction is reverse-biased by V_{cc}. We will again restrict our analysis at this time to a pure resistive load. The development and behavior of hole current within the transistor is blended with the terminal electron current discussion in the following waveform analysis in order to furnish a more complete understanding of the circuit

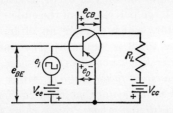

Fig. 4.13   PNP *CE* circuit.

behavior. α_{cb} (or β) is the current amplification factor used in this configuration with typical values of about 50, although 100 or greater is possible.

Figure 4.14a and b (e_i and $e_{B\text{-}E}$). The input signal is an alternating square wave. The polarity of V_{ee} is such that the base is negative with respect to the emitter. e_i is series-opposing to V_{ee} during its positive excursions and series-aiding during its negative excursions. The composite of V_{ee} and e_i, which is $e_{B\text{-}E}$, is the forward bias applied to the base-emitter junction. Notice that the forward bias contains the variations of e_i.

Waveform analysis

Figure 4.14c (i_B). i_B, the external base current, is an electron current which is entering the base and a negative current by the preceding definition. During the positive excursions of e_i, the base becomes less negative with respect to the emitter, and the flow of electrons into the base terminal undergoes a corresponding decrease. When e_i swings in the negative direction, V_{ee} and e_i are series-aiding, thereby making the base more negative with respect to the emitter. The flow of electrons into the base during this e_i excursion undergoes a corresponding increase in magnitude. The electron base current varies in the same manner at the forward voltage, which is the inverse of the e_i variations.

Figure 4.14d (e_D). e_D is the actual potential of the base-emitter junction. The dotted line XX' indicates the relative magnitude of the junction potential prior to the application of any forward bias. With V_{ee} in the circuit and e_i at zero value, e_D is reduced to the value shown as dotted line OQ. Recall that a reduction in forward bias results in an increase in e_D, and e_D experiences a decrease with forward bias increases. e_i reduces the forward bias during its positive swing, and e_D becomes greater (from O to A). Using the same reasoning, e_D is reduced during the negative e_i excursions, since it causes an increase in the forward bias.

Figure 4.14e (i_E).   The electron emitter current leaves the emitter terminal and is a positive current by definition. Electron current within the PNP transistor occurs in the valence energy band. Valence electrons in the emitter region drift from hole to hole toward the emitter terminal, which means that the holes are drifting toward the junction.

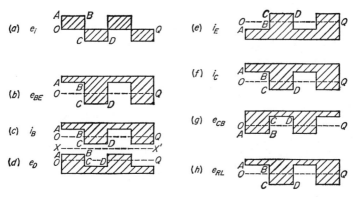

F<small>IG</small>. 4.14   Waveforms for circuit of Fig. 4.13.

Some of the valence electrons within the base region cross the base-emitter depletion region to occupy holes in the emitter region, causing a corresponding hole drift from the emitter to the base regions. The base-emitter potential hill, which is indicated by e_D, is an "uphill" potential for the base electrons. In the absence of forward bias, this action just described is virtually nonexistent. With forward bias, the valence electron drift from the base to the emitter (and hole drift from the emitter to the base) is a direct function of e_D.

When e_D is made to increase (O to A), the probability of a base valence electron moving from its original bond to a bond vacancy (hole) located on the other side of the junction is reduced, since the increment of energy to be acquired by the electron is larger. The number of electrons leaving the emitter terminal at any instant must be equal to the number of electrons drifting from the base across the junction into the emitter at the same instant. Hence i_E decreases in magnitude in accordance with the increase in e_D. A reduction in e_D has the opposite effect, since the increment of energy required of the base valence electrons is smaller, and the probability of acquiring a smaller package of energy is greater; hence, the emitter current increases with e_D reductions.

Figure 4.14f (i_C). As shown in Fig. 4.13, electrons flow into the collector from its terminal, thereby making i_C a negative current by definition. The number of electrons flowing into the collector terminal is determined by the number of valence electrons which cross the collector-base junction to occupy holes in the base region. The valence electron drift across the collector-base junction is relatively unaffected by the voltage across this junction since it is viewed as a "downhill" potential by the collector valence electrons. The number of collector valence electrons that cross the junction, plus the number of electrons entering the base region via the base terminal, equal the valence electrons which cross the base-emitter junction into the emitter region. Because of this relationship, the magnitude of the collector current varies in the same manner as the emitter current. *In the PNP transistor, hole drift occurs from the emitter region through the base into the collector because there is a drift of valence electrons from the collector through the base to the emitter. The electrons entering via the collector and base terminals and leaving via the emitter terminal are determined by the magnitude of the valence electron-hole drift inside the transistor.*

Figure 4.14g (e_{C-B}). The width of the collector-base depletion region varies inversely as the electron current leaving the base. Recall that the base in a PNP transistor is the N-type semiconductor, where the majority mobile carriers are electrons. Each mobile electron balances off a positive ion, thereby reducing the depletion region width. When more electrons are injected into the base via the base terminal, e_D and e_{C-B} undergo a corresponding decrease because more positive ions on the base side of each junction are balanced. When the number of electrons injected into the base is reduced, the junction widths increase, since the balancing action is reduced. In terms of i_C, e_{C-B} increases at the time i_C decreases, and vice versa.

Figure 4.14h (e_{RL}). As in the case of the NPN transistor, R_L is of fixed ohmic value, resulting in e_{RL} varying in the same fashion as the collector current, which passes through it. e_{RL} is a negative voltage, based on the fact that the collector current is assumed to be a negative polarity by definition. Also recall that the following relationship holds for all practical purposes in most transistors:

$$e_{C-B} + e_{RL} \cong V_{cc}$$

where
$$\Delta e_{C-B} = -\Delta e_{RL}$$

in order to maintain a sum always equal to V_{cc}.

Signal inversion occurs with the PNP transistor in this configuration in the sense that e_i and $e_{C\text{-}B}$ vary inversely with respect to each other (when e_i swings positive, $e_{C\text{-}B}$ becomes more negative, and vice versa).

4.9 Input Volt-Ampere Characteristics (PNP). Figure 4.15 illustrates a typical family of input characteristics of the PNP junction transistor. Notice that the base current is plotted on the x axis and base voltage on the y axis with the collector voltage at a fixed value for each curve.

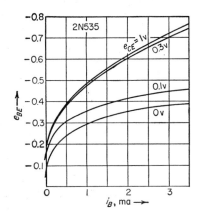

FIG. 4.15 i_B versus $e_{B\text{-}E}$ (2N535). (*Lansdale Tube Co.*)

The dynamic resistance of the input circuit (common-emitter configuration) can be computed from the input characteristic curves by the following relationship:

$$r_i = \frac{\Delta e_i}{\Delta i_i} = \frac{\Delta e_{B\text{-}E}}{\Delta i_B}$$

Example 1: Refer to Fig. 4.15. Find the dynamic input resistance when $e_{CE} = 1$ volt; i_B changes from 0.6 to 1.0 ma.

Solution:

$$r_i = \frac{0.48 - 0.4}{1.0 \text{ ma} - 0.6 \text{ ma}} = 200 \text{ ohms}$$

Example 2: Refer to the same curve as example 1. Find the dynamic input resistance value when i_B changes from 0.2 to 0.6 ma.

Solution:

$$r_i = \frac{0.40 - 0.30}{0.6 \text{ ma} - 0.2 \text{ ma}} = 250 \text{ ohms}$$

The input resistance of the transistor in the common-emitter configuration is greater for smaller values of base current. Also notice that the changes in dynamic resistance are most pronounced at low i_B values.

4.10 Output Volt-Ampere Characteristics (PNP).

Figure 4.16 illustrates the output characteristics of the PNP transistor in the common-emitter configuration. The collector-emitter voltage is plotted on the x axis, and the collector current is on the y axis. Each curve is the e_{CE} versus i_C relationship with a fixed base current. Notice the relative linearity of the curves in the region to the right of the knee. The dynamic

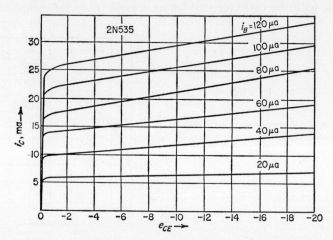

FIG. 4.16   Collector characteristics of the 2N535. $F_{co} = 2$ Mc; $h_{ib} = 33$ ohms; $h_{ob} = 0.4$ μmhos; $h_{fc} = 100$; $h_{rb} \cong 0.25 \times 10^{-3}$. (*Lansdale Tube Co.*)

output resistance of the transistor in the common-emitter configuration may be determined from

$$r_o = \frac{\Delta e_{C\text{-}E}}{\Delta i_C}$$

Example 1: Refer to Fig. 4.16. With i_B fixed at 60 μa, find the dynamic output resistance, using an $e_{C\text{-}E}$ variation of 8 to 10 volts.

Solution:

$$r_o = \frac{10 - 8}{16.4 \text{ ma} - 15.95 \text{ ma}} = 4.44 \text{ kilohms}$$

Example 2: Refer to Fig. 4.16. With i_B fixed at 20 μa, find the dynamic output resistance when $e_{C\text{-}E}$ varies from 8 to 10 volts.

Solution:

$$r_o = \frac{10 - 8}{6.4 \text{ ma} - 6.3 \text{ ma}} = 20 \text{ kilohms}$$

Example 3: Find the dynamic output conductance for the conditions of examples 1 and 2.

Solution 1: In example 1,

$$g_o = \frac{1}{r_o} = \frac{1}{4.44 \times 10^3} = 225 \ \mu\text{mho}$$

Solution 2: In example 2:

$$g_o = \frac{1}{2 \times 10^4} = 50 \ \mu\text{mho}$$

The dynamic output resistance of the transistor in the common-emitter configuration is greater for lower values of base current, and the output conductance is correspondingly less.

4.11 The Common-base Configuration (NPN).

Figure 4.17 illustrates the common-base configuration for the NPN transistor. V_{ee} serves to furnish forward bias for emitter-base junction while V_{cc} supplies reverse bias for the collector-base junction. α_{ce} (or α) is the current amplification factor used for this configuration, with typical values of about 0.98.

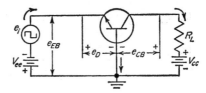

FIG. 4.17 NPN (*CB*) circuit.

Electrons flow into the emitter terminal and out of the base and collector terminals.

The behavior of the transistor is essentially the same regardless of the configuration. In effect, the transistor does not know how it is connected and will follow the behavior patterns stated in prior sections of this chapter. Since this is the case, the waveform analysis of the transistor in the common-base and common-collector configurations need not be as rigorous.

Waveform Analysis

Figure 4.18a (e_i). An alternating square wave is again used as the input voltage waveform for reasons of simplicity.

Figure 4.18b ($e_{E\text{-}B}$). The emitter-to-base voltage is the composite potential of e_i and V_{ee} at every instant. When e_i undergoes its positive excursion (the terminal connected to the emitter is positive with respect to the terminal connected to V_{ee}), the two voltages are series-opposing, resulting in $e_{E\text{-}B}$ undergoing a reduction in magnitude ($O\text{-}A$). $e_{E\text{-}B}$ experiences an increase in magnitude during the e_i negative excursions, since they are then series-aiding (B to C).

Figure 4.18c (e_D). Since $e_{E\text{-}B}$ is the composite forward potential for the emitter-base junction, e_D changes inversely with it. When e_i and

V_{ee} are series-opposing, the resultant forward bias is reduced, and e_D is correspondingly greater (O to A). e_D becomes smaller (B to C) when e_i and V_{ee} are series-aiding, which occurs during the negative swing in e_i. As in the common-emitter configuration, the emitter-base junction potential is modulated by the input signal e_i.

Figure 4.18d (i_C). Based on our previous method for defining current polarity, i_C is positive, since electrons are leaving the collector. Recall that i_E is controlled by changes in e_D. As e_D becomes greater, i_E decreases, and vice versa. Also recall that i_E and i_C vary in the same manner with respect to each other; hence i_C decreases with e_D increases, and vice versa. The waveshape of i_C and e_{RL} are identical.

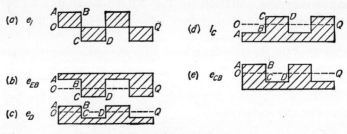

FIG. 4.18   NPN *CB* waveforms.

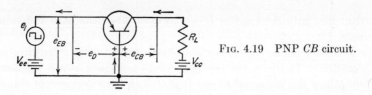

FIG. 4.19   PNP *CB* circuit.

Figure 4.18e ($e_{C\text{-}B}$). $e_{C\text{-}B}$ and i_C vary inversely, as revealed in the common-emitter configuration analysis. There is no signal inversion in the common-base configuration since $e_{C\text{-}B}$ and e_i undergo their positive and negative excursions in step with each other.

4.12 The Common-base Configuration (PNP). The PNP junction transistor in the common-base connection is illustrated in Fig. 4.19. Notice that the polarity of V_{ee} and V_{cc} are reversed as compared to the NPN (*CB*) circuit, in order to apply the proper bias to the junctions. α_{ce} (or α) is the current amplification factor for this connection, again with typical values of 0.95 to 0.98.

Waveform Analysis

Figure 4.20a and b (e_i and $e_{E\text{-}B}$). e_i and V_{ee} are series-aiding during the positive excursions of e_i, resulting in an increase in forward bias. The forward bias is reduced when e_i and V_{ee} are series-opposing, which occurs during the negative swings in e_i.

Figure 4.20c (e_D). e_D is reduced when the composite forward bias $e_{E\text{-}B}$ is made larger, and e_D is increased when $e_{E\text{-}B}$ is decreased. Hence e_D decreases during the positive excursions and increases during the negative excursions of e_i.

Figure 4.20d (i_C). i_E and i_C are greatly determined by the magnitude of e_D. When e_i swings in the positive direction, the increase in forward bias reduces e_D, which permits an increase in i_E and i_C. The negative e_i excursion reduces the forward bias, causing e_D to increase, resulting in a decrease in i_E and i_C.

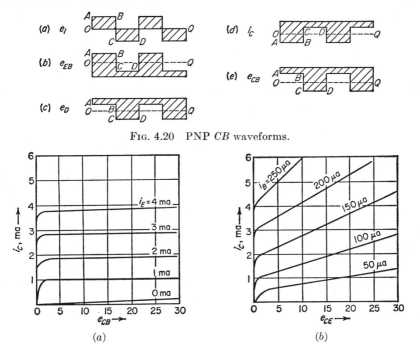

Fig. 4.20   PNP *CB* waveforms.

Fig. 4.21   Comparison of *CB* and *CE* output characteristics: (*a*) common base, (*b*) common emitter.

Figure 4.20e ($e_{C\text{-}B}$). e_D and $e_{C\text{-}B}$, as developed in an earlier section, vary in step. Hence $e_{C\text{-}B}$ decreases when i_C increases, and vice versa. As in the NPN (*CB*) circuit, $e_{C\text{-}B}$ and e_i undergo their positive (graphically upward) and negative (graphically downward) changes in step with each other; hence, there is no signal inversion between these two voltages.

4.13   Output Characteristic Curves for the *CB* Configuration.

Figure 4.21 illustrates the typical output characteristics of the junction transistor in the common-base configuration. e_o is $e_{C\text{-}B}$, i_o is i_C, and the input current i_i is i_E. Compare the common-base characteristics of Fig. 4.21a to the common-emitter characteristics of Fig. 4.21b: The characteristics of the common-base configuration, to the right of the knee region, are more nearly parallel to the x axis (smaller slope values) than is the case with the common-emitter characteristics. This indicates that the collector voltage has a smaller effect upon the collector current in the common-base connection than is the case in the common-emitter configuration. *Since $e_{C\text{-}B}$ has a smaller effect on i_C with i_E constant, relatively larger $e_{C\text{-}B}$ changes are required to produce a given i_C change, resulting in a larger dynamic output resistance in the CB connection as compared to the CE connection.*

Let us examine the common-base characteristic a bit further in order to determine the reason for this difference. Recall the fundamental transistor current relationship:

$$i_E = i_B + i_C$$

hence
$$i_C = i_E - i_B$$

Since i_E is fixed for each CB characteristic curve, any change in i_C must be compensated by a change in i_B. For example, if i_C increases by 50 μa, then i_B must decrease by 50 μa so as to maintain the preceding current relationship. i_B is a small fraction of i_E, while i_C is nearly equal to i_E close to zero collector-base voltage. i_C has the possibility of undergoing only a small change for relatively large changes in collector-base potential, resulting in the output resistance of the transistor in the CB connection being relatively high.

Example 1: Refer to Fig. 4.21. Find the dynamic output resistance for a collector-base voltage change from 5 to 10 volts with i_E fixed at 2 ma.

Solution:

$$r_o = \frac{\Delta e_{C\text{-}B}}{\Delta i_C} = \frac{10 - 5}{1.85 \text{ ma} - 1.82 \text{ ma}} = 167 \text{ kilohms}$$

Example 2: Compute the dynamic output conductance for example 1:

Solution:

$$g_o = \frac{1}{r_o} = \frac{1}{1.67 \times 10^5} = 6.65 \ \mu\text{mho}$$

4.14 The Common-collector Configuration. The NPN junction transistor in the common-collector configuration is shown in Fig. 4.22. The input signal is applied between the base and collector, and the output appears across the emitter and collector. Of special interest is that the emitter-base junction is still under the forward bias condition because of

V_{ee}, and the presence of V_{cc} places the collector-base junction under reverse bias. C_1 is a large bypass capacitor, placing the collector at common potential for the signal variations. *The common-collector circuit displays a high input resistance and a low output resistance.* Typical values for the three configurations are stated in the following section.

Note that V_{ee} and V_{cc} are series-opposing, with V_{cc} generally being the larger of the two. A positive excursion of e_i effectively adds that much potential in series-aiding to V_{ee} and series opposition to V_{cc}. This effectively reduces the collector reverse bias by that amount, thereby causing an increase in the magnitudes of i_E, i_C, and i_B. The negative excursion of e_i produces the opposite effect. This circuit may be used in the opposite way while in the same connection; i.e., the input may be applied between the emitter and collector terminals and the output taken

FIG. 4.22 NPN(*CC*) circuit.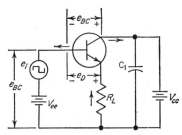

across the base and collector. This feature (alternately reversing the input and output terminals) has found use in many circuit applications.

4.15 Hybrid Parameters. A set of parameters, which can be conveniently obtained by actual measurement or by using the characteristics of the transistor, have been developed. They are the *hybrid parameters.* Several other types of parameters have been developed, but the hybrid parameters are in most popular use because of the ease with which they may be obtained.

Four hybrid parameters are analyzed in this section. Each parameter is given the symbol h and two numbers as its subscript. Only the numbers 1 and 2 are used: number 1 refers to an input circuit measurement, and number 2 indicates an output circuit measurement. Each of the four parameters are obtained from a ratio of two measurements. The first number appearing in the subscript indicates the circuit measurement (input is 1, output is 2) used for the numerator of the ratio, while the second subscript indicates the circuit measurement used for the denominator of the same ratio. For example:

$$h_{22} = \frac{\text{an output circuit measurement}}{\text{an output circuit measurement}}$$

The specific measurements used for each hybrid parameter are analyzed in the following paragraphs.

Figure 4.23 illustrates the hybrid equivalent circuit. Stating the voltage equation for the series input loop, we obtain

$$e_b = h_{11}i_b + h_{12}e_c$$

The current is the prime consideration in the output series loop. Assuming that the current through h_{22} is of the same polarity as the current produced by the generator $h_{21}i_b$, we obtain the following equation

$$i_c = h_{21}i_b + h_{22}e_c$$

The two independent variables i_b and e_c appear in both equations. With these two equations, we shall determine four hybrid parameters, i.e., h_{11}, h_{12}, h_{21}, and h_{22}.

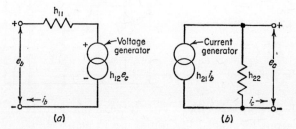

Fig. 4.23   Hybrid equivalent circuit: (a) input, (b) output.

The h_{11} Parameter.   Let e_c of the input-loop equation equal zero, and the equation becomes

$$e_b = h_{11}i_b$$

and

$$h_{11} = \left(\frac{e_b}{i_b}\right)_{E_C=K}$$

Examination of this relationship reveals that h_{11} is the *input resistance* of the equivalent circuit (input voltage divided by the input current) and is expressed in ohms.

h_{11} is most often stated as h_i in manufacturer's data and is further specified with a second subscript which denotes the configuration being considered; i.e.,

$$h_i \text{ or } h_{ie} = \text{input resistance, } CE \text{ configuration}$$
$$h_{ib} = \text{input resistance, } CB \text{ configuration}$$
$$h_{ic} = \text{input resistance, } CC \text{ configuration}$$

NOTE: h_{11} is equal to h_{ib}; the relationship between h_{ib} and the remaining two is given in Sec. 4.17.

This parameter can be determined in an actual transistor by setting e_c to zero in terms of the signal. The correct bias potential is applied to the collector, and it is placed at the common potential (zero volts) in terms of the signal by use of a moderate to large capacitor between collector and common. Precautions should be taken to see that the bypass capacitor is of sufficient capacity to offer negligible impedance to the lowest signal frequency to be handled by the transistor. e_b and i_b are then measured in the conventional manner.

The h_{12} Parameter. Again referring to the input circuit loop equation, set i_b to zero, which results in

$$e_b = h_{12}e_c$$

and

$$h_{12} = \left(\frac{e_b}{e_c}\right)_{I_B=K}$$

Notice that e_b and e_c are both expressed in volts, which results in h_{12} being a pure number with no unit. This relationship is called the *feedback voltage ratio.*

h_{12} most often appears as h_r in manufacturer's data with a second letter added to the subscript to denote the configuration.

h_r or h_{re} = feedback voltage ratio, *CE* configuration
h_{rb} = feedback voltage ratio, *CB* configuration
h_{rc} = feedback voltage ratio, *CC* configuration

NOTE: h_{12} is equal to h_{rb}; the relationship between h_{rb} and the two others is given in Sec. 4.17.

h_{12} is determined for an actual transistor by maintaining a closed circuit for bias voltages but an open circuit for signal currents. The bias potential is applied to the base by way of a high-Q parallel circuit, which resonates at the signal frequency being considered. In this way, there is negligible resistance to the d-c voltage because the coil possesses low resistance, while the tank circuit offers high impedance to the signal because of the high Q value. Once the condition is established, e_b and e_c are measured in the conventional manner, and the feedback voltage ratio is computed. In those cases where a high-Q tank circuit is not readily available, a high impedance choke may be used in its place.

The h_{21} Parameter. Let us now set e_c to zero in the output series loop equation:

$$i_c = h_{21}i_b$$

and

$$h_{21} = \left(\frac{i_c}{i_b}\right)_{E_C=K}$$

Since both i_c and i_b are expressed in amperes, the ratio is a pure number and is the *current amplification* or *forward current ratio* of the circuit. This ratio, when considered with the common-emitter configuration, is also called beta.

h_{21} usually appears as h_f in manufacturers' data with a second subscript added to denote the configuration:

$$h_f \text{ or } h_{fe} = \text{current amplification, } CE \text{ configuration}$$
$$h_{fb} = \text{current amplification, } CB \text{ configuration}$$
$$h_{fc} = \text{current amplification, } CC \text{ configuration}$$

NOTE: h_{21} is equal to h_{fb}; the relationship between h_{fb} and the other two is given in Sec. 4.17.

This parameter is obtained in an actual transistor by setting e_c to zero in the manner described for h_{11} and then measuring i_c and i_b in the conventional manner.

The h_{22} Parameter. Setting i_b to zero in the output series loop equation results in

$$i_c = h_{22}e_c$$

and

$$h_{22} = \left(\frac{i_c}{e_c}\right)_{I_B = K}$$

Notice that this is the ratio of current to voltage, which is conductance, and is expressed in mhos. This ratio is called the *output conductance* or *output admittance*. h_{22} appears as h_o in manufacturers' data with a second subscript added to specify the configuration.

$$h_o \text{ or } h_{oe} = \text{output conductance, } CE \text{ configuration}$$
$$h_{ob} = \text{output conductance, } CB \text{ configuration}$$
$$h_{oc} = \text{output conductance, } CC \text{ configuration}$$

NOTE: h_{22} is equal to h_{ob}; the relationship between h_{ob} and the other two is given in Sec. 4.17.

h_{22} is determined in an actual transistor by setting i_b to zero (in terms of signal) in the manner described earlier and measuring i_c and e_c in the conventional manner.

4.16 Graphical Determination of Hybrid Parameters. In addition to being relatively easy to measure these parameters for an actual transistor, it is equally convenient to determine them by use of the e_B versus i_B and e_C versus i_C characteristics.

h_i (Input Resistance). h_i can be obtained from the e_B versus i_B characteristic by use of the following technique (refer to Fig. 4.24):

1. Select the desired $i_{B,Q}$ value (point A). Draw a line from point A, parallel to the x axis (AB) such that it intersects the curves.

2. Locate the intersection of line AB with the characteristic which has e_C fixed at the desired value (point C). Draw a line tangent to the curve at this point (line DE). Notice that the tangent line is the hypotenuse of a right triangle which has $EF(\Delta i_B)$ for its altitude and $DF(\Delta e_B)$ as its base. The slope of this characteristic is determined by

$$\text{Slope} = \frac{\Delta y}{\Delta x} = \frac{EF}{DF} = \frac{\Delta i_B}{\Delta e_B}$$

3. Recall that each curve is at a constant e_C (zero volts in terms of signal) where the reciprocal of this ratio is h_i:

$$h_i = \left(\frac{\Delta e_B}{\Delta i_B}\right)_{E_C = K}$$

NOTE: The distance between points D and F (which is the length of the tangent line) is arbitrarily selected by the designer.

h_r (*Feedback Voltage Ratio*). The e_B versus i_B characteristic may also be used for the graphical determination of h_r (refer to Fig. 4.25).

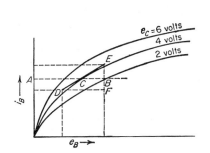

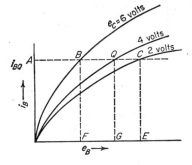

FIG. 4.24 Graphical determination of h_i. FIG. 4.25 Graphical determination of h_r.

1. Select the desired $i_{B,Q}$ (point A). Draw a line parallel to the x axis $(ABQC)$.

2. The line constructed in step 1 will intersect the characteristic which has a fixed e_C that coincides with the desired $e_{C,Q}$. Pick a convenient $e_{C,\text{min}}$ and $e_{C,\text{max}}$ (such as points C and B). Construct perpendicular lines from these points to the x axis (such as CE and BF).

3. Now compute h_r by

$$h_r = \frac{E - F}{B - C} = \frac{\Delta e_B}{\Delta e_C} \qquad \text{with } I_B \text{ constant}$$

h_f (*Forward Current Ratio*).   The parameter h_f can be determined from the collector characteristics (refer to Fig. 4.26).

1. Select the fixed e_C value (point A).   Construct a line perpendicular to the x axis from point A (line AB).

2. Find the characteristic which is fixed at the desired $i_{B,Q}$.   Locate the point of intersection of line AB and this curve (point C).

3. Select convenient values of $i_{B,\min}$ and $i_{B,\max}$ (such as points D and E).

4. Draw lines from points D and E to the y axis and parallel to the x axis (such as lines DF and EG).

5. Using the values read from the characteristics, compute h_f:

$$h_f = \frac{G - F}{E - D} = \frac{\Delta i_C}{\Delta i_B} \quad \text{with } E_C \text{ constant}$$

h_o (*Output Conductance*). h_o can also be computed from the collector characteristics by utilization of the slope technique.   Refer to Fig. 4.27.

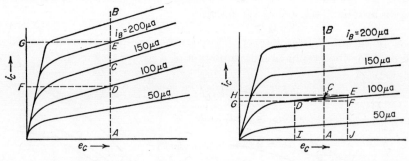

FIG. 4.26   Graphical determination of h_f.   FIG. 4.27   Graphical determination of h_o.

1. Select the desired $e_{C,Q}$ value (point A).

2. Draw a perpendicular line from point A (such as line AB).

3. Locate the intersection of line AB and the curve whose i_B is fixed at the desired $i_{B,Q}$ value (point C).   Draw a line tangent to the curve at point C (line DE).

4. Draw perpendicular lines to each axis from each point (lines DG and DI were constructed from point D, and lines EH and EJ were constructed from point E).

5. Using these values, h_o can now be computed:

$$h_o = \frac{H - G}{J - I} = \frac{\Delta i_C}{\Delta e_C} \quad \text{with } I_B \text{ constant}$$

4.17   Determination of R_i, R_o, and Gains with Hybrid Parameters.   For the common-emitter amplifier:

$$R_i = \frac{h_{ie} + (h_{oe}h_{ie} - h_{fe}h_{re})R_L}{1 + h_{oe}R_L}$$

$$R_o = \frac{h_{ie} + R_g}{h_{oe}h_{ie} - h_{re}h_{fe} + h_{oe}R_g}$$

$$A_v = \frac{-h_{fe}R_L}{R_L(h_{ie}h_{oe} - h_{fe}h_{re}) + h_{ie}}$$

$$A_i = \frac{-h_{fe}}{h_{oe}R_L + 1}$$

$$G = \frac{(h_{fe})^2 R_L}{(h_{oe}R_L + 1)[(h_{ie}h_{oe} - h_{fe}h_{re})R_L + h_{ie}]}$$

The same equations are also true for the common-base amplifier by changing all the subscripts e to b:

$$R_i = \frac{h_{ib} + (h_{ob}h_{ib} - h_{fb}h_{rb})R_L}{1 + h_{ob}R_L}$$

$$R_o = \frac{h_{ib} + R_g}{h_{ob}h_{ib} - h_{rb}h_{fb} + h_{ob}R_g}$$

$$A_v = \frac{-(h_{fb})R_L}{R_L(h_{ib}h_{ob} - h_{fb}h_{rb}) + h_{ib}}$$

$$A_i = \frac{-h_{fb}}{h_{ob}R_L + 1}$$

$$G = \frac{(h_{fb})^2 R_L}{(h_{ob}R_L + 1)[(h_{ib}h_{ob} - h_{fb}h_{rb})R_L + h_{ib}]}$$

And for the common-collector amplifier:

$$R_i = \frac{h_{ic} + (h_{oc}h_{ic} - h_{fc}h_{rc})R_L}{1 + h_{oc}R_L}$$

$$R_o = \frac{h_{ic} + R_g}{h_{oc}h_{ic} - h_{rc}h_{fc} + h_{oc}R_g}$$

$$A_v = \frac{-h_{fc}R_L}{R_L(h_{ic}h_{oc} - h_{fc}h_{rc}) + h_{ic}}$$

$$A_i = \frac{-h_{fc}}{(h_{oc}R_L) + 1}$$

$$G = \frac{(h_{fc})^2 R_L}{(h_{oc}R_L + 1)[(h_{ic}h_{oc} - h_{fc}h_{rc})R_L + h_{ic}]}$$

NOTE: In the preceding equations for the three configurations:

R_g = resistance of the input signal voltage source
R_L = load resistance

It is often necessary to convert a hybrid parameter value from that of one configuration to another. Following are the necessary conversion equations:

1. From CE to CB:

$$h_{ib} = \frac{h_{ie}}{1 + h_{fe}}$$

$$h_{rb} = \frac{h_{ie}h_{oe}}{1 + h_{fe}} - h_{re}$$

$$h_{fb} = \frac{-h_{fe}}{1 + h_{fe}}$$

$$h_{ob} = \frac{h_{oe}}{1 + h_{fe}}$$

2. From CE to CC:

$$h_{ic} = h_{ie}$$
$$h_{rc} = 1 - h_{re}$$
$$h_{fc} = -(1 + h_{fe})$$
$$h_{oc} = h_{oe}$$

3. From CB to CE:

$$h_{ie} = \frac{h_{ib}}{1 + h_{fb}}$$

$$h_{re} = \frac{h_{ib}h_{ob}}{1 + h_{fe}} - h_{rb}$$

$$h_{fe} = \frac{-h_{fb}}{1 + h_{fb}}$$

$$h_{oe} = \frac{h_{ob}}{1 + h_{fb}}$$

4. From CB to CC:

$$h_{ic} = \frac{h_{ib}}{1 + h_{fb}}$$

$$h_{rc} = \frac{h_{ib}h_{ob}}{1 + h_{fb}} + h_{rb}$$

$$h_{fc} = \frac{-1}{1 + h_{fb}}$$

$$h_{oc} = \frac{h_{ob}}{1 + h_{fb}}$$

Example 1: A common-emitter amplifier has the following parameters: $h_{ie} = 1{,}200$ ohms; $h_{fe} = 45$; $h_{re} = 4 \times 10^{-4}$; $h_{oe} = 20$ μmho; $R_L = 2$ kilohms; $R_g = 1$ kilohm. Calculate (a) R_i; (b) R_o; (c) A_v; (d) A_i; (e) G.

Solution:

(a) Calculate R_i:

$$\begin{aligned}
R_i &= \frac{h_{ie} + (h_{oe}h_{ie} - h_{fe}h_{re})R_L}{1 + h_{oe}R_L} \\
&= \frac{1{,}200 + (20 \times 10^{-6} \times 1.2 \times 10^3 - 45 \times 4 \times 10^{-4}) \times 2 \times 10^3}{1 + 20 \times 10^{-6} \times 2 \times 10^3} \\
&= 1{,}165 \text{ ohms}
\end{aligned}$$

(b) Calculate R_o:

$$R_o = \frac{h_{ie} + R_g}{h_{oe}h_{ie} - h_{re}h_{fe} + h_{oe}R_g}$$

$$= \frac{1{,}200 + 1{,}000}{20 \times 10^{-6} \times 1.2 \times 10^3 - 4 \times 10^{-4} \times 45 + 20 \times 10^{-6} \times 1 \times 10^3}$$

$$= 84.6 \text{ kilohms}$$

(c) Calculate A_v:

$$A_v = \frac{-h_{fe}R_L}{R_L(h_{ie}h_{oe} - h_{fe}h_{re}) + h_{ie}}$$

$$= \frac{-45 \times 2 \times 10^3}{2 \times 10^3(1.2 \times 10^3 \times 20 \times 10^{-6} - 45 \times 4 \times 10^{-4}) + 1.2 \times 10^{+3}}$$

$$= 74.2$$

(d) Calculate A_i:

$$A_i = \frac{h_{fe}}{(h_{oe}R_L) + 1} = 43$$

(e) Calculate G:

$$G = \frac{(h_{fe})^2 R_L}{(h_{oe}R_L + 1)[(h_{ie}h_{oe} - h_{fe}h_{re})R_L + h_{ie}]}$$

$$= 3{,}214$$

Example 2: The common-emitter hybrid parameters of a transistor are as follows: $h_{ie} = 2$ kilohms; $h_{re} = 6 \times 10^{-4}$; $h_{fe} = 50$; $h_{oe} = 25$ μmho.   Find the common-base hybrid parameters of this transistor.

Solution:

$$h_{ib} = \frac{h_{ie}}{1 + h_{fe}} = \frac{2 \times 10^3}{1 + 50} = 39.2 \text{ ohms}$$

$$h_{rb} = \frac{h_{ie}h_{oe}}{1 + h_{fe}} - h_{re} = \frac{2 \times 10^3 \times 25 \times 10^{-6}}{1 + 50} - 6 \times 10^{-4}$$

$$= -5.902 \times 10^{-4}$$

$$h_{fb} = \frac{-h_{fe}}{1 + h_{fe}} = \frac{-50}{51} = -0.98$$

$$h_{ob} = \frac{h_{oe}}{1 + h_{fe}} = \frac{25 \times 10^{-6}}{1 + 50} = 4.9 \times 10^{-7} \text{ mho}$$

PROBLEMS

4.1 Find the dynamic resistance of a germanium diode when its forward current is (a) 250 μa; (b) 500 μa; (c) 1 ma

4.2 The dynamic resistance of the diode in Prob. 4.1 changed with the magnitude of the forward current.   Explain.

4.3 A junction transistor has the following parameters: Collector voltage is constant; an emitter voltage change of 0.1 volt is accompanied by an 800-μa change in emitter current.   Find the dynamic resistance of the emitter-base circuit.

4.4 A junction transistor has the following parameters: Collector voltage is constant; an emitter voltage change of 0.5 volt is accompanied by a 4-ma change in emitter current. Find the dynamic resistance of the emitter-base circuit.

4.5 A junction transistor has the following parameters: the emitter current is fixed; a 5-volt change in collector voltage is accompanied by a 5-μa change in collector current. Find the dynamic resistance of the collector-base circuit.

4.6 A junction transistor has the following parameters: With the emitter current fixed, a 10-volt change in collector potential is accompanied by an 8-μa change in collector current. Find the dynamic resistance of the collector-base circuit.

4.7 Find the collector-base circuit dynamic conductance for Prob. 4.5.

4.8 Find the collector-base circuit dynamic conductance for Prob. 4.6.

4.9 With the collector potential fixed, a 500-μa change in emitter current results in a 480-μa change in collector current. Find (a) α; (b) β.

4.10 Assume a junction transistor in a given circuit has $\beta = 100$. Find α.

4.11 A junction transistor has the following parameters: $\alpha = 0.92$; collector voltage is fixed. Find the emitter-current change associated with a 2-ma collector-current change.

4.12 Find β of Prob. 4.11.

4.13 Define emitter efficiency.

4.14 Define transport efficiency.

4.15 Describe the effect of the collector-base voltage upon the emitter and collector currents.

4.16 Refer to Fig. 4.10. Find the dynamic output resistance between collector voltage values of 5 and 10 volts with the base current fixed at 120 μa.

4.17 Refer to Fig. 4.10. Find the dynamic output resistance between collector voltage values of 5 and 10 volts with the base current fixed at 300 μa.

4.18 Refer to Fig. 4.15. With the collector-emitter voltage fixed at 0.1 volt, find the dynamic input resistance for the base current changes from 0.5 to 1 ma.

4.19 Repeat Prob. 4.18 with the collector-emitter voltage fixed at 0.3 volt.

4.20 Refer to Fig. 4.16. i_B is fixed at 40 μa. Find the dynamic output resistance for an e_{CE} variation of 10 to 12 volts.

4.21 Repeat Prob. 4.20 with i_B fixed at 60 μa.

4.22 Repeat Prob. 4.20 with i_B fixed at 80 μa.

4.23 Find the dynamic output conductance of Prob. 4.20.

4.24 Find the dynamic output conductance of Prob. 4.21.

4.25 Find the dynamic output conductance of Prob. 4.22.

4.26 The dynamic output resistance of a junction transistor in the common base configuration is larger than in the common-emitter configuration. Explain.

4.27 Refer to Fig. 4.21. In comparing the CB and CE characteristics, what indicates the relative values of dynamic output resistance?

4.28 The common-collector circuit possesses a high input resistance. Explain.

4.29 The common-collector circuit possesses a low output resistance. Explain.

4.30 Assume that e_e of a certain transistor has set at zero volts (in terms of signal) and the following parameters were measured: $e_B = 0.3$ volt; $i_B = 100$ μa; $i_C = 4$ ma. Find (a) h_{11}; (b) h_{21}.

4.31 Assume that e_C of another transistor was set at zero and the following parameters were measured: $e_B = 1$ volt; $i_B = 500$ μa; $i_C = 30$ ma. Find (a) h_{11}; (b) h_{21}.

4.32 Assume that i_B of a certain transistor was set at zero (in terms of signal) and the following parameters were measured: $e_B = 0.7$ volt; $e_C = 15$ volts: $i_C = 14$ ma. Find (a) h_{12}; (b) h_{22}.

4.33 Assume that i_B of another transistor was set at zero and the following parameters were measured: $e_B = 0.03$ volt: $e_C = 5$ volts; $i_C = 4$ ma. Find (a) h_{12}; (b) h_{22}.

4.34 Refer to Fig. 4.10. $e_{C,Q} = 10$ volts; $i_{B,Q} = 180$ μa. Calculate (a) h_{21} (forward current ratio); (b) h_{22} (output conductance).

4.35 Refer to Fig. 4.10. $e_{C,Q} = 5$ volts; $i_{B,Q} = 300$ μa. Calculate (a) h_{21}; (b) h_{22}.

4.36 Refer to Fig. 4.16. $i_{B,Q} = 60$ μa; $e_{C,Q} = 10$ volts. Calculate (a) h_{21}; (b) h_{22}.

4.37 Refer to Fig. 4.16. $i_{B,Q} = 20$ μa, $e_{C,Q} = 14$ volts. Calculate (a) h_{21}; (b) h_{22}.

4.38 Refer to Fig. 4.15. $i_{B,Q} = 1$ ma; $e_{C,Q} = 0.1$ volt. Calculate (a) h_{11} (input resistance); (b) h_{12} (feedback voltage ratio).

4.39 Refer to Fig. 4.15. $i_{B,Q} = 2.5$ ma; $e_{C,Q} = 0.1$ volt. Calculate (a) h_{11}; (b) h_{12}.

4.40 Refer to Fig. 4.13. R_g (resistance of the input voltage source) $= 1$ kilohm; $R_L = 20$ kilohms; $h_{ie} = 1.5$ kilohms; $h_{re} = 5.8 \times 10^{-4}$; $h_{fe} = 49$; $h_{oe} = 30 \times 10^{-6}$ mho. Find (a) R_i; (b) R_o; (c) A_v; (d) A_i; (e) G.

4.41 Convert the hybrid parameters of Prob. 4.40 to (a) common-base hybrid parameters; (b) common-collector hybrid parameters.

4.42 Refer to Fig. 4.13. $R_g = 2$ kilohms; $R_L = 30$ kilohms; $h_{ie} = 2$ kilohms; $h_{re} = 5.9 \times 10^{-4}$; $h_{fe} = 60$; $h_{oe} = 40$ μmho. Find (a) R_i; (b) R_o; (c) A_v; (d) A_i; (e) G.

4.43 Refer to Fig. 4.19. $R_g = 50$ ohms; $R_L = 50$ kilohms; $h_{ib} = 50$ ohms; $h_{rb} = 6.1 \times 10^{-4}$; $h_{fb} = -0.96$; $h_{ob} = 0.5$ mho. Find (a) R_i; (b) R_o; (c) A_v; (d) A_i; (e) G.

4.44 Convert the hybrid parameters of Prob. 4.43 to (a) common-emitter hybrid parameters; (b) common-collector hybrid parameters.

4.45 Refer to Fig. 4.19. $R_g = 30$ ohms; $R_L = 25$ kilohms; $h_{rb} = 6.5 \times 10^{-4}$; $h_{fb} = 0.98$; $h_{ob} = 6 \times 10^{-7}$ mho. Find (a) R_i; (b) R_o; (c) A_v; (d) A_i; (e) G.

4.46 Convert the hybrid parameters of Prob. 4.45 to common-collector hybrid parameters.

4.47 Using the hybrid parameters in Prob. 4.46 and $R_g = 30$ kilohms; $R_L = 1$ kilohm, find (a) R_i; (b) R_o; (c) A_v; (d) A_i; (e) G.

4.48 Typical hybrid parameters for the 2N333 are: $h_{fe} = 28$; $h_{ib} = 60$ ohms; $h_{ob} = 0.4$ μmho; $h_{rb} = 3 \times 10^{-4}$. Convert h_{fe} to h_{fb}.

4.49 In Prob. 4.48, convert·
   (a) h_{ib} to h_{ie}
   (b) h_{ob} to h_{oe}
   (c) h_{rb} to h_{re}

4.50 A common-emitter amplifier utilizes the 2N333. $R_L = 2$ kilohms; $R_g = 300$ ohms. Using the hybrid parameters found in Prob. 4.49, calculate (a) R_i; (b) R_o; (c) A_v; (d) A_i; (e) G.

4.51 Typical hybrid parameters for the 2N223 are: $h_{ib} = 35$ ohms; $h_{ob} = 1$ μmho; $h_{rb} = 2.5 \times 10^{-4}$, $h_{fe} = 110$. Convert:
   (a) h_{ib} to h_{ie}
   (b) h_{ob} to h_{oe}
   (c) h_{rb} to h_{re}

4.52 A common-emitter amplifier utilizes the 2N223. $R_L = 1$ kilohm; $R_g = 1.5$ kilohms. Using the hybrid parameters made available in Prob. 4.51, calculate (a) R_i; (b) R_o; (c) A_v; (d) A_i; (e) G.

4.53 Typical hybrid parameters for the 2N337 are: $h_{ib} = 50$ ohms; $h_{ob} = 0.2$ μmho; $h_{rb} = 2 \times 10^{-4}$; $h_{fe} = 12.5$. Convert h_{fe} to h_{fb}.

4.54 A common-base amplifier utilizes the 2N337. $R_L = 25$ kilohms; $R_g = 500$ ohms. Using the hybrid parameters made available in Prob. 4.53, calculate (a) R_i; (b) R_o; (c) A_v; (d) A_i; (e) G.

4.55   In Prob. 4.53, convert:

     (a) h_{ib} to h_{ie}

     (b) h_{ob} to h_{oe}

     (c) h_{rb} to h_{re}

4.56   A common-emitter amplifier utilizes the 2N337. $R_L = 1.2$ kilohms; $R_g = 500$ ohms. Calculate (a) R_i; (b) R_o; (c) A_v; (d) A_i; (e) G.

4.57   Typical hybrid parameters for the 2N465 are: $h_{ie} = 1,400$ ohms; $h_{fe} = 45$; $h_{re} = 4.3 \times 10^{-3}$; $h_{oe} = 18$ μmho. Assume this transistor is utilized in a common-emitter amplifier. $R_L = 1.6$ kilohms; $R_g = 300$ ohms. Calculate (a) R_i; (b) R_o; (c) A_v; (d) A_i; (e) G.

4.58   In Prob. 4.57, convert:

     (a) h_{ie} to h_{ic} (b) h_{fe} to h_{fc}

     (c) h_{re} to h_{rc} (d) h_{oe} to h_{oc}

CHAPTER 5

BIAS AND COUPLING TECHNIQUES FOR THE JUNCTION TRANSISTOR

The use of batteries or other sources of emf for bias is customarily restricted to the potential required for the collector-base junction. The bias potential for the emitter-base junction may be obtained by a number of techniques, several of which are discussed in this chapter. Recall that one of the purposes of bias is to set the transistor at a predetermined resistance value at zero signal condition, from which it may vary with application of a signal. The class of operation, to be discussed in a later section, is established by the magnitude of the bias voltage. In those cases where faithful reproduction is desired, the bias sets the operating point such that the output current variations above and below rest are equal and symmetrical (assuming the input variations are also equal and symmetrical).

Another significant purpose of bias is to provide an element of thermal stability in the transistor circuit. The evaluation of the bias techniques in terms of thermal stability is delved into immediately after the presentation of the various bias possibilities.

The latter portion of the chapter is devoted to the analysis of coupling (direct, RC, and transformer). The three basic methods of coupling are examined under actual signal conditions. Design criteria for all coupling techniques are also presented.

5.1 Fixed Bias. Figure 5.1 illustrates a common-emitter amplifier with fixed bias. Notice that a PNP transistor is used in the diagram, making it necessary to maintain $e_{B\text{-}E}$ negative in order to establish forward bias for the input circuit. R_B is connected from the base to the collector side of V_{cc}. By Kirchhoff's series loop law

$$e_{RB} = V_{cc} - e_{B\text{-}E} = e_{RL} + e_{C\text{-}B}$$

or
$$e_{B\text{-}E} = V_{cc} - e_{RB} = V_{cc} - (e_{RL} + e_{C\text{-}B})$$

V_{cc} is a fixed negative potential when using a PNP type transistor, and $e_{B\text{-}E}$ is negative by the magnitude determined by $V_{cc} - e_{RB}$.

When the input signal swings in the positive direction, e_{B-E} becomes less negative, widening the emitter-base and collector-base depletion regions. The emitter and collector currents undergo decreases in magnitude. e_{RL} decreases and e_{C-B} increases by the same amount, such that $e_{RL} + e_{C-B}$ remains fixed. Since $e_{RB} = e_{RL} + e_{C-B}$ and R_B is of fixed ohmic value, then I_B remains fixed. A negative input signal excursion results in an increased collector current, causing e_{RL} to increase and e_{C-B} to decrease. Since $\Delta e_{RL} = \Delta e_{C-B}$, then $e_{RL} + e_{C-B}$ remains constant, as must e_{RB}. Hence a constant bias current is again maintained.

The input signal current may be considered as alternately aiding and opposing this fixed bias current in terms of electron flow into the base terminal of Fig. 5.1. When the signal swings positive, the signal current is opposite to the bias current, resulting in a reduced base current (i_C and e_{RL} increase, e_{C-B} decreases). The value of R_B is determined by the

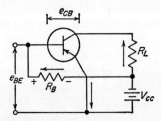

Fig. 5.1　CE circuit with fixed bias.

desired zero signal base current e_{B-E} and magnitude of V_{cc}. In equation form,

$$R_B = \frac{V_{cc} - E_{B-E}}{I_B}$$

where I_B = base current at the operating point
　　　V_{cc} = collector supply voltage
　　　E_{B-E} = base-emitter potential at zero signal condition

NOTE: E_{B-E} is usually much smaller than V_{cc}; hence the preceding equation may be simplified, for all practical purposes, to

$$R_B \cong \frac{V_{cc}}{I_B} \quad \text{and} \quad I_c \cong h_{fe}(I_{co} + I_B) = h_{fe}\left(I_{co} + \frac{V_{cc}}{R_B}\right)$$

NOTE: h_{fe} becomes h_{fb} for the CB connection and h_{fc} for the CC connection.

The fixed-bias technique is not satisfactory in many cases for several reasons. Because of the variations which normally exist between transistors, a bias adjustment is often made necessary by use of another transistor of the same type. The same problems can develop by a change in the temperature of the transistor. The automatic self-bias technique, which is next considered, adjusts to these changes automatically.

Example 1: Refer to Fig. 5.1a. $V_{cc} = 12$ volts; desired zero signal base current = 50 μa; fixed bias is to be used.   Calculate R_B.

Solution:

$$R_B = \frac{12}{50 \times 10^{-6}} = 240 \text{ kilohms}$$

Example 2: Refer to Fig. 5.1a. $V_{cc} = 6$ volts; $I_B = 100$ μa; fixed bias is to be used.   Calculate R_B.

Solution:

$$R_B = \frac{6}{100 \times 10^{-6}} = 60 \text{ kilohms}$$

5.2   Automatic Self-bias.   Refer to Fig. 5.2 for the PNP type common-emitter amplifier with automatic self-bias. R_B is essentially in parallel with the collector-base junction and in this type of bias is connected from

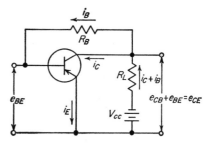

Fig. 5.2   *CE* circuit with automatic self-bias.

the base terminal to the collector terminal.   Since R_B and the collector base junction are in parallel, their voltage variations must coincide, which is an important statement in this analysis.

It will be recalled that an increased collector current is accompanied by a decrease in $e_{C\text{-}B}$.   Since $e_{C\text{-}B} = e_{RB}$, then e_{RB} also decreases with an increase in i_C. R_B is fixed in ohmic value, which means a reduction in e_{RB} results in a reduction in i_{RB}.   This tends to reduce the effect of the input signal variations.

On the positive swing of the input signal, the rate of electron flow into the N base is reduced.   This allows a greater number of unbalanced ions to exist in the base region, which widens the base side of both junction depletion regions.   An enlargement of the depletion region reduces the magnitude of i_C, which reduces e_{RL} and increases $e_{C\text{-}B}$.   Since $e_{RB} = e_{C\text{-}B}$ and R_B is fixed, then i_{RB} increases in accordance with Ohm's law.   This increases the rate of electron flow into the base which tends to bring the transistor back to its operating point condition.

When the input signal swings negative, electron flow into the base is increased, which increases i_C and e_{RL} and decreases $e_{C\text{-}B}$ and e_{RB}.   Hence

i_{RB} decreases, reducing the electron flow into the base to that extent. Again note the tendency of i_{RB} to bring the transistor back to its operating point with changes in the input signal. *This is called stabilization, and negative feedback is incorporated. Negative feedback can be considered as the return of a portion of the output to the input in an inverted form.*

The electron current flowing toward the collector is divided at the collector terminal. When electron flow into the collector is decreased by the internal action of the transistor (depletion-width increase), an increased flow through R_B into the base is developed (which tends to decrease the width of the depletion regions). When the electron flow into the collector is increased (because the depletion widths have been reduced by the negative excursion of the input signal), the electron flow through R_B is reduced (resulting in a tendency to widen the depletion regions since electron flow into the base is reduced). Hence the stabilization effect is achieved.

By Kirchhoff's loop law, we find

$$e_{E\text{-}B} = V_{cc} - e_{RL} - e_{RB}$$

Since

$$e_{RB} = i_{RB}R_B$$

then

$$e_{E\text{-}B} = V_{cc} - e_{RL} - (i_{RB}R_B)$$

Solving for R_B,

$$R_B = \frac{V_{cc} - e_{RL} - e_{E\text{-}B}}{i_{RB}}$$

where i_{RB} = base current at zero signal condition I_B. Since $e_{E\text{-}B}$ is usually small as compared to the other voltages, it may be omitted in many practical cases; hence,

$$R_B \cong \frac{V_{cc} - e_{RL}}{I_B} \qquad \text{and} \qquad I_{C,Q} \cong \frac{I_{co} + V_{cc}/(R_B + R_L)}{1/h_{fe} + R_L/(R_B + R_L)}$$

NOTE: h_{fe} becomes h_{fb} for the CB connection and h_{fc} for the CC connection.

Example 1: Refer to the amplifier of Fig. 5.2. $V_{cc} = 12$ volts; $R_L = 5$ kilohms; $i_C = 1.2$ ma; $I_B = 100$ μa. Calculate R_B.

Solution: We must first determine e_{RL}:

$$e_{RL} = i_{RL}R_L = (i_C + I_B)R_L$$
$$= (1.2 \times 10^{-3} + 0.1 \times 10^{-3}) \times 5 \times 10^3 = 6.5 \text{ volts}$$

Calculating R_B:

$$R_B = \frac{12 - 6.5}{100 \times 10^{-6}} = 55 \text{ kilohms}$$

Example 2: Refer to the amplifier of Fig. 5.2. $V_{cc} = 22$ volts; $R_L = 2$ kilohms; $i_C = 5$ ma; $i_B = 200$ μa. Calculate R_B.

Solution:

$$e_{RL} = (5 \times 10^{-3} + 0.2 \times 10^{-3}) \times 2 \times 10^3 = 10.4 \text{ volts}$$

$$R_B = \frac{22 - 10.4}{200 \times 10^{-6}} = 58 \text{ kilohms}$$

5.3 Automatic Self-bias with Bypass Provisions.

The automatic self-bias arrangement of Fig. 5.2 creates a negative feedback in the amplifier. The advantages of negative feedback are:

1. Increased stability
2. Improved frequency response
3. Improved faithful reproduction (linearity)

The chief disadvantage of negative feedback is the reduction of amplification.

The advantages are weighed against the disadvantage when designing the amplifier, in order to determine whether or not negative feedback is

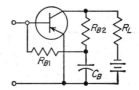

FIG. 5.3   Automatic self-bias with bypassed negative feedback.

desirable. In some cases, the reduction of amplification is more important than the other factors, resulting in a search for a method which would reduce the negative feedback characteristics.   Figure 5.3 is one way in which automatic self-bias can be incorporated with a reduction in negative feedback.

R_B of Fig. 5.2 is replaced with R_{B1} and R_{B2} in Fig. 5.3, such that

$$R_{B1} + R_{B2} = R_B$$

The unidirectional base current path between the collector and base is of the same ohmic value, since R_{B1} and R_{B2} are in series.   The two resistors are often equal in value; hence

$$R_{B1} = R_{B2} = 0.5R_B$$

The sum of R_{B1} and R_{B2} may be computed from the same relationship stated for the automatic self-bias circuit which had no provisions for reduction of negative feedback.

Refer to Fig. 5.4 for the following analysis of negative feedback (degeneration) reduction in the automatic self-bias circuit.   It will be recalled that the input and output signal variations are inverted in the common-emitter configuration, with the greater portion of the voltage appearing across the collector-base (since $e_{B\text{-}E}$ is comparatively small).

Without C_B, e_o appears across $R_{B2} + R_{B1}$. C_B is in series with R_{B2} and in parallel with R_{B1}.

$$X_{CB} = \frac{1}{2\pi f C_B}$$

where f = signal frequency, cps
C_B = farads
X_{CB} = capacitive reactance, ohms

For a given value of C_B, X_{CB} is greatest at the lowest signal frequency. The portion of the output signal variations taken back to the input appears across C_B. *Selecting the value of C_B such that its reactance is 10 per cent or less of R_{B2} at the minimum signal frequency will result in most of e_O appearing across R_{B2}, and only a small portion of e_O will be felt across R_{B1}.* This effectively shunts the output signal away from the input, thereby reducing negative feedback in accordance with the ratio of X_{CB}/R_{B2}.

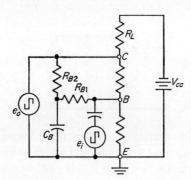

Fig. 5.4   Equivalent circuit of Fig. 5.3.

For example, if only 10 per cent of the output signal appears across C_B at the minimum signal frequency, the negative feedback involves only that portion of the output at that frequency. Recalling that X_C decreases with frequency, we note that X_{CB} becomes much less than $0.1R_{B2}$ at the middle and upper signal frequency range.

Example: Assume the circuit of Fig. 5.4 has the following parameters: $V_{cc} = 12$ volts; $i_C = 2$ ma; $i_B = 75$ μa; signal frequency range is 60 cps − 10,000 cps; $R_L = 3$ kilohms; $R_{B1} = R_{B2}$. Find (a) R_{B1}, R_{B2}; (b) C_B such that X_{CB} at $f_{min} = 0.1R_{B2}$.

Solution:

(a) Find R_{B1} and R_{B2}:

$$R_B = \frac{V_{cc} - e_{RL}}{I_B} = \frac{V_{cc} - (i_C + i_B)R_L}{I_B}$$
$$= \frac{12 - (2 \times 10^{-3} + 0.075 \times 10^{-3})3 \times 10^3}{75 \times 10^{-6}}$$
$$= 77 \text{ kilohms}$$

Since $R_{B1} = R_{B2}$, then

$$R_{B1} = R_{B2} = R_B/2 = 7\frac{7}{2} = 38.5 \text{ kilohms}$$

(b) Find C_B:

$$X_{CB} = 0.1R_{B2} \text{ at } f_{\min}$$

hence

$$X_{CB} = 0.10 \times 38.5 \times 10^3 \text{ ohms at 60 cps} = 3.85 \times 10^3 \text{ ohms at 60 cps}$$
$$C_B = \frac{1}{2\pi f_{\min} X_{CB}} = \frac{1}{6.28 \times 6 \times 10^1 \times 3.85 \times 10^3} = 0.689 \ \mu\text{f}$$

5.4 Voltage Divider and Emitter Bias. Refer to Fig. 5.5 for the following analysis. It will be recalled that the emitter electron current of a PNP type transistor flows out of the emitter terminal. *Connecting a*

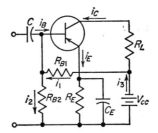

Fig. 5.5 *CE* amplifier utilizing voltage divider and emitter bias.

resistor R_E in the emitter results in a voltage drop with the indicated polarity, which is a reverse bias upon the emitter. An increase in emitter current results in an increase in the reverse bias, which adjusts the current to a reduced magnitude. Likewise, a reduction of emitter current reduces this reverse bias, which enables the current to increase. Therefore R_E introduces negative feedback or degeneration into the circuit. In equation form,

$$R_E = \frac{e_{RE}}{i_E}$$

where e_{RE} is the desired reverse bias.

The signal degeneration introduced by R_E may be minimized by placing a bypass capacitor C_E in parallel with R_E (as shown in Fig. 5.5). C_E is utilized in those cases where the reduction of amplification caused by the unbypassed R_E is undesirable. The negative feedback is minimized most effectively by making X_{CE} as close to $0.10R_E$ at the minimum signal frequency as is possible. In some audio-frequency circuits, where R_E is relatively small and the minimum signal frequency is low, practical considerations may make the ideal relationship difficult to achieve because of the large C_E required.   However, it is still a standard toward which the designer would strive; in equation form,

$$C_E = \frac{1}{2\pi f_{min} X_{CE}}$$

where X_{CE} at $f_{min} = 0.10 R_E$

The use of R_E introduces reverse bias to the input of the amplifier illustrated in Fig. 5.5.   The input circuit of a transistor requires forward bias for operation, which necessitates an additional potential in the input circuit.   Rather than use a separate voltage supply, a voltage-divider network consisting of R_{B1} and R_{B2} may be used for this purpose, with e_{RB2} providing the forward bias to the base-emitter junction.   This forward bias should be greater than e_{RE} by the desired e_{B-E}; hence $e_{RB2} = e_{B-E}$ desired $+ e_{RE}$.   The zero signal base current is determined from the operating point of the characteristics. i_2 is an arbitrarily selected bleeder current, and i_1 is the total of $i_{B,Q}$ and i_2.   For many practical circuits, i_2 is selected to be equal to or less than $i_{B,Q}$ so as not to impose too severe a drain on V_{cc}.

Looking into the base-emitter circuit, the input signal views R_{B2} and R_{B1} as being in parallel.   The equivalent resistance of this parallel combination, which is in shunt with the R_i of the transistor, must be sufficiently high so as not to reduce the input resistance of the transistor to any great extent.   In many practical circuits, R_{B2} should be as much greater than R_i as possible to minimize the shunting effect. R_{B1} is usually greater than R_{B2} since e_{RB1} is many times greater than e_{RB2} while the currents flowing through these resistances are close in value.   Therefore, the equivalent resistance of R_{B1} and R_{B2} must be sufficiently large to minimize the shunting effect they cause across the input circuit.   Consideration of this condition is one of the chief factors in the selection of R_{B1} and R_{B2}.

The collector current for the voltage-divider and emitter-bias network can be reasonably approximated by the following relationship when R_E is many times larger than $R_{B1}R_{B2}/(R_{B1} + R_{B2})$:

$$i_C \cong \frac{[R_{B2}V_{cc}/(R_{B1} + R_{B2})] + [R_{B1}R_{B2}I_{CO}/(R_{B1} + R_{B2})] - e_{B-E}}{R_E + (1/h_{fe}) \times [R_{B1}R_{B2}/(R_{B1} + R_{B2})]}$$

NOTE: h_{fe} becomes h_{fb} for the CB connection and h_{fc} for the CC connection.

Example 1: Assume the circuit of Fig. 5.5 has the following parameters at zero signal condition: $i_E = 2$ ma; $i_B = 50$ μa; desired $e_{RE} = 1$ volt; $V_{cc} = 12$ volts; signal frequency range = 100 to 5,000 cps; $e_{B-E} = 0.5$ volt; $i_{RB2} = i_{B,Q}$.   Find (a) R_E; (b) C_E; (c) R_{B2}; (d) R_{B1}.

Solution:

(a) Find R_E:

$$R_E = \frac{e_{RE}}{i_E} = \frac{1}{2 \times 10^3} = 500 \text{ ohms}$$

(b) Find C_E:

$$X_{CE} \text{ at } f_{\min} = 0.10R_E = 50 \text{ ohms at } 100 \text{ cps}$$
$$C_E = \frac{1}{2\pi f_{\min} X_{CE}} = \frac{1}{6.28 \times 1 \times 10^2 \times 5 \times 10^1}$$
$$= 31.8 \text{ } \mu\text{f}$$

(c) Find R_{B2}:
We must first determine e_{RB2}:

$$e_{RB2} = e_{B\text{-}E} + e_{RE}$$
$$= 0.5 + 1.0 = 1.5 \text{ volts}$$

and

$$i_{RB2} = i_{B,Q} = 50 \text{ } \mu\text{a}$$

Now solving for R_{B2}:

$$R_{B2} = \frac{e_{RB2}}{i_{RB2}} = \frac{1.5}{5 \times 10^{-5}} = 30 \text{ kilohms}$$

(d) Find R_{B1}:

$$e_{RB1} = V_{CC} - e_{RB2}$$
$$= 12 - 1.5 = 10.5 \text{ volts}$$

and
$$i_{RB1} = i_{RB2} + i_{B,Q} = 100 \text{ } \mu\text{a}$$

Solving for R_{B1},

$$R_{B1} = \frac{e_{RB1}}{i_{RB1}} = \frac{10.5}{1 \times 10^{-4}} = 105 \text{ kilohms}$$

Figure 5.6a illustrates the equivalent input circuit of Fig. 5.5 as seen by the input signal. If an emitter bypass capacitor is used, then R_E is

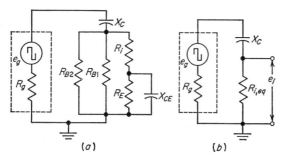

FIG. 5.6   Equivalent input circuit of Fig. 5.5: (a) original, (b) simplified.

effectively removed from the signal circuit because of the low impedance shunt provided by X_{CE}. R_E is sometimes not bypassed in order to increase the input circuit resistance.

In determining the input resistance of the amplifier, the equivalent combination is obtained, yielding the composite value termed $R_{i,eq}$ in Fig. 5.6b. When R_E is not bypassed, the equation becomes

$$R_{i,eq} = \frac{1}{1/R_{B1} + 1/R_{B2} + 1/(R_i + R_E)}$$

where R_i = transistor input $R = \dfrac{h_{ie} + (h_{oe}h_{ie} - h_{fe}h_{re})R_L}{1 + h_{oe}R_L}$

Note that R_i is affected by the value of R_L.

The coupling capacitor C introduces a reactance in series with $R_{i,eq}$, thereby depriving the input circuit of some portion of e_g. *Low-frequency cutoff occurs at the frequency where $e_{Ri,eq} = 0.707e_{g,max}$ (assuming $e_{g,max}$ is always the same).*

X_C is largest at the lower frequencies, and let us assume that $R_{i,eq}$ remains relatively constant. *Since X_C is a quadrature (reactive) component and $R_{i,eq}$ is an in-phase (resistive) component, we recall that when $e_{Ri,eq} = 0.707e_{g,max}$, e_{XC} also equals $0.707e_{g,max}$; hence $e_{Ri,eq} = e_{XC}$ at $f_{co,min}$ and $X_c = R_{i,eq}$ at $f_{co,min}$.* Using this information, the value of C to be used as a coupling capacitor can be readily determined.

Example 2: Refer to the circuit of Fig. 5.5. Assume the hybrid parameters of the transistor are: $h_{ie} = 1,400$ ohms; $h_{fe} = 45$; $h_{re} = 4.3 \times 10^{-4}$; $h_{oe} = 18$ μmho; $R_E = 500$ ohms; $R_{B2} = 2$ kilohms; $R_{B1} = 20$ kilohms; $R_L = 2$ kilohms; $f_{co,min} = 60$ cps. Find (a) R_i; (b) $R_{i,eq}$; (c) C.

Solution:

(a)

$$R_i = \frac{h_{ie} + (h_{oe}h_{ie} - h_{fe}h_{re})R_L}{1 + h_{oe}R_L}$$

$$= \frac{1.4 \times 10^3 + (18 \times 10^{-6} \times 1.4 \times 10^3 - 45 \times 4.3 \times 10^{-4})2 \times 10^3}{1 + 18 \times 10^{-6} \times 2 \times 10^3}$$

$$= 1.363 \text{ kilohms}$$

(b)

$$R_{i,eq} = \frac{1}{1/R_{B1} + 1/R_{B2} + 1/(R_i + R_E)}$$

$$= \frac{1}{1/(2 \times 10^4) + 1/(2 \times 10^3) + 1/[(1.363 \times 10^3) + (5 \times 10^2)]}$$

$$= 917 \text{ ohms}$$

(c) At $f_{co,min}$:

$$X_C = R_{i,eq} = 917 \text{ ohms}$$

$$C = \frac{1}{2\pi f_{co,min}X_C}$$

$$= \frac{1}{6.28 \times 6 \times 10^1 \times 9.17 \times 10^2}$$

$$= 2.9 \ \mu\text{f}$$

5.5 Design of a Voltage-divider and Emitter-bias Network. Let us design the voltage-divider and emitter-bias network for a common-emitter amplifier similar to Fig. 5.5. The 2N465 is to be used; $R_L = 2$ kilohms; $V_{cc} = 12$ volts; the load line for this condition is shown in Fig. 6.8. The operating point is located at the intersection of the -60 μa characteristic curve and the load line (point Q). The hybrid parameters for the 2N465 are: $h_{ie} = 1,400$ ohms; $h_{fe} = 45$; $h_{oe} = 18$ μmho; $h_{re} = 0.43 \times 10^{-3}$; $f_{co,min}$ of the amplifier is to be 40 cps.

Step 1: Construction of the Load Line and Determination of Zero Signal Parameters. The load line is constructed in the usual manner and appears in Fig. 6.8 as BQA. The zero signal parameters read from the characteristics are:

$$i_{B,Q} = 60 \ \mu a$$
$$i_{C,Q} = 3.5 \ \text{ma}$$
$$e_{C,Q} = 5.2 \ \text{volts}$$

Step 2: Determination of R_E. R_E is to be selected so that its voltage drop provides the desired reverse bias at the emitter under zero signal condition. Let $e_{RE,Q} = -1$ volt.

$$i_{RE,Q} = i_{E,Q} \cong i_{C,Q} + i_{B,Q}$$
$$= 3.5 + 0.06 = 3.56 \ \text{ma}$$

Solving for R_E,

$$R_E = \frac{e_{RE,Q}}{i_{RE,Q}} = \frac{1}{3.56 \times 10^{-3}}$$
$$= 280 \ \text{ohms}$$

Step 3: Determination of C_E. At $f_{co,min}$ (40 cps),

$$X_{CE} = 0.1R_E = 28 \ \text{ohms}$$

Solving for C_E,

$$C_E = \frac{1}{2\pi f_{co,min}X_C} = \frac{1}{6.28 \times 4 \times 10^1 \times 2.8 \times 10^1}$$
$$= 142 \ \mu f \ \text{or larger}$$

Step 4: Determination of $e_{B\text{-}E,Q}$. Certain manufacturers furnish $e_{B\text{-}E}$ versus i_C characteristics for some of their transistors. These curves may be used to determine $e_{B\text{-}E}$ by the following:

1. Find $i_{C,Q}$ on the y axis. From $i_{C,Q}$, project a line parallel to the x axis until it intersects the curve which has a fixed $e_C \cong e_{C,Q}$.

2. From the point of intersection, drop a perpendicular to the x axis.

A second graphical method for obtaining $e_{B\text{-}E,Q}$ is by use of the i_B versus $e_{B\text{-}E}$ characteristics.

1. Find the desired $i_{B,Q}$ on the x axis. Project a line parallel to the x axis until it intersects the characteristic curve whose fixed $e_C \cong e_{C,Q}$.

2. From this point, project a line (parallel to the x axis) to the y axis. Read $e_{B\text{-}E,Q}$ on the y axis.

Both of the graphical methods cannot be used in many instances because of the lack of characteristics or an insufficient number of curves within the given characteristics. In such cases, the hybrid parameter h_{ie} may be used in conjunction with $i_{B,Q}$ to obtain a good approximation of $e_{B\text{-}E,Q}$.

The input resistance of the transistor in the CE connection is h_{ie}, while the input resistance of the amplifier is R_i. R_i can be determined by use of the equation in solution a of example 2 in Sec. 5.4, when the hybrid parameters and R_L are known. Substituting values and solving,

$$R_i = 1.36 \text{ kilohms}$$

Notice that R_i is close to the value of h_{ie} for small values of R_L. Even smaller values of R_L will place R_i closer in value to h_{ie}.

Assume that

$$R_i \cong \frac{e_{B\text{-}E,Q}}{i_{B,Q}}$$

and transposing,

$$e_{B\text{-}E,Q} \cong R_i i_{B,Q} \cong 1.36 \times 10^3 \times 60 \times 10^{-6}$$
$$\cong 0.081 \text{ volt}$$

Step 5: Determination of R_{B1} and R_{B2}. The forward bias is developed across R_{B2}. Recall that $e_{RE,Q} = 1$ volt and is a reverse bias. Therefore,

$$e_{RB2,Q} = e_{RE,Q} + e_{B\text{-}E,Q} = 1 + 0.081 = 1.081 \text{ volts}$$

i_{RB2} is the voltage-divider bleeder current and may be arbitrarily selected by the designer. Let $i_{RB2} = 50$ μa, then

$$R_{B2} = \frac{e_{RB2,Q}}{i_{RB2,Q}} = \frac{1.081}{50 \times 10^{-6}} = 21.6 \text{ kilohms}$$

The voltage across R_{B1} is the difference between V_{cc} and e_{RB2}:

$$e_{RB1} = 12 - 1.081 = 10.919 \text{ volts}$$

And i_{RB1} is the sum of $i_{B,Q}$ and the arbitrarily chosen bleeder current:

$$R_{B1} = \frac{e_{RB1}}{i_{RB1}} = \frac{10.919}{110 \times 10^{-6}}$$
$$= 99.26 \text{ kilohms}$$

Note that the use of R_i is permissible when obtaining a reasonable approximation of $e_{B\text{-}E,Q}$ without the use of a second family of characteristic curves (when R_L is small). The use of the computed R_i in place of h_{ie} results in obtaining a reasonably accurate approximation of $e_{B\text{-}E,Q}$.

As in all design problems, the *exact* value of each component is determined when the circuit is constructed. The designed component values are:

$$R_E = 280 \text{ ohms}$$
$$C_E = 142 \ \mu\text{f}$$
$$R_{B2} = 21.6 \text{ kilohms}$$
$$R_{B1} = 99.26 \text{ kilohms}$$

5.6 Variation of the Voltage-divider and Emitter-bias Method. Figure 5.7 illustrates a second method by which the voltage-divider and emitter-bias method can be used. R_E, as in the previous method, provides stability against temperature variations. When desired, R_E may be bypassed by a suitable capacitor to reduce the degenerative effects of R_E upon the signal. In such cases, X_{CE} should be $0.10R_E$ at $f_{co,\min}$.

In the previous voltage-divider arrangement, i_B flowed only through R_{B1}, whereas it is also made to pass through R_{B3} in the network of Fig. 5.7. As can be determined by the polarity of V_{cc}, the transistor used in

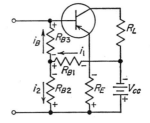

Fig. 5.7 Variation of the voltage-divider and emitter-bias method.

Fig. 5.7 is the PNP type. e_{RB3} acts as a reverse bias (see its polarity in the diagram). This feature adds stability to the circuit because of its negative feedback characteristic. When i_B increases (because of the actual signal or a temperature increase) e_{R3} increases, thereby reducing the forward bias applied to the base by e'_{RB2}. When the base current becomes smaller, the reverse potential of R_{B3} is reduced, thereby allowing a greater amount of forward bias to be felt by the base.

e_{RE} and e_{RB3} are both reverse-bias voltages, and e_{RB2} is a forward potential. e_{RB2} must be of sufficient magnitude to cancel the sum of the two reverse voltages and still have the desired $e_{B\text{-}E,Q}$ as a remainder.

Example: $e_{B\text{-}E,Q} = 0.4$ volt; $e_{RE} = 1$ volt; $e_{RB3} = 0.5$ volt. Find e_{RB2}.
Solution: The loop equation (from the base around the circuit to the emitter) is

$$e_{B\text{-}E} = e_{RB3} + e_{RB2} + e_{RE}$$

and

$$e_{RB2} = e_{B\text{-}E} - e_{RB3} - e_{RE}$$

Referring to Fig. 5.7 for polarity of the voltages,

$$e_{RB2} = 0.4 - (-0.5) - (-1.0)$$
$$= 1.9 \text{ volts}$$

Stating the relationship in words: Since the total reverse potential is 1.5 volts ($e_{RE} + e_{RB3}$), the total forward potential must be 1.9 volts in order to provide the base with the required 0.4 volt of forward bias.

5.7 Design of the Voltage divider and Emitter-bias Variation Network.

The design principles for this type of bias network are basically similar to those of the original voltage-divider and emitter-bias circuit. The variations in the design procedure are the result of introducing the additional stabilizing resistor R_{B3}.

Let us proceed to design the network shown in Fig. 5.7. The 2N465 is to be used; $R_L = 2$ kilohms; $V_{cc} = 12$ volts; the load line is shown in Fig. 6.8. The operating point is at the intersection of the -60 μa characteristic curve and the load line (point Q). The hybrid parameters for the 2N465 are: $h_{ie} = 1,400$ ohms; $h_{fe} = 45$; $h_{oe} = 18$ μmho; $h_{re} = 0.43 \times 10^{-3}$ $f_{co,min} = 60$ cps.

Step 1: Construction of the Load Line and Determination of Zero Signal Parameters. The load line is constructed in the usual manner, as shown in Fig. 6.8 as line BQA. The zero signal parameters available from the characteristics and load line are:

$$i_{B,Q} = 60 \text{ }\mu a$$
$$i_{C,Q} = 3.5 \text{ ma}$$
$$e_{C,Q} = 5.2 \text{ volts}$$

Step 2: Determination of R_E. Let $e_{RE,Q} = 0.5$ volt (the desired reverse emitter bias).

$$i_{RE,Q} = i_{E,Q} \cong i_{C,Q} + i_{B,Q}$$
$$= 3.5 + 0.06 = 3.56 \text{ ma}$$

Solving for R_E,

$$R_E = \frac{e_{RE,Q}}{i_{RE,Q}} = \frac{5 \times 10^{-1}}{3.56 \times 10^{-3}}$$
$$= 140 \text{ ohms}$$

Step 3: Determination of R_{B3}. Let $e_{RB3,Q} = 0.3$ volt. $i_{RB3,Q} = i_{B,Q} = 60$ μa. Solving for R_{B3},

$$R_{B3} = \frac{e_{RB3,Q}}{i_{RB3,Q}} = \frac{3 \times 10^{-1}}{60 \times 10^{-6}}$$
$$= 5 \text{ kilohms}$$

Step 4: *Determination of* $e_{B\text{-}E,Q}$. Since the $e_{B\text{-}E}$ versus i_C or i_B versus $e_{B\text{-}E}$ characteristics are not available, h_{ie} will be used for the computation (since R_L is small):

$$h_{ie} \simeq \frac{e_{B\text{-}E,Q}}{i_{B,Q}}$$

and
$$e_{B\text{-}E,Q} \simeq h_{ie}i_{B,Q}$$
$$= 1.4 \times 10^3 \times 60 \times 10^{-6}$$
$$= 0.084 \text{ volt}$$

Step 5: *Determination of* R_{B1} *and* R_{B2}. The forward bias developed across R_{B2} must exceed the sum of $e_{RE,Q}$ and $e_{RB3,Q}$ by the value of $e_{B\text{-}E,Q}$ (0.084 volt).

$$e_{RB2,Q} = e_{RE,Q} + e_{RB3,Q} + e_{B\text{-}E,Q}$$
$$= 0.5 + 0.3 + 0.084$$
$$= 0.884 \text{ volt}$$

Let the bleeder current i_{RB2} equal 50 μa. Solving for R_{B2},

$$R_{B2} = \frac{e_{RB2,Q}}{i_{RB2,Q}} = \frac{8.84 \times 10^{-1}}{50 \times 10^{-6}}$$
$$= 17.68 \text{ kilohms}$$

The voltage drop of R_{B1} is equal to the difference between V_{cc} and e_{RB2}:

$$e_{RB1,Q} = V_{cc} - e_{RB2,Q}$$
$$= 12 - 0.884$$
$$= 11.116 \text{ volts}$$

and
$$i_{RB1,Q} = i_{\text{bleeder}} + i_{B,Q}$$
$$= 50 + 60$$
$$= 110 \text{ μa}$$

Solving for R_{B1},

$$R_{B1} = \frac{e_{RB1,Q}}{i_{RB1,Q}} = \frac{11.116}{110 \times 10^{-6}}$$
$$\simeq 101 \text{ kilohms}$$

Figure 5.8 illustrates the circuit just designed. As in all designed circuits, the *exact* value of each component is determined when the circuit is actually constructed.

Fig. 5.8 A designed circuit of the voltage-divider and emitter-bias variation method.

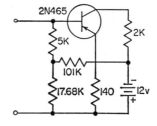

5.8 Fixed Bias with a Decoupling Capacitor. Figure 5.9 illustrates a variation of the fixed-bias method. C_B is sufficiently large so as to effectively shunt R_{B2} at the lowest signal frequency; i.e., X_{CB} should be equal

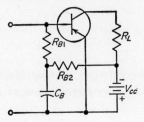

FIG. 5.9   Fixed bias with a decoupling filter.

to $0.1R_{B2}$ at $f_{co,\min}$. The sum of R_{B1} and R_{B2} is computed by

$$R_{B1} + R_{B2} = \frac{V_{cc} - e_{B\text{-}E,Q}}{i_{B,Q}}$$

R_{B1} may be any fraction of $R_{B1} + R_{B2}$, which provides flexibility for impedance matching. An emitter resistor R_E, with or without a bypass capacitor, may be added to provide temperature stabilization.

5.9 Fixed Bias with Emitter Bias. Refer to Fig. 5.10. The forward bias required for the input of the amplifier is provided by R_B. Notice that

FIG. 5.10   Fixed bias with emitter bias.

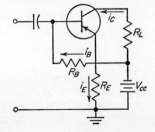

R_B is placed in the same circuit connection as the bias resistor in the fixed-bias arrangement discussed earlier. R_E may or may not be bypassed with a capacitor of appropriate size, depending upon whether or not degeneration is desired. In the circuit of Fig. 5.10:

$$R_{i,eq} \cong \frac{1}{1/R_B + 1/(R_i + R_E)}$$

In many practical circuits, R_B is many times greater than R_i and R_E, in which case

$$R_{i,eq} \cong R_i + R_E$$

5.10 Evaluation of Bias Techniques. Earlier sections have stated the temperature effect upon transistor operation—a temperature rise (with all other factors constant) will produce an increase in collector current. The primary reason for the collector current increase with higher temperature is because of I_{CO}, called the collector saturation current. I_{CO} may be considered as originating from the deathnium center generation of minority carriers in the base and collector regions and is the current which flows through the collector and base when the emitter current is zero. Like the deathnium-center-generated currents discussed in Chap. 3, this current is independent of external collector voltages at magnitudes exceeding about one-half volt. Typical I_{CO} values at room temperature are 5 μa and less. But the value of I_{CO}, in many transistors, doubles for each 10°C rise in temperature. In the common-base configuration, where $i_C = \alpha i_E + I_{CO}$, the I_{CO} increases at higher temperatures do not affect the magnitude of the collector current to a serious extent. In the common-emitter configuration,

$$i_C = \beta i_B + (1 + \beta)I_{CO}$$

The common-emitter connection has the great advantage of reproducing the base current changes in the collector current with variations that are *beta* times greater. The $(1 + \beta)I_{CO}$ portion of the preceding equation indicates that increases in I_{CO} owing to temperature rises produce a substantial increase in the collector current. The increased collector current creates a rise in collector temperature, which further increases I_{CO} (and i_C), resulting in a further temperature rise. This process, if no safeguards are taken, may easily result in the destruction of the transistor.

Example: Assume a common-emitter amplifier has the following parameters: $i_B = 80$ μa; $I_{CO} = 5$ μa at 20°C; assume I_{CO} doubles for each 10°C rise in temperature; $\beta = 25$. Find i_C at the following temperatures: (a) 20°C; (b) 40°C; (c) 60°C.

Solution:

(a) i_C at 20°C:

$$i_C = \beta i_B + (1 + \beta)I_{CO} = 25 \times 80 \times 10^{-6} + 26 \times 5 \times 10^{-6}$$
$$= 2 \times 10^{-3} + 0.13 \times 10^{-3} = 2.13 \text{ ma}$$

(b) i_C at 40°C:
Since I_{CO} doubles for each 10°C,

$$I_{CO} = 10 \text{ } \mu\text{a at 30°C} \quad \text{and} \quad I_{CO} = 20 \text{ } \mu\text{a at 40°C}$$

Now solving for i_C:

$$i_C = 25 \times 80 \times 10^{-6} + 26 \times 20 \times 10^{-6}$$
$$= 2 \times 10^{-3} + 0.5 \times 10^{-3} = 2.52 \text{ ma}$$

(c) i_C at 60°C:

$$I_{CO} = 40 \ \mu\text{a at } 50°C \qquad \text{and} \qquad I_{CO} = 80 \ \mu\text{a at } 60°C$$

Finding i_C:

$$i_C = 25 \times 80 \times 10^{-6} + 26 \times 80 \times 10^{-6}$$
$$= 2 \times 10^{-3} + 2.08 \times 10^{-3} = 4.08 \text{ ma}$$

5.11 Thermal Stability Factor. The preceding example shows that, with the given parameters, the collector current doubled with a 40° rise in temperature. This problem can be even more severe with larger values of beta.

The thermal stability factor of a circuit may be defined as the ratio of the change in collector current to the change in collector saturation current which caused the collector current to change. In equation form,

$$SF = \frac{\Delta i_C}{\Delta I_{CO}}$$

The thermal stability factor indicates the effect a change of I_{CO} has upon i_C. Lower values of SF indicate the circuit has greater thermal stability. The thermal stability of a circuit is greatly determined by the bias arrangement incorporated. The thermal stability factor for several bias arrangements may be determined by the following equations:

Fixed Bias (See Fig. 5.1)

$$SF = \frac{\Delta i_C}{\Delta I_{CO}} = 1 + \beta$$

Emitter Bias with Voltage-divider Arrangement (See Fig. 5.5)

$$SF = \frac{\Delta i_C}{\Delta I_{CO}} = \frac{1 + R_E(1/R_{B1} + 1/R_{B2})}{1 - \alpha + R_E(1/R_{B1} + 1/R_{B2})}$$

Emitter Bias with Fixed Bias (See Fig. 5.10)

$$SF = \frac{\Delta i_C}{\Delta I_{CO}} = \frac{1 + (R_E/R_B)}{1 - \alpha + (R_E/R_B)}$$

Example 1: Assume the amplifier of Fig. 5.1 has $\alpha = 0.96$. Find SF.
Solution: We must first determine beta:

$$\beta = \frac{\alpha}{1 - \alpha} = \frac{0.96}{1 - 0.96} = 24$$
$$SF = 1 + \beta = 25$$

Example 2: The same transistor used in example 1 is to be connected as Fig. 5.5. Assume the following parameters: $R_E = 500$ ohms; $R_{B1} = 20$ kilohms; $R_{B2} = 2$ kilohms. Find SF of the transistor with this bias arrangement.

Solution:

$$SF = \frac{1 + 500(1/20{,}000 + 1/2{,}000)}{1 - 0.96 + 500(1/20{,}000 + 1/2{,}000)}$$
$$= 4.05$$

A comparison of examples 1 and 2 shows that emitter bias is more effective than fixed bias in developing better thermal stability. Larger values of R_E, because of its degenerative characteristics, further reduce the thermal stability factor. The use of smaller R_{B1} and R_{B2} values decreases SF, but also reduces $R_{i,eq}$ of the circuit, which may be detrimental to best circuit performance.

Example 3: Assume the transistor of example 2 is connected as Fig. 5.10 ($\alpha = 0.96$), with the following circuit parameters: $R_B = 30$ kilohms. Find SF if R_E is (a) 500 ohms; (b) 1,000 ohms.

Solution:

(a) When $R_E = 500$ ohms:

$$SF = \frac{1 + (500/30{,}000)}{1 - 0.96 + (1{,}000/30{,}000)}$$
$$= 17.9$$

(b) When $R_E = 1,000$ ohms:

$$SF = \frac{1 + (1{,}000/30{,}000)}{1 - 0.96 + (1{,}000/30{,}000)}$$
$$= 14.2$$

Notice that SF decreased with a larger R_E.

The thermal stability factor is an indication of how effective the circuit is in adjusting to temperature increases without radically altering its operating characteristics. Bias techniques which develop the lowest thermal stability factor are most desirable, since they enable the transistor to be less sensitive to temperature variations. The emitter-bias technique is the most desirable in this respect, and the fixed-bias circuit is least desirable.

5.12 Coupling Considerations. An amplifier is usually associated with other amplifiers. The input signal of the transistor very often is the output signal of another transistor amplifier.

Figure 5.11 illustrates the block diagram of such an arrangement. Each amplifier may be considered a stage. When the output signal of stage 1 is the input of stage 2, the two amplifiers are said to be in *cascade*. In Fig. 5.11, there are three cascaded stages.

Cascading introduces the problem of impedance matching, since R_o of stage 1 and R_i of stage 2 must be reasonably matched for the amplification of stage 1 to be successfully passed on to the input circuit of stage 2. R_{L1} produces an effect on R_{i2}, and R_{i2} affects R_{o1}. The resultant mismatch in some cases is so severe that transformers must be used for interstage

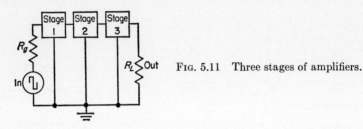

Fig. 5.11 Three stages of amplifiers.

coupling. In other configuration combinations, mismatching is less pronounced, and impedance transformers are not necessary. Since the junction transistor has three external terminals and three configurations are possible, nine cascade arrangements are possible:

1. *CE* to *CE*
2. *CE* to *CB*
3. *CE* to *CC*
4. *CB* to *CB*
5. *CB* to *CE*
6. *CB* to *CC*
7. *CC* to *CC*
8. *CC* to *CE*
9. *CC* to *CB*

Considering the effect of stage 1 upon stage 2 and vice versa, as expressed in the preceding paragraph, only four of the nine possible cascade arrangements can be utilized without incorporating a coupling transformer. These possibilities, in the order of their desirability are:

1. *CE* to *CE*: Voltage and power amplification possibilities are excellent, and the impedance match between stages 1 and 2 are good.

2. *CE* to *CB*: Same possibilities as the *CE* to *CE*, but not quite as good.

3. *CC* to *CE*: Good possibilities. Most desirable where high input impedance is required.

4. *CB* to *CE*: Fair to good possibilities.

5.13 Direct Coupling. In some cases, a unidirectional current path may be allowed between the output of stage 1 and the input of stage 2, which is called *direct coupling*. Figure 5.12*a* illustrates a simple direct-coupled common-emitter amplifier. The load resistance R_L, in a practical circuit, would very likely be some device such as a relay or meter. R_E is used to establish current stability in the event the transistor temperature should undergo changes. R_B provides the necessary forward base bias by means of the fixed-bias technique discussed earlier in this chapter. Automatic self-bias (R_B connected from base to collector) and the voltage-divider arrangement are other methods by which the forward base bias may be obtained.

Since the output is fed directly to the load, the low-frequency response is flat down to nearly zero cps (see Fig. 5.12b). At high frequencies, the output capacitance of the stage 1 transistor appears in parallel with the collector load resistance. When the reactance of the output capacitance X_{co} is sufficiently small, the output will be 0.707 of the mid-frequency output (3 db down); this is $f_{co,\max}$ (see Fig. 5.12b).

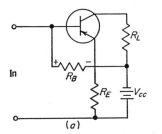
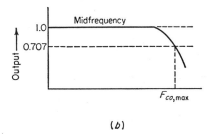

FIG. 5.12   (a) Direct-coupled CE amplifier, (b) frequency response curve for direct coupling.

The chief advantages of direct coupling are:
1. Compactness
2. Economy
3. Good low-frequency response

Since the zero signal and signal circuits are the same, there is only one load line for this type of amplifier, which is determined by R_L. When the stage 1 transistor is coupled directly to a stage 2 transistor, then R_i of the second transistor is the resistance used in determination of the load line (since R_i of transistor 2 is R_L of transistor 1). The load line is illustrated in Fig. 5.13.

Assuming the ideal situation where e_C can be reduced to zero and the output characteristics are linear, the following equations are true:

$$P_o = \frac{e_{C,m} i_{C,m}}{2}$$

$$P_{o,m} = \frac{e_{C,Q} i_{C,Q}}{2}$$

$$P_c = P_{o,m} \quad \text{at maximum signal}$$
$$P_c = 2P_{o,m} \quad \text{at zero signal}$$
$$\text{Eff} = 25\% \quad \text{at full signal}$$

In some applications, a stage 1 transistor is directly coupled to a stage 2 transistor with a separate R_L for the first stage. A practical circuit in which the bias networks are simple to design is the complementary-symmetry type, which is shown in Fig. 5.14.

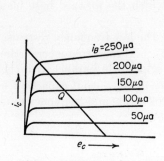

FIG. 5.13 Load line for direct-coupled amplifier.

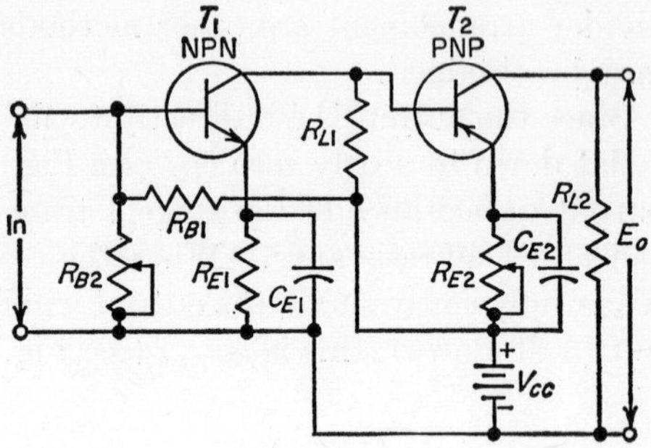

FIG. 5.14 Use of complementary symmetry for direct coupling of two stages.

Notice that opposite-conductivity-type transistors are used (T_1 is NPN, and T_2 is PNP). The load line for T_1 determines the parallel equivalent value of R_{L1} and R_{i2}; i.e.,

$$R_{L1,eq} = \frac{R_{L1}R_{i2}}{R_{L1} + R_{i2}}$$

where

$$R_{i2} = \frac{h_{ie} + (h_{oe}h_{ie} - h_{fe}h_{re})R_{L2}}{1 + h_{oe}R_{L2}}$$

or

$$R_{i2} \cong \frac{\Delta e_{B-E}, T_2}{\Delta i_B, T_2}$$

Since $R_{L1,eq}$ is determined from the constructed load line of T_1 and R_{i2} is found by one of the preceding relationships, R_{L1} can then be calculated:

$$\frac{1}{R_{L1}} = \frac{1}{R_{L1,eq}} - \frac{1}{R_{i2}}$$

NOTE: The selection of R_{L1} whose ohmic value is at least $10R_{i2}$ will make $R_{L1,eq} \cong R_{i2}$.

The remaining stage 1 parameters are designed in accordance with the steps outlined in Sec. 5.5, since the voltage divider with emitter bias is utilized. It is suggested that $e_{C\text{-COMMON}}$ of T_1 at zero signal condition be selected as about one-half of V_{cc}.

Stage 2 is then designed. A load line is selected, thereby determining R_{L2}, and the zero signal parameters are obtained. Special attention must be given to the biasing of stage 2.   Notice that the base of T_2 is at the potential of $e_{C\text{-}COMMON}$ of T_1.   But $e_{B\text{-}E,Q}$ of T_2 must be considerably smaller than this value and can be brought to its desired value by the correct determination of R_{E2}:

$$e_{RE2,Q} = e_{C\text{-}COMMON}T_1 - e_{B\text{-}E,Q}T_2$$

$e_{B\text{-}E,Q}$ of T_2 can be determined by any one of the methods specified in step 4 of Sec. 5.5.

Having determined $e_{B\text{-}E,Q}T_2$ in this manner and $e_{C\text{-}COMMON}$ of T_1 from the load line of T_1, R_{E2} can then be calculated:

$$R_{E2} = \frac{e_{RE2,Q}}{i_{E2,Q}}$$

Expanding, $$R_{E2} = \frac{e_{C\text{-}COMMON}T_1 - e_{B\text{-}E,Q}T_2}{i_{B2,Q} + i_{C2,Q}}$$

where $i_{B2,Q}$ and $i_{C2,Q}$ are determined from the stage 2 load line.

Notice that R_{B2} and R_{E2} are potentiometers.   The correct bias condition for T_1 is achieved by varying R_{B2}.   Increasing R_{B2} increases the forward base bias of T_1, which increases the collector current and decreases the collector potential of T_1.   A voltmeter may be placed between collector and common of T_1 while R_{B2} is varied until this potential is at the desired value.   Upon achievement of this condition, stage 1 is biased in accordance with the design, and attention may now be focused on T_2. R_{E2} is the means by which the stage 2 bias is properly achieved. R_{E2} is varied until e_{RE2} is less than $e_{C\text{-}COMMON}$ of T_1 by the required $e_{B\text{-}E,Q}$ value of T_2.   Therefore the final bias adjustments are quickly made by the use of potentiometers for R_{B2} and R_{E2}.

The direct-coupled complementary-symmetry amplifier of Fig. 5.14 can also be designed with T_1 as a PNP and T_2 as an NPN, in which case the polarity of V_{cc} is reversed.

5.14 RC Coupling.   The signal variations of stage 1 may be passed on to the input circuit of stage 2 by use of one of several coupling techniques. Besides the technique studied in the preceding section, several other methods will be analyzed in this and the following section. RC coupling is commonly used where impedance matching does not present problems of a more serious nature.   The coupling used between two common-emitter connected transistors is often the RC type for this reason, plus its additional advantages of economy and compactness.

Figure 5.15 illustrates a simplified version of an RC-coupled common-emitter amplifier. R_{L1} is the load resistor for T_1, and R_{L2} is the load resistor for T_2.

Referring to the RC coupling network between stages 1 and 2 of Fig. 5.15, the low-frequency equivalent circuit is illustrated in Fig. 5.16a and b. C_2 (assuming it has no leakage current) restricts the unidirectional component of the T_1 collector current to R_{L1}. At frequencies below

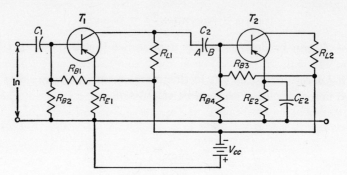

FIG. 5.15 RC-coupled CE amplifier.

$f_{co,\min}$, C_2 appears in series with $R_{i2,eq}$, thereby depriving the input circuit of T_2 of a substantial portion of the input signal. In Fig. 5.16a,

$$R_{o1,eq} = \frac{R_{o1}R_{L1}}{R_{o1} + R_{L1}}$$

Since the amplifier is in the CE configuration, R_{o1} is

$$R_{o1} = \frac{h_{ie} + R_{g1}}{h_{oe}h_{ie} - h_{re}h_{fe} + h_{oe}R_{g1}}$$

where the hybrid parameters are those of transistor 1 and $R_{g1} = R_{i1,eq}$. Refer to Sec. 4.17 for the R_{o1} equation when the CB or CC configuration is used.

The equivalent input resistance of stage 2 is determined by the type of bias network used, as well as the stage 2 transistor. For the stage 2 input circuit of Fig. 5.15,

$$R_{i2,eq} = \frac{1}{1/R_{B4} + 1/R_{B3} + 1/(R_{E2} + R_{i2})}$$

Since the amplifier is in the CE configuration, R_{i2} is

$$R_{i2} = \frac{h_{ie} + (h_{oe}h_{ie} - h_{fe}h_{re})R_{L2}}{1 + h_{oe}R_{L2}}$$

where the hybrid parameters are those of transistor 2, and R_{L2} is $R_{L2,eq}$.

When R_{E2} is shunted by a capacitor C_{E2}, then R_{E2} does not appear in the $R_{i2,eq}$ equation. This is assuming that X_{CE2} is $0.10R_{E2}$ or less at the lowest signal frequency to be handled by the transistor.

As the frequency is reduced, let us assume that $R_{o1,eq}$ and $R_{i2,eq}$ remain fixed, while X_{C2} changes inversely with frequency. The signal voltage applied to the input of stage 2 (e_{i2}) can be determined by

$$e_{i2(1\text{-}f)} = \frac{R_{i2,eq}}{(R_{i2,eq} + R_{o1,eq}) - jX_{C2}} E$$

where E is the stage 2 input voltage if $R_{o1,eq}$ is not shunted by $R_{i2,eq}$.

Notice that this is the voltage-divider relationship and that larger values of X_{C2} (caused by lower frequencies) result in a reduced stage 2 input voltage. When the phase shift of the coupling network (C_2; $R_{o1,eq}$) is 45° (i leads e), the amplifier is at $f_{co,\min}$ and is found by

$$f_{co,\min} = \frac{1}{2\pi(R_{o1,eq} + R_{i2,eq})}$$

The attenuation of $f_{co,\min}$ is somewhat less than 3 db as compared with mid-frequency. Large values of C_2, $R_{o1,eq}$, and $R_{i2,eq}$ result in lower $f_{co,\min}$ values. At $f_{co,\min}$ the sum of $R_{o1,eq}$ and $R_{i2,eq}$ equals X_{C2}. Using the $R_{o1,eq}$ and $R_{i2,eq}$ values obtained from their respective equations, along with the C_2 value, leads to a valid indication of the amplifiers lower-frequency limit.

In examining Fig. 5.15, notice that the stage 1 input signal is also reduced at lower frequencies by C_1. By the same token, if the output signal of T_2 is RC-coupled to a third stage, there would be a reduction due to this coupling capacitor. The value of the coupling capacitors should be sufficiently large so as to present negligible reactance in series with $R_{i,eq}$ at the desired signal frequencies.

At the middle-frequency range, C_2 drops out of the equivalent circuit since X_{C2} becomes considerably smaller than $R_{i2,eq}$. This results in a larger signal voltage applied to the input of stage 2, as determined by

$$e_{i2(m\text{-}f)} = \frac{R_{i2,eq}}{R_{o1,eq} + R_{i2,eq}} E$$

The ratio of the stage 2 input voltage at the low and middle frequencies of the RC coupling circuit, which is useful in evaluating the merit of an RC coupling circuit, can be found by

$$\frac{e_{i2(1\text{-}f)}}{e_{i2(m\text{-}f)}} = \frac{1}{\{1 + [X_{C2}/(R_{o1,eq} + R_{i2,eq})]\}^{\frac{1}{2}}}$$

Recall that power can be determined by the ratio E^2/R. Applying this equation to the stage 2 input of Fig. 5.15 at low and middle frequencies we obtain

$$P_{i(\text{l-f})} = \frac{[e_{i2(\text{l-f})}]^2}{R_{i2,\text{eq}}}$$

and

$$P_{i(\text{m-f})} = \frac{[e_{i2(\text{m-f})}]^2}{R_{i2,\text{eq}}}$$

Expressing the stage 2 input voltage in its root mean square value results in $P_{i,\text{av}}$. The average power ratio of the RC coupling circuit is

$$G_{RC} = \frac{P_{i,\text{av(l-f)}}}{P_{i,\text{av(m-f)}}} = \frac{[e_{i2\text{rms(l-f)}}]^2/R_{i2,\text{eq}}}{[e_{i2\text{rms(m-f)}}]^2/R_{i2,\text{eq}}}$$

where G_{RC} is the average power ratio of the RC coupling circuit.

Notice that $R_{i2,\text{eq}}$ cancels in the numerator and denominator of the preceding equation, simplifying it to

$$\frac{P_{i,\text{av(l-f)}}}{P_{i,\text{av(m-f)}}} = \frac{[e_{i2,\text{rms(l-f)}}]^2}{[e_{i2,\text{rms(m-f)}}]^2}$$

The average input power ratio can also be expressed in decibels:

$$G_{RC(\text{db})} = 20 \log \frac{e_{i2,\text{rms(l-f)}}}{e_{i2,\text{rms(m-f)}}}$$

The RC coupling network loss in decibels can be found by use of the following relationship:

$$\text{db loss} = -10 \log \left[1 + \left(\frac{X_{C2}}{[R_{o1,\text{eq}} + R_{i2,\text{eq}}]^2} \right)^2 \right]$$

Example: Assume the amplifier of Fig. 5.15 has gains of 20 db for stage 1 and 15 db for stage 2. The RC coupling network ahead of stage 1 and between stages 1 and 2 each have a 3-db loss at 60 cps. Find the over-all circuit gain at 60 cps.

Solution:

Over-all amplifier gain $= -3 \text{ db} + 20 \text{ db} - 3 \text{ db} + 15 \text{ db} = 29 \text{ db}$

Returning to the equivalent circuit of Fig. 5.16, let us analyze the manner in which the signal variations are passed from stage 1 to stage 2. When X_{C2} is small, the variations across $R_{i2,\text{eq}}$ are similar to those across $R_{o1,\text{eq}}$. When the collector current of T_1 increases in magnitude, e_{RL1} increases and $e_{C\text{-}B}$ of T_1 decreases. Plate A of C_2 discharges into the collector terminal, with its path completed through $R_{o1,\text{eq}}$, up $R_{i2,\text{eq}}$ to plate B. This places the top of $R_{i2,\text{eq}}$ at a positive potential with respect to common. When the collector current of T_1 decreases in amplitude, $e_{Ro1,\text{eq}}$ becomes smaller, plate B of C_2 discharges down through $R_{i2,\text{eq}}$, up $R_{o1,\text{eq}}$, and back to plate A of C_2. Referring to Fig. 5.16a, h_{ie} is in series

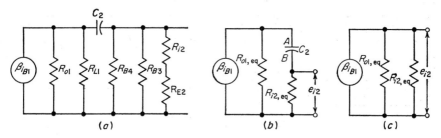

Fig. 5.16   Equivalent circuits for an RC-coupled CE amplifier: (a) low frequency, (b) low frequency simplified, (c) middle frequency.

with R_{E2}.   The top of R_{i2} is of the same polarity as the top of $R_{i2,eq}$. Therefore $e_{Ri2,eq}$ is negative with respect to common when $i_{C,T1}$ decreases and is positive when $i_{C,T1}$ increases (PNP transistor).   Hence the collector-voltage variations of T_1 are coupled to the input circuit of T_2.

At high frequencies, the coupling capacitor drops out of the equivalent circuit, but a number of other capacitances become significant.   Refer to Fig. 5.17 for the high-frequency equivalent circuit of the RC-coupled

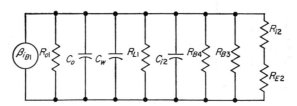

Fig. 5.17   Equivalent high-frequency circuit for an RC-coupled CE amplifier.

amplifier illustrated in Fig. 5.15.   As in the low- and middle-frequency equivalent circuits

$$R_{o1,eq} = \frac{R_{L1}R_{o1}}{R_{L1} + R_{o1}}$$

and
$$R_{i2,eq} = \frac{1}{1/R_{B4} + 1/R_{B3} + 1/(R_{i2} + R_{E2})}$$

C_o, the output capacitance of T_1 in Fig. 5.17, is actually $C_{C\text{-}E}$, since the common-emitter configuration is used.   The output capacitance for the common-base configuration is $C_{C\text{-}B}$ and is often stated on manufacturers' data sheets. $C_{C\text{-}B}$ can be converted to $C_{C\text{-}E}$ value by use of the following relationship:

$$C_{C\text{-}E} = \frac{C_{C\text{-}B}}{1 + h_{fb}}$$

As can be seen in Fig. 5.17, the output capacitance C_o is in shunt with

$R_{o1,eq}$ and $R_{i2,eq}$.   In parallel with C_o are the wiring and stray capacitances of the circuit C_w. C_w is generally small when the two stages are physically close to each other.   A second capacitance in parallel with C_o is the input capacitance of stage 2 (C_{i2}, which is $C_{B\text{-}E}$ in the CE configuration). C_{i2} is considered negligible in many practical cases.   The total shunt capacitance C_s is

$$C_s = C_o + C_w + C_{ie}$$

Figure 5.18 illustrates a simplified version of the high-frequency equivalent circuit.

The computation of the exact total effective capacitance is rather complex because of the many factors involved.   All the internal capacitances of a transistor are dependent upon the dielectric factor, which is a function of the impurity density.   The impurity density is in turn dependent upon the currents (emitter, base, collector), the original impurity level as manufactured, and temperature.   Because of these variable factors, a

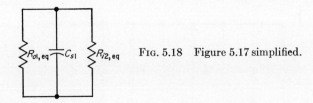

FIG. 5.18   Figure 5.17 simplified.

specific computation would be very complex and applicable to the initial conditions only.

At the high frequencies, X_{Cs} becomes sufficiently small to shunt $R_{ie,eq}$, causing the input signal of T_2 to experience a reduction.   When the input signal of T_2 is 0.707 (-3 db) from its mid-frequency value, $f_{co,\max}$ is achieved.

Most manufacturers provide the frequency at which h_f is reduced to 0.707 of its mid-frequency value (alpha f_{co} for CB, beta f_{co} for CE). Losses within the transistor at high frequencies cause h_f to decrease. The loss in gain, due to high-frequency effects of the transistor, can be computed from the appropriate equation which follows:
For the CE configuration:

$$\text{Gain loss} = \frac{h_{fb}}{[1 + (f/f_{\beta o})^2]^{1/2}}$$

$$\text{db loss} = -10 \log\left[1 + \left(\frac{f}{f_{\beta o}}\right)^2\right]$$

where f = frequency being considered
                          $f_{\beta o}$ = beta frequency cutoff

For the *CB* configuration:

$$\text{Gain loss} = \frac{h_{fb}}{[1 + (f/f_{\alpha o})^2]^{\frac{1}{2}}}$$

$$\text{db loss} = -10 \log \left[1 + \left(\frac{f}{f_{\alpha o}} \right)^2 \right]$$

where $f_{\alpha o}$ is the alpha frequency cutoff.

The input voltage ratio at high frequency to middle frequency is determined by

$$\frac{e_{i2(\text{h-f})}}{e_{i2(\text{m-f})}} = \frac{1}{\left\{ 1 + \left[\frac{R_{o1,\text{eq}} R_{i2,\text{eq}}/(R_{o1,\text{eq}} + R_{i2,\text{eq}})}{X_{Cs}} \right]^2 \right\}^{\frac{1}{2}}}$$

and the *RC* coupling network db loss:

$$\text{db loss} = -10 \log \left\{ 1 + \left[\frac{R_{o1,\text{eq}} R_{i2,\text{eq}}/(R_{o1,\text{eq}} + R_{i2,\text{eq}})}{X_{Cs}} \right]^2 \right\}$$

NOTE: When the equivalent value of $R_{01,\text{eq}}$ and $R_{ie,\text{eq}}$ equals $X_{Cs}, f_{co,\max}$ is achieved.

Looking back at Fig. 5.15, the chief limitation to low-frequency response is the coupling capacitor. There are two major limitations upon high-frequency response; namely, C_s and alpha (or beta) frequency cutoff. The coupling capacitor selected must be sufficiently large so that its reactance is equal to the total of $R_{o1,\text{eq}}$ and $R_{i2,\text{eq}}$ at the desired $f_{co,\min}$. R_{L1} and the bias resistors are selected so that X_{Cs} equals that same total at the desired $f_{co,\max}$. The transistors are selected so that alpha (or beta) f_{co} is at or above the desired $f_{co,\max}$.

5.15 Calculation of $R_{i,\text{eq}}$ and $R_{o,\text{eq}}$ for Several Bias Arrangements.
The manner in which $R_{i,\text{eq}}$ and $R_{o,\text{eq}}$ are computed with the voltage-divider and emitter-bias technique was analyzed in the *RC* coupling section. Let us consider the methods for determining these parameters when other types of bias are incorporated.

Fixed Bias. Refer to Fig. 5.1. In terms of the input signal, R_B is in parallel with R_i of the transistor; therefore

$$R_{i,\text{eq}} = \frac{R_i R_B}{R_i + R_B}$$

And $R_{o,\text{eq}}$ is found in the usual manner:

$$R_{o,\text{eq}} = \frac{R_o R_L}{R_o + R_L}$$

Automatic Self-bias with Bypassed Negative Feedback. Refer to Fig. 5.3. The input signal "sees" R_{B1} in parallel with R_i of the transistor; therefore,

$$R_{i,eq} = \frac{R_{B1}R_i}{R_{B1} + R_i}$$

And R_{B2} appears in parallel with R_L:

$$R_{o,eq} = \frac{1}{1/R_o + 1/R_L + 1/R_{B2}}$$

Variation of the Voltage-divider and Emitter-bias Method. Refer to Fig. 5.7. In terms of the input signal,

$$R_{i,eq} = \frac{1}{\dfrac{1}{R_{B3} + R_{B1}R_{B2}/(R_{B1} + R_{B2})} + \dfrac{1}{R_i + R_E}}$$

and

$$R_{o,eq} = \frac{R_o R_L}{R_o + R_L}$$

Fixed Bias with Decoupling Filter. Refer to Fig. 5.9.

$$R_{i,eq} = \frac{R_{B1}R_i}{R_{B1} + R_i}$$

and

$$R_{o,eq} = \frac{R_o R_L}{R_o + R_L}$$

Fixed Bias with Emitter Bias. Refer to Fig. 5.10.

$$R_{i,eq} = \frac{1}{1/R_B + 1/(R_i + R_E)}$$

and

$$R_{o,eq} = \frac{R_o R_L}{R_o + R_L}$$

Notice that the bias network results in $R_{i,eq}$ being lower than R_i of the transistor because of the shunting effect. The value of the bias resistors must be carefully chosen so as to minimize this shunting effect.

5.16 Power and Signal Efficiency Considerations of the RC-Coupled CE Amplifier. Refer to Fig. 5.16 for the notation associated with the equations which follow. The power output of T_1 may be determined from the following relationship:

$$P_o = \frac{e_{C,m} i_{C,m} R_{L1,eq}}{2(R_{L1,eq} + R_{i2,eq})}$$

The maximum power output of T_1, when ideal conditions have been obtained, is determined from

$$P_o = \frac{e_{C,Q} i_{C,Q} R_{L1,eq}}{2(R_{L1,eq} + R_{i2,eq})}$$

or

$$P_o = \frac{(V_{cc})^2}{8 R_{i2,eq}[(1 + R_{L1,eq})/2R_{I2,eq}]}$$

The collector dissipation is an important factor to consider, in order to avoid excessive transistor temperatures.   For zero signal conditions, the collector dissipation is found by

$$P_c = 2P_{o,m}\left(1 + \frac{R_{i2,eq}}{R_{L1,eq}}\right)$$

For maximum signal conditions, the collector dissipation is

$$P_{c,m} = P_{o,m}\left(1 + \frac{2R_{i2,eq}}{R_{L1,eq}}\right)$$

The full signal efficiency of T_1 is determined by

$$\text{Eff full sig} = \frac{R_{L1,eq} R_{i2,eq}}{(R_{L1,eq} + R_{i2,eq})(R_{L1,eq} + 2R_{i2,eq})} \times 100$$

NOTE: The maximum full signal efficiency for an RC-coupled CE amplifier is 17 per cent, which occurs when the ratio of $R_{L1,eq}/R_{i2,eq}$ is 1.414.

A Trigonometric Method for Determining the Signal Load Line.   When using RC coupling, the construction of the load line involves a zero signal load line and a signal load line, as illustrated in Fig. 5.19.   The zero signal

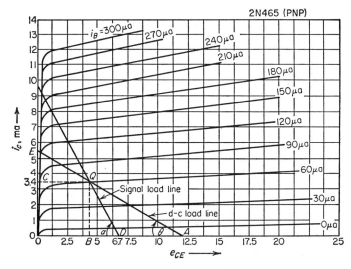

Fig. 5.19   D-c and signal load-line considerations.   (*Characteristics from Raytheon.*)

load line is determined by the actual value of the collector load resistance R_L. The signal load line, which is the one used in actual amplifier design, is determined by the equivalent resistance of R_{L1} and $R_{i2,eq}$ (equivalent resistance of stage 2 input). The slope of the signal load line for T_1 in Fig. 5.15 is determined by

$$R_{L1,eq} = \frac{(R_{L1} + R_{E1})R_{i2,eq}}{(R_{L1} + R_{E1}) + R_{i2,eq}}$$

The unbypassed emitter resistor R_{E1} of stage 1 appears in series with R_{L1} and must be so considered, as shown in the preceding equation. If R_{E1} is bypassed by a capacitor of suitable value, it does not appear in the signal circuit, and the equation becomes

$$R_{L1,eq} = \frac{R_{L1}R_{i2,eq}}{R_{L1} + R_{i2,eq}}$$

Let us determine a technique by which the signal load line may be constructed. $R_{L1,eq}$ can be computed from the preceding equation. The signal load line passes through the same operating point Q as the d-c load line. Since the signal load line, like the d-c load line, is linear, one additional point is required so that it can be constructed. It will be recalled that a load line is determined by the tangent of the angle made by the intersection of the load line and the x axis. Referring to Fig. 5.19, the following relationships are then evident:

$$\tan \theta = \frac{Q - B}{A - B} = -\frac{1}{R_{L1}}$$

and
$$\tan \alpha = \frac{Q - B}{D - B} = -\frac{1}{R_{L1,eq}}$$

where θ is the angle formed by the intersection of the d-c load line and the x axis and α is the angle formed by the intersection of the signal load line and the x axis.

When the y axis is expressed in ma, a factor of 10^3 must be incorporated, expanding the preceding relationships to the following:

$$\tan \theta = \frac{Q - B}{A - B} = \frac{1 \times 10^3}{R_{L1}}$$

and
$$\tan \alpha = \frac{Q - B}{D - B} = \frac{1 \times 10^3}{R_{L1,eq}}$$

In Fig. 5.19, notice that QB is $i_{C,Q}$, which further simplifies the preceding equations:

$$\tan \theta = \frac{i_{c,Q} \times 10^3}{A - B}$$

and
$$\tan \alpha = \frac{i_{c,Q} \times 10^3}{D - B}$$

Signal Load-line Construction

Step 1. When $R_{L1,eq}$ has been determined in accordance with the relationship stated in earlier paragraphs, tan α can be found.

Step 2. Since $i_{C,Q}$ is known from the d-c load line, the value of $D - B$ can be calculated by transposing and substituting values; i.e.,

$$D - B = \frac{i_{C,Q} \times 10^3}{\tan \alpha}$$

Step 3. Notice that $D - B$ is added to $e_{C-E,Q}$ for the intersection of the signal load line and the x axis.

$$x\text{-axis intercept} = e_{C-E,Q} + (D - B)$$

Step 4. Since two points are now available, the signal load line can be constructed.

Example:

Refer to Fig. 5.19 for the characteristics of the 2N465 and the load lines considered in this example. $R_{L1} = 2.2$ kilohms; R_E is bypassed; $R_{i2,eq} = 1$ kilohm; $V_{cc} = 12$ volts.

(*a*) Construct the d-c load line.   (*b*) Construct the signal load line.

Solution:

(*a*) Construct the d-c load line:

R_{L1} and V_{cc} are known, $i_{C(SC)}$ can be found by

$$i_{C(SC)} = \frac{V_{cc}}{R_{L1}} = \frac{12}{2.2 \times 10^3} = 5.45 \text{ ma}$$

Let us select Q to be at the intersection of the 60-μa characteristic and the load line. This is point E where $i_{C,Q} = 3.4$ ma, $e_{C-E,Q} = 4.3$ volts. The line AE is the d-c load line.

(*b*) Construct the signal load line:

We must first determine $R_{L1,eq}$:

$$R_{L1,eq} = \frac{2.2 \times 10^3 \times 1 \times 10^3}{2.2 \times 10^3 + 1 \times 10^3} \cong 700 \text{ ohms}$$

and

$$\tan \alpha = \frac{1,000}{700} = 1.43$$

$D - B$ can then be found:

$$D - B = \frac{3.4 \times 10^{-3} \times 10^3}{1.43} \cong 2.4 \text{ volts}$$

The x-axis intercept is next found:

$$x\text{-axis intercept} = 4.3 + 2.4 = 6.7 \text{ volts}$$

The points D and Q are joined, thereby constructing the signal load line.

A Second Technique for Construction of the Signal Load Line. There is a second method for construction of the signal load line which is more simple than the trigonometric technique. This alternative method is purely graphical and only requires the designer to make a few simple computations. Following are suggested steps for this technique (refer to Fig. 6.11).

Step 1. Compute $R_{L1,eq}$ in accordance with the relationship given at the beginning of this section.

Step 2. Using V_{cc} as the x-axis intercept and the $R_{L1,eq}$ found in step 1, compute $i_{C(SC)}$:

$$i_{C(SC)} = \frac{V_{cc}}{R_{L1,eq}}$$

Join the y-axis point $i_{C(SC)}$ and the x-axis point V_{cc} by a broken line (see line AG in Fig. 6.11).

The slope of line AG is equal to the slope of $R_{L1,eq}$. Because of the RC coupling, the voltage across the transistor under signal conditions cannot increase to the value of V_{cc}. Furthermore, if linear operation is achieved, the operating point Q is the same with and without a signal. Although line AG has the same slope as $R_{L1,eq}$, this load line must pass through the operating point Q. The true signal load line is a line parallel to AG and also passing through Q. Therefore a straight edge may be pivoted around Q until a line parallel to AG is obtained. Line EQ meets these requirements and is the signal load line.

5.17 Maximum Collector Dissipation Curve. In practical design work, the maximum dissipation value of the transistor must not be exceeded during any portion of the input signal cycle. Prior to drawing the load line upon the collector characteristics of the transistor to be used, it is suggested that the maximum power dissipation curve be drawn. The maximum power dissipation (or collector dissipation) value is usually available with the given transistor data. The product of the collector voltage and collector current is equal to the collector dissipation. The maximum collector dissipation curve may be constructed in the following manner:

1. For several values of collector voltage (such as 2, 4, 6, 8, and 10 volts of Fig. 5.20), find the collector current such that the product is equal to the maximum collector power dissipation. In Fig. 5.20, let us find i_C to meet this condition for $e_C = 2, 4, 6, 8,$ and 10 volts. Maximum collector dissipation for the 2N207 is 50 mw.

$$e_C i_C = P_c = 50 \text{ mw}$$

hence
$$i_C = \frac{P_c}{e_C} = \frac{50 \text{ mw}}{e_C}$$

When $e_C = 2$ volts:

$$i_C = \frac{50 \times 10^{-3}}{2} = 25 \text{ ma}$$

When $e_C = 4$ volts:

$$i_C = \frac{50 \times 10^{-3}}{4} = 12.5 \text{ ma}$$

When $e_C = 6$ volts:

$$i_C = \frac{50 \times 10^{-3}}{6} = 8.33 \text{ ma}$$

When $e_C = 8$ volts:

$$i_C = \frac{50 \times 10^{-3}}{8} = 6.25 \text{ ma}$$

When $e_C = 10$ volts:

$$i_C = \frac{50 \times 10^{-3}}{10} = 5 \text{ ma}$$

2. Locate these points on the collector characteristics, and connect them with a dotted line (see Fig. 5.20).

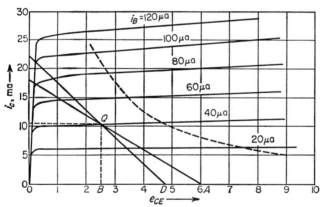

$h_{ib} = 33$ ohms $h_{ob} = 0.4 \times 10^{-6}$ mho $h_{fe} = 100,$ $F_{\alpha co} = 2$ Mc $h_{rb} \cong 0.3 \times 10^{-3}$

Fig. 5.20   Collector characteristics (CE) of the 2N207. (Characteristics from Lansdale Tube Co.)

Having a maximum dissipation curve, we immediately know that the load line must not be to the right of it.   Two additional precautions should be taken prior to selecting a load line: We must see that $e_{C,m}$ and $i_{C,m}$ are kept below their absolute values during the signal excursion.   In the 2N207, $e_{C,m} = 12$ volts, and $i_{C,m} = 20$ ma. $e_C = 12$ volts does not appear on the characteristics of Fig. 5.20.   At $i_C = 20$ ma, a straight dotted line may be drawn from the y axis to the maximum dissipation curve, thereby serving as a reminder.   The load line must be restricted to below the $i_{C,m}$ line as well as to the left of the maximum collector dissipation curve. If maximum power output is desired, the operating point Q should be located somewhere on the maximum collector dissipation curve. The amplifier develops its greatest power output at zero signal condition; therefore, placing the operating point on the maximum collector dissipation curve assures us that the collector dissipation will not exceed its maximum value at any signal condition.

Example: Assume T_1 and T_2 of Fig. 5.15 is a 2N207 whose hybrid parameters are: $h_{ib} = 33$ ohms; $h_{ob} = 0.4$ μmho; $h_{fe} = 100$; and assume $h_{rb} = 2.5 \times 10^{-4}$. The collector characteristics are illustrated in Fig. 5.20. Let $R_{L1} = 333$ ohms; $R_{E2} = 100$ ohms; $R_{B4} = 1$ kilohm; $R_{B3} = 10$ kilohms; $V_{cc} = 6$ volts; $R_{L2} = 2$ kilohms.

(a) Construct the d-c load line.

(b) Select an operating point for faithful reproduction of a maximum input signal excursion of 20 μa.

(c) Construct the signal load line.

Solution:

(a) Construct the d-c load line:

Since V_{cc} and R_{L1} are known, $i_{C(SC)}$ can be found by

$$i_{C(SC)} = \frac{V_{cc}}{R_{L1}} = \frac{6}{3.33 \times 10^2} = 18 \text{ ma}$$

The d-c load line can be drawn from 18 ma on the y axis to 6 volts on the x axis (see Fig. 5.20).

(b) Selection of an operating point to accommodate a 20-μa (maximum) signal:

Consider the intersection of the d-c load line and the 40-μa characteristic for the operating point. When i_B swings above and below its Q value by 20 μa, i_C varies by about 4 ma above and below its rest value. Therefore, this point appears to be suitable for an operating point. The final check as to the suitability of this point is made after the signal load line is constructed.

(c) Construct the signal load line:

In terms of the input signal, $R_{L1,eq}$ is determined by

$$R_{L1,eq} = \frac{1}{1/R_{L1} + 1/R_{B4} + 1/R_{B3} + 1/(R_{E2} + R_{i2})}$$

R_{i2} is not yet known. Since the amplifier is in the CE configuration, the given hybrid parameters must be converted. Using the conversion equations of Sec. 4.17:

$$h_{ie} = \frac{h_{ib}}{1 + h_{fb}} = 3.3 \text{ kilohms}$$

$$h_{re} = \frac{h_{ib}h_{ob}}{1 + h_{fb}} - h_{rb} = 10.7 \times 10^{-4}$$

$$h_{fe} = \text{given as } 100 \quad \text{and} \quad h_{fb} = \frac{-h_{fe}}{1 + h_{fe}} = -0.99$$

$$h_{oe} = \frac{h_{ob}}{1 + h_{fb}} = 0.2 \text{ }\mu\text{mho}$$

With the hybrid parameters determined for the CE connection, we may now calculate R_{i2}:

$$R_{i2} = \frac{h_{ie} + (h_{oe}h_{ie} - h_{fe}h_{re})R_{L2}}{1 + h_{oe}R_{L2}}$$
$$= 5{,}520 \text{ ohms}$$

Substituting into the $R_{L1,\text{eq}}$ equation, and solving,

$$R_{L1,\text{eq}} = \frac{1}{1/(3.33 \times 10^2) + 1/(1 \times 10^3) + 1/(1 \times 10^4) + 1/(100 + 5{,}520)}$$
$$\cong 218 \text{ ohms}$$

Following the steps outlined in a previous paragraph, the signal load line is quickly determined by use of the trigonometric method.

Step 1

$$\tan \alpha = \frac{1 \times 10^3}{R_{L1,\text{eq}}} = \frac{1 \times 10^3}{218} \cong 4.58$$

Step 2

$$D - B = \frac{i_{c.Q} = 10^3}{\tan} = \frac{10.3}{4.58} \cong 2.25 \text{ volts}$$

Step 3

$$x\text{-axis intercept} = e_{c\text{-}E.Q} + (D - B)$$
$$= 2.5 + 2.25 = 4.75 \text{ volts}$$

The signal load line is now drawn between Q and 4.75 volts on the x axis (point D) and extended to the y axis (see Fig. 5.20). In checking the excursions in i_C, while i_B swings between 60 and 20 μa, it is seen that the i_C variations are reasonably symmetrical.

5.18 Transformer Coupling. Another common coupling technique is the use of an impedance matching transformer between the two cascaded stages. Transformer-coupled cascaded common-emitter stages are in popular use, as in the RC coupling technique, although any two configurations can be matched with transformer coupling. Transformer coupling has the advantages of high efficiency, good impedance matching, and better power handling capability. Its chief disadvantages are cost and size.

Figure 5.21a illustrates a single stage transformer-coupled common-emitter amplifier. R_E provides thermal stabilization and may be calculated by

$$R_E = \frac{e_{RE,Q}}{i_{E,Q}} = \frac{e_{RE,Q}}{i_{C,Q} + i_{B,Q}}$$

where $e_{RE,Q}$ is often selected to be about 10 per cent (or less) of V_{cc}, and $i_{C,Q}$ and $i_{B,Q}$ can be determined from the load line.

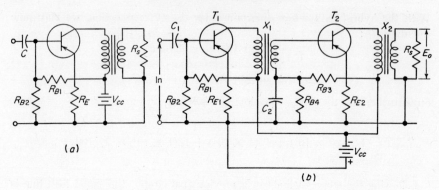

Fig. 5.21   Transformer-coupled CE amplifier: (a) single stage, (b) two stage.

R_{B2} and R_{B1} form a voltage-divider arrangement by which the transistor $e_{B\text{-}E,Q}$ is established.   Recall that in class A operation, the base-to-emitter junction must be placed at a small forward bias at zero signal condition. $e_{B\text{-}E,Q}$ can be determined in one of several ways.   The use of hybrid parameters is one method; i.e.,

$$e_{B\text{-}E,Q} \cong i_{B,Q}R_i$$

where
$$R_i = \frac{h_{ie} + (h_{oe}h_{ie} - h_{fe}h_{re})R_L}{1 + h_{oe}R_L}$$

NOTE: R_L is the reflected primary impedance of the transformer.

A second method is by the combined use of the $e_{C\text{-}E}$ versus i_C and $e_{B\text{-}E}$ versus i_C characteristics.   After determination of $i_{C,Q}$ from the load-line analysis, $i_{C,Q}$ is located on the $e_{B\text{-}E}$ versus i_C characteristics.   A line is then drawn parallel to the $e_{B\text{-}E}$ axis until the characteristic is intersected.   From this point of intersection, a perpendicular line is drawn to the $e_{B\text{-}E}$ axis.   The point at which this line intersects the $e_{B\text{-}E}$ axis renders the $e_{B\text{-}E,Q}$ value that must exist with the selected $i_{C,Q}$.

A third method is by the combined use of the $e_{C\text{-}E}$ versus i_C and $e_{B\text{-}E}$ versus i_B characteristics. $i_{B,Q}$ is determined from the load line and located on the $e_{B\text{-}E}$ versus i_B characteristics.   The corresponding $e_{B\text{-}E}$ for the selected i_B is then found.

The first method is used when the hybrid parameters of the transistor are known and the $e_{B\text{-}E}$ versus i_C or $e_{B\text{-}E}$ versus i_B characteristics are not available.   Method two is recommended if the $e_{B\text{-}E}$ versus i_C characteristics are available and method three if the $e_{B\text{-}E}$ versus i_B characteristics are available.   Occasionally, the designer may be in the position in which the hybrid parameters are unknown, and the $e_{B\text{-}E}$ versus i_C and $e_{B\text{-}E}$ versus i_B characteristics are not on hand.   In such cases, it is recommended that the designer determine the e_{B-E} versus i_C characteristic with e_{C-E} fixed at the $e_{C-E,Q}$ determined from the load-line analysis. Such a characteristic can be determined in short order.

Once $e_{B-E,Q}$ has been determined in one of the four ways stated in the preceding paragraph, the voltage-divider resistors can be computed:

$$R_{B2} = \frac{e_{RB2,Q}}{i_{RB2}} = \frac{e_{RE,Q} + e_{B-E,Q}}{i_{bleeder}}$$

where $i_{bleeder}$ is arbitrarily selected by the designer (about 100 μa is a typical value).

Recall that the forward bias for the base is developed across R_{B2}. Since e_{RE} is a reverse bias, then e_{RB2} must be $e_{B-E,Q}$ volts larger than $e_{RE,Q}$. And,

$$R_{B1} = \frac{V_{cc} - e_{RB2,Q}}{i_{bleeder} + i_{B,Q}}$$

Recall that $i_{bleeder}$ and $i_{B,Q}$ flow through R_{B1}.

Recall that the primary reflected impedance of a perfect transformer is determined from

$$R_L = \left(\frac{N_p}{N_s}\right)^2 R_s$$

where R_L = primary reflected impedance
N_p/N_s = turns ratio
R_s = secondary load impedance

In effect, the transistor will "see" a load impedance equal to R_L, instead of the actual load impedance R_s. In this way, an R_L value properly matched to the transistor amplifier can be established by the selection of the proper turns ratio for a given R_s. The common-emitter amplifier, for example, with a relatively high output resistance, can be used with very low impedance loads.

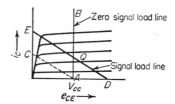

FIG. 5.22 Load lines: transformer-coupled CE amplifier.

Good interstage coupling transformers with efficiencies exceeding 90 per cent are commercially available. These transformers have low values of leakage reactance and primary resistance. For all practical purposes, the zero signal load line (determined by the primary resistance) is vertical (see Fig. 5.22).

With the presence of a signal, the collector current undergoes periodic variations which sets the transformer into action. The collector "sees" the primary as R_L (which is determined by the preceding relationship). This R_L is used to determine the signal load line of the amplifier. Assuming a resistive R_L, the signal load line will appear as shown in Fig. 5.22. A construction technique is discussed shortly. Assuming an ideal coupling transformer and transistor, the following relationships are true:

$$P_o = \frac{e_{C,m} i_{C,m}}{2}$$

$$P_{o,m} = \frac{e_{C,Q} i_{C,Q}}{2}$$

or

$$P_{o,m} = \frac{(V_{cc})^2}{2R_L}$$

$$P_c = P_{o,m} \qquad \text{maximum signal}$$
$$P_c = 2P_{o,m} \qquad \text{zero signal}$$
$$\text{Full signal efficiency} = 50\%$$

Figure 5.23 illustrates the effect of the coupling transformer at the high-, middle-, and low-frequency ranges. At high frequencies, the leakage reactance of the primary coil L_1 contributes to the determination of

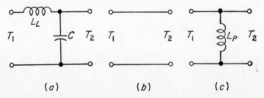

FIG. 5.23   Equivalent transformer circuit: (a) h-f, (b)m-f, (c) l-f.

the high-frequency limit. Effectively in series with the leakage reactance is the equivalent capacitance associated with T_1 output and T_2 input. At some high frequency, these two reactances approach their resonant frequency; being a series resonant circuit, the output of T_1 begins to decrease. The output decreases to 0.707 of the mid-frequency value (3 db down) at $f_{co,\max}$. At mid-frequency, both these reactances are negligible, and best transformer action is obtained. At low frequencies, X_{Lp} becomes small, since X_L is directly proportional to frequency. There is a low frequency where the output is 0.707 of the mid-frequency value (3 db down) because of the reduced X_{Lp}; this is $f_{co,\min}$.

Example: The 2N207 transistor is to be transformer-coupled to a 50-ohm resistive load. R_L is to be 333 ohms (the load line AQ drawn in Fig. 5.20). Using the same operating point Q, find the following: (a) transformer turns ratio required; (b) $P_{o,m}$; (c) P_c at zero signal.

Solution:

(a) Find the required transformer turns ratio:

$$R_L = \left(\frac{N_p}{N_s}\right)^2 R_s$$

hence
$$\frac{N_p}{N_s} = \left(\frac{R_L}{R_s}\right)^{1/2}$$

Substituting values,

$$\frac{N_p}{N_s} = \left(\frac{333}{50}\right)^{1/2} = 2.57/1$$

(b) Find $P_{o,m}$:

$$P_{o,m} = \frac{e_{C,Q} i_{C,Q}}{2} = \frac{2.5 \times 10.5 \times 10^{-3}}{2}$$
$$= 13.12 \times 10^{-3} \text{ watt}$$

(c) Find P_c at zero signal:

$$P_c = 2P_{o,m} = 26.24 \times 10^{-3} \text{ watt}$$

Notice that $P_{o,m}$ is 50 per cent of the collector dissipation, leading to 50 per cent efficiency.

Figure 5.21b shows a two-stage transformer-coupled common-emitter amplifier. The bias networks for each stage are determined in the manner revealed in preceding paragraphs in this section. When the input circuit of the transistor is transformer-coupled, as is the case with the input circuit of stage 2, a capacitor is placed between the secondary coil and common. This capacitor (C_2 in Fig. 5.21) is to serve as a unidirectional current block but to offer negligible impedance to the signal frequencies. Without this capacitor, the input circuit of stage 2 would be shunted (in terms of direct current) by the relatively low resistance of the secondary coil. The design of the remaining components of stage 2 can be performed with the same techniques used for stage 1.

A Technique for Transformer-coupled Signal Load-line Construction. In Fig. 5.21, assume the primary of the output transformer X_2 has negligible resistance. The d-c load line would then correspond to line AB of Fig. 5.22. Once the reflected primary resistance R_L has been determined in the manner revealed in an earlier part of this section, the designer can proceed to the construction of the signal load line.

Step 1. Using the determined R_L and V_{cc}, calculate $i_{C(SC)}$:

$$i_{C(SC)} = \frac{V_{cc}}{R_L}$$

This is point C in Fig. 5.22. Connect these points, thereby forming broken line AC, whose slope is equal to that of R_L.

Step 2. Select an operating point Q somewhere on the d-c load line. Draw a line parallel to broken line AC, which passes through point Q, like line DE in Fig. 5.22.

Line DE is the signal load line. Notice that the collector-base potential of the transistor becomes greater than V_{cc} during the downward excursions of the base current. This is due to the series-aiding effect of the primary induced voltage of the transformer during the time the base current is decreasing, thereby providing additional potential to be dropped across the transistor.

5.19 Theory of Transistor Noise. Transistor noise originates from several sources, some of which are more significant than others. Semiconductor noise, which is due to the properties of semiconductor materials, is most predominant at low frequencies (500 cps and less). In the middle range of frequencies, the transistor exhibits its most quiet behavior. Thermal noise, primarily due to the random motion action of the carriers in the base region, is greater at the middle- and upper-frequency ranges. The carrier random movement across the junction gives rise to shot noise, which is also at its greatest level at the middle- and upper-frequency ranges. A transistor appears less noisy in the upper-frequency range because of its over-all loss in gain which accompanies high-frequency operation.

An indication of the noise quality of a transistor is its noise factor NF which may be defined as the following relationship:

$$NF = \frac{NP_o}{NP_{o,ts}}$$

where NF = noise factor

NP_o = total noise power at the amplifier output

$NP_{o,ts}$ = portion of the output noise power due to the source thermal noise

The noise factor is most frequently stated as a decibel value, which is determined by

$$\text{db } (NF) = 10 \log \frac{NP_o}{NP_{o,ts}}$$

The measurement of the noise factor involves a two-step operation: first the amplifier's total noise output is determined, NP_o, and secondly the portion of the total noise output due to the source thermal noise is found. The noise factor may be found by use of several techniques, including the noise diode method and the single-frequency method.

The noise factor as a function of frequency, e_C, i_C, and source resistance are shown in the graphs of Fig. 5.24 a to d. Refer to Fig. 5.24. The noise factor is highest at low frequencies in well-designed transistors primarily because of semiconductor noise. At the 500 cps to 1 kc range, the semiconductor noise becomes relatively unimportant, with the thermal and shot noise setting the lowest noise level. Figure 5.24b shows that the collector voltage has little effect upon the noise factor up to a value of about 10 volts, at which time the semiconductor noise probably becomes more pronounced. Figure 5.24c illustrates that increases in

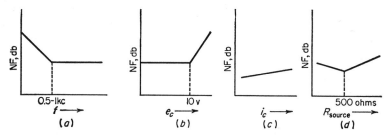

FIG. 5.24   Noise factor as a function of f, e_c, i_c, and R_{source}.

collector current result in an increase in the noise factor. The increased collector current indicates a greater carrier flow intensity across the junction, resulting in an increase in the shot noise level. In Fig. 5.24d, we see that the noise factor is lowest when the resistance of the source is in the neighborhood of 500 ohms. This indicates that transistors perform most quietly when the source resistance is relatively low. Matching the source resistance to the input resistance of the amplifier has no significant effect upon the noise factor. Generally, transformer coupling is utilized when the source resistance is high and low-level noise performance is required. Junction transistors have noise figures that vary from 2 to 25 db.

PROBLEMS

5.1 Refer to Fig. 5.1. $V_{cc} = 12$ volts; $R_L = 1.2$ kilohms; $i_{B,Q} = 100$ μa. Find R_B.

5.2 In Prob. 5.1, assume the transistor hybrid parameters are as follows: $h_{ie} = 1,800$ ohms; $h_{fe} = 30$; $h_{re} = 0.25 \times 10^{-3}$; $h_{oe} = 12$ μmho. Find (a) R_i; (b) $R_{i,\text{eq}}$.

5.3 Refer to Fig. 5.1. $V_{cc} = 22$ volts; $i_{B,Q} = 220$ μa; $R_L = 2$ kilohms. Find R_B.

5.4 In Prob. 3, assume the transistor has the hybrid parameters stated in Prob. 5.2. Find (a) R_i; (b) $R_{i,\text{eq}}$.

5.5 Refer to Fig. 5.2. $V_{cc} = 22$ volts; $R_L = 2$ kilohms; $i_C = 5$ ma; $i_B = 150$ μa. Find R_B.

5.6 Refer to Fig. 5.2. $V_{cc} = 12$ volts; $R_L = 3$ kilohms; $i_C = 1.2$ ma; $i_B = 120\ \mu\text{a}$. Find R_B.

5.7 Refer to Fig. 5.3. $V_{cc} = 12$ volts; $R_L = 4$ kilohms, $i_B = 100\ \mu\text{a}$; $i_C = 2.1$ ma; signal frequency range is 20 cps to 10 kc; $R_{B1} = R_{B2}$. Find (a) R_{B1} and R_{B2}; (b) C_B minimum.

5.8 Find C_B of Prob. 5.7 if the signal frequency range is 10 to 20 kc.

5.9 Refer to Fig. 5.3. $V_{cc} = 22$ volts; $R_L = 4$ kilohms; $i_C = 3$ ma; $i_B = 180\ \mu\text{a}$; $R_{B1} = 0.25\ R_{B2}$; $f = 20$ cps. Find (a) R_{B1}; (b) R_{B2}; (c) minimum C_B.

5.10 Refer to Fig. 5.3. $V_{cc} = 6$ volts; $R_L = 1.2$ kilohms; $i_C = 2.2$ ma; $i_{B,Q} = 40\ \mu\text{a}$; $R_{B1} = 4R_{B2}$; signal frequency range is 20 cps to 10 kc. Find (a) R_{B1}; (b) R_{B2}; (c) minimum C_B.

5.11 Refer to Fig. 5.5. $e_{RE,Q} = 0.9$ volt; $e_{B\text{-}E,Q} = 0.2$ volt; $V_{cc} = 12$ volts; $i_{B,Q} = 120\ \mu\text{a}$; $i_{C,Q} = 6$ ma; $i_{RB2} = 120\ \mu\text{a}$. Find (a) R_E; (b) C_E minimum; (c) R_{B2}; (d) R_{B1}.

5.12 Assume the hybrid parameters of the transistor used in Prob. 5.11 are as follows: $h_{ie} = 1{,}200$ ohms; $h_{fe} = 60$; $h_{re} = 0.5 \times 10^{-3}$; $h_{oe} = 25\ \mu\text{mho}$. Find (a) R_i; (b) $R_{i,\text{eq}}$; (c) R_i with C_E omitted; (d) $R_{i,\text{eq}}$ with C_E omitted.

5.13 In Prob. 5.9, $e_{RE,Q}$ is to be 0.6 volt; $e_{B\text{-}E,Q} = 0.15$ volt; all other parameters are the same. Find (a) R_E; (b) C_E minimum; (c) R_{B2}; (d) R_{B1}.

5.14 Refer to Fig. 5.5. $e_{RE,Q} = 0.5$ volt; $e_{B\text{-}E,Q} = 0.15$ volt; $V_{cc} = 12$ volts; $R_L = 2.2$ kilohms; $i_{B,Q} = 180\ \mu\text{a}$; $i_{C,Q} = 2$ ma; $i_{RB2} = 100\ \mu\text{a}$. Frequency range is 10 to 20 kc. Find (a) R_E; (b) C_E minimum; (c) R_{B2}; (d) R_{B1}.

5.15 Assume the hybrid parameters of the transistor used in Prob. 5.14 are as follows: $h_{ie} = 900$ ohms; $h_{fe} = 15$; $h_{re} = 0.48 \times 10^{-3}$; $h_{oe} = 15\ \mu\text{mho}$. Find (a) R_i; (b) $R_{i,\text{eq}}$; (c) R_i when C_E is omitted; (d) $R_{i,\text{eq}}$ when C_E is omitted.

5.16 Refer to Fig. 5.5. $e_{RE,Q} = 2$ volts; $e_{B\text{-}E,Q} = 0.35$ volt; $V_{cc} = 25$ volts; $R_L = 1.5$ kilohms; $i_{B,Q} = 200\ \mu\text{a}$; $i_{C,Q} = 8$ ma; $i_{RB2} = 100\ \mu\text{a}$. Find (a) R_E; (b) C_E minimum; (c) R_{B2}; (d) R_{B1}.

5.17 Assume the following hybrid parameters are those of the transistor in Prob. 5.16: $h_{ib} = 60$ ohms; $h_{ob} = 4.0\ \mu\text{mho}$; $h_{rb} = 3 \times 10^{-4}$; $h_{fe} = 28$. Find (a) h_{ie}; (b) h_{oe}; (c) h_{re}.

5.18 Using the parameters found in Prob. 5.17 and the stated value of h_{fe}, find the following for the circuit of Prob. 5.16:
(a) R_i (b) $R_{i,\text{eq}}$
(c) R_i with C_E omitted (d) $R_{i,\text{eq}}$ with C_E omitted

5.19 Refer to Fig. 5.7. $e_{RE,Q} = 0.75$ volt; $e_{RB3} = 0.4$ volt; $e_{B\text{-}E,Q} = 0.24$ volt; $i_{C,Q} = 7$ ma; $i_{B,Q} = 180\ \mu\text{a}$; $i_{RB2} = 200\ \mu\text{a}$; $V_{cc} = 12$ volts; $R_L = 1$ kilohm. Find (a) R_E; (b) R_{B3}; (c) R_{B2}; (d) R_{B1}.

5.20 Assume the transistor used in Prob. 5.19 has the following hybrid parameters: $h_{ie} = 1{,}800$ ohms; $h_{oe} = 10\ \mu\text{mho}$; $h_{re} = 0.35 \times 10^{-3}$; $h_{fe} = 35$. Find (a) R_i; (b) $R_{i,\text{eq}}$.

5.21 Compute the C_E required for the circuit of Prob. 5.19 for a signal frequency range of 50 cps to 15 kc.

5.22 Calculate the following for the circuit of Prob. 5.19 with C_E (as found in Prob. 5.21) included in the circuit:
(a) R_i
(b) $R_{i,\text{eq}}$

5.23 Refer to Fig. 5.7. $e_{RE,Q} = 1$ volt; $e_{RB3} = 0.25$ volt; $e_{B\text{-}E,Q} = 0.35$ volt; $i_{C,Q} = 4$ ma; $i_{B,Q} = 90\ \mu\text{a}$; $i_{RB2} = 100\ \mu\text{a}$; $V_{cc} = 25$ volts; $R_L = 2.5$ kilohms. Find (a) R_E; (b) R_{B3}; (c) R_{B2}; (d) R_{B1}.

5.24 Assume the transistor used in Prob. 5.23 has the following hybrid parameter
$h_{ib} = 40$ ohms; $h_{ob} = 1.0$ μmho; $h_{rb} = 4 \times 10^{-4}$; $h_{fe} = 15$. Find (a) h_{ie}
(b) h_{oe}; (c) h_{re}.

5.25 Using the parameters found in Prob. 5.24 and the stated h_{fe}, find the following
for the circuit of Prob. 5.23:
(a) R_i
(b) $R_{i,eq}$

5.26 Calculate the C_E required for the circuit of Prob. 5.23 for a signal frequency
range of 500 to 1,500 kc.

5.27 With C_E (found in Prob. 5.26) in the circuit, calculate the following for the
circuit of Prob. 5.23:
(a) R_i
(b) $R_{i,eq}$

5.28 Refer to Fig. 5.9. $V_{cc} = 12$ volts; $R_L = 1.5$ kilohms; $e_{B\text{-}E,Q} = 0.3$ volt;
$i_{B,Q} = 40$ μa; $R_{B1} = 4R_{B2}$; signal frequency range is 60 cps to 10 kc. Find
(a) R_{B1}; (b) R_{B2}; (c) C_B.

5.29 In Prob. 5.28, assume $R_{B1} = 0.5R_{B2}$. Find (a) R_{B1}; (b) R_{B2}; (c) C_B.

5.30 Assume the hybrid parameters of the transistor used in Prob. 5.28 are: $h_{ie} =$
250 ohms; $h_{fe} = 30$; $h_{re} = 0.5 \times 10^{-3}$; $h_{oe} = 30$ μmho. Find (a) R_i; (b) $R_{i,eq}$.

5.31 Using the hybrid parameters of Prob. 5.30, find the following for the circuit
of Prob. 5.29:
(a) R_i
(b) $R_{i,eq}$

5.32 Refer to Fig. 5.10. $V_{cc} = 22$ volts; $R_L = 2$ kilohms; $i_{C,Q} = 5$ ma; $i_{B,Q} =$
50 μa; $e_{RE,Q} = 1$ volt; $e_{B\text{-}E,Q} = 0.4$ volt. Find (a) R_B; (b) R_E.

5.33 In Prob. 5.32, determine the minimum C_E for a frequency range of 1 to 10 kc.

5.34 Assume the hybrid parameters of Prob. 5.24 are also those for Prob. 5.32.
Find (a) R_i; (b) $R_{i,eq}$.

5.35 Find R_i and $R_{i,eq}$ for Prob. 5.33.

5.36 Assume a CE amplifier has the following parameters: $i_B = 100$ μa; $I_{CO} = 7$ μa
at 20°C; $\beta = 40$. Assume I_{CO} doubles every 10°C rise in temperature.
Find i_C at the following temperatures:
(a) 20°C   (b) 50°C
(c) 60°C   (d) 70°C

5.37 In Prob. 5.36, assume $i_B = 200$ μa. Find i_C at the designated temperatures.

5.38 State the relationship between I_{CO} and temperature.

5.39 State the relationship between i_C and temperature. Explain.

5.40 State the relationship between i_C and i_{CO} at a fixed temperature as i_B is varied.

5.41 In Fig. 5.1, find SF for alpha values of (a) 0.92; (b) 0.95; (c) 0.97; (d) 0.98.

5.42 State the relationship between SF and alpha of the fixed-bias network.

5.43 Refer to Fig. 5.5. $\alpha = 0.98$; $R_E = 1$ kilohm; $R_{B2} = 5$ kilohms; $R_{B1} =$
20 kilohms. Find SF.

5.44 In Prob. 5.43, find SF for the following alpha values:
(a) 0.92   (b) 0.95
(c) 0.97   (d) 0.98

5.45 State the relationship between SF and alpha of the voltage-divider and
emitter-bias circuit.

5.46 In Prob. 5.43, find SF for the following R_E values:
(a) 100 ohms   (b) 500 ohms
(c) 1.5 kilohms   (d) 2 kilohms

5.47 State the relationship between SF and alpha of the fixed bias and emitter-bias circuit.

5.48 Refer to Fig. 5.5. $\alpha = 0.95$; $R_E = 700$ ohms; $R_{B2} = 10$ kilohms; $R_{B1} = 60$ kilohms. Find SF.

5.49 In Prob. 5.43, find SF for R_{B2} values of (a) 2 kilohms; (b) 5 kilohms; (c) 20 kilohms; (d) 30 kilohms.

5.50 In Prob. 5.43, find SF for R_{B1} values of (a) 10 kilohms; (b) 15 kilohms; (c) 40 kilohms; (d) 100 kilohms.

5.51 Refer to the direct-coupled amplifier of Fig. 5.12. The 2N223 is used (see Fig. 5.25b for typical hybrid parameters and collector characteristic curves). $V_{cc} = 15$ volts; $R_L = 1$ kilohm; $e_{RE,Q} = 1$ volt; $i_{B,Q} = 60 \mu a$; $e_{B-E,Q} = 0.2$ volt.
(a) Construct the signal load line.
(b) Determine the Q point and zero signal parameters from the load line.
(c) Calculate the following: R_E, R_B, R_i, $R_{i,eq}$.

5.52 A direct-coupled CE amplifier with the 2N223 is to be designed. The voltage divider with emitter bias is to be used. $R_L = 2$ kilohms; $i_{B,Q} = 40 \mu a$; $V_{cc} = 15$ volts. Find (a) R_i; (b) $e_{B-E,Q}$.

5.53 In the amplifier of Prob. 5.52, $e_{RE,Q} = 0.5$ volt; $f_{co,min} = 20$ cps. Find (a) R_E; (b) C_E.

5.54 Let $i_{bleeder} = 100 \mu a$ in the amplifier of Prob. 5.52. Find (a) R_{B2}; (b) R_{B1}.

5.55 In the amplifier of Prob. 5.54, find (a) $R_{i,eq}$; (b) R_o; (c) $R_{o,eq}$.

5.56 A direct-coupled amplifier with the 2N535 is to be designed. See Fig. 4.16 for hybrid parameters and collector characteristics. $V_{cc} = 15$ volts; $R_L = 500$ ohms; $i_{B,Q} = 60 \mu a$.
(a) Construct the load line.
(b) Determine the Q point and zero signal parameters.
(c) Convert the hybrids into CE values.
(d) Find R_i.

5.57 In the amplifier of Prob. 5.56, the variation of the voltage-divider and emitter-bias network is to be used. $e_{RE,Q} = 1$ volt; $e_{RB3} = 0.4$ volt; $i_{RB2} = i_{B,Q}$. Find:
(a) $e_{B-E,Q}$ (b) R_E
(c) R_{B3} (d) R_{B2}
(e) R_{B1}

5.58 In the amplifier coupled in Prob. 5.57, find (a) $R_{i,eq}$; (b) R_o; (c) $R_{o,eq}$.

5.59 Refer to the RC-coupled amplifier of Fig. 5.15. Assume the stage 1 component values are: $R_{B2} = 2$ kilohms; $R_{B1} = 16$ kilohms; $R_{E1} = 500$ ohms; $R_{L1} = 6$ kilohms. The transistor hybrid parameters are: $h_{fe} = 30$; $h_{ib} = 60$ ohms; $h_{ob} = 0.5 \mu mho$; $h_{rb} = 4 \times 10^{-4}$; and $C_{ob} = 7 \mu\mu f$. Find:
(a) h_{ie} (b) h_{oe}
(c) h_{re} (d) R_{i1}
(e) $R_{i1,eq}$

5.60 Find the following for stage 1 of Prob. 5.59:
(a) R_{o1}
(b) $R_{o1,eq}$

5.61 Assume that stage 2 of the amplifier of Prob. 5.58 has the following parameters: $R_{B4} = 10$ kilohms; $R_{B3} = 60$ kilohms; $R_{E2} = 100$ ohms; $C_2 = 1 \mu f$; $R_{L2} = 1$ kilohm. The transistor hybrid parameters are: $h_{fe} = 100$; $h_{ib} = 30$ ohms; $h_{ob} = 2 \mu mho$; $h_{rb} = 3 \times 10^{-4}$; and $C_{ob} = 90 \mu\mu f$. Find:
(a) h_{ie} (b) h_{oc}
(c) h_{re} (d) R_{i2}
(e) $R_{i2,eq}$

.62 For stage 2 of Prob. 5.61; compute (a) R_{o2}; (b) $R_{o2,eq}$.

5.63 Draw the equivalent low-frequency circuit (see Fig. 5.16a and b). Insert the corresponding values found in Probs. 5.59 to 5.61.

5.64 Draw the simplified low-frequency equivalent circuit and insert the corresponding values.

5.65 In the RC-coupled CE amplifier analyzed in Prob. 5.59 to 5.64, let E (stage 2 input voltage if $R_{o1,eq}$ is not shunted by $R_{i2,eq}$) $= 0.2$ volt; $f = 10$ cps. Find $e_{i2(l-f)}$.

5.66 Repeat Prob. 5.65 for $f = 80$ cps.

5.67 Did $e_{i2(l-f)}$ of Prob. 5.65 differ from $e_{i2(l-f)}$ of Prob. 5.66? Explain.

5.68 Determine $f_{co,min}$ of the RC coupling circuit.

5.69 Using $R_{i2,eq}$ found in Prob. 5.61 and $R_{o1,eq}$ found in Prob. 5.60, find $P_{i2(l-f)}$ at $f = 10$ cps.

5.70 Repeat Prob. 5.69 for $f = 80$ cps.

5.71 Draw the middle-frequency equivalent circuit (as shown in Fig. 5.16). Insert $R_{o1,eq}$ and $R_{i2,eq}$ used in Prob. 5.69.

5.72 In Prob. 5.71, find $e_{i2(m-f)}$ with $E = 0.2$ volt.

5.73 Calculate the ratio of the stage 2 input voltage at the middle and low frequencies (use $f = 10$ cps for low frequency).

5.74 Repeat Prob. 5.73 using $e_{i2(l-f)} = 80$ cps (as found in Prob. 5.66).

5.75 Calculate $P_{i(m-f)}$ of stage 2.

5.76 Using the values obtained in Prob. 5.69 and 5.75, find (a) G_{RC}; (b) $G_{RC(db)}$.

5.77 Find the RC coupling network loss in decibels at 10 cps.

5.78 Repeat Prob. 5.77 at 1 kc.

5.79 Draw the equivalent high-frequency circuit as shown in Fig. 5.17. Insert the resistance values found in the preceding problems. Assume C_w and C_{i2} are negligible.

5.80 C_{ob} (or C_{C-B}) of transistor $1 = 7$ $\mu\mu$f. Find C_{C-E}.

NOTE: C_{C-E} is C_{s1} in the equivalent circuit since C_w and C_{ie} are negligible.

5.81 Refer to Prob. 5.71 for $R_{o1,eq}$ and $R_{i2,eq}$. Find (a) X_{C-E}; (b) $f_{co,max}$ of the RC coupling circuit.

5.82 The parameters of an RC-coupled amplifier are:
$R_{i2,eq} = 1.5$ kilohms; $R_{o1,eq} = 4$ kilohms; C_2 (coupling capacitor) $= 0.4$ μf. Find (a) $f_{co,min}$; (b) X_{C2} at $f_{co,min}$.

5.83 An RC-coupled amplifier has the following parameters:
$R_{i2,eq} = 2$ kilohms; $R_{o1,eq} = 4$ kilohms; $X_{C2} = 1$ kilohm at 200 cps; $E = 0.5$ volt. Find (a) $e_{i2(l-f)}$; (b) $f_{co,min}$; (c) C_2.

5.84 In Prob. 5.83, $R_{i2,eq}$ is reduced to 1 kilohm. Find (a) $e_{i2(l-f)}$; (b) $f_{co,min}$; (c) C_2.

5.85 In Prob. 5.83, $R_{o1,eq}$ is reduced to 2 kilohms. Find (a) $e_{i2(l-f)}$; (b) $f_{co,min}$; (c) C_2.

5.86 Refer to Prob. 5.83 and 5.84. What effect does a change in $R_{i2,eq}$ have upon $e_{i2(l-f)}$? Explain.

5.87 Refer to Prob. 5.83 and 5.85. What effect does a change of $R_{o1,eq}$ have on $e_{i2(l-f)}$? Explain.

5.88 In Prob. 5.83, find the following:
(a) $e_{i2(m-f)}$
(b) ratio of stage 2 input voltage at the middle and low frequencies
(c) G_{RC} (d) $G_{RC(db)}$

5.89 Assume that C_{C-B} of the transistor in Prob. 5.83 $= 15$ $\mu\mu$f. Find (a) C_{C-E}; (b) $X_{C(C-E)}$ at $f_{co,max}$.

5.90 Refer to Fig. 5.20. The 2N207 is directly coupled to a 500-ohm load resistance. Q is the intersection of this load line and the 40-μa characteristic; the signal swing is ±20 μa. Find:

(a) P_o

(b) $P_{o,m}$

(c) P_c (at maximum signal)

(d) P_c (at zero signal)

5.91 Refer to Fig. 5.25 for the collector characteristics of the 2N223. Maximum collector dissipation at room temperature (20°C) is 250 mw. Draw the maximum collector dissipation curve.

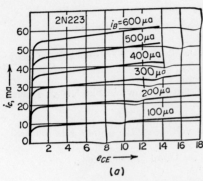 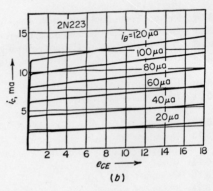

FIG. 5.25 Characteristics of the 2N223; (a) high current region, (b) low current region. $h_{ib} = 35$ ohms, $h_{ob} = 1$ μmho, $h_{fe} = 110$, $h_{rb} = 0.25 \times 10^{-3}$, $f_{aco} = 600$ kc, $C_{ob} = 90$ μμf. (Lansdale Tube Co.)

5.92 Refer to Fig. 5.26 for the collector characteristics of the 2N226. Maximum collector dissipation at room temperature (20°C) is 250 mw. Draw the maximum collector dissipation curve.

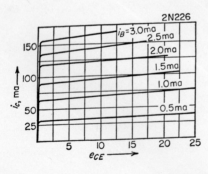

FIG. 5.26 Characteristics of the 2N226: $h_{ib} = 7.5$ ohms, $h_{ob} = 3$ μmhos, $h_{fe} = 60$, $f_{aco} = 400$ kc, $C_{ob} = 140$ μμf. (Lansdale Tube Co.)

5.93 Refer to Fig. 5.21. The transistor is the 2N223; $V_{cc} = 12$ volts; the collector is to "see" a load resistance of 300 ohms, but the actual load resistance is 10 ohms.

(a) Find required turns ratio.

(b) Construct the signal load line.

5.94 In the transformer-coupled amplifier of Prob. 5.93, find a suitable operating point so that ±100-μa input signal variation is faithfully reproduced.

5.95 Based on the signal load line used for Prob. 5.94, find (a) P_o; (b) $P_{o,m}$; (c) P_c (maximum signal); (d) P_c (zero signal).

5.96 Refer to Fig. 5.21. $V_{cc} = 9$ volts; the collector is to "see" a load resistance 1 kilohm; the transistor is the 2N223; the actual load resistance is 14 ohms. (a) Find required turns ratio.
(b) Construct the load line.

5.97 In the amplifier designed in Prob. 5.96, select an operating point so that a maximum input signal of ± 20 μa can be faithfully reproduced.

5.98 From the load line of Prob. 5.97, find (a) P_o; (b) $P_{o,m}$ (c) P_c (maximum signal); (d) P_c (zero signal).

5.99 In an RC-coupled amplifier: $R_{L1,eq} = 2$ kilohms; $R_{i2,eq} = 1.6$ kilohms; $V_{cc} = 12$ volts.   Find (a) $P_{o,m}T_1$; (b) Eff full sig T_1.

5.100 Repeat Prob. 5.99 for $R_{i2,eq} = 1$ kilohm.

CHAPTER 6

LINEAR APPLICATIONS OF TRANSISTORS

The behavior of a control device hinges in great part upon the region of its *E-I* characteristics to which it is restricted. Small signal operation is considered to be the condition where the control device is operated within a small portion of its characteristics. When the small signal behavior is such that the output current and voltage is a faithful reproduction of the input signal variations, operation of the control device is linear. Linear operation is a highly desirable feature in some applications, whereas it is undesirable and/or not important in others. This chapter is devoted exclusively to those applications of the transistor which rely upon linear operation. It will be noted that these applications encompass both low and high frequencies.

This chapter stresses many of the class *A* amplifier possibilities. The first part of the chapter analyzes many of the class *A* voltage amplifiers. The problems associated with *RC* coupling and transformer coupling are dealt with in a practical manner. Class *A* power amplifiers of both low- and high-frequency types are considered. Amplifiers in all configurations are studied (*CB*, *CC*, and *CE*). A careful study of class *A* push-pull amplifiers and phase inverters closes the class *A* amplifier analysis.

6.1 Class *A* Amplifiers. *Class A amplification in transistor work is defined as the condition where the transistor is operated within limits of its E-I characteristics in such a way that the output current flows during the entire input signal cycle.* The common-emitter configuration is most popular, although the common-base connection is sometimes used. The collector is the output terminal in both of these configurations; hence class *A* amplification can be further defined as that condition when collector current flows for the entire input signal cycle. In other classes of operation, the collector current does not flow for the entire input signal cycle, which means that linear applications indicate class *A* operation only. In the other classes of operation, which are discussed in the following chapter, the input signal drives the transistor into the collector cutoff region during some portion of the input signal. *Class A* *operation is the only mode of operation where faithful reproduction can be achieved with a single transistor.* The various advantages and disadvantages of the class A transistor amplifier are discussed in the following sections, which analyze the various applications.

6.2 Voltage Amplifiers. Both the common-emitter and common-base configurations are suitable for voltage amplifiers because of their voltage amplification possibilities. Other considerations, such as the input and output impedance of the transistor, along with the current and power capabilities, lead to the most general use of the common-emitter connection. Since the common-collector amplifier has a voltage ratio of unity or less, it is deemed unsuitable for use as a voltage amplifier. The common-emitter amplifier, as stated earlier, is capable of large voltage, current, and power ratios; possesses low input impedance (but higher than the common-base configuration); and medium to high output impedance.

6.3 *RC*-coupled Voltage Amplifiers. Figure 6.1 is the first stage of a three-transistor hi-fi preamplifier, designed by Texas Instruments

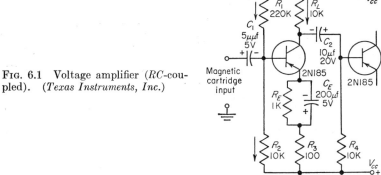

FIG. 6.1 Voltage amplifier (*RC*-coupled). (*Texas Instruments, Inc.*)

Incorporated. This circuit is designed for a GE magnetic cartridge as the input of the stage. The bias and stabilization of the stage are controlled by R_1, R_2, R_3, and R_E. There is little signal degeneration (negative feedback) through R_E since it is bypassed by C_E, which offers very little reactance at the lowest frequency to be handled by the amplifier. Let us verify this with the bypass technique discussed in Chap. 5. The frequency range is 30 to 15,000 cps. Find X_{CE} at 30 cps:

$$X_{CE} = \frac{1}{6.28 \times 30 \times 2 \times 10^{-4}} = 26.5 \text{ ohms}$$

Thus C_E offers a 26.5-ohm impedance path to the lowest input signal frequency (which is less than 10 per cent of R_E), thereby reducing the degenerative effects of R_E to a negligible value. Some negative feedback occurs because of R_3. The placement of R_E and R_3 in the emitter circuit reduces the stability factor of the stage, thereby minimizing the adverse effect of temperature increases upon transistor operation. The base forward bias is supplied by the voltage-divider arrangement of R_1 and R_2.

$$e_{R1} + e_{R2} = V_{cc}$$

where e_{R2} is the base-bias voltage.

Selection of the proper R_1 and R_2 values results in the correct zero signal operating point. In the circuit of Fig. 6.1, R_1, R_2, and R_3 are selected in accordance with the following relationships:

$$R_3 = \frac{\text{cartridge inductance in millihenrys}}{5}$$

$$R_2 = 100R_3$$
$$R_1 = 22R_2$$

The input circuit, as seen by the input signal, appears as shown in Fig. 6.2. R_i, the input resistance of the transistor, is somewhere in the

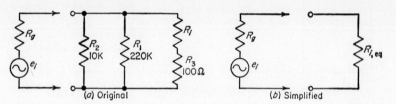

Fig. 6.2   Equivalent input circuit of Fig. 6.1: (a) original, (b) simplified.

neighborhood of 1 kilohm. Since the parallel branch consisting of R_i and R_3 is much smaller in ohmic value than each of the other branches, then

$$R_{i,eq} \cong R_3 + R_i \cong 1 \text{ kilohm}$$

R_g of the equivalent circuit is the impedance of the cartridge, which is about 1.5 kilohms for the GE magnetic type. The cartridge and stage 1 input impedance are reasonably well matched. Using the relationship stated in the preceding paragraph, the new values of R_1, R_2, and R_3 required for other cartridge types can be easily determined.

Notice that C_1, C_2, and C_E are the electrolytic type. Because of the low voltage encountered in this circuit, only low working voltages are required of these capacitors (as indicated in Fig. 6.1).

Figure 6.3 illustrates the complete hi-fi preamplifier, whose first stage has been just analyzed. *RC* coupling is used for all interstage coupling. The three stages are in the common-emitter configuration. The bypassed emitter resistors help to establish the zero signal operating point of each transistor while the unbypassed emitter resistors produce some degeneration (R_3, R_4, and R_{20}). Although these resistors reduce the gain of the respective stages, more faithful reproduction results, which is of vital importance in this type of amplifier.

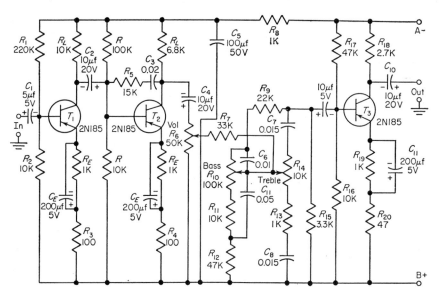

FIG. 6.3   *RC*-coupled amplifier.   (*Texas Instruments, Inc.*)

R_8 and C_5 serve as a *decoupling filter* across the power supply, which is connected from points A to B of Fig. 6.3 with the indicated polarity. Since the power supply has some impedance, a large capacitor across it provides a very low impedance path shunt for the signal frequencies. R_8 must be kept as small as possible, since it reduces the voltage available for the collectors of T_1 and T_2. Using a smaller ohmic value of R_8 necessitates a larger value of C_5, in order to maintain the low impedance shunt for the signal frequencies. As a general rule of thumb,

$$R_8 C_5 > \frac{1}{f_{min}}$$

Substituting values,

$$1 \times 10^3 \times 100 \times 10^{-6} > \frac{1}{30}$$
$$1 \times 10^{-1} > 3.33 \times 10^{-2}$$

which is true.   Hence R_8 and C_5 serve as a good decoupling filter for V_{cc}.
Notice the *base* and *treble* networks, which are located between stages 2
and 3.   When R_{10} is varied, the low-frequency response (bass) is altered,
and the high-frequency response (treble) is changed by varying R_{14}.

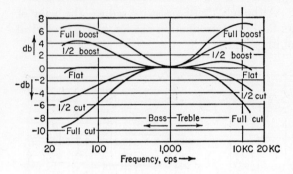

FIG. 6.4 f(cps) versus db
of Fig. 6.3. (*Texas Instruments, Inc.*)

The frequency response of this amplifier with several bass and treble
settings is shown in Fig. 6.4.

6.4   Design of a Single-stage Class A RC-coupled Amplifier.

Let us
consider the selection of components to be used in the class A amplifier
of Fig. 6.5a. R_L is selected as 1.2 kilohms in order to allow a reasonably
faithful reproduction of the input signal.   Forward bias is to be furnished
to the base by means of the voltage-divider network (R_{B1} and R_{B2}).   An
emitter resistance is to be used for thermal stability but is to be bypassed
so as to reduce signal degeneration to a negligible value.   This amplifier
is to handle frequencies from 60 to 15,000 cps.   The supply is 12 volts,
and the transistor is the 2N333 (NPN silicon type).   The input signal is
sinusoidal, with a maximum value of 50 μa.
Absolute maximum values (2N333):

$$E_{C\text{-}E} = 45 \text{ volts}$$
$$E_{C\text{-}B} = 45 \text{ volts}$$
$$E_{E\text{-}B} = 1 \text{ volt}$$

Power dissipation: 150 mw at   25°C
                        100 mw at 100°C
                         50 mw at 150°C

NOTE: Hybrid parameters will not be used in the design of this amplifier.   Recall,
from Sec. 5.5, that the base-to-emitter voltage can be determined from the
i_B versus e_B or i_C versus e_B characteristics if they are available.   The i_C versus e_B
characteristics are furnished in the Transitron data sheet for the 2N333 and will
be used in the following design.   Remember, however, that $e_{B\text{-}E,Q}$ can also be

determined from the relationship:

$$e_{B\text{-}E,Q} \cong i_{B,Q} R_i$$

where, in the CE connection,

$$R_i = \frac{h_{ie} + (h_{oe}h_{ie} - h_{fe}h_{re})R_L}{1 + h_{oe}R_L}$$

Step 1. Draw the load line for $V_{cc} = 12$ volts, and $R_L = 1.2$ kilohms on the characteristics (method is analyzed in Chaps. 2 and 4) in Fig. 6.5b.

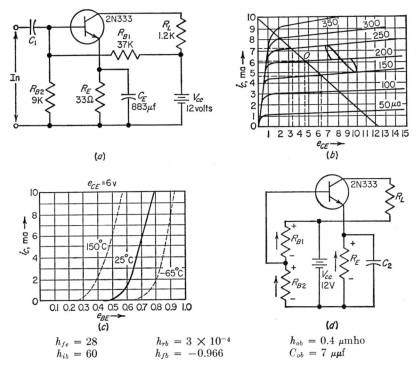

$$h_{fe} = 28 \qquad h_{rb} = 3 \times 10^{-4} \qquad h_{ob} = 0.4 \ \mu\text{mho}$$
$$h_{ib} = 60 \qquad h_{fb} = -0.966 \qquad C_{ob} = 7 \ \mu\mu\text{f}$$

FIG. 6.5 2N333 characteristics: (a) circuit to be designed, (b) $e_{C\text{-}E}$ versus i_C, (c) $e_{B\text{-}E}$ versus i_C, (d) bias circuit. [(b) *and* (c) *Characteristics from Transitron.*]

Step 2. Check to be certain the load line does not intersect the maximum power dissipation curve (25°C). This curve does not fall within the illustrated characteristics.

Step 3. Search for a point on the load line where a 50-μa base current increase and decrease result in a symmetrical variation in the collector current about its rest value. O is such a point.

When i_B swings from 200 to 250 μa, i_C increases from 6 to 7.2 ma, a change of 1.2 ma. When i_B swings from 200 to 150 μa, i_C decreases from

6 to 4.8 ma, also a change of 1.2 ma. Hence the variations of i_C and e_C are symmetrical. This means that the zero signal base current is to be 200 μa.

Step 4: *Determination of Bias Resistor Values.* In step 3, we found i_C at zero input signal to be 6 ma. Referring to the i_C versus $e_{B\text{-}E}$ curve in Fig. 6.5c, the base-to-emitter voltage at rest should be about 0.7 volt. Redrawing the circuit, as in Fig. 6.5d will simplify the following analysis. Taking the series loop from the base to the emitter, we find

$$e_{B\text{-}E} = e_{RB2} + e_{RE}$$

Notice that e_{RB2} and e_{RE} are both positive with respect to common, which means they are series-opposing with respect to each other. *R_E provides reverse bias to the emitter while R_{B2} furnishes forward bias to the base. Since $e_{B\text{-}E}$ is to be 0.7 volt, then e_{RB2} must be 0.7 volt greater than e_{RE}.*

The amount of reverse bias obtained from R_E is optional, so let us arbitrarily select a value of 0.2 volt. This means e_{RB2} will be 0.9 volt. R_E can now be calculated:

$$R_E = \frac{E_{RE}}{I_E} = \frac{0.2}{6 \times 10^{-3}} = 33 \text{ ohms}$$

since $I_E \cong I_C$ for practical purposes.

Turning now to R_{B1} and R_{B2}, we find the following relationships:

$$e_{RB1} + e_{RB2} = 12 \text{ volts}$$

and, at zero signal condition,

$$i_{RB1} = i_B + i_{RB2}$$

and

$$e_{RB2} = \frac{R_{B2}}{R_{B1} + R_{B2}} \times 12 = 0.9 \text{ volt}$$

and

$$e_{RB1} = 12 - e_{RB2} = 12 - 0.9 = 11.1 \text{ volts}$$

We may select i_{RB2}, making sure it does not impose too great a drain upon the power supply. Let $i_{RB2} = 100$ μa. Solving for R_{B2},

$$R_{B2} = \frac{e_{RB2}}{i_{RB2}} = \frac{0.9}{1 \times 10^{-4}} = 9 \text{ kilohms}$$

From the previous relationship, we may determine i_{RB1}:

$$i_{RB1} = i_B + i_{RB2} = 200 \text{ }\mu\text{a} + 100 \text{ }\mu\text{a} = 300 \text{ }\mu\text{a}$$

and

$$R_{B1} = \frac{e_{RB1}}{i_{RB1}} = \frac{11.1}{3 \times 10^{-4}} = 37 \text{ kilohms}$$

Step 5.   Find C_E. X_{CE} should be 10 per cent or less of R_E at the lowest frequency (60 cps), or about 3 ohms:

$$C_E = \frac{1}{6.28 \times 60 \times 3} = 883 \ \mu\text{f}$$

Step 6.   Find C_1. X_{C1} should offer low impedance to the signal variations and should therefore be about the same value as C_E.

Due to the slight variations found between transistors of the same type, the final values of all components are determined when the amplifier is actually constructed.

6.5   Design of a Single-stage Class A Transformer-coupled Amplifier.

The 2N43 (PNP) transistor is used in this design. V_{cc} is 15 volts;

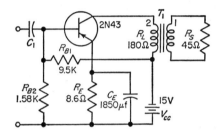

Fig. 6.6   The designed class A transformer-coupled amplifier.

frequency range is 100 cps to 10 kc; R_s is 45 ohms.   The characteristics of the 2N43 are shown in Fig. 6.42a and b.

Step 1: *Construction of the D-C Load Line.*   Refer to Fig. 6.42a and b. Assume the primary of the output transformer possesses negligible resistance, which is a reasonably accurate assumption in many cases. Therefore the d-c load line may be drawn from the value of V_{cc} on the x axis (see line AB).

An operating point Q is selected, which is at the intersection of the 1-ma characteristic and the d-c load line.   The bias network is to be designed such that this operating point is maintained at zero signal condition.

Step 2: *Construction of the Signal Load Line.*   The signal load line may be constructed in one of several ways.   In this circuit design, a value of reflected R_L will be selected first.   Let the reflected R_L be 180 ohms. If a resistor of 180 ohms were connected directly between the collector and the negative terminal of V_{cc}, the load line would appear as AA' of Fig. 6.42a.   However, the operating point is at Q.   Therefore a line parallel to AA' is drawn through $Q(C\text{-}D)$; this is the signal load line.

Notice that $e_{C\text{-}E}$ exceeds V_{cc} when i_B decreases.   This is possible because of the series-aiding effect of the induced voltage of the primary.

Step 3: *Determination of* R_E *and* C_E. Recall that R_E provides thermal stabilization and $e_{RE,Q}$ can be arbitrarily selected. Let $e_{RE,Q} = 0.5$ volt.

$$i_{RE,Q} = i_{C,Q} + i_{B,Q}$$

Obtaining the values of current from Fig. 6.42a,

$$i_{RE,Q} = 57 \times 10^{-3} + 1 \times 10^{-3} = 58 \times 10^{-3} \text{ amp}$$

Now R_E can be determined:

$$R_E = \frac{e_{RE,Q}}{i_{RE,Q}} = \frac{5 \times 10^{-1}}{58 \times 10^{-3}}$$
$$\cong 8.6 \text{ ohms}$$
$$X_{CE} \text{ at } f_{\min} = 0.10 R_E \cong 0.86 \text{ ohm}$$

Since $f_{\min} = 100$ cps, then

$$C_E = \frac{159 \times 10^{-3}}{1 \times 10^2 \times 8.6 \times 10^{-1}}$$
$$= 1,850 \ \mu\text{f}$$

Step 4: *Determination of* R_{B2} *and* R_{B1}. $e_{RB2,Q}$ is to exceed $e_{RE,Q}$ by the desired $e_{B\text{-}E,Q}$. The desired $e_{B\text{-}E,Q}$ can be found in Fig. 6.42c. $i_{C,Q}$ is located on the x axis, and a perpendicular line is projected until it intercepts the characteristic (point Q in Fig. 6.42c). A line parallel to the x axis is projected to the y axis, which is $e_{B\text{-}E,Q}$. From this analysis,

$$e_{B\text{-}E,Q} \cong 0.29 \text{ volt}$$
$$e_{RB2,Q} = e_{RE,Q} + e_{B\text{-}E,Q}$$

Substituting values,

$$e_{RB2,Q} = 0.50 + 0.29 \cong 0.79 \text{ volt}$$

$i_{RB2,Q}$ is a bleeder current and may be arbitrarily selected. **Let**

$$i_{RB2,Q} = 0.5 \text{ ma}$$

R_{B2} may now be computed:

$$R_{B2} = \frac{e_{RB2,Q}}{i_{RB2,Q}} = \frac{7.9 \times 10^{-1}}{5 \times 10^{-4}}$$
$$\cong 1.58 \text{ kilohms}$$

and
$$R_{B1} = \frac{V_{cc} - e_{RB2,Q}}{i_{\text{bleeder}} + i_{B,Q}} = \frac{15 - 0.79}{(0.5 + 1.0) \times 10^{-3}}$$
$$\cong 9.5 \text{ kilohms}$$

Step 5: *Determination of* C_1. Since C_1 is to present negligible impedance at the lowest signal frequency, a value equal to that of C_E may be used.

Step 6: *Determination of the Transformer Ratio*

$$R_L = \left(\frac{N_p}{N_s}\right)^2 R_s$$

where R_L = reflected primary resistance
$\quad\quad R_s$ = actual load resistance
Transposing,

$$\frac{N_p}{N_s} = \left(\frac{R_L}{R_s}\right)^{\frac12}$$

Substituting values,

$$\frac{N_p}{N_s} = \left(\frac{180}{45}\right)^{\frac12} = \frac{2}{1} \quad\quad \text{assuming an ideal transformer}$$

Step 7: *Circuit Construction and Evaluation.* A potentiometer may be used for R_{B2} when constructing the circuit, which is varied until the required $e_{B\text{-}E,Q}$ is obtained. The oscilloscope may be used to check the output waveform.

6.6 Design of a Two-stage RC-coupled (CE) Amplifier. Let us design the two-stage RC-coupled (CE) amplifier illustrated in Fig. 5.15. The 2N465 transistor is to be used for both stages. $V_{cc} = 12$ volts; $f_{co,min} = 60$ cps. Maximum allowable total transistor power dissipation for the 2N465 is determined by

$$P = \frac{(T_{j,\text{max}} - T_a)}{k}$$

where $T_{j,\text{max}}$ = maximum junction temperature = 85°C
$\quad\quad T_a$ = ambient temperature °C
$\quad\quad k$ = dissipation coefficient = 0.4°C/mw in air
Assume the output of the amplifier is fed into a 2-kilohm resistance, and the amplifier input is fed from a 0.5-kilohm resistance. Figure 6.7 illustrates these two statements.

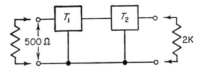

Fig. 6.7 Input and output resistances seen by the two-stage amplifier.

Inasmuch as the signal load line of T_1 is affected by the voltage-divider bias resistors of T_2, the circuit design is started from the output; i.e., the parameters for T_2 are determined first. Following is the step-by-step procedure for the design of this amplifier.

Step 1: *Construction of the Maximum Collector Dissipation Curve.* The relationship stated in the Raytheon data sheet is

$$P = \frac{(T_{j,\max} - T_a)}{k}$$

where $T_{j,\max} = 85°C$
$T_a = 50°C$ in this amplifier
$k = 0.4°C/mw$ in air

therefore $P_{c,\max} = \dfrac{(85° - 50°)}{0.4°/mw} = 87.5 \text{ mw}$

Using the collector characteristics of the 2N465 shown in Fig. 6.8, let

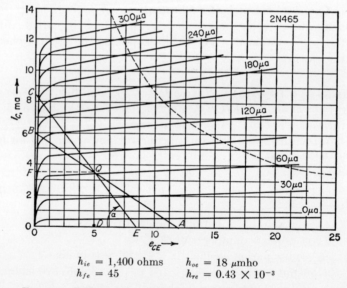

$h_{ie} = 1,400$ ohms $h_{oe} = 18 \ \mu\text{mho}$
$h_{fe} = 45$ $h_{re} = 0.43 \times 10^{-3}$

Fig. 6.8   Collector characteristics of the 2N465.   *(Raytheon.)*

us compute several points and then draw the maximum collector dissipation curve.

$$P_{c,\max} = e_C i_C$$

By arbitrarily selecting i_C values, we may determine the corresponding e_C for each:

$$e_C = \frac{P_{c,\max}}{i_C} = \frac{87.5}{i_C}$$

The following values are obtained:

i_C, ma	e_C, volts
14	6.25
10	8.75
8	10.9
5	17.5
4	21.7

The points are plotted on the collector characteristics and joined by a broken line (see Fig. 6.8).

Step 2: Selection of R_{L2} and Load-line Construction. Let us arbitrarily select $R_{L2} = 2$ kilohms; $V_{cc} = 12$ volts. The d-c load line is easily determined by the open-circuit and short-circuit tests.

If the transistor were shorted, $e_{RL2} = V_{cc} = 12$ volts, and

$$i_C = \frac{V_{cc}}{R_{L2}} = \frac{12}{2 \times 10^3} = 6 \text{ ma (point } B)$$

If the transistor offered infinite ohms, $e_C = V_{cc} = 12$ volts, and

$$i_C = 0 \text{ ma (point } A)$$

The two points are plotted on the characteristics, and the load line is drawn. Notice that R_{E2} does not come into the load-line computations because it is bypassed by C_{E2}. Recall, however, that the output of the amplifier is fed to a 2 kilohm resistance, which is effectively in parallel with R_{L2}:

$$R_{L2,eq} = \frac{2 \times 10^3 \times 2 \times 10^3}{(2 \times 10^3) + (2 \times 10^3)} = 1 \text{ kilohm}$$

Referring to Fig. 6.8,

$$\tan \alpha = -\frac{1}{R_{L2,eq}} = \frac{1 \times 10^3}{1 \times 10^3} = 1$$

$$i_{C,Q} \cong 3.5 \text{ ma} \qquad e_{C\text{-}E,Q} \cong 5 \text{ volts}$$

Substituting values,

$$\tan \alpha = \frac{i_{C,Q} \times 10^3}{D - E}$$

and transposing,

$$D - E = \frac{i_{C,Q} \times 10^3}{\tan \alpha} = \frac{3.5}{1}$$
$$= 3.5 \text{ volts}$$

The x-axis intercept of the signal load line can now be found:

$$x\text{-axis intercept} = e_{C\text{-}E,Q} + (D - E)$$
$$= 5 + 3.5 = 8.5 \text{ volts}$$

The signal load line is now drawn between points CQE (see Fig. 6.8).

Step 3: Selection of T_2 Operating Point, Bias Resistors, and C_{E2}. Find a point of intersection of the signal load line and a characteristic curve

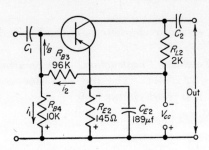

FIG. 6.9   Stage 2 design.

which results in linear reproduction. Q is such a point. At zero signal condition

$$i_{B,Q} = 60 \ \mu a$$
$$i_{C,Q} = 3.5 \text{ ma}$$
$$e_{C,Q} = 5 \text{ volts}$$

Refer to Fig. 6.9 for placement of components. R_{E2} provides stabilization against temperature variations. Let $e_{RE2,Q} = -0.5$ volt.

$$i_{E,Q} = i_{RE,Q} \cong i_{C,Q} + i_{B,Q} = 3.56 \text{ ma}$$

Solving for R_{E2},

$$R_{E2} = \frac{e_{RE2,Q}}{i_{RE2,Q}} = \frac{5 \times 10^{-1}}{3.56 \times 10^{-3}}$$

or
$$R_{E2} \cong 145 \text{ ohms}$$

To find $e_{B\text{-}E,Q}$, recall that

$$e_{B\text{-}E,Q} \cong i_{B,Q} R_i$$

where, in the CE configuration,

$$R_i = \frac{h_{ie} + (h_{oe}h_{ie} - h_{fe}h_{re})R_L}{1 + h_{oe}R_L}$$

where R_L is $R_{L2,eq}$ in this case, which is 1 kilohm.

Substituting the hybrid values given in the data sheet and the calculated value of $R_{L2,eq}$,

$$R_i = \frac{1,400 + (18 \times 10^{-6} \times 1.4 \times 10^{3} - 45 \times 0.43 \times 10^{-3})1 \times 10^{3}}{1 + 18 \times 10^{-6} \times 1 \times 10^{3}}$$

$$\cong 1,380 \text{ ohms}$$

Now solving for $e_{B\text{-}E,Q}$,

$$e_{B\text{-}E,Q} \cong 60 \times 10^{-6} \times 1.380 \times 10^{3}$$
$$\cong 0.083 \text{ volt}$$

NOTE: $e_{B\text{-}E,Q}$ can be determined from the $e_{B\text{-}E}$ versus i_C characteristics, or the i_B versus $e_{B\text{-}E}$ characteristics, if they are available. See Sec. 5.5 and 6.4 for details.

Analyzing the voltages in the series loop from the base to emitter,

$$e_{RB4,Q} + e_{RE,Q} = e_{B\text{-}E,Q}$$

and
$$e_{RB4,Q} = e_{B\text{-}E,Q} - e_{RE2,Q}$$

Now solving,

$$e_{RB4,Q} = 0.083 - (-0.5) = 0.583 \text{ volt}$$

$e_{RE2,Q}$ is a reverse bias, and $e_{RB4,Q}$ is a forward bias. $e_{RB4,Q}$ must be sufficiently large to offset $e_{RE2,Q}$ and still provide 0.083 volt of forward bias.

Let us arbitrarily select $R_{B4} = 10$ kilohms (so as to minimize the shunting effect upon R_{L1}). Knowing $e_{RB4,Q}$, we can determine the bleeder current i_1 of the bias voltage divider:

$$i_{RB4,Q} = i_1 = \frac{e_{RB4,Q}}{R_{B4}} = \frac{5.83 \times 10^{-1}}{1 \times 10^{4}}$$
$$\cong 58 \ \mu a$$

Turning our attention to R_{B3},

$$i_{RB3,Q} = i_2 = i_{B,Q} + i_1$$
$$= 58 + 60 = 118 \ \mu a$$

and
$$e_{RB3,Q} = V_{cc} - e_{RB4,Q} = 12 - 0.583$$
$$= 11.42 \text{ volts}$$

Now solving for R_{B3},

$$R_{B3} = \frac{e_{RB3,Q}}{i_{RB3,Q}} = \frac{11.42}{118 \times 10^{-6}}$$
$$\cong 96 \text{ kilohms}$$

The reactance of C_{E2} at the lowest signal frequency should be 10 per cent of R_{E2} (or 14 ohms); $f_{co,\min} = 60$ cps. Solving for C_{E2},

$$C_{E2} = \frac{1}{2\pi f_{co,\min} X_{CE2}} = \frac{1}{6.28 \times 6 \times 10^{1} \times 14}$$
$$\cong 189 \ \mu f \text{ or as large as practical}$$

Step 4: Determination of $R_{L1,eq}$ and R_{L1}. Figure 6.10a illustrates the manner in which R_{i2}, R_{B3}, and R_{B4} are each in parallel with R_{L1}. The equivalent value of R_{i2}, R_{B3}, and R_{B4} is $R_{i2,eq}$, which is found to be 1,190 ohms in a following paragraph. Figure 6.10b shows $R_{i2,\mathrm{eq}}$ in parallel with R_{L1}.

Let us arbitrarily select $R_{L1,\mathrm{eq}}$ to be 1 kilohm and calculate the R_{L1} value necessary to fulfill these requirements. Note that the $R_{L1,\mathrm{eq}}$

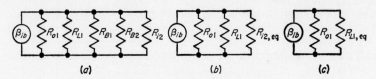

FIG. 6.10   Determination of $R_{L1,\mathrm{eq}}$.

chosen must be less than $R_{i2,\mathrm{eq}}$ because of the parallel resistance relationship which states that the equivalent resistance value $R_{L1,\mathrm{eq}}$ is smaller than the smallest parallel branch resistance $R_{i2,\mathrm{eq}}$.

We must next calculate $R_{i2,\mathrm{eq}}$:

$$R_{i2,\mathrm{eq}} = \frac{1}{1/R_{B4} + 1/R_{B3} + 1/R_{i2}}$$

Note that R_{E2} is not included, since it is shunted by C_{E2}. Solving,

$$R_{i2,\mathrm{eq}} = \frac{1}{1/(1 \times 10^4) + 1/(9.6 \times 10^4) + 1/(1.38 \times 10^3)}$$
$$\cong 1{,}190 \text{ ohms}$$

Since $R_{i2,\mathrm{eq}}$ has been found to be 1,190 ohms, the selection of $R_{L1,\mathrm{eq}}$ of 1 kilohm is permissible.

Referring to the simplified equivalent circuit of Fig. 6.10c, it is seen that

$$R_{L1,\mathrm{eq}} = \frac{R_{L1} R_{i2,\mathrm{eq}}}{R_{L1} + R_{i2,\mathrm{eq}}}$$

By transposing, we may solve for R_{L1}:

$$R_{L1} = \frac{R_{i2,\mathrm{eq}} R_{L1,\mathrm{eq}}}{R_{i2,\mathrm{eq}} - R_{L1,\mathrm{eq}}}$$

Substituting into the preceding equation for R_{L1},

$$R_{L1} = \frac{1{,}190 \times 1{,}000}{1{,}190 - 1{,}000} \cong 6.25 \text{ kilohms}$$

Step 5: Signal Load-line Construction and Zero Signal Parameters of T_1. The d-c load line has a slope determined by the reciprocal of R_{L1}, while the signal load line is determined by $R_{L1,\mathrm{eq}}$. Both load lines are constructed in accordance with the procedure set down in step 2 and are

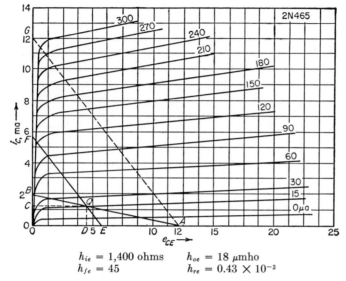

$$h_{ie} = 1{,}400 \text{ ohms} \qquad h_{oe} = 18 \text{ } \mu\text{mho}$$
$$h_{fe} = 45 \qquad h_{re} = 0.43 \times 10^{-3}$$

FIG. 6.11 Collector characteristics of the 2N465. (*Raytheon.*)

shown in Fig. 6.11. The Q-point parameters are taken from the signal load line. Q is selected as the intersection of the signal load line and the 15-μa characteristic:

$$i_{B,Q} \cong 15 \text{ } \mu\text{a}$$
$$e_{C,Q} \cong 4.2 \text{ volts}$$
$$i_{C,Q} \cong 1.2 \text{ ma}$$
$$\tan \alpha = -\frac{1 \times 10^3}{R_{L1,eq}} = \frac{1 \times 10^3}{1 \times 10^3} = 1$$
$$i_{C,Q} \cong 1.2 \text{ ma} \qquad e_{C\text{-}E,Q} \cong 4.2 \text{ volts}$$

Substituting values and transposing,

$$\tan \alpha = \frac{i_{C,Q} \times 10^3}{D - E}$$
$$D - E = \frac{i_{C,Q} \times 10^3}{\tan \alpha} = \frac{1.2}{1} = 1.2 \text{ volts}$$

The x-axis intercept (point E) can now be found:

$$x\text{-axis intercept} = e_{C\text{-}E,Q} + (D - E)$$
$$= 4.2 + 1.2 = 5.4 \text{ volts}$$

The signal load line is now drawn between points FQE (see Fig. 6.11).

Step 6: *Determination of R_{E1}, R_{B2}, and R_{B1}.* Let us assume that R_{i1} and R_{i2} are equal, since the same transistor and $R_{L,eq}$ values are used for both stages; i.e.,

$$R_{i1} = R_{i2} \cong 1{,}380 \text{ ohms}$$

Recall that the input signal applied to the base of T_1 originates from a source whose resistance is 500 ohms. In order to minimize the effect of the input signal source resistance upon T_1, $R_{i1,eq}$ should be made approximately the same value or greater. R_{B2} is the smallest of the two voltage-divider resistors, so let us arbitrarily set its value at 10 kilohms. Recall that

$$\begin{aligned}
e_{B\text{-}E,Q} &\cong i_{B,Q} R_{i1} \\
&\cong 15 \times 10^{-6} \times 1.38 \times 10^3 \\
&\cong 0.02 \text{ volt}
\end{aligned}$$

Let $e_{RE1,Q} \cong 0.70$ volt. Finding $i_{E1,Q}$,

$$\begin{aligned}
i_{E1,Q} = i_{B1,Q} + i_{C1,Q} &= 0.015 + 1.2 \\
&= 1.215 \text{ ma}
\end{aligned}$$

Solving for R_{E1},

$$R_{E1} = \frac{e_{RE1,Q}}{i_{RE1,Q}} = \frac{70 \times 10^{-2}}{1.215 \times 10^{-3}} \cong 570 \text{ ohms}$$

NOTE: It is suggested that the final value of R_{E1} be determined when the circuit is actually constructed. Increase R_{E1} until the output variations are undistorted (place an oscilloscope across R_{L2}).

Again select $R_{B2} = 10$ kilohms; then

$$\begin{aligned}
e_{RB2,Q} &= e_{B\text{-}E,Q} - e_{RE1,Q} \\
&= 0.02 - (-0.70) = 0.72 \text{ volt}
\end{aligned}$$

and the bleeder current i_{RB2} is next found:

$$i_{RB2} = \frac{e_{RB2,Q}}{R_{B2,Q}} = \frac{0.72}{1 \times 10^4} = 72 \ \mu\text{a}$$

Computing R_{B1},

$$\begin{aligned}
R_{B1} &= \frac{V_{cc} - e_{RB2,Q}}{i_{B1,Q} + i_{RB2,Q}} = \frac{12 - 0.72}{15 \ \mu\text{a} + 72 \ \mu\text{a}} \\
&\cong 130 \text{ kilohms}
\end{aligned}$$

Step 7: Determination of the Coupling Capacitors C_1 and C_2
At $f_{co,min}$,

$$X_{C1} = R_{o1,eq} + R_{i2,eq}$$

$R_{i2,eq}$ was found in step 4 to be 1,190 ohms. Recall that

$$R_{o1,eq} = \frac{R_{L1} R_{o1}}{R_{L1} + R_{o1}}$$

We must compute R_{o1}. In the CE connection,

$$R_{o1} = \frac{h_{ie} + R_{i1,eq}}{h_{oe}h_{ie} - h_{re}h_{fe} + h_{oe}R_{i1,eq}}$$

where all the hybrid parameters are of T_1, and $R_{i1,eq}$ is found by

$$R_{i1,eq} = \frac{1}{1/R_{B1} + 1/R_{B2} + 1/(R_{E1} + R_{i1})}$$
$$= \frac{1}{1/(130 \times 10^3) + 1/(10 \times 10^3) + 1/(570 + 1,380)}$$
$$\cong 1.64 \text{ kilohms}$$

Substituting into the R_{o1} equation and solving,

$$R_{o1} = \frac{1,400 + 1,640}{18 \times 10^{-6} \times 1.4 \times 10^3 - 0.43 \times 10^{-3} \times 45 + 18 \times 10^{-6}} \\ \times 1.64 \times 10^3$$
$$\cong 84.2 \text{ kilohms}$$

$R_{o1,eq}$ can now be found:

$$R_{o1,eq} = \frac{6.25 \times 10^3 \times 84.2 \times 10^3}{6.2K \times 10^3 + 84.2 \times 10^3}$$
$$\cong 6 \text{ kilohms}$$

And X_{C1} at 60 cps ($f_{co,min}$),

$$X_{C1} = 6 \times 10^3 + 1.190 \times 10^3 = 7.190 \text{ kilohms}$$

and $\qquad C_1 = \dfrac{1}{2\pi f_{co,min} X_{C1}} \cong 0.369 \ \mu\text{f or larger}$

Let us next determine C_2:

$$X_{C2} \text{ at } f_{co,min} = R_{o2,eq} + R_{i3,eq}$$

where $\qquad R_{i3,eq} = 2$ kilohms $\qquad$ to which T_2 output is fed
Find R_{o2}:

$$R_{o2} = \frac{h_{ie} + R_{i2,eq}}{h_{oe}h_{ie} - h_{re}h_{fe} + h_{oe}R_{i2,eq}}$$
$$\cong 95 \text{ kilohms}$$

and $\qquad R_{o2,eq} \cong \dfrac{95K \times 2K}{95K + 2K} \cong 1.95$ kilohms

therefore $\quad X_{C2}$ at $f_{co,min} = 1.95 \times 10^3 + 2 \times 10^3 = 3.95$ kilohms
Solving for C_2,

$$C_2 = \frac{1}{6.28 \times 6 \times 10^1 \times 3.95 \times 10^3}$$
$$\cong 0.663 \ \mu\text{f or larger}$$

The components required for both stages of the amplifier have been determined and are illustrated in Fig. 6.12.

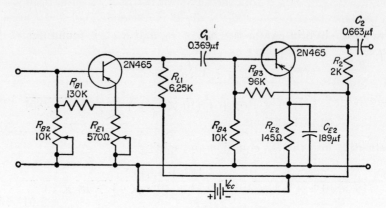

FIG. 6.12   The designed two-stage RC-coupled (CE) amplifier.

Let us briefly summarize the steps that can be taken in the design of this amplifier with some practical considerations.

1. Determination of maximum collector dissipation curve (Fig. 6.8).

2. Selection of R_{L2}.   In many practical cases, R_{L2} can be made 10 times larger than the ultimate load resistance, thereby making $R_{L2,eq}$ a value about equal to the ultimate load resistance.   This procedure saves computation time and is reasonably accurate.

3. Determination of $R_{L2,eq}$ by computation or by the approximation method stated in step 2.   Construction of the signal load line (Fig. 6.8).

4. Selection of $i_{B,Q}$, $i_{C,Q}$ and $e_{C\text{-}E,Q}$ of T_2 from the signal load line.

5. Computation of R_{E2} and C_{E2}.

6. Determination of $e_{B\text{-}E,Q}$ of T_2 (by one of the several graphical methods or by use of hybrid parameters).

7. Computation of R_{B4} and R_{B3}. R_{B4} may be selected 10 times larger than R_E; then i_{RB4} and R_{B3} are calculated.

8. Selection of $R_{L1,eq}$.   Construction of signal load line for T_1 (Fig. 6.11).

9. Computation of R_{L1} and selection of Q parameters.   The technique suggested in step 2 may be used again here.

10. Selection of R_{B2}.

11. Computation of R_{E1} (after the selection of $e_{RE1,Q}$).

12. Determination of $e_{B\text{-}E,Q}$ of T_1 by the same methods used in step 6.

13. Computation of i_{RB2} and R_{B1}.

14. Computation of C_1 and C_2.

The exact value of components is determined when the amplifier is actually constructed. $i_{C1,Q}$ is readily adjusted for the designed value by changing R_{B2} (increasing R_{B2} increases i_{C1}), and $i_{C2,Q}$ is adjusted by changing R_{B4}. For obtaining an output with the least amount of distortion, R_{E1} may be varied until the waveform across R_{L2} appears undistorted. An oscilloscope stage-by-stage analysis is recommended if the output is distorted. In this way, the faulty stage can be localized and then corrected. This amplifier has a gain of at least 40 db.

6.7 Stabistor-coupled Voltage Amplifiers.

Another three-stage preamplifier is illustrated in Fig. 6.13, designed by the Transitron Electronic

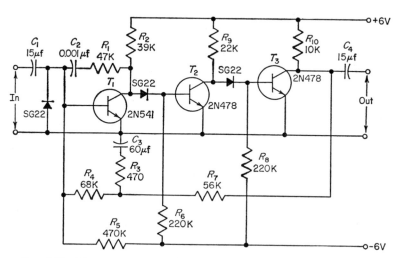

Fig. 6.13 Three-stage preamplifier (stabistor coupling). (*Transitron.*)

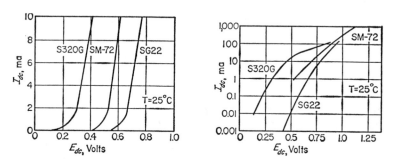

Fig. 6.14 Stabistor characteristics. (*Transitron.*)

Corporation. The interstage coupling of this amplifier differs from RC coupling in that the *interstage stabistors* (SG22) between the collector of T_1 and the base of T_2 and between the collector of T_2 and the base of T_3 eliminate the need for large coupling capacitors. The stabistor behavior is such that it offers high impedance to unidirectional current and very low impedance to bidirectional current.

Stabistors are semiconductor diodes which are especially designed for low-level voltage regulating devices. Both germanium and silicon stabistors are available. Germanium stabistors can be used at temperatures up to 85°C, while the silicon types have a temperature range which extends from −65 to 150°C. Stabistors with current ratings up to 2 amp are available.

The stabistor coupling in the three-stage preamplifier of Fig. 6.13 has the following advantages over *RC* coupling:

1. Reliability is improved with the use of fewer components.
2. Low-frequency response is down to about 0 cps.
3. The transistors are operated at a low average value, which greatly eliminates the problem of I_{co} in these transistors.

The characteristic curves of three stabistors are shown in Fig. 6.14. These stabistors have a maximum reverse potential of 6 volts, which allows for the small reverse potential to which the device may be exposed in certain applications. It is recommended that they not be used as low-voltage rectifiers.

C_3, R_3, and R_7 serve as a *decoupling filter* across the power supply. Since both R_7 and C_3 are large, low-frequency decoupling is ensured. A large value of R_7 is permissible since it is independent of all collector voltages in this circuit. *Negative feedback*, which improves stability and faithful reproduction and increases the input impedance of the amplifier, is used in this circuit. One negative feedback path exists from the collector of T_3 via R_7 and R_4 to the base of T_1. Recall that negative feedback (degeneration) occurs when the signal variations fed back to an earlier stage are inverted with respect to the input signal excursions of the earlier stage. Note that the three stages are in the common-emitter configuration; therefore each stage inverts the signal. Assume a positive signal excursion is applied to the base of T_1. It appears as a negative going variation at the collector of T_1 and base of T_2, positive at the collector of T_2 and base of T_3, and negative at the collector of T_3. Since the collector variations of T_3 are inverted with respect to the base variations of T_1, a return of a portion of these variations from T_3 to T_1 results in negative feedback.

Forward base bias is supplied to T_2 by R_6 and to T_3 by R_8. R_1 and C_2, placed between the collector and the base of T_1, prevent the amplifier from oscillating at high frequencies. The stabistor placed across the base to the emitter of T_1 is to safeguard that transistor from any excessive negative input signal.

The transistors are the NPN silicon type, whose working temperature ranges from −55 to 150°C. A 1-volt rms undistorted output voltage can be obtained from this amplifier.

6.8 Tuned Transformer-coupled Voltage Amplifiers.

Another type of voltage amplifier is one which is designed to accept a particular band of frequencies and reject all frequencies above and below this band. One variation of this type of amplifier is shown in Fig. 6.15. The input signal E_s may possess a wide range of frequencies, which are transformer-coupled to the base of the transistor. These variations are reproduced in the collector. Notice, however, that the collector load impedance is determined by a parallel LC combination (L_3, C_3). Recall that the maximum voltage appears across L and C in a tank circuit at its resonant frequency. Selection of L_3 and C_3 such that they *resonate at the desired frequency* results in voltages of that frequency being large enough to be developed in the secondary circuit L_4. At frequencies below resonance, the coil acts as a low impedance path, and C_3 serves the same purpose for

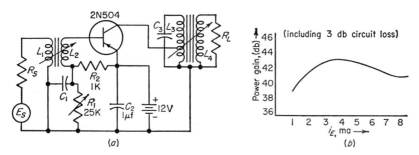

Fig. 6.15 Single-stage voltage amplifier (tuned transformer coupling). (*a*) circuit, (*b*) power gain versus i_E. (*Lansdale Tube Co.*)

the high-frequency variations. Therefore only the variations of the desired frequency are developed across R_L.

The base-bias voltage is obtained by the voltage-divider arrangement of R_1 and R_2, with e_{R2} serving as the forward base-bias voltage. The value of R_1 is varied so the desired bias is obtained. The primary coil L_3 of the output transformer is tapped down so that R_L can be properly matched to the output impedance of the transistor. In this circuit, using the 2N504, maximum power gain occurs with an emitter current of about 4 ma (see Fig. 6.15*b*).

A typical tuned transformer intermediate-frequency (i-f) amplifier for broadcast radio receivers is shown in Fig. 6.16. The circuit is a band-pass amplifier for 455 kc. The input transformer has a tuned primary, as does the output transformer. The base bias may be obtained by automatic gain control (agc) or by the voltage-divider arrangement of R_2 and R_4. Agc is a bias voltage which is made to automatically vary in accordance with the input signal strength. When agc is not used, e_{R2} is the base-bias voltage for the 2N1110.

Since the secondary of the input transformer is of very low resistance, C_2 is placed in series with it to provide a high resistance path for the unidirectional component of the base current. R_3 is an emitter-stabilizing resistor, but signal degeneration does not occur because of its bypass capacitor C_3. In this way, R_3 serves to offset any I_{CO} changes owing to transistor temperature variations, without reducing the over-all amplification of the amplifier.

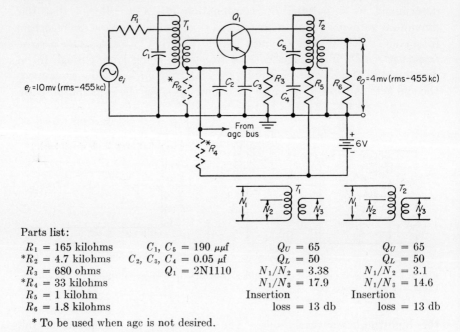

Parts list:

$R_1 = 165$ kilohms	$C_1, C_5 = 190\ \mu\mu\text{f}$	$Q_U = 65$	$Q_U = 65$
$*R_2 = 4.7$ kilohms	$C_2, C_3, C_4 = 0.05\ \mu\text{f}$	$Q_L = 50$	$Q_L = 50$
$R_3 = 680$ ohms	$Q_1 = 2\text{N}1110$	$N_1/N_2 = 3.38$	$N_1/N_2 = 3.1$
$*R_4 = 33$ kilohms		$N_1/N_3 = 17.9$	$N_1/N_3 = 14.6$
$R_5 = 1$ kilohm		Insertion	Insertion
$R_6 = 1.8$ kilohms		loss $= 13$ db	loss $= 13$ db

* To be used when agc is not desired.

FIG. 6.16 Voltage amplifier (tuned transformer coupling) for 455 kc. (*Texas Instruments, Inc.*)

The primary of the output transformer selects the desired frequency (455 kc) which is magnetically coupled to the base-emitter circuit of the 2N1110. The tuned primary of the output transformer also resonates at 455 kc, thereby developing significant voltages of only that frequency, which are felt across the secondary. Notice that the primaries of both transformers are tapped down; T_1 facilitates the proper impedance match between input signal and transistor input, and T_2 serves the same function between the transistor output and the reflected impedance. C_4 and R_5 serve as a decoupling network for the power supply. C_4 provides a low impedance path for the signal variations from the collector back to the emitter. Since the transistor is in the common-emitter configuration, the feedback signal is inverted, thereby providing negative feedback

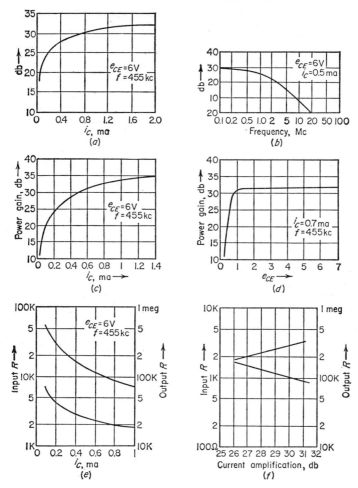

FIG. 6.17   Typical characteristics of the Fig. 6.16 i-f amplifier: (a) typical I amplification versus i_C, (b) typical I amplification versus f, (c) typical power gain versus i_C (includes 14 db i-f transformer losses), (d) typical power gain versus e_{C-E} (includes 14 db i-f transformer losses), (e) typical R_i, R_o versus i_C, (f) typical R_i, R_o versus I amplification.   (*Texas Instruments, Inc.*)

because of R_3.   This negative feedback can be increased by reducing the value of C_3 or adding an unbypassed resistor in series with the emitter lead.   Notice the output voltage is less than the input voltage, primarily because of the transformer losses.

The relationship between current amplification (in decibels) versus collector current with a fixed e_{C-E} of 6 volts and frequency of 455 kc is depicted in Fig. 6.17a.   Figure 6.17b shows that the current amplification decreases with frequency increases.   Power gain as a function of i_C, with e_{C-E} and f constant, is illustrated in Fig. 6.17c. On the other hand, with i_C and frequency constant, the power gain is relatively fixed for collector voltages greater than 1 volt (see Fig. 6.17d). The effect of collector current, with e_{C-E} and frequency fixed, upon the input and output resistance is shown in Fig. 6.17e. An increase in the collector current, which indicates a reduction in the depletion width of the two transistor junctions, results in a reduction of input and output resistance. In Fig. 6.17f, the relationship between current amplification versus input and output resistances are shown. Notice that the input resistance increases from about 2 kilohms at 26 db to 3.5 kilohms at 31 db, while the output resistance decreases from about 180 kilohms at 26 db to 90 kilohms at 31 db.

6.9 Transformer-tuned R-F **Amplifiers.** A typical tuned r-f amplifier utilized in commercial radio receivers is shown in Fig. 6.18. This

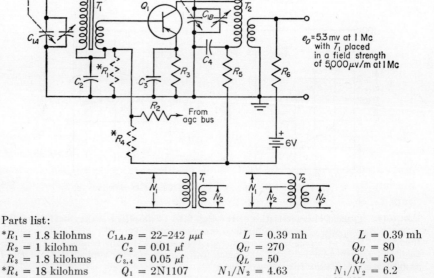

Parts list:

*R_1 = 1.8 kilohms	$C_{1A,B}$ = 22–242 $\mu\mu f$	L = 0.39 mh	L = 0.39 mh
R_2 = 1 kilohm	C_2 = 0.01 μf	Q_U = 270	Q_U = 80
R_3 = 1.8 kilohms	$C_{3,4}$ = 0.05 μf	Q_L = 50	Q_L = 50
*R_4 = 18 kilohms	Q_1 = 2N1107	N_1/N_2 = 4.63	N_1/N_2 = 6.2
R_5 = 470 ohms		K = 0.68	N_1/N_5 = 8.6
R_6 = 5 kilohms			Insertion
			loss = 8.5 db

* To be used when agc is not desired.

Fig. 6.18 Transformer-tuned r-f amplifier (525 to 1,640 kc). (*Texas Instruments, Inc.*)

amplifier differs from the i-f amplifier of Fig. 6.16 in that it is tunable over a relatively wide frequency band (535 to 1,640 kc) instead of one predetermined frequency (such as 455 kc).

The base forward voltage may be obtained from the agc system, or if desired, by use of the voltage-divider arrangement of R_1 and R_4. The forward bias voltage is e_{R1} in such cases.

R_3, placed in the emitter of the 2N1107, ensures circuit reliability even if the ambient temperature should change. The emitter bypass capacitor C_3 reduces degeneration, which would otherwise result. X_{C3} is maximum at the lowest frequency and should be 10 per cent or less of R_3 at that frequency.

$$X_{C3}(535 \text{ kc}) = \frac{1}{6.28 \times 5.35 \times 10^5 \times 5 \times 10^{-8}} = 6 \text{ ohms}$$

C_3 offers a very low impedance to the signal frequencies, thereby preventing negative feedback in the emitter circuit.

The primary portion of the input transformer is tunable from 535 to 1,640 kc by the variable air dielectric capacitor C_{1A}. At full mesh the capacitance C_{1A} is 242 $\mu\mu$f, and 22 $\mu\mu$f when completely out of mesh.

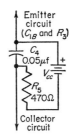

Fig. 6.19   Collector decoupling circuit.

The variable capacitor in parallel with C_{1A} is the semivariable trimmer type, which is adjusted only during alignment of the amplifier. C_2 prevents the base bias from shorting to common (ground), by blocking the unidirectional component of the base current while offering low impedance to the signal variations. The use of a ferrite core increases the coefficient of coupling, enabling weak primary signal currents to be felt in the secondary.

The primary of the output transformer is also tunable from 535 to 1,640 kc, by the variable air dielectric capacitor C_{1B}. Since this tuned circuit is to resonate to the same frequency range as the input transformer, identical component values may be used. The primary of this transformer is tapped down owing to impedance matching considerations. R_5 and C_4 serve as a *decoupling filter* across the power supply. Redrawing this arrangement, as in Fig. 6.19, clarifies how this RC combination serves this purpose.

X_{C4} is maximum at the lowest frequency (535 kc), which is about 6 ohms, resulting in the voltage variations across C_4 being very small. R_5 is effectively in series with V_{cc} and offers more opposition to the signal current than X_{C4}. Therefore the signal variations are bypassed around V_{cc}. For effective decoupling action

$$R_5C_4 > \frac{1}{f_{\min}}$$

Substituting values,

$$470 \times 5 \times 10^{-8} = 2.35 \times 10^{-5}$$

and

$$\frac{1}{f_{\min}} = \frac{1}{5.35 \times 10^5} = 0.187 \times 10^{-5}$$

$$2.35 \times 10^{-5} > 0.187 \times 10^{-5}$$

and effective decoupling is achieved.

Notice that C_{1A} and C_{1B} are "ganged," i.e., mounted on the same shaft so that they are varied together. Varying these capacitors, along with two others associated with the next amplifier stage (not shown), is the manner in which station selection is made in this type of radio receiver.

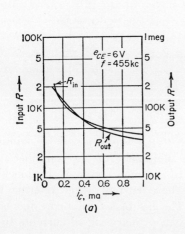

(a)

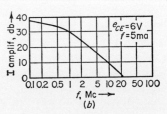

(b)

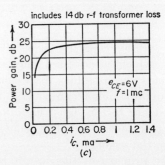

(c)

Fig. 6.20 Typical characteristics of the Fig. 6.18 r-f amplifier. (*Texas Instruments, Inc.*)

Increased amplification could be achieved by reducing the value of R_5, which would increase the collector voltage. Such an arrangement is not desirable in many cases, since greater amplification reduces the *signal noise ratio* of the amplifier.

The typical characteristics of the transformer-tuned amplifier of Fig. 6.18 are shown in Fig. 6.20. With a constant-collector voltage and frequency, both input and output resistances decrease with increases in collector current (see Fig. 6.20a). Since both these resistances in a given transistor hinge directly on the depletion width of the two junctions, the resistances are smaller with greater collector-current values because the depletion regions are more narrow at that time.   Figure 6.20*b* depicts the manner in which current amplification decreases with frequency. There is a sharp rise in power gain for increases of collector current when the collector current is less than 0.5 ma.   Beyond this point, the power gain is relatively independent of changes in collector current.

6.10   Single-ended Power Amplifiers.   In order to achieve faithful reproduction, the single-ended power amplifier must be operated in class *A*.   The full signal efficiency is determined by the type of interstage coupling:

$$\text{Direct coupling} = 25 \text{ per cent}$$
$$RC \text{ coupling} = 17 \text{ per cent}$$
$$\text{Transformer coupling} = 50 \text{ per cent}$$

Bias techniques are very important in the design of power amplifiers. Because of the power dissipated by the transistor, substantial temperature increases may readily occur in normal operation.   The temperature rise can result in an increase in I_{co} (which doubles with each rise of 10°C in many cases) and a reduction of input resistance, especially when proper safeguards are not taken.   In discussing the stability factor in Chap. 5, it was found that certain types of bias were more effective safeguards against thermal runaway than others.   The use of a resistance in the emitter lead, or the voltage-divider arrangement used in Fig. 6.18, results in low stability factors.   Many circuits utilize the combination of both these bias techniques.   Recall that the stability factor of these bias techniques was analyzed in Chap. 5.

A second method in which thermal runaway can be prevented is by use of *thermistors*.   Thermistors possess positive temperature coefficients of resistance.   Coupling the thermistor to the collector in such a way that it remains at the same temperature as the transistor is an effective method of obtaining temperature-compensated bias.   In this way, an increased collector junction temperature due to high dissipation causes the thermistor resistance to increase.   In any of the bias techniques discussed in the preceding chapter, the use of thermistors instead of resistors would result in temperature-compensated bias networks.

The use of a *heat sink* is a third method for the prevention of thermal runaway.   Like any electrical component, a transistor can dissipate heat, by radiation and convection, to an extent which is determined by its size and shape.   Many manufacturers state the maximum dissipation of power transistors without and also with a specified heat sink.   The maximum power dissipation rating without a heat sink is the heat the transistor can dissipate by means of convection and radiation. Placing the transistor in thermal contact with a heat sink, such as a metal chassis, greatly increases the maximum power dissipation of the transistor.

Figure 6.21 illustrates three types of heat sink mountings. Clamping the transistor to the chassis by any one of the three techniques ensures the easy passage of heat from the transistor to the chassis. The chassis

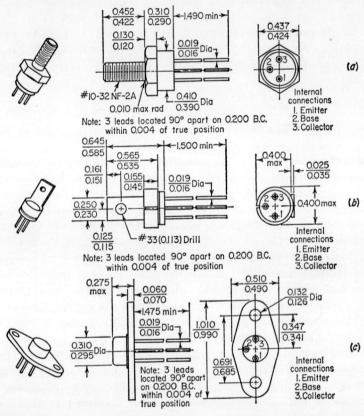

FIG. 6.21 Types of heat-sink mountings: (a) stud, (b) clip, (c) flange. (*Transitron*).

then loses its acquired heat by convection and radiation. If the chassis is sufficiently large, relatively great amounts of heat can be rapidly dissipated. With this type of connection between the transistor and the chassis, the transistor may experience only a 2°C increase in temperature per watt of dissipation. The temperature rise per watt of power dissipation, called the thermal drop or derating, can be determined from the following:

$$\text{Derating} = \frac{\text{junction temperature} - \text{heat sink temperature}}{\text{transistor power rating}}$$

Example: Junction temperature = 100°C; heat sink temperature = 25°C; dissipation is 10 watts. Find the derating.

Solution:

$$\text{Derating} = \frac{100°C - 25°C}{10} = 7.5°C/\text{watt}$$

This means the collector junction temperature of this transistor (obviously silicon) rises at the rate of 7.5°C per watt increase of power dissipation. When dissipating 14 watts, for example, the junction temperature is 100°C + 4(7.5°C) = 130°C.

Some power transistors are also equipped with fins, to facilitate their ability to dissipate heat to the surrounding air. This technique enables the transistor to dissipate a greater amount of heat without being connected to a heat sink.

6.11 A Class *A* Audio Driver Amplifier. The power supply is connected in the emitter circuit, but its polarity is reversed, resulting in no over-all change in its effect upon the transistor. R_1 and R_2 serve as the voltage-divider arrangement across V_{cc} to obtain forward base bias.

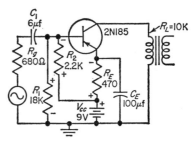

Fig. 6.22 A class *A* audio driver. (*Texas Instruments, Inc.*)

R_E, as in previous amplifiers, provides temperature stability for the amplifier, and C_E greatly eliminates negative feedback. C_1 couples the input variations to the base, while blocking any unidirectional input signal component away from the base. The reflected impedance of the output transformer for this circuit should be 10 kilohms for a power output of 2 mw and a power gain of 40 db. This amplifier, like any driver amplifier, is designed to build the signal power up to a predetermined level, which is delivered to a succeeding power amplifier, such as the push-pull type (discussed in Chap. 7). The input impedance of such amplifiers is generally low, and the transformer turns ratio must be such that this low input impedance is "seen" as 10 kilohms by the driver transistor of Fig. 6.22.

6.12 A 1.5-watt Class *A* Power Amplifier. Figure 6.23 illustrates a class *A* audio power amplifier designed by Texas Instruments Incorporated. The 2N250 transistor is an alloy junction PNP germanium

power transistor with a maximum power dissipation of 25 watts.   Other
maximum values are:

$$e_{C\text{-}B} = -30 \text{ volts}$$
$$i_C = -3 \text{ amp}$$
$$\text{Junction temperature limit} = 85°C$$

At 25°C, the current ratio varies from a minimum of 30 to a maximum
which can exceed 90.

R_E is connected in the emitter circuit to compensate for temperature
variations of the transistor and for differences between transistors of the
same type. R_{B1} and R_{B2} form a voltage-divider arrangement across V_{cc}
to provide forward bias for the base.   Varying R_{B1} alters the preceding
resistance relationship, which changes the forward base voltage.   In
this way, the bias voltage can be varied to the desired value.

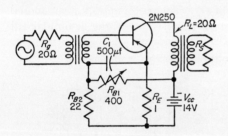

FIG. 6.23   1.5-watt class A power
amplifier.   (*Texas Instruments, Inc.*)

In the amplifier of Fig. 6.23, the zero signal collector current is 0.55
amp.   For all practical purposes, the emitter current is approximately
the same value.   The reverse voltage developed across R_E can be accurately estimated:

$$e_{RE} \cong I_C R_E \cong 0.55 \times 1 \cong 0.55 \text{ volt}$$

Thus e_{RB2} must be greater than 0.55 volt in order to have a forward base-
bias voltage.

The input impedance is 20 ohms at the operating point.   The power
gain, when the input is properly matched, varies from a minimum of 31 db
to a maximum of 40 db.   Frequency cutoff occurs at 8 kc (minimum)
to 12 kc (maximum).   The output transformer primary resistance
should be less than 1.5 ohms.   The transformation ratio of the output
transformer should be such that the reflected primary impedance R_L is
20 ohms.

6.13   A High-frequency Power Amplifier.

Figure 6.24 illustrates a
200-Mc (vhf) amplifier with a minimum power gain of 8 db.   The
Philco 2N502 is a germanium microalloy diffused-base transistor with
polarities similar to PNP junction transistors.   Several absolute maxi-

mum ratings of this transistor are:

$$\text{Junction temperature} = 85°C$$
$$\text{Collector-to-base voltage} = -20 \text{ volts}$$
$$\text{Emitter-to-base voltage} = -0.5 \text{ volt}$$
$$\text{Collector dissipation at } 45°C = 25 \text{ mw}$$

C_1 and C_7 are variable capacitors at the input and output of the amplifier circuit. C_1 is adjusted to match the source and input conditions, while C_7 is adjusted to match the load and output conditions. C_3 provides negative feedback from the output back to the transistor input. This technique is often preferred at high frequencies over the series-RC method, illustrated in some of the previous amplifiers, because of the losses associated with series-RC networks.

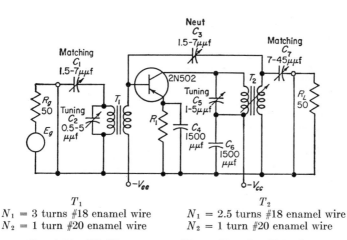

$$T_1 \qquad\qquad\qquad\qquad T_2$$
$$N_1 = 3 \text{ turns } \#18 \text{ enamel wire} \qquad N_1 = 2.5 \text{ turns } \#18 \text{ enamel wire}$$
$$N_2 = 1 \text{ turn } \#20 \text{ enamel wire} \qquad N_2 = 1 \text{ turn } \#20 \text{ enamel wire}$$

Fig. 6.24   200-Mc power amplifier.   (*Lansdale Tube Co.*)

Note that both coupling transformers are the tuned primary, untuned secondary type. The primary of the input transformer can be tuned by C_2 while the output transformer primary is tuned by C_5. The movable cores can be used as a further tuning adjustment for both transformers.

R_1 provides current stabilization against variations in transistor temperature. C_4, the emitter bypass capacitor, effectively reduces degeneration in the emitter circuit. C_6 decouples the signal away from V_{cc}. Some forward bias for the emitter-base junction is required for class A operation. In Fig. 6.24, forward bias is applied to the emitter by the constant current source V_{ee}. The voltage-divider arrangement, discussed in many of the previous amplifiers, could be used instead of V_{ee}. In such an arrangement, the bottom of R_1 would be connected to common. The larger of the two voltage-divider resistors (called R_{B1} in many of the previous amplifiers) would be connected between the negative terminal of V_{cc} and the base, while the smaller resistor R_{B2} would be inserted from base to common. The values of the two resistors would be selected such that e_{RB2} provided the required forward base bias.

Example: Assume the amplifier of Fig. 6.24 is to have the following parameters: zero signal emitter current = 2 ma; zero signal base current = 200 μa; forward base bias = 0.25 volt; V_{cc} = 12 volts. Find R_{B1} and R_{B2}. All calculations are made at zero input signal conditions.

Solution:

(*a*) Determine e_{R1}:
R_1 provides a reverse bias to the emitter, which can be determined by

$$e_{R1} = I_E R_1 = 2 \times 10^{-3} \times 1.5 \times 10^3 = -3 \text{ volts}$$

(*b*) Determine e_{RB2}:
The forward base voltage is determined by

$$e_{B\text{-}E} = e_{RB2} + e_{R1} = -0.25 \text{ volt}$$

But e_{RB2} and e_{R1} are series-opposing, since e_{RB2} is a forward bias and e_{R1} is a reverse bias (see Fig. 6.25).

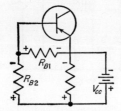

FIG. 6.25 Voltage-divider bias arrangement for Fig. 6.24.

Therefore, e_{RB2} must exceed e_{R1} by the desired base forward bias, which is 0.25 volt. In equation form,

$$e_{RB2} = e_{R1} + \text{desired forward base bias}$$
$$= 3 + 0.25 = 3.25 \text{ volts}$$

(*c*) Determine i_{RB2} and i_{RB1}:
$$i_{RB1} = i_B + i_{RB2}$$

i_B is given as 200 μa. i_{RB2} is to be selected so as not to impose too severe a drain upon V_{cc}. R_{B2} should be large enough to avoid reducing the input resistance of the transistor to too great an extent. Recall that R_{B1} and R_{B2}, in terms of the input signal, are in parallel with each other, and their equivalent value is in parallel with the input impedance of the transistor. The input impedance of this transistor is considerably less than 1 kilohm at this frequency, so R_{B2} and R_{B1} should be of such values that their parallel equivalent is 10 times greater than 1 kilohm. Let us set i_{RB2} = 200 μa. Solving for i_{RB1}:

$$i_{RB1} = i_B + i_{RB2} = 200 \ \mu\text{a} + 200 \ \mu\text{a} = 400 \ \mu\text{a}$$

Note that this imposes a very small drain upon V_{cc}.

(d) Determine R_{B2} and R_{B1}:

$$R_{B2} = \frac{e_{RB2}}{i_{RB2}} = \frac{3.25}{200 \times 10^{-6}} = 16.25 \text{ kilohms}$$

and
$$R_{B1} = \frac{e_{RB1}}{i_{RB1}} = \frac{8.75}{400 \times 10^{-6}} = 21.88 \text{ kilohms}$$

The equivalent parallel value of these two resistors is about 9 kilohms. When placed in the circuit, it results in no appreciable change in the input resistance of the circuit.

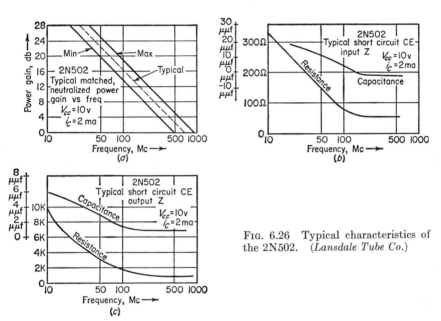

FIG. 6.26 Typical characteristics of the 2N502. (*Lansdale Tube Co.*)

Due to individual differences between transistors of the same type, the final component values are determined when the circuit is actually constructed.

Figure 6.26a illustrates the relationship between the power gain in decibels as a function of frequency of the 2N502 in an amplifier such as Fig. 6.24. The effect of frequency upon the input capacitance and resistance of the transistor in the common-emitter configuration is shown in Fig. 6.26b. Notice that the input resistance levels off to about 60 ohms at 200 Mc. The input capacitance decreases to about zero at 170 Mc, after which the input appears slightly inductive (negative capacitance on the graph is actually inductance). The output resistance also decreases with frequency (see Fig. 6.26c) until 300 Mc, where it levels off to about 1 kilohm. The output capacitance of the transistor is lower than the input capacitance for all frequencies up to about 150 Mc, at which time the input becomes inductive while the output capacitance levels off to about 1 $\mu\mu$f. The characteristics of Fig. 6.26 are for an amplifier like Fig. 6.24.

6.14 The Common-base Amplifier. Figure 6.27 illustrates a transformer-coupled common-base amplifier. R_E may not be necessary in many cases, since the resistance of the secondary coil of the input transformer will serve the same purpose. R_E is added only if the resistance of the secondary coil is very low. In either case, stabilization against I_{CO} variations due to changes in transistor temperature is provided. The voltage-divider bias technique, used for many of the previous common-emitter amplifiers, is also effective for amplifiers in the common-base configuration with e_{RB2} as the forward base bias.

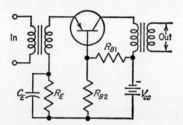

Fig. 6.27 Transformer-coupled CB amplifier.

There is a great similarity between the output characteristics of the CE and CB connections. For the CE configuration, i_C is plotted as a function of e_{C-E}, whereas i_C is a function of e_{C-B} for the CB connection. e_{C-E} and e_{C-B} are nearly the same in magnitude because e_{B-E} is customarily small as compared to e_{C-B}. Because of this similarity, *the common-emitter output characteristics, along with the load line, power output, dissipation and efficiency relationships, can also be used in designing a common-base amplifier (class A).* The results, although slightly inaccurate, are sufficiently correct for many practical cases. The input resistance is determined by

$$R_i \cong \frac{\Delta e_I}{\Delta i_I} = \frac{\Delta e_{E-B}}{\Delta i_E} = \frac{h_{ib} + (h_{ob}h_{ib} - h_{fb}h_{rb})R_L}{1 + h_{ob}R_L}$$

since $i_E = i_B + i_C$

then

$$R_i \cong \frac{\Delta e_{E-B}}{\Delta(i_B + i_C)}$$

Both i_B and i_C changes can be obtained from the common-emitter output characteristics when the load line has been constructed. The input transfer characteristics (e_{B-E} versus i_C) can be used for determining Δe_{E-B}, since

$e_{E-B} = e_{B-E}.$ Or e_{B-E} can be found by

$$e_{B-E} = i_E R_i = \frac{i_E[h_{ib} + (h_{ob}h_{ib} - h_{fb}h_{rb})R_L]}{1 + h_{ob}R_L}$$

Example 1: Refer to Fig. 6.28.   Assume the operating point is at the intersection of the 40 μa characteristic and the load line AB. $\Delta i_B = 10$ μa above and below

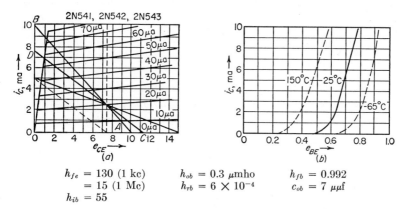

$$
\begin{aligned}
h_{fe} &= 130 \text{ (1 kc)} & h_{ob} &= 0.3 \text{ } \mu\text{mho} & h_{fb} &= 0.992 \\
&= 15 \text{ (1 Mc)} & h_{rb} &= 6 \times 10^{-4} & C_{ob} &= 7 \text{ } \mu\mu\text{f} \\
h_{ib} &= 55
\end{aligned}
$$

FIG. 6.28   Characteristics of the 2N541, 2N542, 2N543.   (*Transitron.*)

the operating point.   Find R_i for the common-base connection, by use of the characteristics.

Solution:

(a) Find Δi_C from the zero signal operating point to $i_{C,\min}$:
Reading the values from Fig. 6.28a,

$$i_{C,Q} = 5.2 \text{ ma} \qquad i_{C,\min} = 3.8 \text{ ma}$$
and
$$\Delta i_C = 5.2 - 3.9 = 1.3 \text{ ma}$$
(b) Find Δi_E:

$$\Delta i_E = \Delta(i_C + i_B) = 1.3 + 0.01 = 1.31 \text{ ma}$$

(c) Find Δe_{E-B} from the input transfer characteristics of Fig. 6.28b.   Using the 25°C characteristic, the input voltage is found to be 0.7 volt at $i_{C,Q} = 5.2$ ma. When i_C is 3.9 ma, the input voltage is 0.66 volt.   Hence

$$\Delta e_{B-E} \cong 0.70 - 0.66 = 0.04 \text{ volt} \cong \Delta e_{E-B}$$

(d) We may now solve for R_i:

$$R_i \cong \frac{4 \times 10^{-2}}{1.31 \times 10^{-3}} \cong 30.5 \text{ ohms}$$

As pointed out in Chap. 4, the same technique can be used to determine the input resistance of a transistor in the common-emitter configuration.

Example 2: Using the load line and parameters of example 1, find R_i for the common-emitter connection, by use of the characteristics.

Solution:

$$R_i \cong \frac{\Delta e_{B\text{-}E}}{\Delta i_B}$$
$$\Delta e_{B\text{-}E} \cong 0.04 \text{ volt}$$
$$\Delta i_B = 10 \ \mu a$$

NOTE: These values were graphically determined in example 1. Substituting values,

$$R_i = \frac{4 \times 10^{-2}}{1 \times 10^{-5}}$$
$$= 4 \text{ kilohms}$$

When the input transfer characteristics are not available, the change of input voltage for a given change of input current may be determined by Ohm's law, if the voltage-divider bias technique is used. Referring to Fig. 6.27,

$$e_{E\text{-}B} = e_{RB2} + e_{RE}$$

The common-base amplifier has a current ratio of less than unity $(\Delta i_C / \Delta i_E)$, resulting in only medium power gain. The input resistance is lower than the common-emitter connection (see examples 1 and 2).

6.15 Design of a Class A Common-base Amplifier. Like the common-emitter amplifier, hybrid parameters can be utilized in the design of the common-base amplifier. Let us design the amplifier illustrated in Fig. 6.29. Assume the input resistance "seen" by the primary of the input

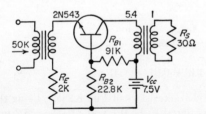

FIG. 6.29 Designed *CB* amplifier.

transformer is 50 kilohms, and the "ultimate load," designated as R_s, is 30 ohms. The output of the 2N543 (NPN-silicon) is to "see" 1.5 kilohms. $V_{cc} = 7.5$ volts; $e_{RE,Q} = 0.5$ volt. The signal load line (constructed in the manner analyzed in Sec. 6.5) is shown in Fig. 6.28a, and $i_{B,Q} = 20 \ \mu a$. As previously stated, the *CE* collector characteristics may be used with reasonable accuracy.

Step 1: Determination of $e_{E\text{-}B,Q}$. Prior to computing the value of R_{B2}, the required forward bias $e_{E\text{-}B,Q}$ must be found. Let us utilize chart (b) of Fig. 6.28.

Finding $i_{C,Q} = 2.5$ ma on the y axis and using the 25°C characteristic:

$$e_{E\text{-}E,Q} \cong 0.64 \text{ volt}$$

Finding R_i:

$$R_i = \frac{h_{ib} + (h_{ob}h_{ib} - h_{fb}h_{rb})R_L}{1 + h_{ob}R_L}$$

$$= \frac{55 + (0.3 \times 10^{-6} \times 55 - (-0.992)(6 \times 10^{-4})1.5 \times 10^3}{1 + 6.3 \times 10^{-6} \times 1.5 \times 10^3)}$$

$$\cong 54 \text{ ohms}$$

Step 2: Determination of R_E. $e_{RE,Q}$ is to be 0.5 volt, and $i_E = 2.52$ ma. Solving,

$$R_E = \frac{e_{RE,Q}}{i_{E,Q}} = \frac{5 \times 10^{-1}}{2.52 \times 10^{-3}}$$

$$\cong 2 \text{ kilohms}$$

Step 3: Determination of R_{B2} and R_{B1}. e_{RB2} is the required forward bias and must exceed $e_{RE,Q}$ by the desired $e_{E\text{-}B,Q}$.

$$e_{RB2,Q} = 0.64 + 0.5 = 1.14 \text{ volt}$$

The bleeder current i_{RB2} is arbitrarily selected. Let $i_{RB2} = 50$ μa. Now R_{B2} may be found:

$$R_{B2} = \frac{e_{RB2,Q}}{i_{RB2,Q}} = \frac{1.14}{50 \times 10^{-6}}$$

$$\cong 22.8 \text{ kilohms}$$

R_{B1} can be readily determined:

$$R_{B1} = \frac{e_{RB1,Q}}{i_{RB1,Q}} = \frac{V_{cc} - e_{RB2,Q}}{i_{B,Q} + i_{\text{bleeder}}}$$

$$= \frac{7.5 - 1.14}{(20 + 50) \times 10^{-6}} \cong 91 \text{ kilohms}$$

Step 4: Determination of the Input Transformer Turns Ratio. The secondary of the input transformer is "loaded down" by $R_{i,\text{eq}}$. Let us find $R_{i,\text{eq}}$; referring to Fig. 6.29,

$$R_{i,\text{eq}} = \frac{1}{1/R_{B1} + 1/(R_{B2} + R_i) + 1/R_E}$$

$$= \frac{1}{1/(2 \times 10^5) + 1/(22,750 + 54) + 1/(2 \times 10^3)}$$

$$\cong 1,820 \text{ ohms}$$

NOTE: The resistance of the secondary winding should be considered as a part of R_E. The actual value of R_E is decreased accordingly.

Recall that the reflected primary resistance R_p is found by the following relationship in a 100 per cent efficient transformer:

$$R_p = \left(\frac{N_p}{N_s}\right)^2 R_s$$

R_s is $R_{i,eq}$ in this case, and R_p is 50 kilohms. Transposing and solving for the required turns ratio,

$$\frac{N_p}{N_s} = \left(\frac{R_p}{R_s}\right)^{\frac{1}{2}} = \left(\frac{5 \times 10^4}{1.82 \times 10^3}\right)^{\frac{1}{2}}$$
$$= \frac{5.25}{1}$$

Step 5: Determination of the Output Transformer Turns Ratio. R_s is given as 30 ohms, and R_p is 1.5 kilohms; solving,

$$\frac{N_p}{N_s} = \left(\frac{1.5 \times 10^3}{3 \times 10^1}\right)^{\frac{1}{2}}$$
$$= \frac{7.07}{1}$$

The exact values of the components are determined by practical considerations. Increasing R_{B2} increases all current values, and a reduction in R_{B2} produces the opposite effect. Increasing R_E reduces all current values. Final adjustment of the values of R_E and R_{B2} is best made when the circuit is constructed.

6.16 The Common-collector Amplifier. Figure 6.30 illustrates the transformer-coupled *CC* amplifier. Notice that degeneration is provided to the emitter by the ohmic value of the primary of the output transformer.

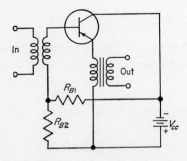

Fig. 6.30 Transformer-coupled *CC* amplifier.

Forward base bias may be obtained by the voltage-divider arrangement (R_{B1} and R_{B2}). The common-collector amplifier has a voltage ratio of less than unity, high current gain, low power gain (because of the low voltage ratio), high input resistance, and low output resistance. The input voltage is generally high, since it is slightly greater than the output voltage. Since the input resistance is high, a relatively small input current will produce the required input voltage.

The base current is the input current, as in the CE connection. i_C and i_E are rather close in value in most cases. The input voltage $e_{B\text{-}C}$ is nearly identical to $e_{C\text{-}E}$. In many practical cases, $e_{C\text{-}E}$ can be taken as the input voltage of the amplifier. In such cases, the common-emitter output characteristics can be used as the approximate input characteristics of the common-collector stage:

$$R_i \cong \frac{\Delta e_{C\text{-}E}}{\Delta i_B} \qquad \text{or} \qquad R_i = \frac{h_{ic} + (h_{oc}h_{ic} - h_{fc}h_{rc})R_L}{1 + h_{oc}R_L}$$

Also, in those cases where $i_E \cong i_C$, the common-emitter relationships for power output, dissipation, and efficiency are reasonably accurate for the common-collector configuration.

Example: Refer to Fig. 6.29a. Using the intersection of the load line AB with the 40 μa characteristic as the input operating point, find R_i of the transistor in the common-collector connection with an i_B variation of 10 μa.
Solution: Reading from the characteristics:

$$R_1 = \frac{6 - 4.7}{10 \times 10^{-6}} = 130 \text{ kilohms}$$

6.17   Design of a Class A Common-collector Amplifier.

Refer to Fig. 6.5 for the CE collector characteristics for the 2N333 (NPN-silicon), which may be used as the input characteristics for the amplifier of Fig. 6.31, since $e_{C\text{-}E} \cong e_{C\text{-}B}$ and $i_C \cong i_E$. Let $R_L = 500$ ohms, $i_{B,Q} = 200$ μa, and $i_{E,Q} = 6$ ma.

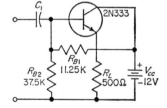

FIG. 6.31   Designed CC amplifier.

Step 1: *Determination of* $e_{B\text{-}C,Q}$ *(Zero Signal Input Voltage).* Let us select $i_{RL,Q}$ as 6 ma. $i_{RL,Q}$ is $i_{E,Q}$, which is approximately equal to $i_{C,Q}$. Locate $i_C = 6$ ma in Fig. 6.5b, project this point horizontally until it intersects the 200-μa characteristic, then read the corresponding $e_{C\text{-}E}$ value on the x axis. This value is about 4.5 volts. Since $e_{C\text{-}E} \cong e_{B\text{-}C}$, 4.5 volts can be chosen as $e_{B\text{-}C,Q}$. For class A work, a maximum input signal of 4.5 volts may be used.

Step 2: Determination of R_{B2} and R_{B1}. $e_{RB2,Q}$ is to furnish the forward bias to the base-collector junction and should be greater than $e_{RL,Q}$ by the desired value of $e_{B\text{-}C,Q}$; therefore

$$e_{RB2,Q} = e_{B\text{-}C,Q} + e_{RL,Q}$$

where
$$e_{RL,Q} = i_{E,Q}R_L = 6 \times 10^{-3} \times 5 \times 10^2$$
$$\cong 3 \text{ volts}$$

Solving for $e_{RB2,Q}$,

$$e_{RB2,Q} = 4.5 + 3 = 7.5 \text{ volts}$$

Let i_{RB2} (the bleeder current) = 200 μa. Then

$$R_{B2} = \frac{e_{RB2,Q}}{i_{RB2,Q}} = \frac{7.5}{2 \times 10^{-4}} \cong 37.5 \text{ kilohms}$$

and R_{B1} is now found:

$$R_{B1} = \frac{V_{cc} - e_{RB2,Q}}{i_{B,Q} + i_{\text{bleeder}}} = \frac{12 - 7.5}{(200 + 200) \times 10^{-6}}$$
$$\cong 11.25 \text{ kilohms}$$

Note that this amplifier was designed with only the *CE* collector characteristics of the transistor. The exact value of R_{B2} is best selected when the circuit is actually constructed.

6.18 The Split-load Amplifier. Figure 6.32 illustrates the class *A* split-load amplifier. Notice that a portion of the output is taken from the collector while the remainder of the output is taken from the emitter.

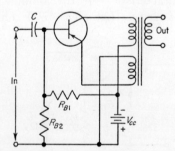

Fig. 6.32 The split-load amplifier.

This amplifier is, in effect, a combination of the common-emitter and common-collector configurations. The emitter portion of the output transformer primary provides negative feedback. The degeneration can be increased by placing a greater portion of the primary in the emitter circuit. Because of the common-collector characteristic of taking a portion of the output off of the emitter, the input resistance is substantially higher than would be the case in the straightforward common-emitter connection. On the other hand, the common-emitter characteristic of taking a portion of the output off of the collector increases the output resistance to a higher value than in the ordinary common-collector configuration. The emitter, because a portion of the output transformer primary is located there, can be used as a take-off point for applying negative feedback to a preceding stage. The volt-ampere characteristics and relationships of the common-emitter amplifier can be utilized in analyzing this amplifier.

6.19 Push-Pull Amplifiers. There are many instances when the required input signal to an amplifier is too large for a single transistor to faithfully reproduce it. Figure 6.33 illustrates the effect of such a situation. The collector current variations from rest to maximum are greater than the variations from rest to minimum. The waveforms of i_C and e_C

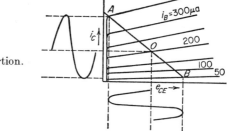

Fig. 6.33   Harmonic distortion.

are distorted as a result. Since i_C and e_{RL} have identical waveforms, e_{RL} also has one of its peaks flattened by this action. Such a waveshape, since it resembles the composite waveform of a fundamental and a 90° out-of-phase second harmonic, is often called *second harmonic distortion*. This problem can be easily resolved by using two transistors for handling the large signal, called a push-pull amplifier.

Figure 6.34 illustrates a conventional class *A* push-pull amplifier. Two reasonably well-matched PNP transistors are shown, although two matched NPN transistors could be used by merely reversing V_{cc}. The positive swing of the input signal is felt as a negative pulse at the base of T_1 (owing to the inverting action of the transformer), driving that transistor into heavy conduction. At the same time, the collector current of T_2 is reduced, since this portion of the input signal is felt as a positive pulse at the base of T_2. The reverse action occurs when the input signal swings positive. In this way, the characteristics of T_1 need only accommodate the negative portion of the input signal, while T_2 reproduces the positive half in a faithful manner. In class *A*, both transistors are always conducting; but T_1 conducts heavily while i_C of

T_2 is relatively small, and vice versa. Each transistor distorts in the manner shown in Fig. 6.33. If the transistors are well matched, T_1 distorts on the positive input signal swing by the same amount that T_2 distorts on the negative input signal excursion. Equal amounts of distortion during each portion of the input signal result in a comparatively undistorted output, as shown in Fig. 6.35.

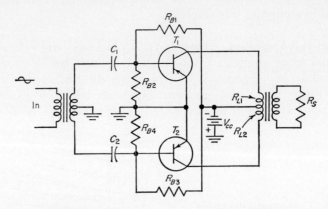

FIG. 6.34   Conventional push-pull amplifier.

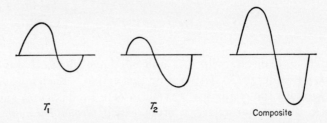

FIG. 6.35   Signal collector currents in a class A push-pull amplifier.

Assuming the two transistors are well matched, the operating point of each is located on the same point of their respective output characteristics. The center-tapped secondary of the input coupling transformer inverts the input signal to T_2, as compared to T_1. The load resistance of each transistor should be equal, and is the resistance reflected back to each half of the output transformer primary, which is also center-tapped. Each transistor has half of the output transformer primary turns across it, ensuring the same reflected resistance back to each transistor output. R_{B2} and R_{B4} of Fig. 6.34, which provide forward base bias, should also be equal in value in order to maintain the matched relationship. C_1 and C_2 are also the same value. A common-emitter resistance (not shown) may be incorporated in the push-pull amplifier of

Fig. 6.34, if provisions for temperature compensation are necessary. In Fig. 6.34,

$$R_{L1} = R_{L2} = \frac{1}{4}\left(\frac{N_p}{N_s}\right)^2 R_s$$

From the preceding relationship the turns ratio of the output transformer required for proper matching can be readily found.

The load-line analysis of a class A push-pull amplifier may be approached in several ways. One technique is illustrated in Fig. 6.36. Only the portion of each family of characteristics concerned with the heavy conduction region is considered in this type of analysis. The slope of the load line, which is the reciprocal of each R_L, and the operating point are determined in the usual manner. Two sets of characteristics

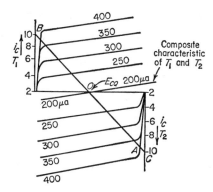

FIG. 6.36 Load-line analysis of a class A push-pull amplifier.

are used, with the heavy conduction portion of the second family placed under the second set in the manner shown in Fig. 6.36. The load line and operating point are drawn on these characteristics. If i_C of T_1, during its heavy conduction half cycle, varies from 2 to 6 ma, then T_2 performs in an identical fashion during the next half cycle of the input signal.

A second load-line technique is to analyze push-pull operation from a single set of characteristics, such as Fig. 6.33. The load line and operating point are obtained in the usual manner. This type of analysis is sufficient in many cases.

The use of a push-pull class A amplifier, as compared to a single transistor, has a number of advantages:

1. Larger signal handling capability while still maintaining faithful reproduction

2. Larger power output

3. Reduction of certain types of distortion, particularly the "second harmonic" type discussed in this section

The disadvantages of this type of amplifier are:

1. Cost of an additional transistor along with input and output transformers

2. Matched transistors

The problem of selecting well-matched transistors is no longer as severe as was the case in the past. Many manufacturers can provide matched transistors with a minimum of difficulty. Matched transistors in one compact unit are becoming commonplace. Furthermore, unmatched transistors can be balanced by biasing. The cost of transformers, like that of transistors, is steadily decreasing as manufacturing techniques are improved.

6.20 Design of a Class A Push-Pull Amplifier. Let us consider the approach which may be used in the design of a class A push-pull amplifier similar to the one illustrated in Fig. 6.34.

Step 1: Determination of R_{L1} and R_{L2}. The type of load-line analysis shown in Fig. 6.36 may be used to determine suitable values of R_{L1}, R_{L2}, and zero signal parameters. However, the construction of a load line upon a single family of collector characteristics (see Fig. 6.33) is permissible. The desired R_{L1} and R_{L2} values can be found by use of either type of load-line construction. Precautions should be taken to be sure the maximum power dissipation curve of the transistor is not intersected.

Step 2: Determination of the Output Transformer Turns Ratio. Notice that the primary of the output transformer is center-tapped. One of the chief purposes of this transformer is to properly match each transistor to the "ultimate" load resistance R_s. In other words, each transistor should "see" the R_L determined in step 1; i.e., R_s should look like R_L to each transistor. By the selection of the proper turns ratio, this requirement can be met.

$$R_{L1} = R_{L2} = \frac{1}{4}\left(\frac{N_p}{N_s}\right)^2 R_s$$

Transposing,

$$\frac{N_p}{N_s} = \left(\frac{R_{L1}}{0.25R_s}\right)^{1/2}$$

and

$$\frac{N_p}{N_s} = \left(\frac{R_{L2}}{0.25R_s}\right)^{1/2}$$

since $R_{L1} = R_{L2}$.

Step 3: Determination of $e_{B\text{-}E,Q}$. $e_{B\text{-}E,Q}$ is the same for each transistor. Since the CE connection is used, then

$$e_{B\text{-}E,Q} \cong R_i i_{B,Q}$$

where R_i may be determined by either of the following methods:

$$R_i \cong \frac{\Delta e_{B\text{-}E,Q}}{\Delta i_B}$$

or

$$R_i = \frac{h_{ie} + (h_{oe}h_{ie} - h_{fe}h_{re})R_L}{1 + h_{oe}R_L}$$

where $R_L = R_{L1} = R_{L2}$.

NOTE: As in previous amplifiers, $e_{B\text{-}E,Q}$ may be determined by use of the i_C versus $e_{B\text{-}E}$ or i_B versus $e_{B\text{-}E}$ characteristic in conjunction with the collector characteristics.

Step 4: Determination of the bias resistors. The voltage-divider type of bias is shown in Fig. 6.34, and the values of the resistors are computed in the usual manner. Since the transistors are considered matched, and when there is no R_E,

$$R_{B2} = R_{B4} = \frac{e_{RB2,Q}}{i_{RB2,Q}} = \frac{e_{B\text{-}E,Q}}{i_{\text{bleeder}}}$$

where i_{bleeder} is arbitrarily selected and $e_{B\text{-}E,Q}$ is determined in step 3, and

$$R_{B1} = R_{B3} = \frac{V_{cc} - e_{B\text{-}E,Q}}{i_{\text{bleeder}} + i_{B,Q}}$$

where $i_{B,Q}$ and i_{bleeder} are of one transistor only. If R_E is included in the circuit, then

$$R_{B2} = R_{B4} = \frac{e_{B\text{-}E,Q} + e_{RE,Q}}{i_{\text{bleeder}}}$$

and

$$R_{B1} = R_{B3} = \frac{V_{cc} - (e_{B\text{-}E,Q} + e_{RE,Q})}{i_{\text{bleeder}} + i_{B,Q}}$$

Step 5: Determination of the Input Transformer Turns Ratio. One of the chief functions of the input transformer is to ensure a proper match between the output of the driver stage and the input of each push-pull transistor:

$$R_{i,\text{eq}} = \frac{R_{L(d)}}{0.25(N_p/N_s)^2}$$

Transposing,

$$\frac{N_p}{N_s} = 2\left(\frac{R_{L(d)}}{R_{i,\text{eq}}}\right)^{1/2}$$

where $R_{L(d)}$ = load resistance of the driver stage
$R_{i,\text{eq}}$ = equivalent input resistance of each push-pull transistor circuit

When the voltage-divider type of bias and matched transistors are used,

$$R_{i1,eq} = R_{i2,eq} = \frac{1}{1/R_{B1} + 1/R_{B2} + 1/(R_E + R_{i1})}$$

R_E does not appear in the preceding equation if it is bypassed or not included in the actual circuit.

Step 6: Determination of C_1 and C_2. C_1 and C_2 prevent the secondary of the input transformer from shunting the d-c input circuit of the two push-pull transistors. They should present negligible impedance to the lowest signal frequency.

At $f_{co,\min}$,

$$X_{C1} = X_{C2} = R_{i1,eq}$$

and
$$C_1 = C_2 = \frac{1}{2\pi f_{co,\min} X_{C1(f_{co,\min})}}$$

6.21 Phase Inverters. In many cases, it is desirable to furnish a signal to a double-ended input of a push-pull amplifier without the use of an

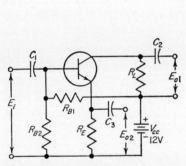

Fig. 6.37 Split-load phase inverter.

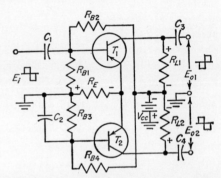

Fig. 6.38 The common-emitter phase inverter.

input coupling transformer. Recall that the input of the push-pull transistors must be equal in amplitude but opposite in phase, if most faithful reproduction is desired. The use of a phase-inverter circuit can serve this function.

The Split-load Phase Inverter. One version of a phase-inverter circuit, called the split-load phase inverter, is illustrated in Fig. 6.37. The double-ended output of this amplifier, when properly balanced, has $E_{o1} = -E_{o2}$. R_{B1} and R_{B2} form the voltage-divider arrangement by which forward bias is applied to the base and establish the zero signal operating point of the circuit. Since the collector current is slightly less than the emitter current (which is the sum of collector and base currents), R_L must be slightly larger than R_E in order to make $E_{o1} = -E_{o2}$. The magnitude of the zero signal base current determines the difference between the R_E and R_L values. Notice that this circuit connection is intermediate between the CE and CC configurations.

The Common-emitter Phase Inverter. Figure 6.38 illustrates a second type of phase-inverter circuit which has found popular use. Two transistors of the same type are used (either PNP or NPN), with a resistor common to both emitters (R_E in Fig. 6.38). R_E provides reverse bias to both transistors. Consider the input portion of the circuit when E_i undergoes its positive variation. Being a PNP type, a positive pulse to the base of T_1 results in a decrease in its emitter current, which passes through R_E. This results in a reduction of e_{RE}, thereby decreasing the reverse emitter bias of T_2. As a result, the emitter current of T_2 increases with the same variation that i_E of T_1 decreases, which effectively inverts the input signal to T_2. The opposite effect is produced by the negative pulse of E_i. T_1 is driven to heavier conduction, producing an increase of emitter reverse bias at the emitter of T_1, causing i_E of T_1 to decrease in the same manner that i_E of T_1 increased.

For the proper operation of the CE phase inverter, the zero signal emitter current of T_1 should be slightly larger than the zero signal emitter current of T_2. The gain of T_1 is low, while T_2 gain is high. The correct selection of T_2 gain/T_1 gain will result in E_{o1} and E_i, E_{o2} and $-E_i$ being identical in waveshape. The load-line analysis and the determination of the zero signal operation point are done by use of the techniques discussed in preceding sections. R_{B1} and R_{B2} are selected to ensure the correct forward bias at the base of T_1; R_{B3} and R_{B4} serve the same purpose for T_2. The values of R_{L1} and R_{L2} hinge on the magnitude of i_{C1} and i_{C2}, since $e_{RL1} = e_{RL2}$ for correct operation. All coupling capacitors (C_1, C_2, C_3, and C_4) must be sufficiently large in order not to cause any significant signal reduction at the low end of the signal frequency band.

6.22 Design of a Split-load Phase Inverter.

Let us design a split-load phase inverter, as shown in Fig. 6.37. Assume the input is fed from a transistor whose $R_{o,eq} = 2$ kilohms. Frequency range is 60 cps to 10 kc. The output is fed to an ultimate load of 1 kilohm.

Step 1: *Determination of the Load Line.* The load line is determined by the sum of R_E and R_L, since R_E is unbypassed. Let $R_E + R_L = 1,200$ ohms; $V_{cc} = 12$ volts. The 2N333 transistor is to be used. Hence the load line is identical to that of Fig. 6.5b.

Step 2: *Determination of R_E and R_L.* The value of R_L and R_E is determined by the current passing through each.

$$e_{RE} = i_{RE}R_E$$

and

$$e_{RL} = i_{RL}R_L$$

Using the operating point of Fig. 6.5b:

$$i_{RL,Q} = i_{C,Q} = 6 \text{ ma}$$

and

$$i_{RE,Q} = i_{C,Q} + i_{B,Q} = 6 + 0.2 = 6.2 \text{ ma}$$

and

$$\frac{R_L}{R_E} = \frac{i_{RE,Q}}{i_{RL,Q}} = \frac{6.2}{6} = \frac{1.033}{1}$$

The sum of R_L and R_E is 1.2 kilohms. Let $R_E = x$ ohms, and $R_L = 1.033x$ ohms; then

$$x + 1.033x = 1,200$$
$$2.033x = 1,200$$
$$x = 590 \text{ ohms} = R_E$$

and

$$R_L = 610 \text{ ohms}$$

NOTE: Make R_E a potentiometer. Upon circuit construction, R_E is adjusted so that $e_{RE,Q} = e_{RL,Q}$.

Step 3: Determination of $e_{B\text{-}E,Q}$. $e_{B\text{-}E,Q}$ can be determined by use of the i_C versus $e_{B\text{-}E}$ characteristics when they are available, as stated previously. These characteristics for the 2N333 are shown in Fig. 6.5b. Since they are not readily available in many cases, let us determine $e_{B\text{-}E,Q}$ from the following relationship:

$$e_{B\text{-}E,Q} \cong R_i i_{B,Q}$$

where R_i is to be computed with the hybrid parameters in this example. The typical h_b and h_{fe} values are given in Fig. 6.5. Let us convert them to h_e values:

$$h_{ie} = \frac{h_{ib}}{1 + h_{fb}} = \frac{60}{1 + (-0.966)} \cong 1,765 \text{ ohms}$$

$$h_{re} = \frac{h_{ib}h_{ob}}{1 + h_{fe}} - h_{rb} = \frac{60 \times 0.4 \times 10^{-6}}{1 + 28} - 3 \times 10^{-4} \cong 3 \times 10^{-4}$$

$$h_{oe} = \frac{h_{ob}}{1 + h_{fb}} = \frac{0.4 \times 10^{-6}}{1 + (-0.966)} \cong 11.8 \times 10^{-6} \text{ mho}$$

$$h_{fe} = 28$$

Now solving for R_i,

$$R_i = \frac{h_{ie} + (h_{oe}h_{ie} - h_{fe}h_{re})R_L}{1 + h_{oe}R_L}$$

where R_L is actually $R_L + R_E = 1.2$ kilohms. Substituting values and solving,

$$R_i \cong 1,740 \text{ ohms}$$

Knowing R_i and $i_{B,Q}$, we may now compute $e_{B\text{-}E,Q}$:

$$e_{B\text{-}E,Q} \cong 1.74 \times 10^3 \times 2 \times 10^{-4}$$
$$\cong 0.35 \text{ volt}$$

Step 4: *Determination of* R_{B1} *and* R_{B2}.   Recall that $e_{RB2,Q}$ is the forward base-bias voltage and must exceed $e_{RE,Q}$ by the desired value of $e_{B\text{-}E,Q}$; i.e.,

$$e_{RB2,Q} = e_{RE,Q} + e_{B\text{-}E,Q}$$

where
$$e_{RE,Q} \cong i_{E,Q}R_E \cong (i_{C,Q} + i_{B,Q})R_E$$
$$\cong 6.2 \times 10^{-3} \times 5.9 \times 10^2$$
$$\cong 3.66 \text{ volts}$$

and
$$e_{RB2,Q} \cong 3.66 + 0.35 \cong 4.0 \text{ volts}$$

Let $i_{RB2,Q}$ (bleeder current) $= 100 \ \mu a$.   Find R_{B2}:

$$R_{B2} = \frac{e_{RB2}}{i_{RB2,Q}} = \frac{4.0}{1 \times 10^{-4}} \cong 40 \text{ kilohms}$$

R_{B1} can now be determined:

$$R_{B1} = \frac{V_{cc} - e_{RB2}}{i_{B,Q} + i_{\text{bleeder}}} = \frac{12 - 4.0}{(100 + 200) \times 10^{-6}}$$
$$\cong 27 \text{ kilohms}$$

Step 5: *Determination of Coupling Capacitors* C_1, C_2, *and* C_3.   Recall that the reactance of a coupling capacitor should equal the sum of the output resistance of the preceding stage (or generator) and the input resistance of the following stage (or generator) at $f_{co,\min}$.   The input is coupled from a transistor whose $R_{o,eq} = 2$ kilohms. $R_{i,eq}$ of the split-load phase inverter must be determined:

$$R_{i,eq} \cong \frac{1}{1/R_{B1} + 1/R_{B2} + 1/(R_E + R_i)}$$
$$\cong 2.04 \text{ kilohms}$$

Therefore,
$$X_{C1} \text{ at } f_{co,\min} = 2.04\text{K} + 2\text{K} = 4.04 \text{ kilohms}$$

$f_{co,\min}$ is 60 cps.

and
$$C_1 = \frac{1}{6.28 \times 60 \times 4.04 \times 10^3}$$
$$\cong 0.65 \ \mu\text{f or larger}$$

X_{C2} at $f_{co,\min}$ should be equal to the sum of R_E and the ultimate load of 1 kilohm; i.e.,

$$X_{C2} \text{ at } f_{co,\min} = 590 + 1\text{K} \cong 1.59 \text{ kilohms}$$

and
$$C_2 \cong \frac{1}{6.28 \times 60 \times 1.59 \times 10^3}$$
$$\cong 1.67 \ \mu\text{f or larger}$$

Since $R_L \cong R_E$ and the ultimate load connected to C_3 is equal to the ultimate load associated with C_2, then C_3 should be the same value as

C_2; i.e.,

$$C_3 \cong 1.67 \ \mu\text{f or larger}$$

The exact values of all components are determined by practical considerations upon construction of the circuit. E_{o1} and E_{o2} may be precisely matched by changing R_E; increasing R_E increases E_{o2} and decreases E_{o1}. If E_{o1} is to be increased, R_E should be reduced. E_{o1} and E_{o2} undergo opposite changes in magnitude with variations in R_E. The exact value of R_E is best determined by use of two voltmeters, one across R_E and the second across R_L. When the voltages are identical, the correct R_E value has been found.

PROBLEMS

6.1   Refer to Fig. 6.1. What is the type of bias network used in this amplifier? State the resistors that form the bias network and the function of each.

6.2   Refer to Fig. 6.1. State the function of C_E, R_E, and R_3.

6.3   Refer to Fig. 6.3. State the type of bias network and the component resistors for stages 2 and 3.

6.4   Refer to the class A amplifier of Fig. 6.5. The 2N333 is to be used. $V_{cc} = 15$ volts; $R_L = 1.5$ kilohms.

    (a) Construct the load line.

    (b) Select an operating point which results in minimum distortion of a base current variation of 100 μa(pp).

6.5   In Prob. 6.4, calculate (a) the h_e parameters from the given h_b values; (b) R_i; (c) $e_{B\text{-}E,Q}$.

6.6   In Prob. 6.4, let $e_{RE,Q} = 1$ volt, calculate:

    (a) R_E (b) C_E

    (c) R_{B1} (d) R_{B1}

    (e) $R_{i,\text{eq}}$

6.7   In Prob. 6.4, assume $R_{o,\text{eq}}$ of the preceding stage = 3 kilohms, and the frequency range of the amplifier is 20 cps to 15 kc. Find (a) X_{C1} at $f_{co,\text{min}}$; (b) C_1.

6.8   In the class A amplifier designed in Probs. 6.4 to 6.7, calculate:

    (a) R_o (b) $R_{o,\text{eq}}$

    (c) A_v (d) A_i

    (e) G

6.9   Refer to the class A amplifier of Fig. 6.6. $V_{cc} = 9$ volts. The 2N223 is to be used (see Fig. 5.25a for the characteristics and hybrid parameters). $R_L = 250$ ohms; the "ultimate" load resistance R_s is 50 ohms.

    (a) Construct the signal load line.

    (b) Select an operating point which results in minimum distortion of a base current variation of 200 μa(pp).

6.10   For the amplifier of Prob. 6.9, calculate:

    (a) output transformer turns ratio

    (b) the h_e parameters from the given h_b values

    (c) R_i (d) $e_{B\text{-}E,Q}$

6.11   For the amplifier of Probs. 6.9 and 6.10, let $e_{RE,Q} = 0.5$ volt. Calculate:

    (a) R_E (b) C_E

    (c) R_{B2} (d) R_{B1}

    (e) $R_{i,\text{eq}}$

6.12 In the amplifier of Probs. 6.9 to 6.11, assume $R_{o,eq}$ of the preceding stage = 1.4 kilohms, and the frequency range of the amplifier is 60 cps to 15 kc. Find (a) X_{C1} at $f_{co,\min}$; (b) C_1.

6.13 In the class A amplifier designed in Probs. 6.9 to 6.12, find:

(a) R_o (b) $R_{o,eq}$

(c) A_v (d) A_i

(e) G

6.14 Refer to the class A amplifier of Fig. 6.5a. The 2N207 is to be used (see Fig. 5.20 for collector characteristics and hybrid parameters). V_{cc} = 10 volts; R_L = 500 ohms.

(a) Construct the load line.

(b) Select an operating point which results in minimum distortion of a base current variation of 40 μa(pp).

6.15 For the amplifier of Prob. 6.14, calculate (a) the h_e parameters from the given h_b parameters; (b) R_i; (c) $e_{B\text{-}E,Q}$.

6.16 For the amplifier of Probs. 6.14 and 6.15, let $e_{RE,Q}$ = 0.6 volt. Calculate:

(a) R_E (b) C_E

(c) R_{B2} (d) R_{B1}

(e) $R_{i,eq}$

6.17 In the amplifier of Probs. 6.14 to 6.16, assume $R_{o,eq}$ of the preceding stage = 3.8 kilohms, and the frequency range of the amplifier is 40 cps to 20 kc. Find (a) X_{C1} at $f_{co,\min}$; (b) C_1.

6.18 Find the following for the amplifier designed in Probs. 6.14 to 6.17:

(a) R_o (b) $R_{o,eq}$

(c) A_v (d) A_i

(e) G

6.19 Refer to Fig. 6.12. The 2N223 transistor is to be used for both stages. V_{cc} = 12 volts; $f_{co,\min}$ = 40 cps. Refer to Fig. 5.25b for the collector characteristics and hybrid parameters of the 2N223. $R_{L2,eq}$, as seen by T_2, is to be equal to 1.2 kilohms; the ultimate load resistance of stage 2 is 2 kilohms. For stage 2:

(a) Compute R_{L2}.

(b) Construct the zero signal load line.

(c) Construct the signal load line.

6.20 For stage 2 of the amplifier in Prob. 6.19, select $i_{B,Q}$ = 80 μa; $e_{RE2,Q}$ = 1.5 volts. Find:

(a) $i_{C2,Q}$ (b) $e_{C2,Q}$

(c) R_{E2} (d) C_{E2}

6.21 Referring to stage 2 of the amplifier in Probs. 6.19 and 6.20, calculate:

(a) h_e parameters from the given h_b values   (b) R_{i2}

(c) $e_{B\text{-}E,Q}$ of stage 2   (d) R_{B4} (let $i_{\text{bleeder}} = i_{B2,Q}$)

(e) R_{B3} (f) $R_{i2,eq}$

6.22 $R_{L1,eq}$ of the amplifier in Probs. 6.19 to 6.21 is to be 500 ohms.

(a) Find R_{L1}.

(b) Construct the zero signal load line.

(c) Construct the signal load line.

6.23 Referring to stage 1 considered in Prob. 6.22, let $i_{B1,Q}$ = 40 μa; $e_{RE1,Q}$ = 0.6 volt. Find (a) $i_{C1,Q}$; (b) $e_{C1,Q}$; (c) R_{E1}.

6.24 Find the following for stage 1 (see Probs. 6.22 and 6.23):

(a) R_{i1} (b) $e_{B\text{-}E,Q}$

(c) R_{B2} (d) R_{B1}

(e) $R_{i1,eq}$

6.25 Find the following for the two-stage RC-coupled CE amplifier designed in Probs. 6.19 to 6.24:

(a) R_{o1} (b) $R_{o1,eq}$

(c) X_{C1} at $f_{co,min}$ (d) C_1

6.26 Find the following for the amplifier considered in Prob. 6.25:

(a) R_{o2} (b) $R_{o2,eq}$

(c) X_{C2} at $f_{co,min}$ (d) C_2

6.27 Refer to Fig. 5.15. The 2N465 transistor is to be used for both stages. Refer to Fig. 6.11 for the collector characteristics and hybrid parameters for the 2N465. $V_{cc} = 22$ volts; $f_{co,min} = 60$ cps; $R_{L2,eq}$, as seen by T_2, is to be 2 kilohms; the ultimate load resistance is 10 kilohms. For stage 2:

(a) Compute R_{L2}.

(b) Construct the zero signal load line.

(c) Construct the signal load line.

6.28 For stage 2 of the amplifier in Prob. 6.27, select $i_{B2,Q} = 90$ μa; $e_{RE2,Q} = 0.5$ volt. Find:

(a) $i_{C2,Q}$ (b) $e_{C2,Q}$

(c) R_{E2} (d) C_{E2}

6.29 Referring to stage 2 of the amplifier in Probs. 6.25 and 6.28, calculate:

(a) R_{i2} (b) $e_{B-E,Q}$

(c) R_{B4} (let $i_{bleeder} = 2i_{B2,Q}$)   (d) R_{B3}

(e) $R_{i2,eq}$

6.30 $R_{L1,eq}$ of the amplifier in Probs. 6.27 to 6.29 is to be 1.8 kilohms.

(a) Compute R_{L1}.

(b) Construct the zero signal load line.

(c) Construct the signal load line.

6.31 Referring to stage 1 considered in Prob. 6.30, let $i_{B1,Q} = 60$ μa; $e_{RE1,Q} = 1$ volt. Find (a) $i_{C1,Q}$; (b) $e_{C1,Q}$; (c) R_{E1}.

6.32 Find the following for stage 1 (see Probs. 6.30 and 6.31):

(a) R_{i1} (b) $e_{B-E,Q}$

(c) R_{B2} (let $i_{bleeder} = 100$ μa)   (d) R_{B1}

(e) $R_{i1,eq}$

6.33 Find the following for the two-stage RC-coupled CE amplifier designed in Probs. 6.27 to 6.32:

(a) R_{o1} (b) $R_{o1,eq}$

(c) X_{C1} at $f_{co,min}$ (d) C_1

6.34 Find the following for the amplifier considered in Prob. 6.33:

(a) R_{o2} (b) $R_{o2,eq}$

(c) X_{C2} at $f_{co,min}$ (d) C_2

6.35 Refer to Fig. 6.16. Assume the base-bias voltage is obtained by use of R_2 and R_4. Which resistor provides the forward base voltage? Explain.

6.36 Refer to Fig. 6.16. What is the reactance of C_3 at the (a) lowest signal frequency; (b) highest signal frequency?

6.37 What is the purpose of C_3 in Fig. 6.16?

6.38 Refer to Fig. 6.18. R_5 and C_4 are to bypass all signal variations around V_{cc}. C_4 is predetermined at the stated value. Find R_5 for the following signal frequency ranges:

(a) 88 to 108 mc

(b) 25 to 750 mc

6.39 Refer to Fig. 6.39. Draw the maximum collector dissipation curve for (a) 25°C (ambient); (b) 150°C (ambient).

6.40 In Prob. 6.39, why are the two maximum collector dissipation curves not alike?

6.41 What type of bias system is used in the class A audio driver of Fig. 6.22? List the components used in this bias network and the purpose of each.

6.42 Refer to Fig. 6.23. Explain how the forward base bias is varied by changes in R_{B1}.

6.43 Refer to Fig. 6.29. The 2N475 (NPN-silicon) is to be used (see Fig. 6.39 for the characteristics and hybrid parameters). $V_{cc} = 15$ volts; the input resistance "seen" by the primary of the input transformer is 5 kilohms; the ultimate load R_s is 10 ohms. The transistor output is to "see" 2 kilohms. Let $i_{B,Q} = 150$ μa. Construct the signal load line.

6.44 In the amplifier considered in Prob. 6.43, find (a) $i_{C,Q}$; (b) $e_{C,Q}$.

6.45 Refer to the amplifier of Probs. 6.43 and 6.44. Compute (a) R_i; (b) $e_{B-E,Q}$; (c) R_E.

6.46 In Prob. 6.45, what C_E value would serve as a suitable R_E bypass if the signal frequency range is 60 cps to 15 kc?

6.47 In the amplifier of Probs. 6.43 to 6.45, calculate (a) R_{B2} (let $i_{\text{bleeder}} = 200$ μa); (b) R_{B1}.

6.48 Using the parameters determined in Probs. 6.43 to 6.47, compute (a) $R_{i,\text{eq}}$; (b) input transformer turns ratio; (c) output transformer turns ratio.

6.49 Refer to Fig. 6.29. The 2N465 is to be used (refer to Fig. 6.11 for the collector characteristics and hybrid parameters). The input resistance "seen" by the primary of the input transformer is 2.5 kilohms; and the ultimate load resistance R_s is 75 ohms. The transistor output is to "see" 500 ohms; $V_{cc} = 6$ volts; let $i_{B,Q} = 90$ μa. Construct the signal load line.

6.50 In the amplifier considered in Prob. 6.49, find (a) $i_{C,Q}$; (b) $e_{C,Q}$.

6.51 Refer to the amplifier of Probs. 6.49 and 6.50. Find (a) R_i; (b) $e_{B-E,Q}$; (c) R_E.

6.52 In Prob. 6.51, what C_E value would serve as a suitable R_E bypass if the signal frequency range is 60 cps to 15 kc?

6.53 In the amplifier of Probs. 6.49 to 6.52, calculate (a) R_{B2} (let $i_{\text{bleeder}} = i_{B,Q}$); (b) R_{B1}.

6.54 Using the parameters determined in Probs. 6.49 to 6.53, compute (a) $R_{i,\text{eq}}$; (b) input transformer turns ratio; (c) output transformer turns ratio.

6.55 Why can the common-emitter characteristics be used in the design of a common-base amplifier?

6.56 Refer to Fig. 6.31. The 2N475 (NPN-silicon) is to be used. $R_L = 500$ ohms; $i_{B,Q} = 250$ μa; $i_{E,Q} = 7.5$ ma. See Fig. 6.39 for the common-emitter characteristics of the 2N475. Using the CE characteristics, find $e_{B-C,Q}$.

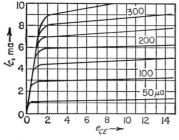
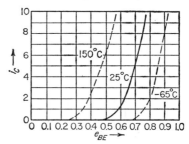

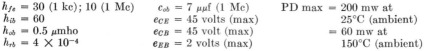

$h_{fe} = 30$ (1 kc); 10 (1 Mc) $c_{ob} = 7$ $\mu\mu$f (1 Mc)   PD max $= 200$ mw at
$h_{ib} = 60$ $e_{CE} = 45$ volts (max)   25°C (ambient)
$h_{ob} = 0.5$ μmho $e_{CB} = 45$ volt (max) $= 60$ mw at
$h_{rb} = 4 \times 10^{-4}$ $e_{EB} = 2$ volts (max)   150°C (ambient)

Fig. 6.39   2N475 characteristics.   (*Transitron.*)

6.57 In the CC amplifier of Prob. 6.56, find (a) $e_{RL,Q}$; (b) R_{B2}; (c) R_{B1}.

6.58 Refer to Fig. 6.31. The 2N541 (NPN-silicon) is to be used. $R_L = 400$ ohms; $i_{B,Q} = 40$ μa; $i_{E,Q} = 5.5$ ma. See Fig. 6.28 for the common-emitter characteristics of the 2N541. Using the CE characteristics, find $e_{B-C,Q}$.

6.59 In the CC amplifier of Prob. 6.58, find (a) $e_{RL,Q}$; (b) R_{B2}; (c) R_{B1}.

6.60 Why can the common-emitter characteristics be used in the design of a common-collector amplifier?

6.61 Refer to Fig. 6.34. The class A push-pull amplifier is to have $R_{L1} = R_{L2} = 33$ ohms, where R_{L1} and R_{L2} are the load resistances "seen" by the transistors. The ultimate load resistance $R_s = 4$ ohms; $V_{cc} = 12$ volts. The 2N1117 (NPN-silicon) is to be used (see Fig. 6.40 for the characteristics).
(a) Using a single set of collector characteristics, construct the signal load line.
(b) Let $i_{B,Q} = 4$ ma; find $i_{C,Q}$ and $e_{C,Q}$.

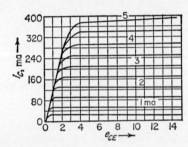
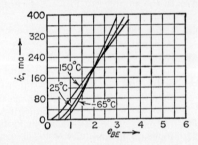

$h_{fe} = 65$

Absolute Ratings

$e_{CE} = 60$ volts	PD $= 5$ watts at 100°C (case)
$e_{CB} = 60$ volts	$= 0.5$ watts at 200°C (case)
$e_{EB} = 10$ volts	$= 0.6$ watts at 100°C (ambient)
	$= 0.05$ watts at 200°C (ambient)

FIG. 6.40   Characteristics of the 2N1117.   (*Transitron.*)

6.62 In the push-pull amplifier of Prob. 6.61, determine the required output transformer turns ratio.

6.63 In the push-pull amplifier of Probs. 6.61 and 6.62, find:
(a) $e_{B-E,Q}$ (by use of the given characteristics)   (b) R_E (let $e_{RE,Q} = 0.5$ volt)
(c) R_{B2} (let $i_{bleeder} = i_{B1,Q}$)   (d) R_{B4}
(e) R_{B1} (f) R_{B3}

6.64 In the push-pull amplifier of Probs. 6.61 to 6.63, let $R_{L,eq}$ of the driver stage $= 1$ kilohm. Find:
(a) R_i of each push-pull transistor   (b) $R_{i,eq}$ of each push-pull transistor
(c) input transformer turns ratio   (d) C_1 and C_2

6.65 Refer to Fig. 6.34. The class A push-pull amplifier is to have $R_{L1} = R_{L2} = 15$ ohms, where R_{L1} and R_{L2} are the load resistances "seen" by the transistors. The ultimate load resistance $R_s = 3$ ohms; $V_{cc} = 12$ volts; the 2N548 (NPN-silicon) is to be used (see Fig. 6.41 for the characteristics).
(a) Using a single set of collector characteristics, construct the signal load line.
(b) Let $i_{B,Q} = 4$ ma; find $i_{C,Q}$ and $e_{C,Q}$.

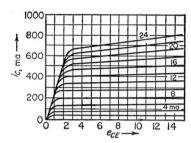
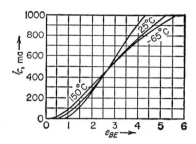

$h_{fe} = 35$

Absolute Maximum Values

$e_{CE} = 30$ volts	PD = 5 watts at 100°C (case)
$e_{CB} = 30$ volts	= 0.5 watts at 200°C (case)
$e_{EB} = 10$ volts	= 0.6 watts at 100°C (ambient)
	= 0.05 watts at 200°C (ambient)

FIG. 6.41 Characteristics of the 2N548. (*Transitron.*)

6.66 In the push-pull amplifier of Prob. 6.65, determine the required output transformer turns ratio.

6.67 In the push-pull amplifier of Probs. 6.65 and 6.66, find:
(*a*) $e_{B\text{-}E,Q}$ (by use of the given characteristics) (*b*) R_E (let $e_{RE,Q} = 0.5$ volt)
(*c*) R_{B2} (let $i_{\text{bleeder}} = 1$ ma) (*d*) R_{B4}
(*e*) R_{B1} (*f*) R_{B3}

6.68 In the push-pull amplifier of Probs. 6.65 to 6.67, let $R_{L,\text{eq}}$ of the driver stage = 500 ohms. Find:
(*a*) R_i of each push-pull transistor (*b*) $R_{i,\text{eq}}$ of each push-pull transistor
(*c*) input transformer turns ratio (*d*) C_1 and C_2

6.69 Refer to Fig. 6.34. The class A push-pull amplifier is to have $R_{L1} = R_{L2} = 200$ ohms, where R_{L1} and R_{L2} are the load resistances "seen" by the transistors. The ultimate load $R_s = 10$ ohms; $V_{cc} = 20$ volts. The 2N43 (PNP) is to be used (see Fig. 6.42 for the characteristics, hybrid parameters, and total transistor power dissipation).
(*a*) Compute the maximum power dissipation of the 2N43 at 25°C with an infinite heat sink.
(*b*) Construct the maximum power dissipation curve for (*a*).
(*c*) Using a single set of collector characteristics, construct the signal load line.
(*d*) Let $i_{B,Q} = 1$ ma; find $i_{C,Q}$ and $e_{C,Q}$.

6.70 In the push-pull amplifier of Prob. 6.69, determine the required output transformer turns ratio.

6.71 In the push-pull amplifier of Probs. 6.69 and 6.70, calculate:
(*a*) h_e parameters from the given h_b values (*b*) R_i of each transistor
(*c*) $e_{B\text{-}E,Q}$ (*d*) R_E (let $e_{RE,Q} = 0.5$ volt)

6.72 In the push-pull amplifier of Prob. 6.71, compute:
(*a*) R_{B2} (*b*) R_{B4}
(*c*) R_{B1} (*d*) R_{B3}
(*e*) $R_{i,\text{eq}}$

6.73 Referring to Probs. 6.69 to 6.72, assume $R_{L,\text{eq}}$ of the driver stage = 1.2 kilohms $F_{i,\text{min}} = 60$ cps; find (*a*) input transformer turns ratio; (*b*) C_1 and C_2.

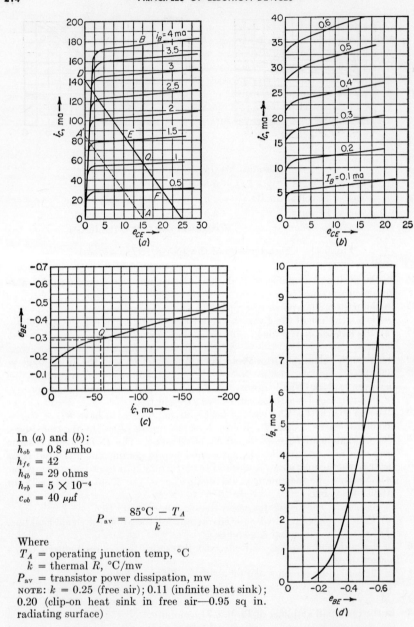

In (a) and (b):
$h_{ob} = 0.8$ μmho
$h_{fe} = 42$
$h_{ib} = 29$ ohms
$h_{rb} = 5 \times 10^{-4}$
$c_{ob} = 40$ μμf

$$P_{av} = \frac{85°C - T_A}{k}$$

Where
T_A = operating junction temp, °C
k = thermal R, °C/mw
P_{av} = transistor power dissipation, mw
NOTE: $k = 0.25$ (free air); 0.11 (infinite heat sink);
0.20 (clip-on heat sink in free air—0.95 sq in.
radiating surface)

FIG. 6.42   Characteristics of the 2N43.   (*General Electric Co.*)

6.74 Refer to the collector characteristics of the 2N43 in Fig. 6.42. Compute the maximum transistor power dissipation when (*a*) free air is used for heat dissipation; (*b*) clip-on heat sink in free air (0.95 sq in. radiating surface) is used.

6.75 Construct the maximum dissipation curves for Prob. 6.74*a* and *b*.

6.76 Compare the maximum transistor dissipation curves constructed in Probs. 6.69*a* and 6.74*a* and *b*. Explain why they are different.

6.77 Refer to the split-load phase inverter of Fig. 6.37. Let $R_E + R_L = 2$ kilohms; $V_{cc} = 15$ volts. The 2N475 (NPN-silicon) is to be used (refer to Fig. 6.39 for the characteristics and hybrid parameters).
(*a*) Construct the load line.
(*b*) Let $i_{B,Q} = 100$ μa; find $i_{C,Q}$ and $e_{C,Q}$.

6.78 In the amplifier of Prob. 6.77, find (*a*) R_E; (*b*) R_L.

6.79 In the amplifier of Probs. 6.77 and 6.78, determine (*a*) the h_e parameters from the given h_b values; (*b*) R_i; (*c*) $e_{B\text{-}E,Q}$.

6.80 Refer to the split-load phase inverter of Probs. 6.77 to 6.79. Find (*a*) R_{B2} (let $i_{\text{bleeder}} = 2i_{B,Q}$); (*b*) R_{B1}; (*c*) $R_{i,\text{eq}}$.

6.81 In Prob. 6.80, assume $f_{co,\min} = 20$ cps and the transistor $R_{o,\text{eq}}$ ahead of this amplifier $= 1.5$ kilohms; ultimate R_L for each output $= 500$ ohms. Find (*a*) C_1; (*b*) C_2; (*c*) C_3.

6.82 Refer to the split-load phase inverter of Fig. 6.37. Let $R_E + R_L = 1$ kilohm; $V_{cc} = 10$ volts. The 2N541 (NPN-silicon) is to be used (refer to Fig. 6.28 for the characteristics and hybrid parameters).
(*a*) Construct the load line.
(*b*) Let $i_{B,Q} = 40$ μa; find $i_{C,Q}$ and $e_{C,Q}$.

6.83 In the amplifier of Prob. 6.82, find (*a*) R_E; (*b*) R_L.

6.84 In the amplifier of Probs. 6.82 and 6.83, determine (*a*) the h_e parameters from the given h_b values; (*b*) R_i; (*c*) $e_{B\text{-}E,Q}$.

6.85 Refer to the split-load phase inverter of Probs. 6.82 to 6.84. Find (*a*) R_{B2} (let $i_{\text{bleeder}} = i_{B,Q}$); (*b*) R_{B1}; (*c*) $R_{i,\text{eq}}$.

6.86 In Prob. 6.85, assume $f_{co,\min} = 60$ cps, and the transistor $R_{o,\text{eq}}$ ahead of this amplifier $= 2$ kilohms. Find (*a*) C_1; (*b*) C_2; (*c*) C_3.

6.87 In Prob. 6.85, let $i_{\text{bleeder}} = 240$ μa. Find (*a*) R_{B2}; (*b*) R_{B1}; (*c*) $R_{i,\text{eq}}$.

6.88 Refer to Probs. 6.85 and 6.87. What effect did the change in bleeder current have on $R_{i,\text{eq}}$? Explain.

6.89 In Prob. 6.86, let $f_{co,\min} = 20$ cps and the transistor $R_{o,\text{eq}}$ ahead of the amplifier $= 2$ kilohms. Find (*a*) C_1; (*b*) C_2; (*c*) C_3.

6.90 What relationship exists between the C values and $f_{co,\min}$? Explain.

CHAPTER 7

NONLINEAR TRANSISTOR AMPLIFIERS AND SINUSOIDAL OSCILLATORS

Nonlinear operation indicates the transistor is experiencing large signal operation, which alternately drives the transistor far into the heavy conduction and cutoff regions. In nonlinear work, faithful reproduction of the input signal by a single transistor is not possible. Power is one of the chief considerations in this type of circuit. Faithful reproduction can be achieved by use of push-pull circuits, as studied in a following section, with the development of large output power as well.

The class of operation is determined by the portion of the input signal at which the transistor is at cutoff or zero collector signal current condition. The three nonlinear classes of operation (AB, B, and C) are analyzed in this chapter.

Many of the common transistor oscillators are studied in the latter portion of the chapter. The distinction between sinusoidal and non-sinusoidal oscillators is carefully made, with the requirements for each type studied along with a number of typical applications.

7.1 The Class AB Amplifier. *The class AB amplifier is one in which collector signal current flows for more than half but less than the entire input signal cycle.* A single transistor class AB amplifier is a nonlinear amplifier since faithful reproduction cannot be achieved. Class AB linear push-pull operation is possible, however, since one of the two units is conducting when the second transistor is in its cutoff region. This mode of push-pull operation is more desirable than class A, since higher maximum efficiency can be achieved.

The load-line analysis whereby the output characteristics of both transistors are combined is an effective design technique for the class AB amplifier. The manner in which this may be done is best described with the use of Fig. 7.1. As in the class A push-pull amplifier, two matched transistors are used; therefore the output characteristics of the two transistors are identical. The load line (whose slope is the reciprocal of R_{L1}) is drawn upon the characteristics of transistor 1, which is the line AQE in Fig. 7.1a. Point B, the point on the load line where the base current of transistor 1 is zero, is the collector cutoff point. Transistor 1 is at i_{co} at this point. The operating point Q is above cutoff to the extent that a portion of the input signal equal to less than a half cycle will drive transistor 1 to cutoff, as required in class AB operation. From cutoff to the remainder of that pulse, the operating point moves along the zero base current curve to point B'.

Since $R_{L2} = R_{L1}$ and matched transistors are used, the same characteristics and load line represent the action of transistor 2 as well as transistor 1. In order to interpret their joint behavior as a class AB

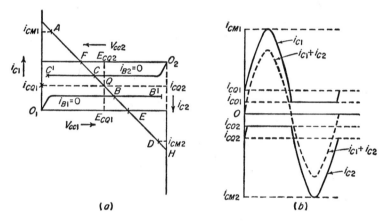

FIG. 7.1   Class AB amplifier analysis: (a) load line, (b) waveform.

amplifier, a load line which is continuous and common to both transistors is suggested. Rotating the characteristics of transistor 1 by 180°, about the operating point Q, results in a common load line for both transistors. The origin for the characteristics of transistor 2 is O_2, with its collector current values plotted on line O_2H. The operating point of transistor 2 moves from Q at zero signal to D (heavy conduction) to cutoff at point C, and then to C' for the remainder of the pulse.

Recall that the same input signal is applied to both push-pull transistors, but the input to T_1 is inverted with respect to the T_2 input. Assume a matched pair of PNP transistors is used in the push-pull circuit of Fig. 7.2. A negative pulse of the input signal drives the operating point of T_1 from Q to A and back again. This waveform, as a function of the operating point, is shown in Fig. 7.1b. At the same time, T_2 feels a positive pulse, identical in waveshape and magnitude to that pulse applied to T_1. The operating point of T_2 moves from Q to cutoff at point C and along the zero base current line to point C' and back. Since T_2 was in the cutoff region for a substantial portion of the pulse, it did not reproduce the input signal (see Fig. 7.1b). Note that the collector current is not actually zero because of i_{CO2}, which cancels a portion of i_{C1}, rendering a resultant collector current waveform as illustrated by the dotted lines of Fig. 7.1b. When the positive pulse is applied to T_1, its operating point moves from Q to B to B' and then back to Q. Note that i_{CO1} flows for a good portion of this pulse. T_2, at the same time, is subjected to a similar but negative pulse. T_2's operating point moved from Q to D and back to Q, which reproduces the pulse. The resultant current waveform $(i_{CO1} + i_{C2})$ is slightly less than that of i_{C2}, since i_{CO1} cancels a portion of it (see dotted lines in Fig. 7.1b).

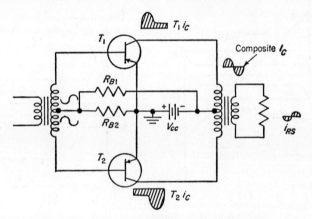

FIG. 7.2 Class AB (CE) push-pull operation.

The composite results for the entire signal cycle, assuming a good operating point was selected, are a faithful reproduction of the input signal.

A practical design technique for the class AB push-pull amplifier is a combination of the load-line analysis and final determination of components by previously discussed methods. The construction of several load lines upon the output characteristics will yield a good approximation of the reflected impedance needed for each transistor. Knowing the ultimate load impedance (such as R_s in Fig. 7.2), the required output transformer turns ratio can be determined.

$$R_L = \frac{1}{4} \times \left(\frac{N_p}{N_s}\right)^2 R_s$$

then

$$\frac{N_p}{N_s} = \left(\frac{4R_L}{R_s}\right)^{1/2}$$

where R_L = load resistance of 1 transistor

N_p/N_s = output transformer turns ratio

R_s = ultimate load resistance

The maximum power dissipation curve should always be drawn upon the output characteristics in order to prevent operation which would damage or destroy the unit. Selection of the operating point determines the zero signal base current of each transistor. When this current is known, the voltage-divider base-bias resistors can be determined. During zero signal condition,

$$i_{RB1} = 2i_{B,Q1} + i_{RB2}$$

Recall that i_{RB2} can be arbitrarily selected, keeping in mind that a heavy drain is not to be imposed on V_{cc}. $i_{B,Q1} = i_{B,Q2}$, and both are obtainable from the output characteristics once the load line has been constructed.

The zero signal input voltage $e_{B-E,Q}$ is required and may be obtained by use of the hybrid parameters when they are available:

$$e_{B-E,Q} \cong i_{B,Q}R_i$$
$$\cong i_{B,Q} \frac{h_{ie} + (h_{oe}h_{ie} - h_{fe}h_{re})R_L}{1 + h_{oe}R_L}$$

$e_{B-E,Q}$ may also be determined by use of the i_C versus e_{B-E} or i_B versus e_{B-E} characteristics in conjunction with the collector characteristics.

When the i_C versus e_{B-E} characteristics are available:

1. Find $i_{C,Q}$ on the collector characteristics.

2. Locate the $i_{C,Q}$ value on the i_C versus e_{B-E} characteristics, and find the corresponding e_{B-E} value.

When the i_B versus e_{B-E} characteristics are available:

1. Find $i_{B,Q}$ on the collector characteristics.

2. Locate the $i_{B,Q}$ value on the i_B versus e_{B-E} characteristics, and locate the corresponding e_{B-E} value.

A fourth method for the determination of $e_{B-E,Q}$ is possible when the manufacturer states the typical voltage amplification ratio of the transistor.

Since

$$E_{ratio} = \frac{\Delta e_o}{\Delta e_i}$$

then

$$\Delta e_i = \frac{\Delta e_o}{E \text{ ratio}}$$

where Δe_o is determined from the load line, and the E ratio is given by the manufacturer.

A fifth method for the determination of $e_{B\text{-}E,Q}$ is analyzed in the oscillator section of this chapter (page 249).

The values of e_{RB1} and e_{RB2} are next found. When no emitter resistance is used,

$$e_{RB2,Q} = e_{B\text{-}E,Q}$$

and when an emitter resistance is incorporated,

$$e_{RB2,Q} = e_{B\text{-}E,Q} + e_{RE,Q}$$

$e_{RB1,Q}$, in either case, is determined by

$$e_{RB1,Q} = V_{cc} - e_{RB2,Q}$$

NOTE: It should be pointed out that an emitter resistance of small ohmic value is frequently used to provide a degree of thermal stability. Other methods for provision of temperature compensation are analyzed in the class B section of this chapter.

Knowing the voltage and current of both resistors, they can be determined by Ohm's law:

$$R_{B1} = \frac{V_{cc} - e_{RB2,Q}}{i_{B,Q1} + i_{B,Q2} + i_{RB2}}$$

and

$$R_{B2} = \frac{e_{RB2}}{i_{RB2}}$$

where i_{RB2} is the arbitrarily determined bleeder current of the voltage-divider bias network.

The final step in the design of any amplifier is its actual construction and test operation. Examination of the output waveforms with an oscilloscope readily indicates the accuracy of the design work. Placing an ammeter in series with V_{cc} such that it measures the composite collector current is a valuable testing technique. The average composite collector current is the sum of $i_{C,Q1} + i_{C,Q2}$. If the measured current exceeds the value indicated from the load-line analysis, the forward base-bias voltage is too large, which calls for a reduction in R_{B2}. On the other hand, if the measured current is below the desired value, a larger R_{B2} value is required. It is not considered unusual to end up with an R_{B2} value slightly less or more than the designed value because of the small difference in the output characteristics of the particular transistors as compared to the general output characteristics made available by the manufacturer.

After the amplifier has been constructed, the power output, power input, and efficiency can be approximated.

Power Output

$$P_o \cong \frac{i_{C,m1}(V_{cc} - e_{C,\min 1})}{2}$$

$$\cong \frac{i_{C,m1}V_{cc}(1 - e_{C,\min 1}/V_{cc})}{2}$$

(These relationships are most accurate when $i_{C,m1}$ is much larger than $I_{co,1}$.)

Power Input. The average current drawn from V_{cc} must first be determined. We may assume without too much inaccuracy that the current waveform of each transistor is a half sine wave above its operating point and a rectangular wave below the operating point.

The average value of the half sine wave over a complete cycle is

$$I_{av,\sin} = \frac{i_{C,m1} - i_{C,Q}}{3.14}$$

The average value of the rectangular wave over a complete cycle, which is also a half cycle long, is

$$I_{av,rect} = \frac{i_{C,Q}}{2}$$

The average current value for one transistor can be approximated by use of the following relationship:

$$I_{av} \cong I_{av,\sin} + I_{av,rect}$$

$$\cong \frac{i_{C,m1} - i_{C,Q}}{3.14} + \frac{i_{C,Q}}{2}$$

$$\cong \frac{i_{C,m1}}{3.14}\left(1 + \frac{0.6i_{C,Q}}{i_{C,m1}}\right)$$

The total input power, which considers the average current of both transistors, is

$$P_i \cong \frac{2i_{C,m1}V_{cc}}{3.14}\left(1 + \frac{0.6i_{C,Q}}{i_{C,m1}}\right)$$

Efficiency:

$$\text{Eff} = \frac{P_o}{P_i} \cong 78\left(1 - \frac{e_{C,\min 1}}{V_{cc}} - \frac{0.6i_{C,Q}}{i_{C,m1}}\right)$$

These equations are most accurate when $i_{C,m}$ is much larger than $i_{C,Q}$. Maximum efficiency approaches 78 per cent, as compared to 50 per cent with a class A transformer-coupled amplifier.

PROBLEMS

7.1 Refer to the class AB (CE) amplifier of Fig. 7.2. The 2N656 (NPN-silicon) is to be used. $V_{cc} = 14$ volts. The actual load resistance to be "seen" by

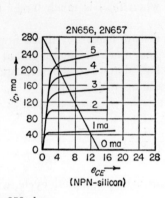

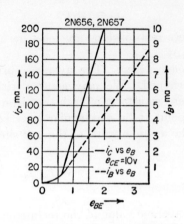

$h_{ie} = 350$ ohms
$h_{fe} = 60$
$h_{re} = 400 \times 10^{-6}$
$h_{oe} = 90$ μmho
$I_{CO} = 60$ μa at 150°C, 10 μa at 25°C

e_{C-B} (breakdown) $= 60$ volts (2N656), 100 volts (2N657)
e_{B-E} (breakdown) $= 8$ volts for both types
PD (max) $= 4$ watts at 25°C (derate 22.8 mw/°C)

FIG. 7.3   Characteristics of the 2N656, 2N657. (*Texas Instruments, Inc.*)

each transistor (R_{L1} for transistor 1 and R_{L2} for transistor 2) is to be 50 ohms. The ultimate load resistance $R_s = 4$ ohms; let $i_{B,Q} = 1$ ma.

(a) Construct the signal load line.

(b) Find $i_{C,Q}$ and $e_{C,Q}$ of each transistor.

7.2 In the amplifier of Prob. 7.1, determine the required turns ratio of the output transformer.

7.3 In the amplifier of Probs. 7.1 and 7.2, using the i_B versus e_{B-E} characteristics in conjunction with the collector characteristic, determine $e_{B-E,Q}$.

7.4 Assume an R_E common to both emitters is to be used in the amplifier of Probs. 7.1 to 7.3. Let $e_{RE,Q} = 0.5$ volt. Find R_E. Note: $i_{RE,Q} = i_{E,QT1} + i_{E,QT2}$.

7.5 In the amplifier of Probs. 7.1 to 7.4, find (a) R_{B2} (let $i_{bleeder} = 100$ μa); (b) R_{B1}.

7.6 Assume the output resistance of the driver stage (referring to the amplifier of Probs. 7.1 to 7.5) $= 2$ kilohms. Determine:

(a) R_i of each transistor

(b) $R_{i,eq}$ of each transistor

(c) turns ratio of the input transformer

7.7 The 2N656 is to be used in a class AB amplifier, and its ambient temperature is to be 50°C. Find the maximum power dissipation of the transistor.

7.8 In the amplifier considered in Prob. 7.7, why was the maximum power dissipation of the transistor at 50°C less than its stated value at 25°C?

7.9 Refer to the class AB (CE) amplifier of Fig. 7.2. The 2N497 (NPN-silicon) is to be used. $V_{cc} = 20$ volts; the actual load resistance to be "seen" by each transistor = 100 ohms; the ultimate load resistance R_s = 12 ohms; let $i_{B,Q}$ = 1 ma.

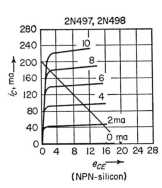

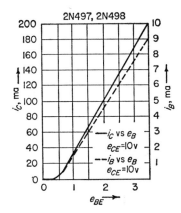

(NPN-silicon)

FIG. 7.4   Characteristics of the 2N497, 2N498.   (*Texas Instruments, Inc.*)

h_{ie} = 250 ohms

h_{fe} = 30

h_{re} = 200 × 10^{-6}

h_{oe} = 70 μmho

I_{CO} = 60 μa at 150°C, 10 μa at 25°C

$e_{C\text{-}B}$ (breakdown) = 60 volts (2N497), 100 volts (2N498)

$e_{B\text{-}E}$ (breakdown) = 8 volts for both types

PD (max) = 4 watts at 25°C (derate 22.8 mw/°C)

(*a*) By interpolation, draw the characteristic curve with i_B = 1 ma on the collector characteristics shown in Fig. 7.4.

(*b*) Construct the signal load line.

(*c*) Determine $i_{C,Q}$ and $e_{C,Q}$ of each transistor.

7.10 In the amplifier of Prob. 7.9, determine the required turns ratio of the output transformer.

7.11 In the amplifier of Probs. 7.9 and 7.10, using the i_C versus $e_{B\text{-}E}$ characteristics in conjunction with the collector characteristics, determine $e_{B\text{-}E,Q}$.

7.12 Assume an R_E common to both emitters is to be used in the amplifier of Probs. 7.9 to 7.11.   Let $e_{RE,Q}$ = 1 volt.   Find R_E.

7.13 In the amplifier of Probs. 7.9 to 7.12, let the bleeder current of the voltage divider = 1 ma.   Find (*a*) R_{B2}; (*b*) R_{B1}.

7.14 Assume the output resistance of the driver stage (referring to the amplifier of Probs. 7.9 to 7.13) = 1.2 kilohms.   Determine (*a*) R_i of each transistor; (*b*) $R_{i,\text{eq}}$ of each transistor; (*c*) turns ratio of the input transformer.

7.15 Refer to the amplifier of Prob. 7.6.   Assume R_E is bypassed by a suitable capacitor.   Find (*a*) R_i of each transistor; (*b*) $R_{i,\text{eq}}$ of each transistor; (*c*) turns ratio of the input transformer.

7.16 Refer to the amplifier of Prob. 7.6.   Find (*a*) power output; (*b*) power input; (*c*) efficiency.

7.17 Refer to the amplifier of Prob. 7.14.   Find (*a*) power output; (*b*) power input; (*c*) efficiency.

7.2 The Class B Amplifier.   *A class B amplifier is one in which collector signal current flows for one-half of the input signal cycle.*   During the second half of the signal cycle, the collector current is not truly zero but at the I_{CO} value. The class B amplifier, like the class A and AB types, can be designed in any of the three configurations, as well as intermediate connections (the split load is an example which is intermediate between CE and CC). Any of the previously discussed coupling techniques may be incorporated. The type of configuration and coupling hinges on the intended use of the amplifier. Several possibilities are studied in this section.

The composite characteristics are obtained in the same manner described for the class AB amplifier (see Fig. 7.1). The slope of the load line, as with all amplifiers, is determined by the load resistance of each transistor. The maximum power dissipation line should be observed, to ensure safe operation of the transistors.

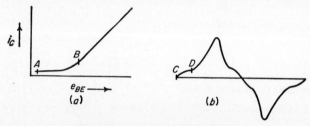

FIG. 7.5 Nonlinear distortion.

One of the problems which must be considered when using class B amplifiers is *nonlinear distortion*. Figure 7.5a illustrates the portion of base-to-emitter voltage versus collector-current characteristic (points A to B) which would result in nonlinear distortion. The input characteristic of a class B amplifier resembles that of Fig. 7.5a when the driver amplifier presents low impedance to the input circuit of the class B amplifier. This results in the output current waveform shown in Fig. 7.5b, which is called *crossover distortion*.

If the driver amplifier is transformer-coupled to the class B amplifier, the presentation of high impedance to its input (particularly at the low portion of the frequency band) is a difficult problem. Since X_L decreases with frequency, the primary inductance must be high, requiring a costly and large transformer. This is not practical in most cases. Unless other precautions are taken, such an amplifier would produce crossover distortion for a substantial portion of its low-frequency range.

One technique which successfully reduces nonlinear distortion is the use of class AB instead of true class B operation. Refer to the input characteristics of Fig. 7.5a. As $e_{B\text{-}E}$ swings from A to B, there is no substantial increase in collector current, resulting in the crossover distortion shown between points C and D of Fig. 7.5b. This distortion reoccurs each time e_{B-E} is driven back to the low values between points A and B. To eliminate this effect, the bias point of the transistor must be located at point B of Fig. 7.5a instead of point A (cutoff). Since the bias point B is not at collector cutoff, the condition is between true class B and A, which is class AB. The specific bias point, which hinges on the input characteristics of the transistor to be used, is to be located just above the nonlinear portion of the input characteristics. In Fig. 7.1, the faithful reproduction possibilities of class AB operation are shown.

7.3 A Conventional Class B (CE) Push-Pull Amplifier. Like many class B amplifiers in transistor circuitry, the push-pull amplifier of Fig. 7.6 is not operated at true class B conditions. R_{B1} and R_{B2} form

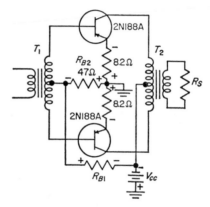

Fig. 7.6 A conventional class B (CE) push-pull amplifier. (*General Electric Co.*)

the voltage-divider type of bias network. R_{B2}, which is 47 ohms in this circuit, provides a forward base bias of about 0.1 volt. The zero signal collector current, instead of being at the I_{co} value, is about 3 ma. R_{B2} places the zero signal operating point just above the nonlinear region of the base voltage versus collector current characteristics, thereby avoiding the cross-over distortion problem.

Notice that thermal stability is enhanced by placing 8.2 ohms in series with each emitter. The possibility of thermal runaway, which tends to occur as the junction temperature increases, is thereby minimized. The use of emitter resistances results in a reduction in gain, but the degenerative features also add circuit stability and more faithful reproduction.

Signal Power Output. The following relationship is often useful in designing this type of amplifier, since the output signal power is frequently the beginning point of the design.

$$P_{o,m,\text{sig}} = \frac{2(e_{C-E,Q})^2}{R_{c-c}} = \frac{2(e_{C-E,Q})^2}{4R_L}$$

where $e_{C\text{-}E,Q}$ = collector-to-emitter voltage at zero signal condition

R_L = load resistance of one collector

$R_{c\text{-}c}$ = collector-to-collector impedance, which is equal to $4R_L$

The output impedance of the transistor is generally not matched to the load impedance because the low load impedance customarily used in power output amplifiers is much lower than the output impedance of the transistor.

Signal Power Gain Ratio

$$\text{Power gain ratio} = \frac{P_o}{P_i} = \frac{2(\beta)^2(e_{C\text{-}E,Q})^2}{(R_{b\text{-}b})P_{o,m,\text{sig}}}$$

where β = current gain ratio for the common-emitter configuration

$R_{b\text{-}b}$ = base-to-base input resistance

Class B Push-Pull (CE) Amplifier with Temperature Compensation Diode. Figure 7.7 illustrates a class B (common-emitter) push-pull

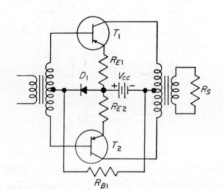

Fig. 7.7 Class B (CE) push-pull amplifier with temperature compensation diode.

amplifier which utilizes a diode instead of R_{B2} in the voltage-divider base-bias arrangement. The diode has a temperature coefficient of resistance which is similar to that of the emitter-base junction. When an increase of current through the emitter-base junction is encountered, this junction experiences a temperature rise. By providing tight thermal coupling between the diode and the transistors, this action also causes the diode to increase in temperature. Since the diode has a negative coefficient of resistance, as it rises in temperature its resistance decreases. Reducing R_{D1} results in a corresponding reduction in e_{D1}, which is the forward base-bias voltage for both transistors. The currents of the two transistors undergo a corresponding reduction because of the decreased forward base bias. When the transistors return to a lower temperature, R_{D1} increases accordingly, and the normal magnitude of forward base bias is provided.

Recall that in Sec. 4.3, the resistance of a germanium junction (or diode) can be approximated by

$$r_j \cong \frac{1}{39 I_f}$$

where I_f is the diode forward current in amperes.

I_f, the diode forward current, is the bleeder current of the voltage-divider bias network and may be arbitrarily selected. The chief precaution to be observed in an actual design problem is the selection of a bleeder current well within the safe forward current range of the diode, such that the diode is not damaged.

When using the temperature compensation diode, the bias network may be designed in the following manner:

1. Select the bleeder current (which is I_f in the preceding equation), and solve for r_j. This is the equivalent of R_{B2} in the straightforward voltage-divider bias network.

2. Calculate R_{B1}:

$$R_{B1} = \frac{V_{cc} - e_{D1}}{i_{B,Q} + I_f}$$

This technique is very effective in helping to prevent thermal runaway. A thermistor, since it also has a negative coefficient of resistance, may be used in place of the diode. As in the case of the temperature compensation diode, tight thermal coupling between the thermistor and transistor must be provided for most effective operation. A thermistor is selected, which presents a sufficiently low resistance at some temperature below the critical temperature of the transistor, to reduce the forward bias to near zero volts. In this manner, the reverse bias provided by the emitter resistor(s) will reduce all transistor currents until the transistor has cooled to a safe temperature.

In actual practice, a thermistor is selected; $e_{B\text{-}E,Q}$ is determined. The room temperature resistance of the thermistor and $e_{B\text{-}E,Q}$ determine the bleeder current:

$$i_{\text{bleeder}} = \frac{e_{B\text{-}E,Q}}{R_{\text{thermistor}} \text{ at } 25°C}$$

where R of the thermistor at 25°C replaces R_{B2} in the conventional voltage-divider bias network. In those cases where the amplifier is to operate at higher ambient temperatures, the resistance of the thermistor at that ambient temperature would be used in the determination of the bleeder current. R_{B1} is determined by

$$R_{B1} = \frac{V_{cc} - e_{\text{thermistor}} \text{ at desired temperature}}{i_{\text{bleeder}} + i_{B,Q}}$$

PROBLEMS

7.18 Refer to Fig. 7.6. The 2N657 (NPN-silicon) is to be used (see Fig. 7.3). True class B operation is to be achieved. $V_{cc} = 22$ volts; the load resistance "seen" by each transistor = 100 ohms; the ultimate load resistance $R_s = 5$ ohms.

(a) Construct the signal load line for each transistor.

(b) Determine $i_{C,Q}$ and $e_{C,Q}$ of each transistor.

7.19 In the amplifier of Prob. 7.18, determine the required turns ratio of the output transformer.

7.20 Using the i_C versus $e_{B\text{-}E}$ characteristic in conjunction with the collector characteristics, find $e_{B\text{-}E,Q}$ of each transistor.

7.21 Let $e_{RE,Q}$ of each transistor = 0.1 volt. Find R_{E1} and R_{E2}.

7.22 In the amplifier of Probs. 7.18 to 7.21, let $i_{\text{bleeder}} = 150$ μa. Find (a) R_{B2}; (b) R_{B1}.

7.23 The output resistance of the driver stage (referring to the amplifier of Probs. 7.18 to 7.22 = 2.4 kilohms. Find:

(a) R_i of each transistor

(b) $R_{i,eq}$ of each transistor

(c) required turns ratio of the input transformer

7.24 In the amplifier designed in Probs. 7.18 to 7.23, find (a) $P_{o,m,\text{sig}}$; (b) signal power gain ratio.

7.25 State the difference between class AB and B operation in terms of (a) collector current; (b) base current.

7.26 Why is nonlinear distortion a problem in true class B operation?

7.27 What is the most practical remedy for nonlinear distortion in a class B amplifier? Explain.

7.28 Refer to the class B push-pull (CE) amplifier with a temperature compensation diode of Fig. 7.7. The 2N498 (NPN-silicon) transistor is to be used. The load resistance "seen" by each transistor = 50 ohms; $V_{cc} = 12$ volts; the ultimate load resistance $R_s = 3$ ohms; let $i_{B,Q} = 1$ ma.

(a) Is this class B or AB operation?

(b) Construct the signal load line.

(c) Find $i_{C,Q}$ and $e_{C,Q}$.

7.29 In the amplifier of Prob. 7.28, calculate the required turns ratio of the output transformer.

7.30 In the amplifier of Probs. 7.28 and 7.29, using the i_C versus $e_{B\text{-}E}$ characteristic in conjunction with the collector characteristics, find $e_{B\text{-}E,Q}$ of each transistor.

7.31 Let $e_{RE,Q}$ of each transistor = 0.2 volt. Find R_{E1} and R_{E2}.

7.32 In the amplifier of Probs. 7.28 to 7.31, let $i_{\text{bleeder}} = 100$ μa. A temperature compensation diode is to be used. Find (a) r_j of the diode; (b) R_{B1}.

7.33 The output resistance of the driver stage (referring to the amplifier of Probs. 7.28 to 7.32 = 800 ohms. Find:

(a) R_i of each transistor

(b) $R_{i,eq}$ of each transistor

(c) required turns ratio of the input transformer

7.34 In Prob. 7.32, a thermistor is to be used in place of the diode. Find the required thermistor resistance at room temperature for the stated conditions.

7.35 When using a temperature compensation diode or thermistor, why is tight thermal coupling important? Explain.

7.4 Class B Push-Pull (CE) Amplifier (Single-ended). Figure 7.8 illustrates a push-pull possibility which does not use an output transformer. The input transformer has two separate windings, which establishes the necessary phase inversion. Each transistor obtains forward base-bias voltage by its own voltage-divider arrangement. As indicated by the arrows at R_L, the two transistor collector currents flow through R_L in opposite directions in an alternate manner. T_1 conducts on the positive half of the input cycle. At this condition, the collector current of T_2 is at its cutoff value. During the negative portion of the input signal cycle, T_2 conducts and T_1 is at its cutoff value.

The two transistors are properly matched if both collector currents cancel at zero signal condition, which is indicated by $e_{RL} = 0$ at that

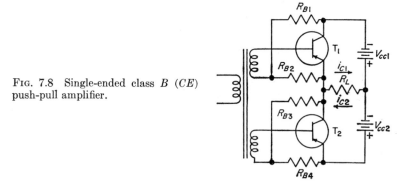

Fig. 7.8 Single-ended class B (CE) push-pull amplifier.

time. Mismatched transistors can be balanced in many cases by adjusting the forward bias voltage resistors (R_{B2} and R_{B4}) so the zero signal collector currents do cancel. The combined action of the two transistors is a bidirectional voltage variation across R_L. This circuit arrangement has the advantage of not requiring an output transformer.

The input transformer of any of the push-pull circuits analyzed up to this point could be replaced with a phase-inverter circuit. Depending upon the specific application, either the split-load or common-emitter phase inverter could be utilized.

7.5 Analysis and Design of a Class B Complementary-symmetry (Common-collector) Amplifier. The complementary-symmetry amplifier is illustrated in Fig. 7.9. Opposite conductivity type transistors are used. T_1 is a PNP while T_2 is an NPN. Notice that R_L is associated with both emitters, resulting in this complementary-symmetry amplifier being in the common-collector configuration. The same input signal is applied to both bases in the same phase. When the input signal undergoes a positive excursion, T_1 is at cutoff and T_2 conducts, with its collector current passing through R_L in the direction indicated by the arrow i_{C2}. The negative input signal swing cuts off T_2 and T_1 conducts; and its collector current passes through R_L in the direction shown by i_{C1}. Notice that no over-all unidirectional current flows through R_L, resulting in only signal power being developed across the output. The low value of direct current which does flow, since neither transistor is brought to true cutoff, flows through the two transistors in series.

The complementary amplifier is especially well suited for RC coupling. i_{B1} flows only when T_1 is conducting, and this is an electron flow into the base of T_1. While this action is occurring, i_{B2} is essentially zero. Note that i_{B2} is an electron flow out of the T_2 base. $i_{B1} \cong i_{B2}$ when a symmetrical output is obtained, resulting in zero unidirectional base current

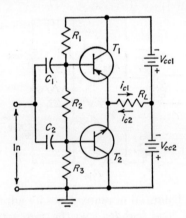

Fig. 7.9 Complementary-symmetry (CC) amplifier.

for the entire signal cycle. Since the unidirectional base current due to the signal variations is zero, the operating point is not altered, which is an important advantage of this amplifier.

Being a common-collector amplifier, the output voltage is essentially the same as the input voltage. The preceding stage, called the driver, must develop a large output voltage to be used as the input of the complementary-symmetry arrangement. The power gain is less than would be obtained with the common-emitter configuration. But the reduction in power gain is due to negative feedback, which means the amplifier also has less distortion and high-frequency phase shift (a significant advantage). If the driver amplifier is RC-coupled to the output stage, the over-all gain of the entire amplifier is just as good as would be the case with a common-emitter output stage. The reason for this fact is that the driver and output stage are better matched with a CC output than would be the case with a CE output stage. Thus a CE-CC cascade arrangement results in approximately the same gain as a CE-CE cascade.

R_1, R_2, and R_3 form a combined voltage-divider arrangement from which the forward bias for T_1 and T_2 are obtained. Recall that a small amount of forward bias is desirable so as to place each transistor in class AB (rather than in true class B) operation. A bleeder current flows through the three resistors, since they are directly across V_{cc1} and V_{cc2}. The bleeder current is arbitrarily selected by the designer, remembering that the power dissipation of these resistors should be kept at a low value. e_{R2} is the forward base bias for both transistors. Notice the polarity of e_{R2}: the top of R_2 is negative with respect to its bottom, which is felt as forward base bias by the PNP transistor (T_1 of Fig. 7.9); and the bottom of R_2 is positive with respect to its top, which is felt as forward base bias by the NPN transistor (T_2 of Fig. 7.9). This forward base bias should be very small, often about 0.01 to 0.1 volt, so that the class AB condition is closer to class B than to class A. In this way, higher efficiency is achieved. R_1 and R_3 are approximately equal in value, and R_2 is customarily small. R_1 and R_3 can be potentiometers which encompass the designed values. Upon construction of the circuit, R_1 and R_3 are varied until the d-c voltage across R_L is zero, which indicates the desired balanced output signal condition. It is also suggested that R_2 be divided into two equal resistors, whose sum is equal to the designed R_2. In this way, R_1 and R_3 may be adjusted so the voltage from V_{cc1} negative to the center of R_2 is equal to the voltage from V_{cc2} positive to the center of R_2. When this condition is achieved, the two transistors are balanced.

The e_{CE} versus i_C (common-emitter) characteristics can be used to determine the input characteristics of the common-collector stage in most cases. Since $R_1 \cong R_3$, the input load line for each transistor can be determined by

$$R_1 + R_2 \cong R_3 + R_2$$

and
$$R_i \cong \frac{\Delta e_{CE}}{\Delta i_B}$$

Both Δe_{CE} and Δi_B can be obtained from the load line. With class B operation, the driver amplifier sees T_1 for one half cycle and T_2 for the other half cycle. The input resistance of the complementary-symmetry common-collector amplifier is sufficiently high to properly match it to the output resistance of the driver stage. The output impedance of the output stage is low:

$$R_o \cong \frac{\Delta e_{EB}}{\Delta i_E}$$

Recall that the emitter and collector currents are nearly identical in value in most cases; therefore the $e_{E\text{-}B}$ versus i_C characteristics may be used in determining the approximate value of R_o. The output load line, determined by R_L, may be constructed on the $e_{E\text{-}B}$ versus i_C characteristics. Since $i_E \cong i_C$, the best R_L and the operating point can be selected with a reasonable degree of accuracy. As in most design work, the final component values are determined upon construction of the circuit.

Figure 7.10 shows a designed complementary-symmetry common-collector stage identical to Fig. 7.9, with a typical common-emitter driver stage. The voltage gain of the driver stage must be sufficient to develop the required input voltage to the output stage. This usually

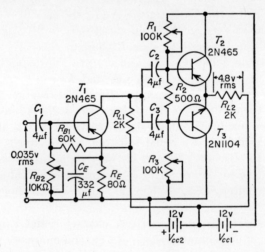

FIG. 7.10   A designed complementary symmetry (CC) and driver stage.

is not a difficult problem because of the high input resistance of the output stage, thereby permitting a good impedance match between the driver output and complementary-symmetry input.

Design Criterion for a Complementary-symmetry and Driver Circuit. The driver stage of Fig. 7.10 is operated in the class A mode and is designed in the conventional manner. The design techniques and values of steps 2 and 3 of Sec. 6.5 were used. All coupling capacitors were made large (4 μf) so as to offer negligible impedance to the signal frequencies (100 cps to 15 kc).

The voltage-divider network for the complementary-symmetry stage was designed next. The current flow in the voltage-divider network is

$$i_{R1,Q} = i_{\text{bleeder}} + i_{B2,Q}$$
$$i_{R2,Q} = i_{\text{bleeder}}$$
$$i_{R3,Q} = i_{\text{bleeder}} + i_{B3,Q}$$

$i_{B2,Q}$ and $i_{B3,Q}$ were selected to be 10 μa, and i_{bleeder} was selected as 100 μa. The forward base bias for each complementary-symmetry transistor was selected as 0.05 volt. R_2 was then computed:

$$R_2 = \frac{e_{B-E,Q}}{i_{\text{bleeder}}} = \frac{5 \times 10^{-2}}{100 \times 10^{-6}} = 500 \text{ ohms}$$

Notice that

$$e_{R1} = e_{C-B} \text{ of } T_2$$

and

$$e_{R3} = e_{C-B} \text{ of } T_3$$

At zero signal condition, each transistor should be very close to cutoff; therefore e_{C-B} of T_2 and T_3 was selected to be 11 volts. Since the current through each resistor is known, their values can then be found:

$$R_1 = \frac{e_{C-B,Q}T_2}{i_{\text{bleeder}} + i_{B2,Q}} = \frac{11}{110 \times 10^{-6}} = 100 \text{ kilohms}$$

and

$$R_3 = R_1 = 100 \text{ kilohms}$$

From the collector-emitter characteristics of the 2N465 (Fig. 7.39), it is seen that $i_{c,Q}$ at the imposed conditions is about 0.5 ma, which is nearly identical to $i_{E,Q}$. Since $e_{C-B,Q} = 11$ volts, then $e_{RL,Q} = 1$ volt if only one of the transistors is connected, and

$$R_L = \frac{e_{RL,Q}}{i_{E,Q}} \simeq \frac{e_{RL,Q}}{i_{c,Q}} = \frac{1}{5 \times 10^{-4}} = 2 \text{ kilohms}$$

The final steps in the design were constructing the circuit, balancing, and evaluating. A sinusoidal signal less than $e_{B-E1,Q}$ is connected to the input circuit of T_1. A d-c voltmeter is connected across R_{L2}; R_1 and R_3 are varied until the d-c voltage across R_{L2} is zero. At this condition, the two complementary transistors are balanced, and the signal voltage across R_{L2} (checked with an oscilloscope) is a faithful reproduction of the input voltage. The over-all gain of the circuit is approximately equal to that obtained from T_1, since A_v of the complementary-symmetry stage is slightly less than unity (common-collector configuration).

The following parameters were obtained from the designed circuit of Fig. 7.10:

$$A_v(T_1) \cong 140$$
$$A_v \text{ (over-all)} \cong 135$$
$$P_{o,\text{sig}} = \frac{(e_{RL2\text{rms}})^2}{R_{L2}} \cong 12 \text{ mw}$$

This type of amplifier has the great advantage of eliminating the need for all coupling transformers, which results in a substantial cost reduction. Also, because of the high input resistance of the output stage, the cross-over distortion problem is virtually nonexistent. Although both Figs. 7.9 and 7.10 indicate two power supplies, one supply with a center tap will serve the purpose.

Complementary-symmetry common-emitter amplifiers are also possible, but offer no advantages over the common-collector design and introduce the problem of power supply ripple. The circuit of Fig. 7.9 can also be directly coupled from its driver.

This involves the use of opposite conductivity type transistors for the driver as well as for the output amplifier. Such an arrangement is most worthy of consideration only when the conventional driver cannot produce the proper drive for the output stage.

7.6 Class B Common-base Push-Pull Amplifier.   Figure 7.11 illustrates a class B common-base push-pull amplifier with a transformer-coupled

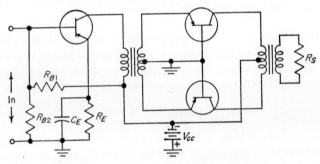

Fig. 7.11   Class B (CB) push-pull amplifier.

driver. If transformer coupling is to be used between the driver and the output stage, the common-base push-pull circuit has many desirable features. This configuration, because of feedback, has less distortion than the common-emitter connection. The feedback possibility is especially desirable in helping to balance the two output transistors.

7.7 Class B Common-collector Push-Pull Amplifier.   The common-collector push-pull amplifier of Fig. 7.12 serves especially well in those cases where the driver stage must be under light load conditions. Since the push-pull amplifier is in the CC configuration, its input resistance is very high, thereby imposing a relatively light load upon the output of the preceding driver stage. Because of the high input resistance, RC coupling can be effectively used in this arrangement. Low distortion is one of the chief advantages of the common-collector push-pull amplifier. The diodes placed between the base and emitter of the two output transistors are optional; they help to prevent i_B from charging the

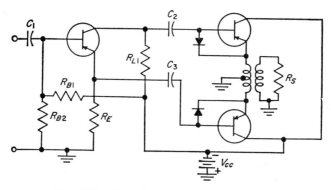

FIG. 7.12 Class B (CC) push-pull amplifier.

coupling capacitors, which may bias the output transistors beyond cutoff. Several low-value resistors could be used in their place.

PROBLEMS

7.36 Refer to the single-ended class B (CE) push-pull amplifier of Fig. 7.8. Let $i_{B,Q}$ of each transistor $= 0$ μa; the 2N656 is to be used. $V_{cc} = 24$ volts, with a center tap, thereby permitting a V_{cc} of 12 volts for each transistor; $R_L = 50$ ohms; let the bleeder current of each voltage-divider bias arrangement $= 100$ μa. (*a*) Construct the signal load line. (*b*) Find $i_{C,Q}$ and $e_{C,Q}$ of each transistor.

7.37 In the amplifier of Prob. 7.36, find $e_{B\text{-}E,Q}$ of each transistor by use of the i_C versus $e_{B\text{-}E}$ characteristic in conjunction with the collector characteristics.

7.38 In the amplifier of Probs. 7.36 and 7.37, find (*a*) R_{B2}, R_{B4}; (*b*) R_{B1}, R_{B3}.

7.39 In the amplifier of Probs. 7.36 to 7.38, find (*a*) R_i of each transistor; (*b*) $R_{i,eq}$ of each transistor.

7.40 Assume $R_{o,eq}$ of the preceding driver stage of the amplifier in Probs. 7.36 to 7.39 $= 2.4$ kilohms. Find the input transformer turns ratio (primary/one of the secondaries).

7.41 Refer to the complementary-symmetry (CC) amplifier of Fig. 7.9. Assume the bleeder current which flows through R_1, R_2, and R_3 is to be 100 μa; $e_{B\text{-}E,Q}$ of each transistor $= 0.5$ volt. Find R_2.

7.42 In Prob. 7.41, assume $i_{B,Q}$ of each transistor $= 0.25$ ma. Let $e_{C\text{-}B,Q}$ of each transistor $= 11$ volts. Find R_1 and R_3.

7.43 Assume the bleeder current of Prob. 7.41 is to be 300 μa, and $e_{B\text{-}E,Q}$ of each transistor $= 0.5$ volt. Find R_2.

7.44 In Prob. 7.43, assume $i_{B,Q}$ of each transistor is still to be 0.25 ma. Let $e_{C\text{-}B,Q}$ of each transistor $= 11$ volts. Find R_1 and R_3.

7.45 Why did the change in the value of bleeder current alter the designed values of R_1, R_2, and R_3 in Probs. 7.43 and 7.44? Explain.

7.46 Refer to the complementary-symmetry (CC) amplifier and driver stage of Fig. 7.10. State the relationship for determining $R_{i,eq}$ of (*a*) T_1; (*b*) T_2; (*c*) T_3.

7.47 Refer to Fig. 7.10. Assume $R_{i,eq,T2} = R_{i,eq,T3} = 4$ kilohms; $R_{L1,eq} = 2$ kilohms. Determine the required R_{L1}.

7.48 In Prob. 7.47, assume $f_{co,min} = 60$ cps. Find (*a*) C_2; (*b*) C_3.

7.49 In Prob. 7.47, assume $R_{i,eq,T1} = 1.2$ kilohms; $R_{o,eq}$ of the stage preceding the driver $= 2$ kilohms. Compute the value of C_1.

7.50 In Prob. 7.47, assume R_E of the driver $= 130$ ohms. Compute the value of C_E.

7.8 The Class C Amplifier. *The class C amplifier is one in which the collector signal current flows for less than one-half of the input signal cycle.* Since the collector current is at its cutoff value for the majority of the signal cycle, faithful reproduction of the input signal is not possible, and this mode of operation falls under the category of nonlinear amplification. The chief purpose of class C amplifiers is to furnish energy to oscillator or switching circuits.

Figure 7.13a illustrates the collector characteristics and load line of a transistor operated in class C. Notice that Q, the zero signal operating

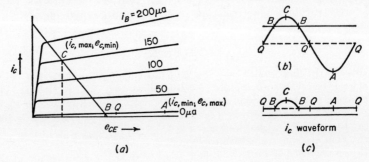

Fig. 7.13 Class C characteristics: (a) load line, (b) input signal riding on a reverse bias, (c) i_C waveform.

point, is well beyond cutoff since only I_{co} flows at that time. Assume the transistor is the NPN type. The transistor must be under reverse bias at zero signal in order to place Q in the cutoff region. A positive excursion of the input signal (see Fig. 7.13b and c) overcomes the reverse bias at point B (series-opposing), at which time the transistor begins to conduct. After reaching a peak value at point C, resulting in maximum collector current, the positive input signal decreases. When the magnitude of the positive swing is equal to the reverse bias, the transistor cuts off. The effect of the reverse bias is sufficient to keep the transistor at cutoff condition for the remainder of the positive signal pulse. When the input signal undergoes its negative excursion, it is series-aiding to the reverse bias, which effectively drives the operating point further into the cutoff region. The operating point, because of the reverse bias, is in the cutoff region for more than half the signal cycle, which meets the prerequisite of class C operation.

As stated in class AB and B operation, the transistor does not achieve true cutoff because of I_{co}.   For good class C operation, a transistor with a low I_{co} should be used; otherwise operating efficiency is seriously lowered.

7.9   Principles of Feedback.   Refer to Fig. 7.14a.   Beta is that fraction of E_o which is fed back to the input.   The total input voltage is equal to the sum of the original input voltage plus beta; i.e.,

$$E_{i,t} = E_i + \beta$$

where $E_{i,t}$ = total input voltage with feedback
E_i = original input voltage without feedback
β = fractional portion of E_o which is fed back to the input

The transistor amplifier, which amplifies its input signal by the factor A_v, provides an output voltage $E_{o,t}$, when feedback is provided.   The

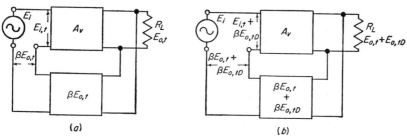

(a)   **(b)**

Fɪɢ. 7.14   Transistor circuit feedback: (a) voltage feedback, (b) voltage and distortion feedback.

voltage fed back to the input is $\beta E_{o,t}$.   The relationship between the input and output voltages in an amplifier with provisions for feedback is

$$(E_i + \beta)A_v = E_{o,t} = E_{i,t}(A_v)$$

Recall that the voltage ratio of an amplifier is

$$A_v = \frac{E_o}{E_i}$$

And the voltage ratio for an amplifier with feedback becomes

$$A_{v,t} = \frac{E_{o,t}}{E_i}$$

and

$$A_{v,t} = \frac{A_v}{1 - \beta A_v}$$

where $A_{v,t}$ = amplifier gain with feedback
A_v = amplifier gain without feedback
βA_v = feedback factor

Positive Feedback. Refer to Fig. 7.14a. The phase relationship between $\beta E_{o,t}$ and E_i determines whether $E_{i,t}$ is larger or smaller than E_i. Beta is positive when the feedback voltage is in phase with E_i, resulting in $E_{i,t}$ being larger than E_i. The use of a feedback voltage $\beta E_{o,t}$, which is in phase with E_i, is called *positive* or *regenerative* feedback. Positive feedback results in increased amplifier gain and also an increase in output distortion. The advantage of increased gain must be weighed against the disadvantage of increased distortion when considering the use of positive feedback.

It should be noted that $A_{v,t}$ becomes infinite when the feedback factor βA_v becomes $+1$. In effect, this means that the amplifier is capable of supplying its own input signal, and self-sustained operation is achieved. *This condition of self-sustained operation is achieved when positive feedback with a feedback factor of $+1$ is made available, and the amplifier is an oscillator.*

Example 1: Refer to Fig. 7.14a. Assume $A_v = 20$; find the required beta for oscillations to occur.
Solution: Oscillations will occur when the feedback factor $= +1$, i.e.,

$$\beta A_v = +1$$

Substituting the given value of A_v, and solving,

$$\beta = \frac{+1}{A_v} = \frac{+1}{20} = +0.05$$

That is, 5 per cent of E_o must be fed back to the input in phase so that the amplifier will oscillate.

Example 2: Refer to Fig. 7.14a. Assume $A_v = 20$. Find $A_{v,t}$ for the following beta values: (a) 0.01; (b) 0.02; (c) 0.03.
Solution: Substituting the beta value into the following equation,

$$A_{v,t} = \frac{A_v}{1 - \beta A_v}$$

(a) When beta $= 0.01$:

$$A_{v,t} = \frac{20}{1 - (0.01)(20)} = 25$$

(b) When beta $= 0.02$:

$$A_{v,t} = \frac{20}{1 - (0.02)(20)} = 33.3$$

(c) When beta $= 0.03$:

$$A_{v,t} = \frac{20}{1 - (0.03)(20)} = 50$$

Example 1 illustrates the manner in which the beta value required for oscillations may be obtained. Example 2 bears out the fact that an increase in positive feedback results in an increase in amplifier gain

$A_{v,t}$. Because of the increased distorted output, positive feedback is not often used in general amplifier work. On the other hand, positive feedback is utilized in many oscillator circuits, where the property of self-sustained oscillations is of prime consideration and distortion is of lesser importance.

Negative Feedback. When the feedback voltage $\beta E_{o,t}$ returned to the input is 180° out of phase with the input signal, the resultant input voltage $E_{i,t}$ is reduced in magnitude. This is called *negative feedback* or *degeneration.* Since $E_{i,t}$ is less than E_i, then $E_{o,t}$ is less than E_o; and the over-all gain of the amplifier $A_{v,t}$ is reduced by the application of negative feedback.

Example 3: Refer to Fig. 7.14a. $A_v = 20$. Find $A_{v,t}$ for the following beta values: (a) -0.01; (b) -0.10; (c) -0.5.

Solution: Substituting the appropriate beta value into the following equation,

$$A_{v,t} = \frac{A_v}{1 - \beta A_v}$$

(a) When $\beta = -0.01$:

$$A_{v,t} = \frac{20}{1 - (-0.01)(20)} = 16.7$$

(b) When $\beta = -0.1$:

$$A_{v,t} = \frac{20}{1 - (-0.1)(20)} = 6.7$$

(c) When $\beta = -0.5$:

$$A_{v,t} = \frac{20}{1 - (-0.5)(20)} = 1.8$$

Example 3 illustrates the fact that negative feedback does reduce the over-all gain of the amplifier.

Distortion Feedback. Refer to Fig. 7.14b. Recall that in positive feedback a portion of the output voltage is fed back to the input such that it is in phase with the input signal. Since the amplifier develops a distortion of the output voltage $E_{o,tD}$, a fraction of the distortion $\beta E_{o,tD}$ is also fed back to the input. The total distortion is actually the sum of the distortion due to the characteristics of the amplifier and the amplified value of $\beta E_{o,tD}$; i.e.,

$$E_{o,tD} = E_{o,D} + (\beta E_{o,tD})A_v$$

where $E_{o,tD}$ = total distortion with feedback

$E_{o,D}$ = distortion due to the characteristics of the amplifier

$\beta E_{o,tD}$ = fraction of the total distortion fed back to the input

A_v = voltage gain of the amplifier without feedback

and $\quad E_{o,tD} = \dfrac{E_{o,D}}{1 - \beta A_v}$

An examination of the preceding equations reveals that when beta is positive (which results in positive feedback), the total distortion with feedback is greater than without feedback. Furthermore, with positive values of beta, the distortion increases as beta is made larger.

Example 4: Refer to Fig. 7.14b. $A_v = 20$; $E_{o,D} = 2$ volts; find the total distortion with feedback for the following beta values: (a) 0.01; (b) 0.02; (c) 0.03. *Solution:* Substituting into the preceding equation:
(a) When beta = 0.01:

$$E_{o,tD} = \frac{2}{1 - (0.01)(20)} = 2.5 \text{ volts}$$

(b) When beta = 0.02:

$$E_{o,tD} = \frac{2}{1 - (0.02)(20)} = 3.3 \text{ volts}$$

(c) When beta = 0.03:

$$E_{o,tD} = \frac{2}{1 - (0.03)(20)} = 5 \text{ volts}$$

Example 4 reveals that the total distortion is increased with larger values of positive beta. As pointed out in a preceding paragraph, positive feedback is rarely used in linear amplifier service, but is commonly used for oscillator service.

The use of negative feedback (beta is negative) has the opposite effect upon distortion. When beta is negative, increasing the amount of feedback results in a reduction of distortion as well as reduced over-all gain. Recall that the negative feedback is 180° out of phase with the input signal, resulting in $E_{i,t}$ being smaller than E_i. This means a smaller portion of the feedback distortion is magnified by the amplifier.

Example 5: Refer to Fig. 7.14b. $A_v = 20$; $E_{o,D} = 2$ volts; find the total distortion with feedback for the following beta values: (a) −0.01; (b) −0.05; (c) −0.1.

Solution:

(a) When beta = −0.01:

$$E_{o,tD} = \frac{2}{1 - (-0.01)(20)} = 1.67 \text{ volts}$$

(b) When beta = −0.05:

$$E_{o,tD} = \frac{2}{1 - (-0.05)(20)} = 1 \text{ volt}$$

(c) When beta = −0.1:

$$E_{o,tD} = \frac{2}{1 - (-0.1)(20)} = 0.67 \text{ volt}$$

Example 5 reveals that the distortion is reduced with larger negative values of beta.

Negative feedback is utilized in many amplifier circuits for the primary purpose of minimizing distortion at the expense of some reduction in gain. Both voltage (shunt) feedback and current (series) feedback are utilized in transistor circuitry.

Shunt Feedback. Figure 5.1, a common-emitter circuit with fixed bias, is an example of negative shunt feedback. R_B is the resistor through which the degenerative feedback current flows from the collector to the base. When i_C increases (because of temperature increase or some other reason), e_{RL} increases and e_{C-B} decreases. i_{RB}, which is i_B, undergoes a decrease, thereby reducing i_C.

Figure 6.13 illustrates the use of negative shunt feedback in a three-stage amplifier. R_7 and R_4 are connected from the collector of $T3$ to the base of $T1$. Since this amplifier has all stages in the common-emitter configuration, an odd number of stages must be included in the feedback loop. An increase of i_{C3} results in a reduction in the increase of i_{B1}, thereby adding stability to the circuit.

In general, a feedback loop involving several stages is more preferable than several stages with individual feedback loops.

Example 6: Consider a three-stage amplifier. The over-all gain of the amplifier is 3,000, without negative feedback; and the inherent distortion of the amplifier is 20 per cent. The distortion is to be reduced to 5 per cent. Find the required beta and the new over-all gain.

Solution: The ratio of $E_{o,D}/E_{o,tD} = 20/5 = 4/1$.

$$E_{o,tD} = \frac{E_{o,D}}{1 - \beta A_v}$$

Transposing and substituting values,

$$1 - \beta A_v = \frac{E_{o,D}}{E_{o,tD}} = 4$$

and

$$-\beta = \frac{3}{3,000} = -0.001$$

or -0.1 per cent over-all feedback.
And the new over-all amplifier gain

$$A_{v,t} = \frac{A_v}{1 - \beta A_v} = \frac{3,000}{1 - (-0.001)(3,000)}$$
$$= 750$$

Example 6 shows that a reduction of distortion is achieved at the expense of amplifier gain. It should be pointed out, however, that the gain would have been further reduced if the negative shunt feedback was applied to each individual stage instead of the entire amplifier.

Series Feedback. Series feedback is frequently utilized in transistor circuits. Refer to the common-base circuit of Fig. 4.17. The placement of a resistor in the base lead would produce negative feedback. Recall that the emitter current is always greater than the collector current in a junction transistor. The emitter current tends to pass through the base resistor, producing a voltage drop which places the base at a reverse potential. The collector current tends to flow through the base resistor in the opposite direction, which is positive feedback and reduces the reverse base potential. Since i_C cannot exceed i_E, the base feedback is always degenerative and the junction transistor is always stable.

Figure 5.5 also illustrates the utilization of series feedback which is degenerative. The emitter current passing through R_E produces reverse bias. A rise in i_E increases the reverse potential, and a drop in i_E results in a reduced reverse potential. Series (current) feedback is most commonly utilized within a given stage whereas shunt (voltage) feedback is incorporated in feedback for several stages.

7.10 Fundamentals of Transistor-oscillator Action. Figure 7.14a illustrates, by block diagram, a transistor circuit with provisions for positive external feedback. The box labeled $\beta E_{o,t}$ is the feedback loop, where beta is that fractional portion of the output that is returned to the input. Recall that the feedback signal, in order to create regeneration, must be in phase with the original input signal. A second requirement is that the feedback loop must return sufficient energy to the input such that the power gain around the closed loop is equal to unity; i.e., the feedback factor βA_v must be equal to $+1$.

When these two conditions are met, the energy dissipated by the various circuit components is replaced by an equal amount of energy returned to the input via the feedback loop. Since the power output of the transistor is many times greater than the power input, only a small fraction of P_o needs to be returned to the input. This is the chief factor which determines the value of beta in the feedback loop.

Since positive feedback is one of the chief prerequisites for oscillator action, oscillators may be classified in accordance with the manner in which the feedback is obtained:

   1. External feedback
   2. Internal feedback

The external feedback type of oscillator is most frequently sinusoidal in waveform, although this is not always the case. A resonant LC circuit is often used in the feedback loop. The internal feedback oscillator utilizes the properties of negative resistance, and the output waveform is most often nonsinusoidal. Several oscillators of both types are analyzed in the sections which immediately follow.

Prior to studying several types of oscillators, let us briefly review the behavior of a tank circuit, such as the one illustrated in Fig. 7.15. A pulse of collector current flowing through the transformer primary will develop induced voltage across the secondary coil. The secondary induced voltage acts like a generator in series with the secondary, thereby driving current, which charges plate A of the tank capacitor. As the

Fig. 7.15 Tank circuit with transformer-coupled input.

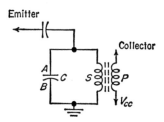

induced voltage decays to zero, the current likewise decays. But plate A of C is now negative with respect to plate B, as shown in Fig. 7.16a.

When the induced voltage has decayed to a magnitude smaller than e_c, plate A will discharge and electrons flow from plate A to plate B. Since e_c exponentially decays as a function of its time constant, this current flow also decays. As the current flow (with the direction shown in Fig. 7.16a) begins to decay, the secondary develops an induced voltage which tends to maintain the current flow (see Fig. 7.16b). Recall that induced voltage opposes *changes* in current. Since the current tends to decay, the induced voltage strives to maintain it. This enables plate B to become negative with respect to plate A of the capacitor. The current decays in a manner determined by e_c and the induced emf.

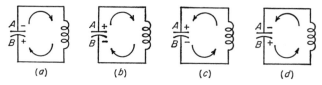

Fig. 7.16 Circulating tank circuit action: (a) plate A discharging, (b) e_{in} maintaining i flow, (c) plate B discharging, (d) e_{in} maintaining i flow.

At the instant e_c is greater than the induced emf, plate B then discharges, as shown in Fig. 7.16c. As in the plate A discharge, the current tends to exponentially decay as e_c decreases. An induced voltage builds up which maintains the current flow long enough to allow plate A to charge, as can be seen in Fig. 7.16d. The current finally decays as the induced emf and e_c decreases, and the action depicted by Fig. 7.16a is repeated.

The tank circulating current would continue indefinitely if no power was dissipated. Since the secondary coil Q is a finite value and other resistive elements are present, the circulating current continues until the energy injected by the original primary current pulse is dissipated. In a tank circuit containing a reasonably low value of resistance, a primary current pulse of less than a half cycle duration for each tank cycle is sufficient to replenish the energy dissipated by the tank circuit. Furthermore, energy can also be drawn from the tank circuit and fed to the input circuit of the transistor. The current can be just sufficient to drive the transistor into conduction for less than half of a signal cycle. The collector current will be at its I_{co} value for more than half the time, which is class C operation. The brief but comparatively large pulse of collector current during the time it is conducting is sufficient to replenish the energy "lost" by the tank circuit. Therefore class C operation is well suited for many types of oscillator operation. Regardless of the shape of the primary waveform, the tank-circuit circulating current will be sinusoidal if its loaded Q is reasonably high. The frequency of the circulating current is determined by the resonant frequency of the tank circuit:

$$f_o = \frac{1}{2\pi(LC)^{1/2}} \qquad \text{when tank } Q \text{ is high}$$

7.11 A Common-base Oscillator. Figure 7.17 illustrates a simple common-base oscillator. R_{B1} and R_{B2} form a voltage-divider arrangement, with e_{RB2} providing forward base bias for the transistor. R_{B2} must be of the correct ohmic value to provide class A bias condition, since the

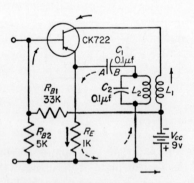

Fig. 7.17   A designed common-base oscillator.

oscillator will start in class A and promptly swing into class C upon energizing the tank circuit. The values of R_{B1} and R_{B2} are obtained from the information found by the load-line technique.

Class C Self-biasing. The manner in which the class C bias value is achieved is of importance. Upon closing the circuit, since the oscillator starts in class A, the transistor conducts.   Plate A of C_1 is charged (takes on electrons) in its attempt to achieve a potential equal to the top of the tank circuit.   As plate A charges, electrons also flow through R_E.   The sum of these two currents flows toward $+V_{cc}$.   From the $-V_{cc}$ terminal, the majority of this current flows to the collector, with the base receiving its proportionate share.   Note that electron flow of this direction through R_E develops reverse bias.   The charging time constant of C_1 is very short; therefore plate A becomes charged in about one-tenth the time of a half cycle.   Upon becoming fully charged in short order, plate A proceeds to discharge through R_E (see broken arrows).   This drives the transistor into cutoff because of the large reverse bias developed across R_E by the discharge current of C_1. C_1 has a long discharge time constant, thereby passing sufficient current through R_E to keep the transistor at cutoff for over half the signal cycle.   While C_1 discharges, its current decreases until the reverse bias is small enough to be offset by the forward bias of

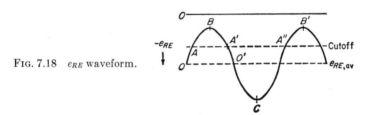

Fig. 7.18 e_{RE} waveform.

e_{RB2}.   The transistor again briefly conducts, charging up plate A, which completes one cycle of operation. i_{RE}, although always flowing in one direction, is sinusoidal if the tank circuit output is sinusoidal, as shown in Fig. 7.18.

Notice that the transistor conducts only from A to A' of each cycle. Plate A of C_1 is charging until point B, at which time it begins to discharge through R_E.   This action drives e_{RE} to cutoff A' and down to the value at point C.   The current due to the C_1 discharge begins to decrease in a sinusoidal manner, allowing e_{RE} to decrease back to A''.   The transistor begins to conduct at this value, and C_1 charges until B'.   At this instant, the voltage from plate A to common is maximum, and plate A begins its discharge action again at B'.   The waveform across R_E has the same shape as that across the tank circuit, but is slightly smaller in magnitude. This technique of a *short charge* and *long discharge* time constant of C_1 is a common bias technique for oscillators, and is called *self-biasing*.

Class C Efficiency and Voltage Ratio.   Recall that the complete sinusoidal cycle can be described as 360°.   The *angle of collector current* is the portion of 360° during which the transistor conducts.   The efficiency of a class C oscillator increases as the angle of collector current is reduced, up to the point where insufficient power is returned to the tank circuit. Class C efficiency can be made to approach 90 per cent without too much difficulty. The voltage ratio in class C operation is considerably smaller than is the case with the same configuration in class A operation. Although the efficiency increases with smaller angles of collector current, the voltage ratio decreases from 30 to 40 per cent at a 300° angle of collector current to 5 per cent or less of the class A value at a 100° angle of collector current.

7.12 Design of a Common-base Oscillator. Let us design the common-base oscillator of Fig. 7.19, which oscillates at 1 kc.

Step 1: *Determination of V_{cc} and Transformer.* V_{cc} is arbitrarily selected to be suitable for the transistor to be used. Nine volts is

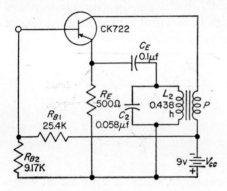

FIG. 7.19   A designed CB oscillator (1 kc).

selected here, although any value from 6 to 12 volts would have served the purpose. The oscillator frequency is determined by the resonant frequency of the tank circuit (C_2 and L_2). A transformer with about a 2/1 turns ratio and with good frequency response at the desired oscillator frequency is next selected. When a suitable transformer has been found,

FIG. 7.20   Determination of L_2.

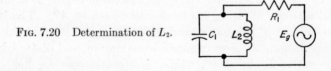

the following procedure may be used in the determination of the remaining parameters.

Step 2: *Determination of the Secondary Inductance of the Transformer.* This value is sometimes stated by the manufacturer. When not stated, the following procedure may be used to approximate its value.

1. Construct the circuit shown in Fig. 7.20.

2. C_1 should be known within close tolerance if the results are to be reasonably accurate. R_1 is to be large enough to permit its voltage drop to be accurately measured. E_g must be constant in magnitude for all frequencies.

3. Vary the frequency of E_g, keeping its output voltage fixed, until the voltage across the tank circuit is maximum. This is the resonant frequency of the tank circuit.

With the selected transformer secondary, $C_1 = 0.1 \ \mu f$; $R_1 = 5$ kilohms; E_g was kept constant at 4 volts. The maximum tank voltage (2.4 volts) was developed at 760 cps.

4. Working on the assumption that $X_L = X_C$ at resonance, these reactances can be found. This relationship is most accurate for high-Q coils.

$$X_L = X_C = \frac{1}{2\pi f C} = \frac{15.9 \times 10^{-2}}{7.6 \times 10^2 \times 1 \times 10^{-7}} = 2.09 \text{ kilohms}$$

5. Now solving for the value of L_2:

$$L_2 = \frac{X_L}{2\pi f} = \frac{2.09 \times 10^3}{6.28 \times 7.6 \times 10^2} = 0.438h$$

Step 3: *Determination of C_2 (Tank Capacitance).* Knowing the desired resonant frequency and the coil value, the required C_2 can be readily found. X_L of the coil is first calculated:

$$X_L = 2\pi f L = 6.28 \times 1 \times 10^3 \times 4.38 \times 10^{-1}$$
$$= 2.75 \text{ kilohms}$$
and $\qquad X_{C2} \cong 2.75 \text{ kilohms}$

We may now compute the value of C_2:

$$C_2 = \frac{1}{2\pi X_{C2} f_o} = \frac{15.9 \times 10^{-2}}{2.75 \times 10^3 \times 1 \times 10^3}$$
$$= 0.05782 \ \mu f$$

Step 4: *Construction of the Tank Circuit*
1. Construct the tank circuit as shown in Fig. 7.21.

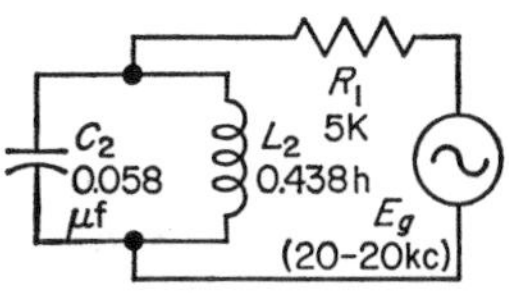

FIG. 7.21   Determination of f_o.

2. Vary the frequency of E_g, keeping its output voltage constant. F_o occurs at that frequency where E_{tank} is maximum.

3. Because of the tolerance of the stated C_2 value, f_o may occur above

or below the desired 1 kc. Recall that

$$f_o = \frac{1}{2\pi(LC)^{1/2}}$$

Increasing C_2 reduces f_o, and reducing C_2 increases f_o. This final solution of C_2 is empirical. When the correct C_2 is obtained, E_{tank} of Fig. 7.21 is maximum at 1 kc.

Step 5: Determination of the Secondary Load Impedance. Assuming L_2 has a reasonably high Q, the impedance of the tank circuit at resonance is very large. R_E and C_E are in parallel with the tank circuit. C_E is to offer negligible reactance at the resonant frequency. R_E (customarily

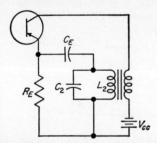

FIG. 7.22   Determination of secondary load Z.

in the 1-kilohm range) has an ohmic value many times smaller than the tank circuit impedance at resonance. Therefore

$$R_s \cong R_E$$

where R_s is the secondary circuit load impedance.

Step 6: Determination of the Primary Reflected Impedance R_L and R_E.
Recall that the reflected primary impedance is found by

$$R_L = \left(\frac{N_p}{N_s}\right)^2 R_s$$

where R_L = reflected primary impedance
$N_p N_s$ = transformer turns ratio $\cong 2/1$
R_s = secondary load impedance $\cong R_E$
therefore, $R_L = (2)^2 R_E = 4R_E$
R_L should be such that none of the absolute values of the transistor are exceeded. This is conveniently determined by examining the collector characteristic curves of the transistor, which are illustrated in Fig. 7.23.

It can be seen that an R_L value of 2 kilohms meets these requirements. R_E can now be found, since

$$R_L = 4R_E$$

then $$R_E = \frac{R_L}{4} = 500 \text{ ohms}$$

Step 7: *Selection of a Class A Operating Point for Starting the Oscillator.* The class *A* operating point (*Q* of Fig. 7.23) can be arbitrarily determined. The intersection of the load line and the 150-μa characteristic was selected in this case.

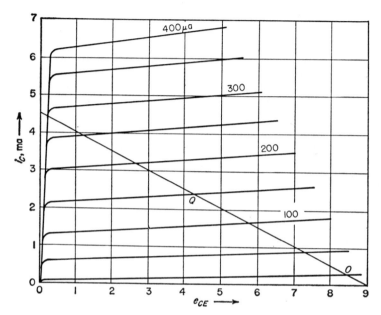

Fig. 7.23   Characteristics of CK722.   (*Raytheon.*)

Step 8: *Determination of R_{B1} and R_{B2}.*   The $e_{B\text{-}E}$ when $I_{C,Q}$ = 2.4 ma (see Fig. 7.23) must next be obtained, which can be found by one of the four methods described in the preceding sections.   As a fifth method, $e_{B\text{-}E}$ can be quickly determined from the following circuit (Fig. 7.24):

Fig. 7.24   Test circuit for approximating $e_{B\text{-}E}$.

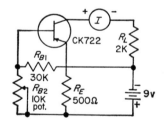

The values found above for R_E, R_L, and V_{cc} are those values in the oscillator that is being designed.   Vary R_{B2} until $i_C = i_{C,Q}$, which is 2.4 ma in this example.   Measure $e_{B\text{-}E}$ with the voltmeter.   This value was found to be about 0.1 volt with the particular transistor used.

Recall that

$$e_{B\text{-}E} = e_{RB2} - e_{RE}$$
$$e_{RE} = i_{RE}R_E = (i_{B,Q} + i_{C,Q})R_E$$
$$= 2.55 \times 10^{-3} \times 5 \times 10^2 = 1.275 \text{ volts}$$

and $e_{RB2} = e_{RE} + e_{B\text{-}E} = 1.275 + 0.1 = 1.375 \text{ volts}$

i_{RB2}, the bleeder current of the voltage divider, is arbitrarily determined by the designer. In this case, $i_{RB2} = 150 \ \mu\text{a}$. R_{B2} can now be found:

$$R_{B2} = \frac{e_{RB2}}{i_{RB2}} = \frac{1.375}{1.5 \times 10^{-4}} = 9.17 \text{ kilohms}$$

We may now find R_{B1}:

$$i_{RB1} = i_{B,Q} + i_{RB2} = 300 \ \mu\text{a}$$
$$e_{RB1} = V_{cc} - e_{RB1} = 9 - 1.375 = 7.625 \text{ volts}$$
$$R_{B1} = \frac{7.625}{3 \times 10^{-4}} = 25.42 \text{ kilohms}$$

Step 9: *Construction of the Circuit.* Upon construction of the circuit, the oscillator may not function for one of several reasons. Assuming a proper transformer has been selected, the primary may be phased improperly. This possibility is quickly checked by reversing the primary

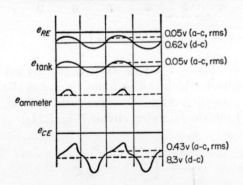

FIG. 7.25   Waveforms of Fig. 7.19.

lead connections. Another common reason for the oscillator being non-operative is the magnitude of e_{RB2}. For this reason, a potentiometer ranging below and above the calculated value of R_{B2} should be used. Varying this potentiometer, if all the components are functioning normally, will result in the correct e_{RB2} being obtained. In some cases, a larger R_E may be required. An oscilloscope connected across the tank circuit will verify whether or not the oscillator is functioning. The frequency of the oscillator can also be verified with the oscilloscope. The waveforms obtained in this particular oscillator are shown in Fig. 7.25.

The waveforms of e_{RE} and e_{tank}, being reasonably sinusoidal, verify that the tank Q is sufficiently high. The waveform obtained across the ammeter inserted in the collector circuit indicates collector current flows for less than 180°, a requirement for class C operation. The angle of collector current, which is seen by this waveform, is altered by varying R_{B2}. As the angle of collector current is reduced, the tank circuit voltage decreases until an insufficiently large angle of collector current is obtained, which fails to energize the tank circuit.

7.13 Several Variations of the CB Oscillator

CB Oscillator with a Collector Tank Circuit. Another common-base oscillator is shown in Fig. 7.26. The oscillation frequency is determined by the resonant frequency of the collector tank circuit. Energy from

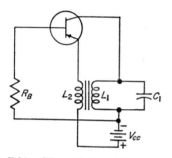

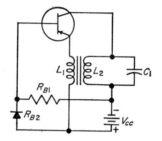

Fig. 7.26 *CB* oscillator with a collector tank circuit.

Fig. 7.27 *CB* oscillator with a varistor starter.

the tank circuit is transformer-coupled to the emitter. R_B is computed so that the base-to-emitter forward bias is sufficiently high to set the transistor into conduction, thus providing class A starting.

CB Oscillator with a Varistor Starter. Figure 7.27 illustrates a common-base oscillator with a nonlinear varistor in place of R_{B2} in the conventional voltage-divider base-bias arrangement. A varistor is a nonlinear resistive device possessing a negative temperature coefficient of resistance. When i_{RB2} is low in magnitude, the varistor (designated as R_{B2}) is relatively cool and its resistance is high. Until the circuit begins to oscillate, i_{RB2} is small, resulting in R_{B2} being of high ohmic value. This places the base under sufficient forward bias to start the circuit action in class A. As the oscillations begin, the current through the varistor becomes greater, which causes its temperature to rise, resulting in a reduction in R_{B2}. This produces a reduction in e_{RB2}, causing a reduction in forward base voltage. The net result is a more stable oscillator.

CB Oscillator with a Capacitor Starter. Figure 7.28 illustrates a common-base oscillator with a capacitor starter. The action of the oscillator is triggered by closing the switch, which causes a momentary surge of current because of C_s. The pileup of electrons at plate B of C_s results in a flow of electrons from plate A. This electron flow, whose direction is shown by the arrows, develops a forward base voltage across R_{B2}. The sudden surge of forward base bias is sufficient to drive the transistor into conduction, thereby starting the oscillatory action of the circuit. Notice that i_{RB1} and i_{RB2}, after the start of oscillations, are primarily determined by the leakage resistance of C_s, since V_{cc} is a unidirectional source of potential. A large value of C_s, with small leakage losses, provides the oscillator with a high degree of efficiency. This method is superior to the customary voltage-divider arrangement, since the power losses associated with the voltage-divider resistors are minimized.

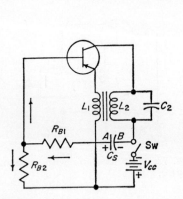
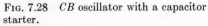
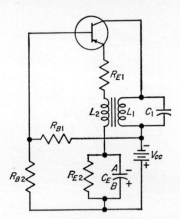

FIG. 7.28 *CB* oscillator with a capacitor starter.

FIG. 7.29 *CB* oscillator with a collector tank circuit and self-bias.

CB Oscillator with a Collector Tank and Self-bias. A common-base oscillator with a collector tank circuit and provisions for self-biasing are shown in Fig. 7.29. The oscillator is started in class A, with R_{B1} and R_{B2} forming the conventional voltage-divider arrangement. Their values are computed by use of the technique described earlier. The starting forward base bias is provided by R_{B2} and must be sufficiently high to set the oscillator into action. The collector tank circuit (L_1 and C_1) is made to resonate at the desired frequency, and a portion of its energy is magnetically coupled to L_2. When the polarity of the emf across L_2 is plus at top and minus at bottom, e_{RE2} becomes large, and plate A of C_E takes on electrons. The charging action of C_E has a very short time constant, enabling plate A to become quickly charged. Plate A then discharges through R_{E2}, which is a long time constant. Notice that this is similar to the self-bias action of Fig. 7.17. Because of the short charge and long discharge time constants of C_E, the transistor is driven into class C operation. For most dependable operation, the product of R_{E2} and C_E should be made as small as possible. An excessively long discharge time constant may result in reducing the angle of collector current to such a low value that oscillations cannot be maintained.

Recall that a long time constant may be defined as that condition where $RC = 20$ half-cycle times or 10 full-cycle times. Using this definition, we can approximate $R_{E2}C_E$:

$$R_{E2}C_E \cong 10T_p$$

where T_p is the time of 1 cycle.
Since

$$f = \frac{1}{T_p}$$

then

$$T_p = \frac{1}{f}$$

and

$$R_{E2}C_E \cong \frac{1}{10f_0}$$

C_E should offer negligible reactance to the oscillator frequency; using the rule of thumb,

$$X_{CE} \cong 0.10R_{E2}$$

and

$$C_E \cong \frac{1}{2\pi f_0(0.1)R_{E2}}$$

R_{E1} serves the usual function of providing thermal stability and reduced distortion in the output. R_{E1} is customarily much smaller than R_{E2}.

7.14 The Colpitts Oscillator. One version of the Colpitts oscillator is shown in Fig. 7.30. Class A starting is utilized, with the proper for-

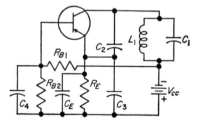

FIG. 7.30   A Colpitts oscillator.

ward bias being supplied by the voltage-divider arrangement of R_{B1} and R_{B2}.

Let us further consider the operation of the Colpitts oscillator with the help of Fig. 7.31a and b.　Refer to Fig. 7.31a, which illustrates the oscillator circuit at the time the transistor is at cutoff (designated by the open switch).　At this time, plate A of C_3 is sufficiently negative to cause it to discharge through R_E and back to plate B of C_3.　This discharge action drives the transistor far into its cutoff region, since e_{RE} is much larger than the forward potential of e_{RB2} at this time.　The transistor remains in the cutoff region as long as e_{RE} exceeds e_{RB2}.　By selecting a suitable R_E ($R_E = 10X_{C3}$ at f_o), the transistor will remain at cutoff for over half of the signal cycle, which is a requirement for class C operation.　In other words, $R_E C_3$ should be a long time-constant arrangement ($R_E C_3 = 10T_p$ or more).

As the signal cycle time progresses, the discharge current of C_3 decreases and e_{RE} decreases.　Eventually, e_{RE} is less than e_{RB2}, at which time the

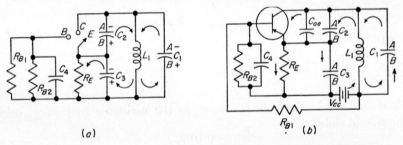

(a)　　　　　　　　　　　　　　　　　　　　　(b)

FIG. 7.31　Colpitts oscillator action: (a) nonconducting, (b) conducting.

transistor conducts.　Refer to Fig. 7.31b, which illustrates the oscillator circuit at the time the transistor conducts.　Plate A of C_3 now charges, with plate B of C_3 discharging through the inductor, transistor, and back to plate A of C_3.　This time constant, $C_3(R_L + R_o)$, should be short as compared to the signal cycle time; i.e.,

$$C_3(R_L + R_o) = 0.1T_p \text{ or less}$$

The feedback voltage is e_{C2}, and beta is

$$\beta = \frac{X_{C2}}{X_{C2} + X_{L1}}$$

In the initial design of the Colpitts oscillator, it is convenient to first use a C_3 which is equal to C_2.　This usually provides sufficient feedback to ensure operation.　At this condition, the angle of collector current will be reasonably small, and the efficiency is high.　Reducing the size of C_3 or increasing C_2 decreases the amount of feedback (since a larger C results in a smaller X_C), and the final value of C_2 and C_3 may be determined empirically.　Assuming a high-Q coil is used, the resonant frequency is determined by

$$f_o \cong \frac{1}{2\pi(LC_T)^{1/2}}$$

where

$$C_T = \frac{C_2 C_3}{C_2 + C_3} + C_1$$

C_1 may be omitted in this circuit, since the equivalent values of C_2 and C_3, in conjunction with the inductor, form a tank circuit. In such cases,

$$C_T = \frac{C_2 C_3}{C_2 + C_3}$$

At higher frequencies, the output capacitance of the transistor C_{oe} should be considered. Notice that C_{oe} is in parallel with C_2. When C_1 is not used, C_T is determined by

$$C_T = \frac{(C_{oe} + C_2) \times C_3}{C_{oe} + C_2 + C_3}$$

When C_1 is used in the circuit, then C_T is found by

$$C_T = C_1 + \frac{(C_{oe} + C_2) \times C_3}{C_{oe} + C_2 + C_3}$$

As stated previously, it is important that R_E be large as compared to X_{C3} at the resonant frequency; otherwise nonoperation will result. C_4 serves as a bypass capacitor to R_{B1} and R_{B2} at the resonant frequency; i.e.,

$$X_{C4} \text{ at } f_o = 0.1 \frac{R_{B1}R_{B2}}{R_{B1} + R_{B2}}$$

The values of R_{B1} and R_{B2}, the conventional voltage-divider bias resistors' are designed in the usual manner so as to provide the oscillator with class A starting.

After the construction of the designed oscillator, the angle of collector current can be studied by placing a small resistor in the collector and observing its waveform with an oscilloscope. The waveform of the tank circuit can also be analyzed with an oscilloscope.

7.15 Criteria for Designing a Colpitts Oscillator

1. *Selection of the Transistor and V_{cc}.* The transistor selected should have a cutoff frequency above the desired oscillator frequency. V_{cc} should be of such a value that $e_{C\text{-}E,\max}$ does not exceed the maximum rating of the transistor.

2. *Determination of $R_{L,\text{eq}}$.* Because of feedback, the equivalent circuit becomes much more complicated than is the case with a class A amplifier

without feedback. If a high-Q tank circuit is used, its impedance when unloaded is high and resistive. Such is the condition when the transistor is cut off. When the transistor conducts, the tank circuit is loaded down as shown in Fig. 7.32a. R_o and R_i are very small at this time, as indicated by the low $e_{C\text{-}E}$ value when the transistor conducts, and the equivalent value of the resistance loading down the tank circuit approaches the value of R_E. For a first approximation for use in the design of the Colpitts oscillator, R_E may be considered as $R_{L,\text{eq}}$, as shown in Fig. 7.32b.

3. *Load-line Construction and Selection of a Class A Operating Point.* Using R_E as a first approximation for $R_{L,\text{eq}}$, select R_E such that its load line does not exceed the maximum power dissipation rating of the transistor. Select an operating point well into the active region of the characteristics. Then determine $i_{B,Q}$, $i_{C,Q}$, and $e_{C,Q}$.

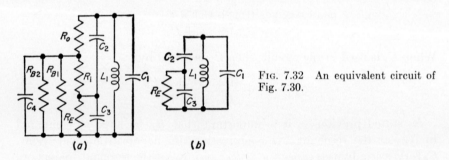

FIG. 7.32 An equivalent circuit of Fig. 7.30.

(a)   (b)

4. *Determine $e_{B\text{-}E,Q}$ for the Selected Class A Operating Point of Step 3.* Any one of the techniques used in previous sections may be used in the determination of $e_{B\text{-}E,Q}$.

5. *Computation of the Voltage-divider Bias Resistors.* Arbitrarily select a bleeder current, then

$$R_{B2} = \frac{e_{RB2,Q}}{i_{\text{bleeder}}}$$

and

$$R_{B1} = \frac{V_{cc} - e_{RB2,Q}}{i_{B,Q} + i_{\text{bleeder}}}$$

6. *Calculation of C_4.* C_4 is to serve as a bypass for R_{B1} and R_{B2} at the oscillator frequency; therefore

$$X_{C4} \text{ at } f_0 = 0.1 \frac{R_{B1}R_{B2}}{R_{B2} + R_{B1}}$$

7. *Determination of C_T and L for the Desired Resonant Frequency.* Recall that

$$f_o \cong \frac{1}{2\pi(LC)^{1/2}}$$

When using a specified coil, C_T may be found by rearranging the preceding equation:

$$C_T = \frac{1}{4(\pi)^2 L(f_o)^2}$$

Also recall that

$$C_T = \frac{C_2 C_3}{C_2 + C_3} + C_1$$

As a first approximation, let $C_2 = C_3$, with the remainder of C_T made up by C_1. Select C_2 and C_3 such that $X_{C3} = 0.1 R_E$ or less.

NOTE: The C_T/L ratio should be kept as high as possible for higher tank voltages.

8. *Circuit Construction.* The final step in any designed circuit is its construction and final determination of component values. For an increased angle of collector current, R_{B2} may be made larger or C_3 may be made smaller. A decreased angle of collector current may be achieved

FIG. 7.33   A designed Colpitts oscillator.

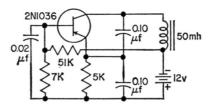

by increasing C_3 or increasing R_E. The effects created by these changes occur until a certain point, of course. It should be noticed that the ratio of C_2/C_3 may be varied but C_T must remain the same value, unless inductor adjustments can be made so that the tank resonates at the same f_o.

Figure 7.33 illustrates a Colpitts oscillator, with C_1 omitted, which was designed by use of the preceding criteria. The Colpitts oscillator of Fig. 7.33 oscillates at a frequency of 3.2 kc with an angle of collector current of about 50°. Oscillations are maintained at C_3 values of about 0.005 μf, but with an increased angle of collector current.

In high-frequency work, more attention should be given to a high L/C ratio, a high-Q coil, and the output capacitance of the transistor.

7.16   The Hartley Oscillator.   Figure 7.34 illustrates a designed Hartley oscillator, and an approximate equivalent circuit of this oscillator is shown in Fig. 7.35. R_{B1} and R_{B2} do not appear in the equivalent circuit, since they are shunted at the oscillator frequency by C_4, and R_E does not appear because it is shunted by C_E. The collector load in the equivalent circuit, for an approximation, may be viewed as a parallel combination where C_1 and L_1 form one series branch and L_2 is the second leg.

Since C_1 and L_1 are in series, the total reactance of this leg is equal to the arithmetic difference of X_{L1} and X_{CT}. The reactance of the other leg is equal to X_{L2}. The collector load may be approximated by

$$Z_L \cong \frac{X_{L2}(-X_{CT} + X_{L1})}{X_{L2} + (-X_{CT} + X_{L1})}$$

where Z_L is very large with a high-Q tank circuit. The frequency of oscillation may be determined by

$$f_o = \frac{1}{2\pi[C_1(L_1 + L_2 + 2M) + (h_{oe}/h_{ie})(L_1L_2 - M^2)]^{1/2}}$$

where M = mutual inductance = $k(L_1L_2)^{1/2}$.

Since the coil is tapped (see Fig. 7.34), the mutual inductance between L_1 and L_2 must be considered. L_1 and L_2 are series-aiding, resulting in

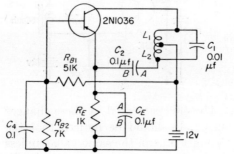

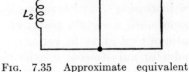

FIG. 7.34   A designed Hartley oscillator.   FIG.   7.35   Approximate   equivalent circuit of Fig. 7.34.

a total inductance which is larger than that of L_1 and L_2 if they were magnetically isolated. The parameters h_{oe} and h_{ie} are available in the manufacturers' data or can be graphically determined in the manner described in Chap. 4.

The amount of feedback is determined by the ratio of X_{L1} to the sum of X_{L1} and X_{CT}; i.e.,

$$\beta = \frac{X_{L1}}{X_{L1} + X_{CT}}$$

The feedback must be sufficiently large so that the product of the transistor gain and β, which is $A_v\beta$, equals unity. Also, the feedback must be positive, which means a 360° phase shift must occur between the transistor input circuit and the end of the feedback loop. The transistor furnishes a 180° phase shift between its input and collector, and a second 180° phase shift is produced by the presence of X_{L1} and C_1.

C_E should be such a value that its reactance at the oscillator frequency is 10 per cent or less than the ohmic value of R_E. The self-bias required for class C operation is achieved from the action of R_E and C_E. R_{B2} and R_{B1} are voltage-divider bias resistors for placing the oscillator at a class A operating point when it is first turned on. C_4 is to be sufficiently large so that its reactance at the frequency of oscillations is 10 per cent or less than the equivalent value of R_{B1} and R_{B2}. The transistor selected must have a cutoff frequency above the desired oscillator frequency.

The angle of collector current can be further reduced by increasing R_E and C_4, up to a point. Adjustments in the angle of collector current can also be made by varying R_{B2}.

Class C bias is achieved by R_E and C_E. When the transistor conducts, plate A of C_E takes on electrons, and this charging action should require only one-tenth of the cycle time. Upon being charged, sufficient reverse bias is present in the emitter to cut off transistor conduction. When the transistor stops conducting, plate A of C_E discharges through R_E, and the discharging time constant $(R_E C_E)$ should be 10 or more times longer than the cycle time. In this way, the transistor can be maintained at cutoff for more than half the signal cycle, thereby achieving class C operation.

7.17 Design Criteria for the Hartley Oscillator

1. *Selection of V_{cc} and Transistor.* V_{cc} is determined by practical considerations, and a transistor whose cutoff frequency is above the oscillator frequency is selected.

2. *Determination of R_E and C_E.* As a first approximation, select R_E such that the load line to be constructed (whose slope $= 1/R_E$) is a good class A load line on the collector characteristics. Select C_E such that X_{CE} at $f_o = 0.1R_E$ (or less).

3. *Determination of a Class A Operating Point and Bias.* For a first approximation, R_E and V_{cc} may be used in the construction of the zero signal load line. Select a class A operating point toward the center of the active region. Then calculate R_{B2} and R_{B1}. Recall that

$$R_{B2} = \frac{e_{RB2,Q}}{i_{\text{bleeder}}}$$

and
$$R_{B1} = \frac{V_{cc} - e_{RB2,Q}}{i_{B,Q} + i_{\text{bleeder}}}$$

It is suggested that a potentiometer whose maximum value is greater than the computed R_{B2} be first placed in the circuit. The potentiometer is varied until the desired operating conditions are achieved.

4. *Selection of Tank Circuit Components.* As a first approximation,

$$f_o = \frac{1}{2\pi[C_1(L_1 + L_2 + 2M)]^{1/2}}$$

The capacitor and inductor are selected in accordance with the desired f_o.

5. *Selection of C_2 and C_4.* Select C_4 such that

$$X_{C4} \text{ at } f_o \cong 0.1 \frac{R_{B2}R_{B1}}{R_{B2} + R_{B1}} \qquad \text{or less}$$

In this way, C_4 will serve as a low impedance shunt for the two voltage-divider resistors at the oscillator frequency. C_2 is to prevent a d-c path from the tank to the emitter via the feedback loop. X_{C2} should be reasonably low at the oscillator frequency.

6. *Construction and Testing of the Circuit.* A small resistor may be placed in series with the collector to be used for an oscilloscope analysis of the angle of collector current. After all adjustments have been made, the resistor in the collector circuit is removed, and a fixed R_{B2} (corresponding to the setting of the potentiometer) is inserted as a replacement for the potentiometer.

The Hartley oscillator of Fig. 7.34 was designed in accordance with the preceding criteria.

7.18 The Clapp Oscillator. Figure 7.36a illustrates the Clapp oscillator, which is similar to the Colpitts oscillator, except a series resonant circuit (L_1 and C_1) is used in the collector circuit. Feedback is provided

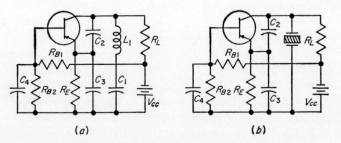

Fig. 7.36 The Clapp oscillator: (*a*) with series resonant circuit, (*b*) with a crystal.

by the voltage dividing capacitors C_2 and C_3. L_1 and C_1 are selected to resonate at the desired frequency of oscillations. R_L limits the d-c collector current of the transistor to safe values. The forward bias required for class A starting is provided by the voltage-divider network of R_{B2} and R_{B1}.

The series resonant circuit in the collector circuit can be replaced with

a crystal, as shown in Fig. 7.36b. The frequency of the crystal determines the frequency of oscillations in this circuit.

7.19 The Wien Bridge Oscillator. There are a number of sinusoidal
oscillators which do not incorporate the principles of resonance. Recall
that the use of resonant circuits in the preceding oscillators made it
possible for the circuit to possess the required positive feedback at a
predetermined frequency. An RC circuit which is frequency selective
is the Wien bridge, which is illustrated in Fig. 7.37.

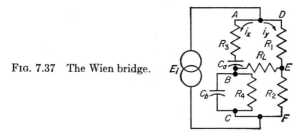

FIG. 7.37   The Wien bridge.

Let us examine the basic Wien bridge circuit. At the condition of
balance, the potential across R_L is zero, and the following relationships
exist:

$$e_{AB} = e_{DE}$$

and
$$e_{BC} = e_{EF}$$

where
$$e_{AB} = (e_{R3} - je_{XCa}) = i_x(R_3 - jX_{Ca})$$

and
$$e_{DE} = e_{R1} = i_yR_1$$

and
$$e_{BC} = e_{XCb} = e_{R4} = i_x\frac{R_4(-jX_{Cb})}{R_4 - jX_{Cb}}$$

and
$$e_{EF} = e_{R2} = i_yR_2$$

Substituting into the original equations,

$$i_x(R_3 - jX_{Ca}) = i_yR_1$$

and
$$i_yR_2 = i_x\frac{R_4(-jX_{Cb})}{(R_4 - jX_{Cb})}$$

The ratio of R_1/R_2 is obtained by dividing the two equations, resulting in

$$\frac{R_1}{R_2} = \frac{(R_3 - jX_{Ca})R_4(-jX_{Cb})}{R_4 - jX_{Cb}} = \frac{C_b}{C_a} + \frac{R_3}{R_4}$$

and
$$f_b = \frac{0.159}{(R_3R_4C_aC_b)^{\frac{1}{2}}}$$

where f_b is the frequency at which balance exists.

The preceding equations are true when the Wien bridge is balanced. If the circuit is such that $C_a = C_b$, and $R_3 = R_4$, then

$$R_1 = 2R_2$$

and
$$f_b = \frac{0.159}{RC}$$

where
$$R = R_3 = R_4$$
$$C = C_a = C_b$$

The Wien bridge analyzed in the preceding paragraphs can be used in conjunction with two transistors to produce a Wien bridge transistor oscillator, as illustrated in Fig. 7.38.

The Wien bridge is associated with the input of T_1. R_1 must be at least $2 + 1/A_{v,t}$ times greater than R_2 in order for oscillations to occur. The required forward bias for T_1 is obtained by the automatic self-bias

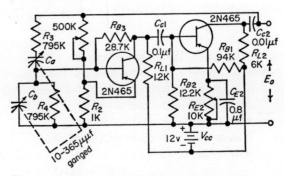

Fig. 7.38   A designed Wien bridge oscillator.

provided by R_{B3}. R_2 is in the emitter circuit of T_1, thereby providing temperature stabilization. R_3 is equal to R_4, C_a is equal to C_b, and they are ganged together. This provides a means for varying the frequency within the designed range of the oscillator. The remaining parameters of T_1 and T_2 are designed in accordance with the conventional class A design principles.

Refer to the Wien bridge circuit of Fig. 7.38 and notation of Fig. 7.37. Let $C_a = C_b = C$, and $R_3 = R_4 = R$, then

$$Z_{ABC} = R - jX_C + \frac{R(-jX_C)}{R - jX_C}$$

The total current in branch ABC is

$$i_{ABC} = \frac{e_{ABC}}{Z_{ABC}}$$

And the voltage from base to common is

$$e_{B\text{-COMMON}} = i_{ABC}\frac{R(-jX_C)}{R - jX_C}$$

and

$$e_{B\text{-COMMON}} = \frac{Z_{BC}}{Z_{ABC}}e_{ABC}$$

Substituting values,

$$e_{B\text{-COMMON}} = \frac{R(-jX_C)/(R - jX_C)}{(R - jX_C) + [R(-jX_C)/(R - jX_C)]}e_{ABC}$$

Since the resistances R_3 and R_4 and the reactances are equal at the frequency of oscillations,

$$e_{B\text{-COMMON}} = \frac{-jRX_C}{-j3RX_C}e_{ABC} = \frac{e_{ABC}}{3}$$

Therefore one-third of the bridge voltage is applied to the base of T_1 at the frequency of oscillations. At frequencies above and below this value, the potential applied to the T_1 base is lower in magnitude and too small to allow the two-stage amplifier to feed back a signal of sufficient magnitude to maintain oscillations. The feedback loop is from the collector of T_2 back to point D of the Wien bridge.

7.20 Design of a Wien Bridge Oscillator.

The Wien bridge oscillator illustrated in Fig. 7.38 is designed in this section. The collector characteristics of the 2N465 are shown in Fig. 7.39. Let $V_{cc} = 12$ volts. Frequency range is 2 to 20 kc; $C_a = C_b = 365$ to 10 $\mu\mu$f.

Step 1: Design of the Wien Bridge Components. Let us select the special Wien bridge condition where $R_3 = R_4 = X_{Ca} = X_{Cb}$ at the frequency of oscillations. Recall that the oscillator is at its maximum frequency (20 kc) when C_a and C_b are at their minimum setting; i.e.,

$$C_a = C_b = C = 10 \ \mu\mu\text{f when } f = 20 \text{ kc}$$

Also recall that

$$f_{o,\max} = \frac{0.159}{RC_{\min}}$$

where

$$R = R_3 = R_4$$

Transposing, and solving for R,

$$R = \frac{0.159}{f_{o,\max}C_{\min}} = \frac{0.159}{2 \times 10^4 \times 10 \times 10^{-12}}$$
$$= 795 \text{ kilohms}$$

therefore,

$$R_3 = R_4 = 795 \text{ kilohms}$$

Assume the two-stage amplifier is to have some nominal over-all gain

$A_{v,t}$, such as 5. Recall that

$$\frac{R_1}{R_2} = \frac{2 + (1/A_{v,t})}{R_2}$$

Substituting, $\quad\quad R_1 = 2.2R_2 \quad$ minimum

From the circuit of Fig. 7.38, it is seen that R_2 serves as the emitter resistor of the stage 1 transistor. Since the stages are to operate in class

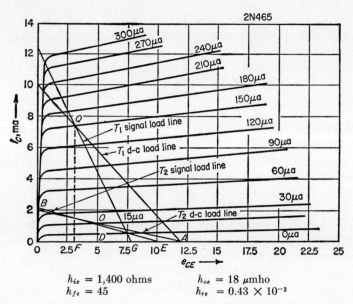

$$
\begin{aligned}
h_{ie} &= 1{,}400 \text{ ohms} & h_{oe} &= 18 \ \mu\text{mho} \\
h_{fe} &= 45 & h_{re} &= 0.43 \times 10^{-3}
\end{aligned}
$$

FIG. 7.39 Collector characteristics with Wien bridge load lines. (*Characteristics from Raytheon.*)

A, R_2 should not be too large. Let $R_2 = 1$ kilohm as a first approximation, and

$$R_1 = 2.2(1\text{K}) = 2.2 \text{ kilohms} \quad\quad \text{minimum}$$

It is suggested that a 500-kilohm potentiometer be used for R_1 in the initial circuit, which can be varied for the desirable feedback magnitude. The determination of the bridge components is now complete.

Step 2: Determination of Stage 2 Parameters. The Wien bridge designed in step 1 appears as a shunt across the output of stage 2. The equivalent value of the Wien bridge in Fig. 7.38 is approximately equal to the value of R_1. Assume a conservative value for R_1, such as 20 kilohms.

Let $R_{L2,eq} = 4.6$ kilohms. R_{L2} can then be computed from

$$R_{L2} = \frac{1}{1/R_{L2,eq} - 1/R_{\text{bridge,eq}}}$$

Substituting values and solving,

$$R_{L2} = \frac{1}{1/(4.6 \times 10^3) - 1/(20 \times 10^3)} \cong 6 \text{ kilohms}$$

An operating point in the center of the active region is selected (point O), and the zero signal parameters are:

$$i_{B2,Q} = 15 \ \mu\text{a}$$
$$i_{C2,Q} \cong 1.2 \text{ ma}$$
$$e_{C\text{-}E,Q} \cong 5 \text{ volts}$$

The signal load line for T_2 can next be determined:

$$\tan \alpha_2 = -\frac{1 \times 10^3}{R_{L2,eq}} = \frac{1 \times 10^3}{4.6 \times 10^3}$$
$$= 0.217$$

The distance between $e_{C\text{-}E,Q}$ and the point where the signal load line intersects the x axis $(D - E)$ is next determined by

$$D - E = \frac{i_{C,Q} \times 10^3}{\tan \alpha_2}$$

Substituting values and solving,

$$D - E = \frac{1.2}{0.217} \cong 5.5 \text{ volts}$$

and the x-axis intercept of the signal load line is now found:

$$x\text{-axis intercept (point } E) = e_{C\text{-}E,Q} + (D - E)$$
$$= 5 + 5.5 \cong 10.5 \text{ volts}$$

The signal load line for T_2 is then drawn on the characteristics (see Fig. 7.39).

The bias network of stage 2 may next be determined. Let

$$e_{RE2,Q} = 1.2 \text{ volts}$$

Solving for R_{E2},

$$R_{E2} = \frac{e_{RE2,Q}}{i_{E,Q}} = \frac{1.2}{1.2 \times 10^{-3}} = 1 \text{ kilohm}$$

NOTE: R_{E2} should be a 10-kilohm potentiometer in the initial circuit, and is to be adjusted for the value which results in the best waveform.

$X_{CE2} = 100$ ohms at the lowest frequency, which is 2 kc, and

$$C_{E2} = \frac{0.159}{2 \times 10^3 \times 1 \times 10^2} \cong 0.8 \ \mu\text{f} \qquad \text{minimum}$$

The required $e_{B\text{-}E,Q}$ can be determined from

$$e_{B\text{-}E,Q} \cong R_{i,2} i_{B,Q}$$

where $\quad R_{i2} = \dfrac{h_{ie} + (h_{oe}h_{ie} - h_{fe}h_{re})R_{L2,eq}}{1 + h_{oe}R_{L2,eq}}$

$$= \frac{1{,}400 + (18 \times 10^{-6} \times 1{,}400 - 45 \times 0.43 \times 10^{-3})6 \times 10^3}{1 + 18 \times 10^{-6} \times 6 \times 10^3}$$

$$\cong 1{,}420 \text{ ohms}$$

and $\qquad e_{B\text{-}E,Q} = 1.42 \times 10^3 \times 15 \times 10^{-6} = 0.0213$ volt

The required voltage across R_{B2} must be greater than $e_{RE2,Q}$ by the required $e_{B\text{-}E,Q}$; i.e.,

$$e_{RB2} = 1.2 + 0.021 = 1.221 \text{ volts}$$

Let the bleeder current $= 100 \ \mu\text{a}$, and R_{B2} can now be determined:

$$R_{B2} = \frac{1.22}{100 \times 10^{-6}} \cong 12.2 \text{ kilohms}$$

and $\qquad R_{B1} = \dfrac{V_{cc} - e_{RB2}}{i_{\text{bleeder}} + i_{B,Q}} = \dfrac{12 - 1.22}{1.15 \times 10^{-4}}$

$$\cong 94 \text{ kilohms}$$

The interstage coupling capacitor is determined after the design of stage 1.

Step 3: Determination of Stage 1 Parameters. In order to make a wise selection of $R_{L1,eq}$ the value of $R_{i2,eq}$ should first be found:

$$R_{i2,eq} = \frac{1}{1/R_{B1} + 1/R_{B2} + 1/R_{i2}}$$

NOTE: R_{E2} is not included in the $R_{i2,eq}$ computations because it is bypassed. Substituting values and solving,

$$R_{i2,eq} = \frac{1}{1/(12.2 \times 10^3) + 1/(94 \times 10^3) + 1/(1.42 \times 10^3)}$$

$$\cong 1.25 \text{ kilohms}$$

Let $R_{L1} = 1.2$ kilohms and the d-c load line can be drawn (see Fig. 7.39). The operating point Q is selected, and

$$i_{B,Q} = 150 \ \mu\text{a}$$
$$i_{C,Q} = 7.3 \text{ ma}$$
$$e_{C\text{-}E,Q} = 3.25 \text{ volts}$$

$R_{L1,eq}$ is next determined:

$$R_{L1,eq} = \frac{R_{L1}R_{i2,eq}}{R_{L1} + R_{i2,eq}}$$
$$= \frac{(1.2 \times 10^3)(1.25 \times 10^3)}{(1.2 \times 10^3) + (1.25 \times 10^3)}$$
$$= 612 \text{ ohms}$$
$$\tan \alpha_1 = -\frac{1 \times 10^3}{612} \cong 1.63$$

The distance between $e_{C\text{-}E,Q1}$ and the point where the signal load line intersects the x axis $(F - G)$ is next found:

$$F - G = \frac{i_{C,Q1} \times 10^3}{\tan \alpha_1} = \frac{7.3}{1.63} \cong 4.5 \text{ volts}$$

and the x-axis intercept of the signal load line for T_1 is now found:

$$x\text{-axis intercept (point } G) = e_{C\text{-}E,Q1} + (F - G)$$
$$= 3.25 + 4.5 \cong 7.75 \text{ volts}$$

The signal load line for T_1 is then drawn on the characteristics (see Fig. 7.39).

The reverse bias supplied by R_2, which serves as the emitter resistance for T_1, can now be determined,

$$e_{R2,Q} = i_{E,Q}R_2$$
$$= (i_{C,Q} + i_{B,Q})R_2 = 7.45 \times 10^{-3} \times 1 \times 10^3$$
$$= 7.45 \text{ volts}$$

The required $e_{B\text{-}E,Q}$ can be determined from

$$e_{B\text{-}E,Q} = R_{i1}i_{B,Q}$$

where

$$R_{i2} = \frac{h_{ie} + (h_{oe}h_{ie} - h_{fe}h_{ie})R_{L1,eq}}{1 + h_{oe}R_{L1,eq}}$$
$$= \frac{1{,}400 + (18 \times 10^{-6} \times 1.4 \times 10^3 - 45 \times 0.43 \times 10^{-3})1.2 \times 10^3}{1 + (18 \times 10^{-6} \times 1.2 \times 10^3)}$$
$$\cong 1.38 \text{ kilohms}$$

and

$$e_{B\text{-}E,Q} = 1.38 \times 10^3 \times 1.5 \times 10^{-4} \cong 0.21 \text{ volt}$$

R_{B3}, the automatic self-bias resistor for the stage 1 transistor, can now be determined by

$$R_{B3} = \frac{V_{cc} - (e_{B\text{-}E,Q} + e_{R2,Q})}{i_{B,Q}}$$

Substituting values and solving,

$$R_{B3} = \frac{12 - (0.21 + 7.45)}{1.5 \times 10^{-4}}$$
$$\cong 29 \text{ kilohms}$$

Step 4: *Determination of* C_{C1} *and* C_{C2}. In order to maintain a negligible phase shift in the RC coupling network, X_{CC1} at the minimum frequency should be no greater than $R_{o1,eq} + R_{i2,eq}$. In actual practice, it is suggested that C_{C1} be larger than the value designed (which is the minimum acceptable value).

Let us first find $R_{o1,eq}$:

$$R_{o1,eq} = \frac{1}{1/R_{o1} + 1/R_{L1}}$$

R_{o1} may be determined graphically. Refer to the collector characteristics of Fig. 7.39:

$$R_o \cong \frac{\Delta e_{C\text{-}E}}{\Delta i_C} \qquad \text{with } i_B \text{ fixed}$$

Selecting values near the operating point,

$$R_o = \frac{3.75 - 2.5}{(7.4 - 7.2) \times 10^{-3}} \cong 6.25 \text{ kilohms}$$

Substituting values and solving for $R_{o1,eq}$,

$$R_{o1,eq} = \frac{1}{1/(6.25 \times 10^3) + 1/(1.2 \times 10^3)}$$
$$\cong 1.04 \text{ kilohms}$$

and $R_{i2,eq}$ was determined in step 3 as 1.25 kilohms; therefore

$$X_{CC1} \text{ at 2 kc} = R_{o1,eq} + R_{i2,eq}$$
$$= 1.04 \times 10^3 + 1.25 \times 10^3 \cong 2.29 \text{ kilohms}$$

Solving for C_{C1},

$$C_{C1} = \frac{0.159}{2 \times 10^3 \times 2.29 \times 10^3} = 0.035 \ \mu\text{f} \qquad \text{minimum}$$

Let us select $C_{C1} = 0.1 \ \mu\text{f}$.

The value of C_{C2} is determined by the same type of analysis, as used for C_{C1}:

$$X_{CC2} \text{ at 2 kc} = R_{o2,eq} + R_{\text{bridge,eq}}$$

In step 2, the equivalent value of the bridge resistance was approximated at 20 kilohms, and

$$R_{o2,eq} = \frac{1}{1/R_{o2} + 1/R_{L2}}$$

R_{o2} is graphically determined from points taken near the stage 2 operating point O in Fig. 7.39:

$$R_{o2} \cong \frac{\Delta e_{C\text{-}E}}{\Delta i_C} \quad \text{with } i_B \text{ fixed}$$

$$= \frac{6.25 - 3.75}{(1.25 - 1.1) \times 10^{-3}} \cong 16.7 \text{ kilohms}$$

R_{L2} was determined to be 6 kilohms in step 2. Substituting values and solving,

$$R_{o2,\text{eq}} = \frac{1}{1/(16.7 \times 10^3) + 1/(6 \times 10^3)}$$
$$\cong 4.4 \text{ kilohms}$$

and X_{CC2} at 2 kc $= 20 \times 10^3 + 4.4 \times 10^3 = 24.4$ kilohms

Solving for C_{C2},

$$C_{C2} = \frac{0.159}{2 \times 10^3 \times 24.4 \times 10^3}$$
$$\cong 0.0033 \ \mu\text{f} \quad \text{minimum}$$

Let us arbitrarily select $C_{C2} = 0.01 \ \mu\text{f}$.

Step 5: Construction and Testing of the Circuit. Upon construction of the circuit, R_1 (500-kilohm potentiometer) and R_{E2} (10-kilohm potentiometer) are adjusted. R_1 is adjusted for the desired amount of feedback, and R_{E2} is adjusted until the oscillator waveform is sinusoidal. It is suggested that these adjustments be made in conjunction with the oscilloscope, which may be placed from the base of transistor 1 to common, or across the designated output terminals of the oscillator. R_2, the Wien bridge resistor which serves as the emitter resistor for stage 1, may be adjusted within a small range (about 20 per cent) for improvement of the waveform. When the circuit is first assembled, set R_1 at its maximum value, set R_{E2} at the calculated value of 1 kilohm, and begin the adjustments from these values.

7.21 The Phase-shift Oscillator.

A two-transistor phase-shift oscillator is shown in Fig. 7.40. T_1 is designed to give a high A_v value. T_2 is utilized for matching the output back into the base of T_1. Positive feedback occurs from T_2 back to the base of the first transistor. The circuit will oscillate at the frequency where the phase shift around the loop is 360°. T_1 provides a 180° phase shift; and the RC network of R_1, R_2, R_3, C_1, C_2, and C_3 undergo 180° phase shift at the frequency of oscillations.

The frequency of oscillations is determined by

$$f_o = \frac{0.159}{(6)^{1/2}RC}$$

and

$$e_o = e_i/29$$

where

$$R = R_1 = R_2 = R_3$$
$$C = C_1 = C_2 = C_3$$

The phase-shift network should possess low impedance, and R_{i2} should be low as compared to R_{i1}.

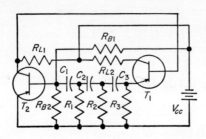

FIG. 7.40   The phase-shift oscillator.

R_{B2} and R_{B1} form the conventional voltage-divider bias network for class A starting of transistor 1.   The class A starting point of transistor 1 may be determined by the load-line design techniques employed for class A amplifiers, with an aim toward obtaining the highest possible A_v. The purpose of R_{L2} is to provide a means to keep $e_{C\text{-}B}$ of T_2 at lower values than e_{RL1}. e_{RL2} may be used as the output terminals of this oscillator.

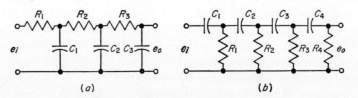

FIG. 7.41   Additional phase-shift networks.

There are several other possible phase-shift networks which may be used in phase-shift oscillators, two of which are illustrated in Fig. 7.41. In Fig. 7.41a,

$$f_o = \frac{(6)^{1/2}}{6.28RC} \qquad \text{and} \qquad e_o = e_i/29$$

In Fig. 7.41b,

$$f_o = \frac{0.159}{(19\!\!\;/\!\!\;7)^{1/2}RC} \qquad \text{and} \qquad e_o \cong e_i/18$$

where $$R = R_1 = R_2 = R_3 = R_4$$
$$C = C_1 = C_2 = C_3 = C_4$$

7.22 Design of a Phase-shift Oscillator. The phase-shift oscillator of Fig. 7.42 is designed in this section. Let $V_{cc} = 12$ volts, and the 2N465 is to be used for T_1 and T_2. The frequency of oscillation $= 5$ kc.

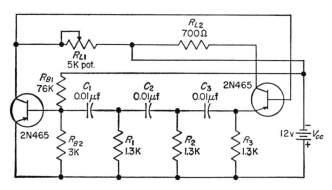

Fig. 7.42 The designed phase-shift oscillator.

Step 1: Design of the RC Phase-shift Network (R_1, R_2, R_3, C_1, C_2, and C_3).
Let us arbitrarily select a convenient value for the three capacitors in the RC phase-shift network. Let $C_1 = C_2 = C_3 = 0.01$ μf. The frequency of oscillations is determined by

$$f_o = \frac{0.159}{(6)^{1/2}RC}$$

and transposing,

$$R = \frac{0.159}{(6)^{1/2}Cf_o}$$

where $$R = R_1 = R_2 = R_3$$

Substituting values and solving,

$$R = \frac{0.159}{2.45 \times 1 \times 10^{-8} \times 5 \times 10^3}$$
$$= 1.3 \text{ kilohms}$$

Step 2: Design of T_2 Parameters. $e_{C\text{-}B}$ of T_2 is to be maintained at lower values than e_{RL1}. Notice that R_3 serves as an unbypassed emitter resistor for T_2; therefore the slope of the T_2 load line is determined by $R_{L2} + R_3$. Since R_3 has already been determined as 1.3 kilohms, $R_{L2} + R_3$ must be greater than this value.

Let $$R_{L2} + R_3 = 2 \text{ kilohms}$$
then $$R_{L2} = 2\text{K} - R_3 = 2\text{K} - 1.3\text{K}$$
$$= 700 \text{ ohms}$$

The load line for $R_{L2} + R_3$ is shown as AQ_2 in Fig. 7.43. The operating point Q_2 is selected at the intersection of the load line and the 90-μa characteristic. The zero signal parameters for T_2 are:

$$e_{C\text{-}E,Q} = 2.9 \text{ volts}$$
$$i_{C,Q} = 4.6 \text{ ma}$$
$$i_{B,Q} = 90 \ \mu\text{a}$$

The reverse bias provided by R_3 for T_2 may be computed by

$$e_{R3,Q} = i_{E,Q}R_3 = (i_{C,Q} + i_{B,Q})R_3$$
$$= 4.69 \times 10^{-3} \times 1.3 \times 10^3 \cong 6.1 \text{ volts}$$

The base to emitter must obtain a forward bias value slightly greater than 6.1 volts to enable T_2 to conduct. Notice that the base of T_2 is

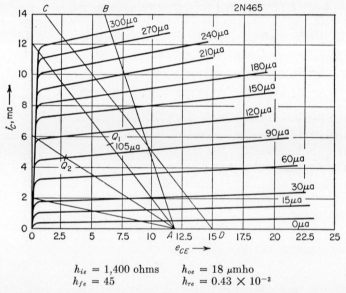

$$h_{ie} = 1{,}400 \text{ ohms} \qquad h_{oe} = 18 \ \mu\text{mho}$$
$$h_{fe} = 45 \qquad\qquad h_{re} = 0.43 \times 10^{-3}$$

Fig. 7.43 Collector characteristics with phase-shift oscillator load lines. (*Characteristics from Raytheon.*)

directly connected to the collector of T_1; therefore $e_{C\text{-}E}$ of T_1 is applied to the base of T_2. When designing the T_1 parameters in step 3, $e_{C\text{-}E,Q}$ of T_1 will be set at a value slightly larger than 6.1 volts.

Step 3: Design of T_1 Parameters. As stated in step 2, $e_{C\text{-}E,Q}$ of T_1 should be slightly larger than 6.1 volts so as to enable T_2 to conduct at zero signal. The reverse bias of 6.1 volts is provided by R_3 to the emitter of T_2. The $e_{B\text{-}E,Q}$ of T_2 can be determined by

$$e_{B\text{-}E,Q}T_2 \cong i_{B,Q}R_{i2}$$

where

$$R_i = \frac{h_{ie} + (h_{oe}h_{ie} - h_{fe}h_{re})(R_{L2} + R_3)}{1 + h_{oe}(R_{L2} + R_3)}$$

Substituting values and solving,

$$R_{i2} = \frac{1{,}400 + (18 \times 10^{-6} \times 1{,}400 - 45 \times 0.43 \times 10^{-3})(2 \times 10^3)}{1 + 18 \times 10^{-6} \times 2 \times 10^3}$$

$$\cong 1{,}360 \text{ ohms}$$

and $\qquad e_{B\text{-}E,Q}T_2 = 90 \times 10^{-6} \times 1.36 \times 10^3 = 0.12 \text{ volt}$

Since the $e_{C\text{-}E}$ of T_1 is directly connected to the base of T_2, then $e_{C\text{-}E}$ of T_1 should be 0.12 volt greater than 6.1 volts so that T_2 is at its correct forward bias value; i.e.,

$$e_{C\text{-}E,Q} = e_{R3,Q} + e_{B\text{-}E,Q}T_2$$
$$= 6.1 + 0.12 \cong 6.2 \text{ volts}$$

A number of R_{L1} values could be selected. Let us select $R_{L1} = 1$ kilohm, whose load line is shown as AQ_1 in Fig. 7.43. Select the operating point Q_1 such that the previously determined $e_{C\text{-}E,Q}$ of 6.2 volts is obtained. The zero signal parameters of T_1 are:

$$e_{C\text{-}E,Q} = 6.2 \text{ volts}$$
$$i_{C,Q} = 5.7 \text{ ma}$$
$$i_{B,Q} = 105 \text{ } \mu\text{a}$$

Let us next determine $e_{B\text{-}E,Q}T_1$:

$$e_{B\text{-}E,Q}T_1 = i_{B1,Q}R_{i1}$$

where

$$R_{i1} = \frac{h_{ie} + (h_{oe}h_{ie} - h_{fe}h_{re})R_{L1}}{1 + h_{oe}R_{L1}}$$

$$= \frac{1{,}400 + (18 \times 10^{-6} \times 1{,}400 - 45 \times 0.43 \times 10^{-3})(1 \times 10^3)}{1 + 18 \times 10^{-6} \times 1 \times 10^3}$$

$$= 1{,}380 \text{ ohms}$$

and $\qquad e_{B\text{-}E,Q}T_1 = 105 \times 10^{-6} \times 1.38 \times 10^{+3} \cong 0.15 \text{ volt}$

The voltage-divider bias network will be incorporated. Let the bleeder current through $R_{B2} = 50$ μa, then

$$R_{B2} = \frac{e_{B\text{-}E,Q}T_1}{i_{\text{bleeder}}} = \frac{1.5 \times 10^{-1}}{50 \times 10^{-6}} = 3 \text{ kilohms}$$

and $\qquad R_{B1} = \frac{V_{cc} - e_{B\text{-}E,Q}T_1}{i_{B1,Q} + i_{\text{bleeder}}} = \frac{12 - 0.15}{(105 + 50) \times 10^{-6}}$

$$\cong 76 \text{ kilohms}$$

Step 4: Circuit Construction and Evaluation. The *RC* phase-shifting network used in this design reduces the output voltage to $\frac{1}{29}$. In order for oscillations to occur, the voltage gain of T_1 must be at least 29. In actual practice, A_{v1} may have to be greater than 29 because of other losses in the circuit.

Consider the gain of T_1 when i_{B1} varies from 105 to 60 μa.

$$R_{i1} = 1{,}380 \text{ ohms}$$

as determined in step 3. The corresponding values of e_{B-E} for each i_B can be readily determined. When $i_B = 105$ μa:

$$e_{B-E,Q} = R_{i1}i_{B,Q} = 1.38 \times 10^3 \times 105 \times 10^{-6}$$
$$\cong 0.15 \text{ volt}$$

When $i_B = 60$ μa:

$$e_{B-E,Q} = 1.38 \times 10^3 \times 60 \times 10^{-6} \cong 0.083 \text{ volt}$$
and $$\Delta e_i = 0.15 - 0.083 = 0.067 \text{ volt}$$

The corresponding values of e_{C-E} are obtained from the load line of Fig. 7.43.

When $i_B = 105$ μa: $\qquad e_{C-E} = 6.2$ volts
When $i_B = 60$ μa: $\qquad e_{C-E} = 8.25$ volts
and $$\Delta e_{C-E} = 8.25 - 6.20 = 2.05 \text{ volts}$$
and $$A_{v1} = \frac{\Delta e_o}{\Delta e_i} = \frac{\Delta e_{C-E}}{\Delta e_{B-E}} = \frac{2.05}{.067}$$
$$\cong 31$$

The output of the oscillation (across R_{L2}) should result in a good sinusoidal waveform and can be checked with the oscilloscope. The frequency of oscillations can also be checked with the oscilloscope.

Let R_{L1} be a 5-kilohm potentiometer. Adjust R_{L1} for the required value, as indicated by circuit oscillation, and e_{RL1} should be sinusoidal.

PROBLEMS

7.51 Define class C operation in terms of collector current.

7.52 At what bias condition is the transistor at zero signal condition in class C operation?

7.53 Why is true collector current cutoff not achieved in class C operation? Explain.

7.54 What is the angle of collector current?

7.55 State the relationship between the angle of collector current and (a) A_v; (b) efficiency.

7.56 Define beta in the feedback loop of Fig. 7.14.

7.57 Assume an amplifier has $A_v = 50$ without feedback. Find $A_{v,t}$ for the following beta values:
(a) 0.001
(b) 0.005
(c) 0.01

7.58 In the amplifier of Prob. 7.57, find the beta value required for the circuit to function as an oscillator.

7.59 In the amplifier of Prob. 7.57, find $A_{v,t}$ for the following beta values:
    (*a*) −0.001
    (*b*) −0.05
    (*c*) −0.10

7.60 In Prob. 7.57, state the relationship between positive values of beta and $A_{v,t}$. Explain.

7.61 In Prob. 7.59, state the relationship between negative values of beta and $A_{v,t}$. Explain.

7.62 Assume an amplifier has $A_v = 25$ without feedback. Find $A_{v,t}$ for the following beta values:
    (*a*) 0.001
    (*b*) 0.005
    (*c*) 0.01

7.63 Compare the results of Probs. 7.57 and 7.62. What is the relationship between $A_{v,t}$ and A_v of an amplifier for a given positive beta value? Explain.

7.64 Assume an amplifier has $A_v = 25$ without feedback. Find $A_{v,t}$ for the following beta values:
    (*a*) −0.001
    (*b*) −0.05
    (*c*) −0.10

7.65 Compare the results of Probs. 7.59 and 7.64. What is the relationship between $A_{v,t}$ and A_v of an amplifier for a given negative beta value? Explain.

7.66 State the relationship between distortion and (*a*) positive feedback; (*b*) negative feedback.

7.67 Assume an amplifier has $A_v = 30$ without feedback, and $E_{o,D} = 1$ volt. Find $E_{o,tD}$ for the following beta values:
    (*a*) 0.002
    (*b*) 0.005
    (*c*) 0.01

7.68 Repeat Prob. 7.67 for the following beta values:
    (*a*) −0.002
    (*b*) −0.05
    (*c*) −0.10

7.69 Refer to Prob. 7.67. State the relationship between positive beta values and distortion. Explain.

7.70 Refer to Prob. 7.68. State the relationship between negative beta values and distortion. Explain.

7.71 Assume a three-stage amplifier has a gain of 5,000 without feedback. The inherent distortion of the amplifier is 10 per cent, and is to be reduced to 5 per cent. Find (*a*) required beta; (*b*) $A_{v,t}$.

7.72 In Prob. 7.71, assume the distortion is to be reduced to 2 per cent. Find (*a*) required beta; (*b*) $A_{v,t}$.

7.73 Compare the results of Probs. 7.71 and 7.72. State the relationship between the amount of distortion reduction and $A_{v,t}$ for a given amplifier. Explain.

7.74 State the two basic feedback types of oscillators.

7.75 Why must positive feedback be utilized in oscillators?

7.76 Refer to the circuit of Fig. 7.15. What is the purpose of the transformer primary?

7.77 In Fig. 7.15, what determines the oscillator frequency?

7.78 In Fig. 7.15, what relationship exists between the tank circuit Q and (*a*) angle of collector current required to maintain oscillations; (*b*) oscillator efficiency; (*c*) waveform of the tank circulating current.

7.79 Refer to Fig. 7.15. f_o is to be 100 kc; $L_s = 50$ mh. Compute the required C.

7.80 In Prob. 7.79, assume $R_{Ls} = 5$ ohms. Determine the unloaded tank circuit Q.

7.81 In Fig 7.17, why is the oscillator started in class A?

7.82 In Fig. 7.17, describe the manner in which C_1 develops the bias required for class C operation.

7.83 What circuit efficiency is possible in class C operation?

7.84 Why is class C efficiency higher than in other classes of operation? Explain.

7.85 Refer to the common-base oscillator of Fig. 7.19. The 2N597 (PNP) will be used (see Fig. 7.44). $V_{cc} = 12$ volts; a transformer with a 2/1 turns ratio, with $L_s = 50$ mh is to be used; $f_o = 10$ kc. Find (a) C_2; (b) secondary load impedance (assume L_2 has a high Q).

7.86 In the CB oscillator of Prob. 7.85, let $R_L = 1$ kilohm:
(a) Construct the load line.
(b) Determine R_E.
(c) Determine C_E.

7.87 In Prob. 7.86, let the starting $i_{B.Q} = 0.1$ ma; assume the starting $e_{B\text{-}E.Q} = 0.2$ volts. Find (a) R_{B2} (let $i_{bleeder} = 100$ µa); (b) R_{B1}.

7.88 Refer to the designed common-base oscillator of Fig. 7.19. The 2N475 (NPN-silicon) is to be used (see Fig. 6.39). $V_{cc} = 12$ volts; a transformer with a 2/1 turns ratio, with $L_s = 10$ mh is to be used; the oscillator frequency is to be variable from 100 to 500 kc. Find (a) C_2 (minimum and maximum); (b) secondary load impedance (assume L_2 has a high Q).

7.89 In Prob. 7.88, let $R_L = 2$ kilohms.
(a) Construct the load line.
(b) Determine R_E.
(c) Determine C_E.

7.90 In Prob. 7.89, let the starting $i_{B.Q} = 100$ µa. Using the i_C versus $e_{B\text{-}E}$ characteristic in conjunction with the collector characteristics, find $e_{B\text{-}E.Q}$.

7.91 In Prob. 7.90, let $i_{bleeder} = 100$ µa. Calculate (a) R_{B2}; (b) R_{B1}.

7.92 Refer to the designed common-base oscillator of Fig. 7.19. The 2N465 (PNP) is to be used (see Fig. 6.11). $V_{cc} = 15$ volts; a transformer with a 1/1 turns ratio is to be used, with $L_s = 40$ mh; $f_o = 500$ kc. Find (a) C_2; (b) secondary load impedance (assume L_2 has a high Q).

7.93 In Prob. 7.92, let $R_L = 1.5$ kilohms.
(a) Construct the load line.
(b) Determine R_E.
(c) Determine C_E.

7.94 In Prob. 7.93, let the starting $i_{B.Q} = 90$ µa. Using the hybrid parameters, compute (a) R_i; (b) $e_{B\text{-}E.Q}$.

7.95 In Prob. 7.94, let $i_{bleeder} = 50$ µa. Calculate (a) R_{B2}; (b) R_{B1}.

7.96 Refer to Fig. 7.42. What components determine the oscillator frequency?

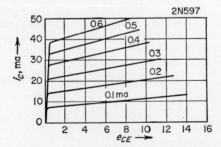

Absolute values (25°C):

$e_{cE} = 45$ volts
$i_c = 400$ ma
$i_B = 50$ ma
PD $= 250$ mw

FIG. 7.44 Characteristics (low I region) of 2N597. (*Lansdale Tube Co.*)

7.97 In Fig. 7.26, state the behavior of the varistor.

7.98 In Fig. 7.27, explain how C_s enables the oscillator to be self-starting.

7.99 In Fig. 7.29, state the purpose of L_1.

7.100 In Fig. 7.34, what determines the magnitude of the feedback?

7.101 Refer to the Colpitts oscillator of Fig. 7.30. $C_1 = 200$ $\mu\mu$f; $C_2 = C_3 = 500$ $\mu\mu$f; $L_1 = 50$ mh. Find (a) C_t; (b) f_o.

7.102 Using the parameters and frequency of Prob. 7.101, find beta.

7.103 Refer to the Hartley oscillator of Fig. 7.34. $L_1 = 30$ mh; $L_2 = 10$ mh; k between L_1 and $L_2 = 1$; $h_{oe} = 20 \times 10^{-6}$ mho; $h_{ie} = 1,200$ ohms; C_1 can be varied from 50 to 360 $\mu\mu$f. Find (a) M; (b) $f_{o,\max}$; (c) $f_{o,\min}$.

7.104 In the Hartley oscillator of Prob. 7.103, Find (a) beta at $f_{o,\max}$; (b) beta at $f_{o,\min}$.

7.105 Refer to the Wien bridge oscillator of Fig. 7.39. Assume the oscillator frequency can be varied from 50 to 500 cps. $R_3 = R_4 = 600$ kilohms; $C_a = C_b$ and they are ganged together. Find (a) $C_{a,\min}$; (b) $C_{a,\max}$.

7.106 In the Wien bridge oscillator of Prob. 7.105, assume the over-all two-stage gain is 10. Let $R_2 = 800$ ohms. Find the minimum R_1.

7.107 Refer to the phase-shift oscillator of Fig. 7.40. Assume R_1, R_2, and R_3 are potentiometers and are ganged together. Let $C_1 = C_2 = C_3 = C = 0.02$ μf. The potentiometers can be varied from 500 ohms to 5 kilohms. Find (a) $f_{o,\max}$; (b) $f_{o,\min}$.

7.108 Refer to the phase-shift oscillator of Fig. 7.40. Assume C_1, C_2, and C_3 are ganged together and can be varied from 50 to 500 $\mu\mu$f. $R_1 = R_2 = R_3 = 50$ kilohms. Find (a) $f_{o,\max}$; (b) $f_{o,\min}$.

7.109 What is the minimum gain required in the designed phase-shift oscillator of Fig. 7.42?

7.110 What class of operation is utilized in the designed phase-shift oscillator of Fig. 7.42?

CHAPTER 8

TRANSISTOR MULTIVIBRATORS AND THE UNIJUNCTION TRANSISTOR

The types of multivibrators analyzed in the following three sections also utilize external positive feedback for the maintenance of oscillations. They differ from the previously discussed oscillators in that the frequency of oscillations is determined by the time constant of one or two capacitors instead of a resonant LC circuit. Also, two junction transistors are used instead of one. An additional distinction between multivibrators and the LC type is that nonsinusoidal waveforms are obtained in this type.

Junction transistor multivibrators fall into three general categories:
1. Astable
2. Monostable
3. Bistable

The theory of operation of the unijunction transistor, along with its applications as a relaxation oscillator, pulse generator, and multivibrator, are given separate treatment in the latter sections of this chapter.

8.1 The Astable Multivibrator. This section is devoted to the analysis of the astable multivibrator, which may also be called a *relaxation* or *free-running* oscillator. Both transistors, without the cross-coupling capacitors (see Fig. 8.1) are biased as class A amplifiers. A square-wave output $e_{C\text{-}E}$ is obtained when the circuit is symmetrical. A short pulse output and saw-tooth base-to-emitter voltage is obtainable by making the circuit asymmetrical.

The astable multivibrator is frequently called a relaxation oscillator because the circuit has two conditions of operation, neither of which can be maintained permanently. In other words, the two circuit states are both *unstable*, hence the name *astable*. Let us consider condition 1 as that state where T_1 is ON and T_2 is OFF. Condition 2 is then the state where T_1 is OFF and T_2 is ON. When a square-wave output waveform is desired, both transistors and their parameters must be symmetrical. Referring to Fig. 8.1, this means:

1. T_1 and T_2 are reasonably well matched.

2. $R_{L1} = R_{L2}; R_{E1} = R_{E2}; R_{B1} = R_{B3}; R_{B2} = R_{B4}.$

3. $C_1 = C_2$.

When circuit symmetry has been achieved, T_1 ON $= T_2$ ON; T_1 OFF $= T_2$ OFF; and a square-wave voltage from collector to emitter of either transistor is available.

Prior to delving further into the actual theory of operation of this circuit, it is necessary to define the conditions ON and OFF, which is most conveniently done with the use of the collector characteristic curves (see Fig. 8.2). Without the cross-coupling capacitors C_1 and C_2, both transistors would function as class A amplifiers. If the entire circuit is symmetrical, both transistors have the same operating point, which is Q in Fig. 8.2. Because of the action of the cross-coupling capacitors, several voltages and currents are made to change so that one transistor is driven to ON and the other transistor goes to OFF at the same time.

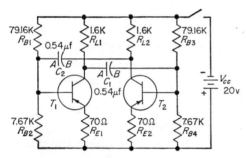

FIG. 8.1 A designed astable multivibrator.

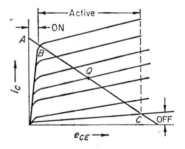

FIG. 8.2 ON-OFF conditions of the multivibrator.

Notice that the ON transistor now has its operating point to the left of the knee region of the characteristics, which is called the *saturation* region. The chief characteristic of the ON condition, as can be read from Fig. 8.2, is that $e_{C\text{-}E}$ is very low, i_C is very high, and i_B is likewise very high.

The OFF transistor is at *cutoff*, and its parameters can be read from the characteristics. Note that $e_{C\text{-}E}$ is very high, i_C is very low, and i_B is zero. The region of operation between ON and OFF, through which the operating point of both transistors must pass when going from one condition to the other, is called the *active* region.

Theory of Operation of the Astable Multivibrator. Let us arbitrarily refer to T_1 ON, T_2 OFF as condition 1, and T_1 OFF, T_2 ON as condition 2.

Condition 1 (T_1 ON, T_2 OFF). With T_1 ON, T_2 OFF, $e_{C\text{-}E}T_1$ is a small negative value, since T_1 is in its saturation region; and $e_{C\text{-}E}T_2$ is a large negative value because T_2 is in its cutoff region. As T_2 moves toward cutoff, plate A of C_2 discharges electrons through R_{B2}, V_{cc}, R_{L2}, and back to plate B of C_2 (since $e_{C\text{-}E}T_2$ becomes more negative). The current flowing through R_{B2} places the base of T_1 at a more negative potential. Being a PNP transistor, this is forward base bias and drives T_1 to its saturation region.

Condition 2 (T_1 OFF, T_2 ON). While T_1 is conducting heavily, $e_{C\text{-}E}T_1$ is low. The discharge current of C_2 decreases exponentially, reducing e_{RB2}, and the operating point of T_1 begins to move from saturation toward the active region. $e_{C\text{-}E}T_1$ becomes more negative, and so must plate A of C_1. Plate B of C_1 discharges, and the path of electron flow is from plate B of C_1, R_{B4}, V_{cc}, R_{L1}, and back to plate A of C_1. Notice that e_{RB4} places the base of T_2 at a negative potential. Being a PNP transistor, this brings T_2 out of its cutoff region into the active region. Plate B of C_2 and $e_{C\text{-}E}T_2$ are at the same potential. Since T_2 is now conducting, $e_{C\text{-}E}T_2$ becomes less negative, which reduces the potential across C_2 by that amount. The result of this change is a reduction in e_{RB2}, causing T_1 to conduct less. $e_{C\text{-}E}T_1$ becomes more negative, as does plate A of C_1, and plate B of C_1 discharges more heavily. e_{RB4} becomes greater, driving T_2 into heavier conduction, bringing T_1 closer to cutoff, and T_2 moves toward saturation. Eventually, T_2 arrives at saturation, and T_1 is cut off. Once T_2 is full ON and T_1 is OFF, the action described in condition 1 again takes place.

It is important to note that the period time (ON or OFF) hinges directly upon the charge time of the capacitor which drives the base of the ON transistor. The time constant for each condition (using the symbol notation of Fig. 8.1) is:

$$\text{tc (condition 1)} \cong (R_{B2} + R_{L2})C_2$$

and
$$\text{tc (condition 2)} \cong (R_{B4} + R_{L1})C_1$$

If the circuit is symmetrical, then $R_{B2} = R_{B4}$; $R_{L1} = R_{L2}$; and $C_1 = C_2$; hence

$$\text{tc (condition 1)} = \text{tc (condition 2)}$$

Waveform Analysis of the Symmetrical Astable Multivibrator. A study of the waveforms of the symmetrical astable multivibrator is useful in determining the value of components to be used in its design. Figure 8.3 shows the waveforms of the circuit in Fig. 8.1.

Notice that $e_{C\text{-}E}$ of both transistors, in order to obtain a reasonably good square-wave output, must undergo its change from ON to OFF or OFF to ON in a relatively short period of time. Let us restrict the discussion to T_1, since the requirements of T_2 are identical. When T_1 is ON, e_{RB2} is a relatively large negative potential (point A) in order to drive the transistor to its saturation region. e_{RB2}, at the beginning of the ON period, instantaneously increases because of the discharge current due to plate A of C_2. This current flow should be sufficient to drive T_1 well into its saturation region (e_{RB2} is at point A). e_{RB2} should be sufficiently large to keep T_1 in its saturation region for the entire ON period. This is important, since allowing T_1 to pass into its active region prior to the end of ON time would result in distortion of the square wave of $e_{C\text{-}E}$. e_{RB2} can therefore decrease from A to B of the illustrated waveform and still enable $e_{C\text{-}E}T_1$ ON to appear as shown. The discharge current of C_2's plate A is through R_{B2}, V_{cc}, R_{L2}. Assuming V_{cc} has only negligible resistance, the charging time constant is

$$\text{tc (charging plate } A \text{ of } C_2) \cong (R_{B2} + R_{L2})C_2$$

A handy rule of thumb, to ensure a reasonably good square-wave output

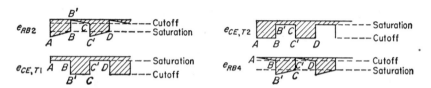

FIG. 8.3 Waveforms of the Fig. 8.1 circuit.

for T_1 ON, is to make this time constant 10 times longer than the time of T_1 ON:

$$\text{tc (charging plate } A \text{ of } C_2) \cong (R_{B2} + R_{L2})C_2 \cong 10T_1 \text{ ON}$$

In this way, the discharge current of C_2's plate A will decrease to not less than 90 per cent of its initial value.

At point B, plate A of C_2 begins to charge and plate B discharges, since T_2 is now conducting heavily. The plate B discharge path is through T_2 ON, R_{E2}, V_{cc}, and R_{B1}. This results in an instantaneous increase in e_{RB1}. Since $e_{RB1} + e_{RB2} = V_{cc}$, then e_{RB2} undergoes an instantaneous reduction (B to B'), bringing T_1 to well within its cutoff region. As the discharge current of C_2's plate B decreases, e_{RB1} decreases and e_{RB2} increases. Although the increase in e_{RB2} brings T_1 closer to its active region, this change can be made relatively small by providing a long time constant for this action. In this way, e_{RB2} can be at the threshold of its cutoff value at the end of the T_1 OFF period, at which time plate A of C_2 discharges, thereby repeating the T_1 ON sequence described earlier. The time constant of plate A discharge should be about 10 times longer than the T_1 OFF time.

$$\text{tc (discharging plate } A \text{ of } C_2) \cong (RT_2 \text{ ON} + R_{E2} + R_{B1})C_2 \cong 10T_1 \text{ OFF}$$

C_1 is the capacitor which drives T_2 into heavy conduction. As T_1 cuts off, $e_{C-E}T_1$ becomes highly negative, which causes plate A of C_1 to take on electrons. Plate B discharges electrons via R_{B4}, V_{cc}, and R_{L1}. e_{RB4} instantaneously increases from B to B', driving T_2 from cutoff at point B to far into the saturation region at B'. Providing plate A of C_1 with a time constant which is about ten times longer than the time of T_2 ON will allow e_{RB4} at the end of T_2 ON (point C) to be about 90 per cent of its value at the beginning of that period. T_1 fires ON at this instant, since e_{RB2} crossed over its cutoff point, thereby reversing the action of the capacitors. Plate A of C_1 discharges through T_1 ON, R_{E1}, V_{cc}, and R_{B3}, which instantaneously drives e_{RB4} to beyond the cutoff value of T_2 (C to C'). The discharge of plate A of C_2 at this same time instantaneously brings e_{RB2} to well within the saturation region of T_1 (C to C').

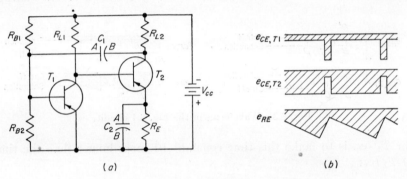

FIG. 8.4   A nonsymmetrical astable multivibrator.

As pointed out in the preceding analysis, the period time hinges directly upon the size of the capacitors. Larger values increase the period time, which is a lower oscillation frequency, and vice versa.

Nonsymmetrical Astable Multivibrator. In the nonsymmetrical circuit, in which a pulse output is desired (or a saw-tooth across the base-emitter circuit), the relationship between tc1 and tc2 is determined by the size of the involved resistors and the capacitors. Figure 8.4 illustrates a nonsymmetrical astable multivibrator. T_1 is biased by the voltage-divider arrangement of R_{B1} and R_{B2} in such a way that T_1 is ON (in its saturation region) most of the time and T_2 conducts only for a fraction of the time.

When T_2 is at cutoff, plate A of C_1 discharges electrons through V_{cc}, R_{L2}, and plate B of C_1. When T_2 conducts, plate B of C_2 discharges electrons through V_{cc}, R_{L2}, T_2, and back to plate A of C_2. Hence plate A of C_2 is negative, which places the emitter of T_2 at a negative potential. Being a PNP transistor, this is felt as a reverse bias. When plate A

becomes sufficiently negative, T_2 will cut off, at which time plate A of C_2 discharges electrons through R_E to plate B of C_2. As C_2 discharges in this manner, e_{RE} exponentially decreases until T_2 can again conduct, and C_2 then charges again. The time during which T_2 remains at cutoff is largely determined by the discharge time constant of C_2, which is

$$tc\ C_2\ (\text{discharge}) \cong R_E C_2$$

When T_2 conducts, plate B of C_1 becomes less negative and plate A of C_1 takes on electrons. The electron flow is from plate B of C_1 through T_2, R_E, and back to plate A of C_1. T_1 is cut off by this action until T_2 again cuts off, at which time plate A of C_1 discharges in the manner previously described.

8.2   Design of a Symmetrical Astable Multivibrator. Let us design a symmetrical astable multivibrator similar to Fig. 8.1. T_1 and T_2 are to be Raytheon 2N465 transistors; $V_{cc} = 20$ volts; frequency $= 1$ kc.

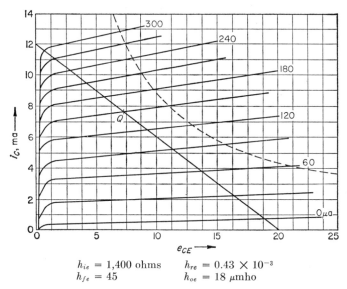

$$h_{ie} = 1{,}400 \text{ ohms} \qquad h_{re} = 0.43 \times 10^{-3}$$
$$h_{fe} = 45 \qquad\qquad h_{oe} = 18 \ \mu\text{mho}$$

Fig. 8.5   Typical collector characteristics of the 2N465. (*Characteristics from Raytheon.*)

Since the circuit is to be symmetrical, corresponding components associated with each transistor will be of the same value. The typical collector characteristics of the 2N465 are illustrated in Fig. 8.5.

The first step is to construct the maximum power-dissipation curve for an ambient temperature which will not be exceeded by the circuit. An $R_L + R_E$ value will then be selected, and a portion of this value will be assigned to R_E. By careful evaluation of the base voltage for both ON and OFF conditions, the voltage-divider resistors can be determined. The final step of the design computations is the selection of the coupling capacitor value, which is determined in great part by the desired frequency of oscillations.

Step 1: Construction of the Maximum Allowable Total Transistor Power-dissipation Curve. The maximum allowable total transistor power dissipation hinges on the ambient temperature. Assume the ambient temperature is to be 50°C, which is a safe assumption for this circuit. Using the equation stated in the Raytheon technical information sheet for the 2N465, $P_{c,\text{max}}$ can be found:

$$P_{c,\text{max}} = \frac{(T_{j,\text{max}} - T_a)}{k}$$

where $T_{j,\text{max}}$ = maximum junction temperature (85°C)
$\quad\ T_a$ = ambient temperature (50°C)
$\quad\ k$ = dissipation coefficient (0.40°C/mw)
Substituting into the equation,

$$P_{c,\text{max}} = \frac{85 - 50}{0.4} \cong 90 \text{ mw}$$

Recall that $P_{c,\text{max}} = e_c i_c$, and transposing,

$$e_c = \frac{P_{c,\text{max}}}{i_c}$$

Substituting several values of i_c yields the corresponding values of e_c required for maximum power dissipation. Connecting these points by a broken line results in the maximum power-dissipation curve for 50°C ambient temperature (see Fig. 8.5).

Step 2: Selection of $R_L + R_E$. The selection of $R_L + R_E$ is made such that the maximum power-dissipation curve is not intersected. Assume $V_{cc} = 20$ volts; and $i_{C,M} = 12$ ma. Connecting these two points produces a load line which meets these requirements. The value of $R_L + R_E$ can now be found:

$$R_L + R_E = \frac{V_{cc}}{i_{C,M}} = \frac{20}{12 \times 10^{-3}} = 1.67 \text{ kilohms}$$

Let us make $R_L = 1.6$ kilohms and $R_E = 70$ ohms.

Step 3: Selection of an Operating Point Q. Q is to be so selected that both transistors would operate as class A amplifiers if the cross-coupling capacitors were omitted. Using this factor as a basis, Q can be located on a number of points along the load line in the active region. The selected Q is at the intersection of the load line and the 150-μa characteristic.   The transistor parameters at this point are:

$$i_{C,Q} = 7.8 \text{ ma}$$
$$e_{C\text{-}E,Q} = 7.13 \text{ volts}$$
$$i_{B,Q} = 150 \ \mu\text{a}$$

Step 4: *Determination of* $e_{B\text{-}E,Q}$.   This value is read directly from the $e_{B\text{-}E}$ versus i_C or $e_{B\text{-}E}$ versus i_B characteristics, when available.   Since neither of these characteristics is included in the specification folder, one of the other possible methods for approximating this value must be used. Since the hybrid parameters are given, R_i can be easily computed.   Using the $i_{B,Q}$ value determined in step 3 (150 μa), we can arrive at a good approximation of $e_{B\text{-}E,Q}$: Since

$$R_i \cong \frac{e_{B\text{-}E,Q}}{i_{B,Q}}$$

then, by transposing,

$$e_{B\text{-}E,Q} \cong R_i i_{B,Q}$$
$$\cong \frac{h_{ie} + (h_{oe}h_{ie} - h_{fe}h_{re})R_L}{1 + h_{oe}R_L} i_{B,Q}$$

Substituting values, and solving,

$$e_{B\text{-}E,Q} \cong \frac{1{,}400 + (18 \times 10^{-6} \times 1.4 \times 10^3 - 45 \times 0.43 \times 10^{-3})}{1 + (18 \times 10^{-6})(1.67 \times 10^3)} 1.67 \times 10^3 \times 1.5 \times 10^{-4}$$
$$\cong 0.21 \text{ volt}$$

Step 5: *Determination of* R_{B2}. R_{B1} and R_{B2} form the conventional voltage-divider arrangement, with e_{RB2} being the forward base-bias voltage to place T_1 at the operating point Q.   Note, however, that e_{RE1} is providing reverse bias to the emitter, which is series-opposing to e_{RB2}. Hence, e_{RB2} must be large enough to overcome this reverse bias and still furnish 0.21 volt to the base.   In equation form,

$$e_{RB2,Q} \cong e_{RE1,Q} + e_{B\text{-}E,Q}$$

$e_{RE1,Q}$ must first be computed:

$$e_{RE1,Q} \cong (i_{C,Q} + i_{B,Q})R_{E1} \cong (7.8 + 0.15) \times 10^{-3} \times 70$$
$$\cong 0.557 \text{ volt}$$

We may now find $e_{RB2,Q}$:

$$e_{RB2,Q} \cong 0.557 + 0.21 \cong 0.767 \text{ volt}$$

Recall that $i_{RB2,Q}$ is a bleeder current with a value arbitrarily selected.

In this circuit, let us use a bleeder current of 100 μa. Solving for R_{B2},

$$R_{B2} = \frac{e_{RB2,Q}}{i_{RB2,Q}} = \frac{7.67 \times 10^{-1}}{1 \times 10^{-4}} = 7.67 \text{ kilohms}$$

Step 6: Determination of R_{B1}. $i_{RB1,Q}$ is the composite of $i_{B,Q}$ and $i_{RB2,Q}$, and is

$$i_{RB1,Q} = i_{B,Q} + i_{RB2,Q} = (0.15 + 0.10) \times 10^{-3}$$
$$= 250 \ \mu\text{a}$$

e_{RB1} can now be found, since

$$e_{RB1} = e_{C\text{-}B,Q} + e_{RL,Q} = V_{cc} - e_{RB2}$$

Substituting values,

$$e_{RB1} = 20 - 0.767 \cong 19.23 \text{ volts}$$

R_{B1} can now be computed by use of the current and voltage values just determined:

$$R_{B1} = \frac{e_{RB1}}{i_{RB1}} = \frac{19.233}{2.5 \times 10^{-4}} \cong 80 \text{ kilohms}$$

Step 7: Determination of C_2. Adopting the rule of thumb stated in the preceding discussion,

$$\text{tc (charging plate } A \text{ of } C_2) \cong (R_{B2} + R_{L2})C_2 \cong 10T_1 \text{ ON}$$

T_1 ON should be one-half of each cycle. The time of one cycle for a frequency of 1 kc is

$$T_p = \frac{1}{f} = \frac{1}{1 \times 10^3} = 0.001 \text{ sec}$$

and $$T_1 \text{ ON} = 0.5T_p = 0.0005 \text{ sec}$$

The charging time constant of C_2's plate A should be

$$10 \times 0.0005 = 0.005 \text{ sec}$$

Recall that $R_{L1} = R_{L2}$, hence

$$R_{B2} = 7.67 \text{ kilohms}$$
$$R_{L1} = 1.6 \text{ kilohms}$$
and $$R_{B2} + R_{L2} = 9.27 \text{ kilohms}$$

Transposing the tc equation, and solving for C_2,

$$C_2 = \frac{10T_1 \text{ ON}}{(R_{B2} + R_{L2})} = \frac{5 \times 10^{-3}}{9.27 \times 10^3}$$
$$\cong 0.54 \ \mu\text{f}$$

Step 8: Circuit Construction and Checkout. The required circuit components have been determined to be

$$R_{L1}, R_{L2} = 1.6 \text{ kilohms}$$
$$R_{E1}, R_{E2} = 70 \text{ ohms}$$
$$R_{B2}, R_{B4} = 7.67 \text{ kilohms}$$
$$R_{B1}, R_{B3} = 80 \text{ kilohms}$$
$$C_1, C_2 = 0.54 \text{ } \mu\text{f}$$

Troubleshooting Hints. Because of individual variations of transistors and components, a number of adjustments may be required so that oscillations will occur. If the circuit fails to oscillate, the following procedure will help to locate the problem.

Remove the coupling capacitors that are connected between each base and collector. Using a d-c voltmeter, measure $e_{C\text{-}E}$ of each transistor. The collector-to-emitter voltage of each transistor should be that Q value selected on the collector characteristics.

If $e_{C\text{-}E}$ is too high, this indicates the collector current of that transistor is too low. Increase the forward bias of the transistor (R_{B2} for T_1, and R_{B4} for T_2) until $e_{C\text{-}E}$ decreases to the selected Q value. If $e_{C\text{-}E}$ cannot be reduced to the desired value, the other voltage-divider bias resistor may have to be reduced (R_{B1} for T_1, and R_{B3} for T_2).

This establishes the designed class A operation point of each transistor. The coupling capacitors are then placed back into the circuit.

The output waveforms are affected by R_E. Increasing R_E reduces the output-voltage magnitude but improves the linearity of the waveform. The final value of R_E can be decided by compromise of these two factors.

The final adjustment for the frequency of oscillations is conveniently made on an empirical basis, recognizing that a smaller C value increases the frequency and vice versa. Another convenient technique for making changes in the frequency is to alter R_{B2} and R_{B4}. Raising the values of these resistors increases the period (frequency reduction), and the frequency can be increased by reducing the values of R_{B2} and R_{B4}.

NOTE: The characteristics of several transistors suitable for multivibrator circuits are given in Figs. 8.26 to 8.28 to enable the reader to design the preceding circuit.

8.3 Design Criteria for the Nonsymmetrical Astable Multivibrator.

A nonsymmetrical multivibrator similar to the circuit illustrated in Fig. 8.4 can be designed with a minimum of difficulty. Recall that T_1 is ON most of the time and T_2 conducts for only a fraction of the period time.

Step 1. Select an appropriate V_{cc} and transistors.

Step 2. By using the load-line technique, select R_{L1} and the operating point of T_1 near its saturation region.

Step 3. Determine R_{B2} and R_{B1} so as to achieve the proper bias for the selected Q value of T_1 in step 2.

Step 4. Determine C_1. Plate A of C_1 discharges through V_{cc}, R_{L2}, and to plate B of C_1 when T_1 is conducting. This time constant should be about 10 times longer than T_1 ON.

Step 5. Determine C_2 so that the ratio of the plate A charging time (plate B, to V_{cc}, to R_{L2}, to T_2, and back to plate A of C_2) to the plate B charging time (plate A, to R_E, back to plate B) is equal to the desired ratio of T_1 OFF/T_1 ON.

Raising the value of R_{B2} increases T_1 ON (and vice versa); increasing R_E also increases T_1 ON, since it decreases T_2 OFF time.

PROBLEMS

8.1 Refer to the astable multivibrator of Fig. 8.1. How would T_1 and T_2 operate if C_1 and C_2 were excluded from the circuit? Explain.

8.2 In Fig. 8.1, why are both conditions of operation unstable?

8.3 In Fig. 8.1, state the conditions required for a square-wave output in terms of (a) transistors; (b) resistances; (c) capacitances.

8.4 In Fig. 8.1, describe the action of C_2 in bringing T_2 toward cutoff and T_1 toward saturation.

8.5 In Fig. 8.1, describe the action of C_1 in bringing T_1 to cutoff and T_2 to saturation.

8.6 In Fig. 8.2, describe what occurs in the saturation region.

8.7 In Fig. 8.4, describe how R_{B1} and R_{B2} can cause T_1 to be ON most of the time.

8.8 What is the fundamental difference between a symmetrical and nonsymmetrical astable multivibrator?

8.9 Refer to the symmetrical astable multivibrator of Fig. 8.1. The 2N465 transistor is to be used (see Fig. 8.5). $V_{cc} = 25$ volts; frequency = 10 kc; let $R_{L1} = R_{L2} = 2.2$ kilohms; $R_{E2} = R_{E1} = 300$ ohms; $i_{B.Q} = 120$ μa.
(a) Construct the load line.
(b) Determine $i_{C.Q}$ and $e_{C.Q}$ of each transistor.

8.10 In the astable multivibrator of Prob. 8.9, find (a) R_i of each transistor; (b) $e_{B-E.Q}$ of each transistor.

8.11 Referring to Probs. 8.9 and 8.10, find (a) e_{RE} of each transistor; (b) e_{RB2} and e_{RB4}; (c) R_{B2} and R_{B4} (let $i_{bleeder} = 100$ μa); (d) R_{B1} and R_{B3}.

8.12 Referring to the astable multivibrator of Probs. 8.9 to 8.11, find (a) time of 1 cycle; (b) T_1 ON; (c) charging time constant of plate A of C_2; (d) C_2 and C_1.

8.13 Refer to the symmetrical multivibrator of Fig. 8.1. The 2N337 transistor is to be used (see Fig. 8.26). $V_{cc} = 30$ volts; $R_{L1} = R_{L2} = 1.9$ kilohms; $R_{E1} = R_{E2} = 100$ ohms; frequency = 100 kc; $i_{B.Q} = 200$ μa.
(a) Construct the load line.
(b) Determine $i_{C.Q}$ and $e_{C.Q}$ of each transistor.

8.14 In the astable multivibrator of Prob. 8.13, by using the collector characteristics in conjunction with the i_C versus e_{B-E} characteristic, find $e_{B-E.Q}$.

8.15 Referring to Probs. 8.13 and 8.14, find (a) e_{RE} of each transistor; (b) e_{RB2} and e_{RB4}; (c) R_{B2} and R_{B4} (let $i_{bleeder} = 200$ μa); (d) R_{B1} and R_{B3}.

8.16 Referring to the astable multivibrator of Probs. 8.13 to 8.15, find (a) time of 1 cycle; (b) T_1 ON; (c) charging time constant of plate A of C_2; (d) C_2 and C_1.

8.4 The Monostable (One-shot) Multivibrator.

A monostable multivibrator has one stable and one unstable state. In the circuit of Fig. 8.6, T_2

is normally ON and T_1 is OFF. In the normal state, e_{RB2} is less than e_{RE2}, so that reverse bias is applied to T_1, which holds this transistor at cutoff. Plate A of C_1 is at the same high negative potential as the collector of T_1. The potential across C_1 at this time is $e_{C-E}T_1 + e_{RE} - e_{RB4}$. T_2, in the normal condition, is biased so that e_{RB4} exceeds e_{RE} sufficiently to maintain this transistor at saturation. Notice that the forward base-bias voltage for both transistors is obtained by use of the conventional voltage-divider arrangement.

The normal condition of T_1 OFF, T_2 ON is upset by a negative pulse applied to the base of T_1 (when NPN transistors are used, this activating pulse is positive). This pulse must be sufficiently large to drive T_1 into its saturation region. Plate A of C_1 instantaneously begins to discharge since $e_{C-E,T1}$ has suddenly decreased to a very low negative potential (T_1

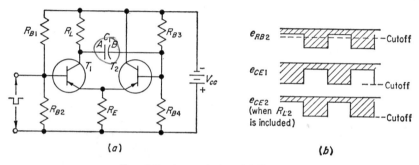

Fig. 8.6   A one-shot multivibrator.

is now in saturation). This discharge path is through T_1 ON, R_E, V_{cc}, R_{B3}, to plate B of C_1. Recall that the discharge current instantaneously rises to a value determined by the change in voltage and series resistance. In equation form,

$$i_{C1} \text{ (plate } A \text{ discharge)}_{0 \text{ time}} \cong \frac{\Delta e_{C-E,T1}}{R_{T1,ON} + R_E + R_{B3}}$$

Since T_1 is at saturation, its resistance is very low (100 ohms or less in most cases). The instantaneous current rise results in a corresponding increase in e_{RB3}. e_{RB4} experiences an instantaneous decrease which brings T_2 into its cutoff region. T_2 is maintained at cutoff until the capacitor discharge current through R_{B3} has decayed to the point where e_{RB3} is small enough to allow e_{RB4} to be sufficiently great to drive T_2 into its ON condition. Once T_2 turns back ON, e_{RE} will increase enough to cut off T_1 again. This is the stable condition and will be maintained until the next negative pulse is applied to the base of T_1.

The time during which T_1 is ON is determined by the discharge current of C_1's plate A. This type of multivibrator is especially useful in standardizing trigger pulses of nonuniform widths. Regardless of the trigger pulse width, T_1 will remain ON for a predetermined amount of time. C_1, $R_{T1,\text{ON}}$, R_E, and R_{B2} determine the time of T_1 ON, making its operation independent of the input pulse width. The one-shot circuit may be used for obtaining a square-wave output of predetermined frequency or as a time-delayed pulse output. T_2 may have an R_{L2} in its collector circuit if a negative-going output pulse is desired (as shown in Fig. 8.7).

8.5 Design of a One-shot Multivibrator. Let us design the one-shot multivibrator illustrated in Fig. 8.7. A resistor R_{L2} will be in the collector circuit of T_2 in the event a negative-going square-wave pulse is desired. T_1 and T_2 are to be Raytheon 2N465 transistors; V_{cc} is 20 volts; frequency of the output square-wave pulse from either collector to

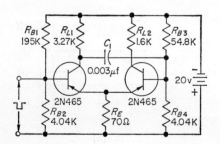

Fig. 8.7 A designed one-shot multivibrator.

emitter is to be 1 kc. The parameters of each transistor will be designed to achieve the normal condition of T_1 OFF, T_2 ON. Then C_1 will be selected to maintain the unstable condition of T_1 ON, T_2 OFF for a period such that the output pulse frequency is 1 kc. We will first determine the parameters of T_2, which is normally ON. Since V_{cc} is reasonably high (20 volts), it is not necessary to drive T_2 to saturation during its ON time.

Step 1: Construction of T_2 Load Line and Determination of R_{L2} and R_E. The slope of the load line is selected in the usual manner. Let $i_{C,M} = 12$ ma. This results in $R_{L2} + R_E = 1.67$ kilohms. We will use $R_E = 70$ ohms and $R_{L2} = 1.6$ kilohms. See Fig. 8.8 for this load line BQ_2.

Step 2: Selection of Q_2. Q_2 is selected at the intersection of the 150-μa characteristic and the load line. Parameters at this point are:

$$i_{B2,Q} = 150 \ \mu a$$
$$i_{C2,Q} \cong 7.8 \ ma$$
$$e_{C2,Q} \cong 7.125 \ volts$$
$$e_{RL2,Q} \cong 12.48 \ volts$$

Step 3: Determination of R_{B4} and R_{B3}. R_{B4} can be found once e_{RB4}

and i_{RB4} are known:

$$e_{RB4,Q} = e_{B-E2,Q} + e_{RE,Q}$$

Both of these voltages must next be determined:

$$e_{B-E2,Q} \cong R_i i_{B2,Q}$$
$$\cong \frac{h_{ie} + (h_{oe}h_{ie} - h_{fe}h_{re})R_L}{1 + h_{oe}R_L} i_{B2,Q}$$

Substituting values and solving,

$$e_{B-E,Q} \cong \frac{1{,}400 + (18 \times 10^{-6} \times 1.4 \times 10^3 - 45 \times 0.43 \times 10^3)}{1 + (18 \times 10^{-6} \times 1.67 \times 10^3)} \cdot 1.67 \times 10^3 \times 1.5 \times 10^{-4}$$
$$\cong 0.21 \text{ volt}$$

and

$$e_{RE,Q} \cong (i_{E1,Q} + i_{E2,Q})R_E$$

$i_{C1,Q}$ is at cutoff and this value, which is $i_{E1,Q}$, is approximated by the collector current that flows with the condition of zero base current and a

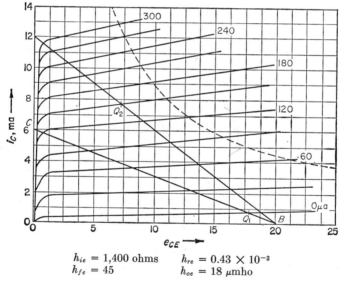

$$h_{ie} = 1{,}400 \text{ ohms} \qquad h_{re} = 0.43 \times 10^{-3}$$
$$h_{fe} = 45 \qquad h_{oe} = 18 \text{ } \mu\text{mho}$$

Fig. 8.8 Typical collector characteristics of the 2N465. (*Characteristics from Raytheon.*)

collector voltage equal to V_{cc}. $i_{E2,Q}$ is the composite of $i_{C2,Q}$ and $i_{B2,Q}$. Substituting these values into the preceding equation,

$$e_{RE,Q} = (0.7 + 7.8 + 0.15) \times 10^{-3} \times 70$$
$$= 8.65 \times 10^{-3} \times 7 \times 10^1 \cong 0.606 \text{ volt}$$

We are now in the position to compute e_{RB4}:

$$e_{RB4} = 0.210 + 0.606 = 0.816 \text{ volt}$$

R_{B4} can be found next. i_{RB4} is an arbitrarily selected bleeder current, which is 200 μa in this case:

$$R_{B4} = \frac{e_{RB4,Q}}{i_{RB4,Q}} = \frac{8.16 \times 10^{-1}}{2 \times 10^{-4}} = 4.04 \text{ kilohms}$$

and

$$R_{B3} = \frac{V_{cc} - e_{RB4,Q}}{i_{RB4,Q} + i_{B2,Q}} = \frac{20 - 0.816}{(200 + 150) \times 10^{-6}}$$
$$= 54.8 \text{ kilohms}$$

Now that the parameters for T_2 have been calculated, we next turn our attention to T_1, which is normally OFF.

Step 4: Construction of T_1 Load Line and Determination of R_{L1}. T_1 can be maintained at cutoff more easily by use of a larger R_L. Let $i_{C1,M} = 6$ ma. Joining points C and B with a straight line renders a load line of the resistance equal to 3.34 kilohms. R_E was computed in step 1; hence

$$R_E = 70 \text{ ohms}$$
$$R_{L1} = 3.27 \text{ kilohms}$$

Step 5: Selection of Q_1. As stated earlier, T_1 is OFF under normal conditions; therefore Q_1 is at the intersection of the T_1 load line and the 0-μa collector characteristic. The parameters at this point are:

$$i_{C1,Q} = 0.7 \text{ ma}$$
$$e_{C1,Q} = 17.805 \text{ volts}$$
$$e_{RL1,Q} = 1.589 \text{ volts}$$
$$e_{RE,Q} = 0.606 \text{ volt}$$
$$i_{B1,Q} = 0 \text{ } \mu\text{a}$$

Step 6: Determination of R_{B2} and R_{B1}. $e_{B\text{-}E,Q1}$ is first determined:

$$e_{B\text{-}E,Q1} = i_{B1,Q}R_i$$

since $i_{B1,Q}$ is 0 μa, then $e_{B\text{-}E,Q1} = 0$ volts or some reverse bias value.

$e_{RE,Q}$ was earlier found to be 0.606 volt; thus e_{RB2} need only be equal or less than this value in order to place T_1 at cutoff. Let e_{RB2} be equal to 0.406 volt, thereby assuring a reverse bias of 0.2 volt at the base of T_1, because

$$e_{B\text{-}E,Q} = e_{RE} + e_{RB2} \qquad \text{algebraic sum}$$

Since e_{RE} and e_{RB2} are series-opposing, then

$$e_{B\text{-}E,Q} = 0.606 - 0.406 \cong 0.2 \text{ volt}$$

Note that this is a reverse bias. As in the previous voltage-divider computations, i_{RB2} is a bleeder current, and 100 μa will be used in this instance. We may now determine R_{B2}:

$$R_{B2} = \frac{e_{RB2,Q}}{i_{RB2,Q}} = \frac{4.06 \times 10^{-1}}{1 \times 10^{-4}} \cong 4.0 \text{ kilohms}$$

and
$$R_{B1} = \frac{V_{cc} - e_{RB2,Q}}{i_{RB2,Q} + i_{B1,Q}} = \frac{20 - 0.406}{(100 + 0) \times 10^{-6}}$$
$$\cong 195 \text{ kilohms}$$

Step 7: *Determination of Pulse Amplitude Required to Turn T_1* ON. T_1 will turn ON when e_{RB2} is increased by a minimum of 0.2 volt (which is the difference between $e_{RE,Q}$ and $e_{RB2,Q}$). Assume a negative pulse is applied to the base of T_1 which can drive T_1 to saturation condition of $i_{B1} = 120$ μa and $e_{C-E,T1} = 0.75$ volt, resulting in i_{C1} of 5.8 ma. T_2 is cut off at this instant, and e_{RE} is

$$e_{RE} = (i_{E,T1,\text{sat}} + i_{C,T2,\text{co}})R_E$$
$$= (5.8 + 0.12 + 0.7) \times 10^{-3} \times 70$$
$$= 0.463 \text{ volt}$$

$e_{B-E,T1}$ at this condition can be approximated:

$$e_{B-E,T1} \cong i_{B1}R_i$$
$$\cong 1.2 \times 10^{-4} \times 14 \times 10^2 \cong 0.168 \text{ volt}$$

Thus e_{RB2} is to be 0.168 volt greater than e_{RE} at this condition:

$$e_{RB2(\text{sat})} \cong 0.463 + 0.168 \cong 0.631 \text{ volt}$$
but $$e_{RB2,Q} \cong 0.406 \text{ volt}$$

hence a negative pulse of $0.631 - 0.406 \cong 0.225$ volt is sufficient to drive T_1 ON to saturation.

Step 8: *Determination of C_1.* The desired frequency is 1 kc. The period time is

$$T_p = \frac{1}{f} = \frac{1}{1 \times 10^3} = 1 \times 10^{-3} \text{ sec}$$

T_1 and T_2 should each be ON half this time and OFF half this time, which is 5×10^{-4} sec. Recall that T_1 is turned ON by the application of a negative pulse to its base, which turns T_2 OFF. The time T_2 is OFF is determined by the time e_{RB3} is large enough to deprive the base of T_2 of sufficient forward voltage. T_2 will conduct when $e_{B-E,T2}$ is negative, but $e_{B-E,T2}$ is equal to the algebraic sum of e_{RE} and e_{RB4}. When T_2 is OFF, T_1

is on, and
$$e_{RE} = 0.463 \text{ volt}$$

hence e_{RB4} is less than 0.463 volt for the time T_2 is off. Since

$$e_{RB3} = V_{cc} - e_{RB4}$$

then e_{RB3} is greater than 19.537 volts for 5×10^{-4} sec.

The resistance of T_1 on is very small because T_1 is in its saturation region and R_E is small as compared to the value of R_{B3} (54.8 kilohms). C_1 should be large enough to ensure a sufficient capacitor discharge current through R_{B3} to maintain e_{RB3} greater than 19.537 volts for 5×10^{-4} sec.

When T_2 is off, i_{RB3} is a composite of the voltage-divider bleeder current and the capacitor discharge current. Using the superposition technique, let us determine e_{RB3} with the capacitor discharge current:

$$i_{RB3(\text{bleeder})} = \frac{20}{54.8\text{K} + 4.06\text{K}} = 339 \ \mu\text{a}$$

and e_{RB3} is

$$3.39 \times 10^{-4} \times 5.48 \times 10^4 = 18.577 \text{ volts}$$

Therefore e_{RB3} is $19.537 - 18.577 = 0.96$ volt short of maintaining T_2 at off without the capacitor discharge current.

Let us next determine what capacitor discharge current must be flowing at 5×10^{-4} sec to maintain T_2 at its cutoff threshold:

$$i_{RB3(\text{capacitor})} = \frac{e_{RB3} \ (\text{needed})}{R_{B3}}$$
$$= \frac{9.6 \times 10^{-1}}{5.48 \times 10^4} = 17.5 \ \mu\text{a}$$

The zero time capacitor current is

$$i_{C1(0 \text{ time})} = \frac{\Delta e_{C\text{-}E,T1}}{R_{B3}} = \frac{17.805 - 0.75}{5.48 \times 10^4}$$
$$= 311 \ \mu\text{a}$$

i_{C1} is to exponentially decrease from 311 to 17.5 μa in 5×10^{-4} sec. In other words, i_{C1} must decay to about 5 per cent of its initial value in that time. It will be recalled that the capacitor discharge current exponentially decays to about 5 per cent of its initial value in 3 tc's. Hence

$$3 \text{ tc} = 5 \times 10^{-4} \text{ sec}$$

and

$$\text{tc} = 1.67 \times 10^{-4} \text{ sec}$$

Using R_{B3} for the total resistance and knowing the time constant, we can determine C_1:

$$tc = R_{B3}C_1$$

and
$$C_1 = \frac{tc}{R_{B3}} = \frac{1.67 \times 10^{-4}}{5.48 \times 10^4} = 0.003047 \ \mu f$$

Step 9: *Circuit Construction.* The final value of C_1 is determined upon completion of circuit construction. Recall that enlarging C_1 reduces the frequency, and vice versa. R_{B3} can also be changed within limits with the same effect (an increase in R_{B3} reduces the frequency, and vice versa). The determined component values are:

$$R_{B1} = 195 \text{ kilohms}$$
$$R_{B2} \cong 4.0 \text{ kilohms}$$
$$R_{B3} = 54.8 \text{ kilohms}$$
$$R_{B4} \cong 4.0 \text{ kilohms}$$
$$R_{L1} = 3.27 \text{ kilohms}$$
$$R_{L2} = 1.6 \text{ kilohms}$$
$$R_E = 70 \text{ ohms}$$
$$C_1 = 0.003047 \ \mu f$$

PROBLEMS

8.17 Refer to the monostable multivibrator of Fig. 8.6. What is the relationship between e_{RB2} and e_{RE} in the normal state?

8.18 In Fig. 8.6, how is T_1 fired ON?

8.19 In Fig. 8.6, what determines the time of T_1 ON?

8.20 Why is T_1 ON time independent of the trigger pulse width?

8.21 Refer to the one-shot multivibrator of Fig. 8.7. T_1 and T_2 are to be 2N465 transistors; $V_{cc} = 25$ volts; frequency of the output square wave from either collector to emitter is to be 10 kc; the normal condition is to be T_1 OFF and T_2 ON. Let $R_{L2} = 2.4$ kilohms; $R_E = 100$ ohms; $i_{B2,Q} = 150 \ \mu a$.
(*a*) Construct the T_2 load line.
(*b*) Determine $i_{C2,Q}$, $e_{C2,Q}$, and $e_{RL2,Q}$.

8.22 In Prob. 8.21, find (*a*) R_{i2}; (*b*) $e_{B\text{-}E2,Q}$; (*c*) $e_{RE,Q}$.

8.23 Referring to Prob. 8.22, compute (*a*) $e_{RB4,Q}$; (*b*) R_{B4} (let $i_{\text{bleeder}} = 100 \ \mu a$); (*c*) e_{RB3}; (*d*) R_{B3}.

8.24 Referring to Probs. 8.21 to 8.23, let $R_{L1} = 5$ kilohms; $i_{B1,Q} = 0 \ \mu a$.
(*a*) Construct the T_1 load line.
(*b*) Determine $i_{C1,Q}$, $e_{C1,Q}$, and $e_{RL1,Q}$.

8.25 In Prob. 8.24, assume 0.3 volt of reverse bias at the base of T_1 is desired for the time T_2 is ON. Find:
(*a*) $e_{RE,Q}$ (*b*) e_{RB2} (let $i_{RB2} = 100 \ \mu a$)
(*c*) R_{B2} (*d*) R_{B1}

8.26 Referring to the one-shot multivibrator of Probs. 8.17 to 8.25, find the following when T_2 is OFF and T_1 is ON:
(*a*) e_{RE} (*b*) $e_{B\text{-}E,T1}$
(*c*) e_{RB2} (*d*) required pulse to drive T_1 ON to saturation

8.27   In Probs. 8.17 to 8.26, find (a) T_p; (b) T_1 ON time, T_2 ON time; (c) i_{C1} (0 time); (d) C_1.

8.28   Refer to the one-shot multivibrator of Fig. 8.7. T_1 and T_2 are to be the 2N337 (see Fig. 8.26); $V_{cc} = 20$ volts; frequency of the output square wave from either collector to emitter is to be 50 kc; the normal condition is to be T_1 OFF and T_2 ON. Let $R_{L2} = 1.9$ kilohms; $R_E = 100$ ohms; $i_{B2,Q} = 200$ μa.
(a) Construct the T_2 load line.
(b) Determine $i_{C2,Q}$, $e_{C2,Q}$, and $e_{RL2,Q}$.

8.29   In Prob. 8.28, find (a) $e_{B\text{-}E2,Q}$ (by use of the i_C versus $e_{B\text{-}E}$ characteristics); (b) $e_{RE,Q}$.

8.30   Referring to Probs. 8.28 and 8.29, compute (a) $e_{RB4,Q}$; (b) R_{B4} (let $i_{\text{bleeder}} = 200$ μa); (c) e_{RB3}; (d) R_{B3}.

8.31   Referring to Probs. 8.28 to 8.30, let $R_{L1} = 3.9$ kilohms; $i_{B1,Q} = 0$ μa.
(a) Construct the T_1 load line.
(b) Determine $i_{C1,Q}$, $e_{C1,Q}$, and $e_{RL1,Q}$.

8.32   In Prob. 8.31, assume 0.4 volt of reverse bias at the base of T_1 is desired for the time T_1 is ON. Find:
(a) $e_{RE,Q}$ (b) e_{RB2}
(c) R_{B2} (let $i_{\text{bleeder}} = 100$ μa)   (d) R_{B1}

8.33   Refer to the one-shot multivibrator of Probs. 8.28 to 8.32. Find the following when T_2 is OFF and T_1 is ON:
(a) e_{RE} (b) $e_{B\text{-}E,T1}$
(c) e_{RB2} (d) required pulse to drive T_1 ON to saturation

8.34   In Probs. 8.28 to 8.33, find (a) T_p; (b) T_1 ON time, T_2 ON time; (c) $i_{C1(0\text{ time})}$; (d) C_1.

8.6   The Bistable Multivibrator.

Bistable multivibrators can be divided into two general classifications:

1. The flip-flop
2. All ON or all OFF

Both of these types are analyzed in this section.

The Symmetrical Flip-Flop. The bistable flip-flop circuit, as implied by its name, has two stable states:

1. T_1 ON, T_2 OFF
2. T_1 OFF, T_2 ON

The circuit is switched from one stable state to the other by the application of an input signal. The flip-flop circuit may be symmetrical or asymmetrical. The asymmetrical variety is treated later in this section. Figure 8.9 illustrates several symmetrical flip-flop possibilities.

The circuit of Fig. 8.9a is of great use where low-power operation is a chief requisite. Notice the direct coupling between collector 1, base 2 and collector 2, base 1. This circuit must be operated as a saturated flip-flop, since the collector-emitter voltage of the ON transistor must be less than the base-emitter voltage of the same transistor. The direct-coupled flip-flop is most effectively designed with transistors which display low collector voltages in their typical operation. At saturation, the base current is close in value to the collector current of that transistor. T_1 is turned OFF by the application of a positive going pulse to the base-emitter circuit of T_1, and a positive going pulse at the base-emitter circuit of T_2 will turn off T_2.

A more conventional type of flip-flop is illustrated in Fig. 8.9b. Base bias is obtained by the voltage-divider method, as shown in Fig. 8.10. Refer to Fig. 8.10a, which shows the voltage-divider network for T_1.

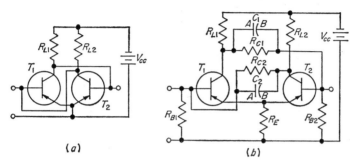

FIG. 8.9   Symmetrical flip-flop circuits: (a) direct coupled, (b) RC-coupled.

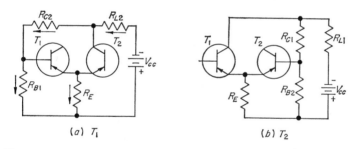

FIG. 8.10   Voltage-divider network for each transistor: (a) T_1, (b) T_2.

When T_1 is ON, e_{RB1} should be sufficiently greater than e_{RE} to drive the operating point of T_1 to its ON condition. When designing a saturated flip-flop circuit, e_{RB1} is such that e_{B-E} is that value required to bring the transistor to saturation. On the other hand, e_{RB2} is made not as large if T_1 ON is to be nonsaturated, since a smaller e_{B-E} is required for an ON condition which is less than saturation. Allowing the bleeder current i_{RB1} to be larger than i_{B1} can restrict e_{RB1} to a value which is too small to develop the saturation e_{B-E} ON value and therefore is a useful approach to the design of a non-saturated flip-flop.

A convenient method for designing a saturated flip-flop is as follows:

1. Determine R_L of each transistor and the common R_E by the usual load-line technique used in preceding sections. Do not exceed the

maximum dissipation rating of the transistor. Since the flip-flop is also symmetrical, all computations may be based on one of the two similar transistors.

2. Next select R_c and C, keeping in mind that R_c functions as part of the voltage-divider network for obtaining base bias. The value of C determines the maximum frequency (a smaller C results in a higher maximum repetition rate). C also permits a high base current during the transient interval.

3. Determine R_B. The value of e_{RB} and i_{RB} with the transistor ON must be found. Recall that e_{RB} ON should be sufficiently larger than e_{RE} to bring the transistor to saturation. In Fig. 8.10a,

$$e_{RB1} \text{ ON} = e_{RL2} + e_{RC2} - V_{cc}$$
$$i_{RB1} \text{ ON} = i_{RC2} - i_{B1} \text{ ON} = i_{bleeder}$$

and
$$R_{B1} = \frac{e_{RB1} \text{ ON}}{i_{RC2} - i_{B1} \text{ ON}}$$

Also of interest in the voltage-divider circuit in Fig. 8.10 is the bias change when T_1 is turned OFF. T_2 is at saturation when T_1 is OFF, resulting in $R_{T2,ON}$ being very low. i_{RL2} increases since i_{C2} flows through this resistor, and e_{RL2} increases to a value close to V_{cc}. e_{RC2} and e_{RB1} undergo a correspondingly large reduction. This reduction places e_{RB1} at a value less than e_{RE}, thereby placing the T_1 base-emitter circuit under reverse bias. Therefore, both the collector-base and base-emitter junctions of the OFF transistor are reverse-biased. This adds stability to the circuit, since the OFF transistor cannot create a regeneration process under this condition. In this way, the frequency of the flip-flop is more directly determined by the externally applied trigger.

The repetition rate of this circuit is decided by the applied trigger rate up to a frequency determined by the cross-coupling capacitors C_1 and C_2. The use of smaller capacitors increases the maximum repetition rate.

8.7 Design of a Symmetrical Saturated Flip-Flop. The circuit to be designed is that shown in Fig. 8.9b. The symmetrical saturated flip-flop is most conveniently designed with the use of the collector characteristics of the transistors to be used. From the load-line technique, the R_L and R_E values can be computed. The next step is to assume a suitable cross-coupling network (C and R_C). The base current ON and base voltage OFF are then calculated, from which the R_B values can be found. In the illustrated design which follows, Raytheon 2N465 transistors are used.

Step 1: Selection of R_L, R_E, and Load-line Construction. $V_{cc} = 20$ volts. The load line to be drawn has a slope equal to the reciprocal of $R_E + R_L$.

Select a load line such that the maximum allowable transistor power dissipation is not exceeded. The load line drawn between $i_C = 12$ ma and $e_{C\text{-}E} = 20$ volts was selected. This is shown in Fig. 8.5.

$$R_L + R_E = \frac{20}{12 \times 10^{-3}} = 1.67 \text{ kilohms for each transistor}$$

Arbitrarily let

$$R_L = 1.6 \text{ kilohms}$$
$$R_E = 70 \text{ ohms}$$

Step 2: Assume a Crossover Coupling Network. Let $C = 220$ μf; $R_C = 33$ kilohms.

Step 3: Determination of R_B. In order to compute R_B, we must first determine the values of e_{RB} ON and i_{RB} ON:

$$e_{RB} \text{ ON} = e_{RE} + e_{B\text{-}E} \text{ ON}$$

Since the i_C versus $e_{B\text{-}E}$ or i_B versus $e_{B\text{-}E}$ characteristics are not available, the $e_{B\text{-}E}$ ON is found by the following technique

$$e_{B\text{-}E} \text{ ON} \cong R_i i_B \text{ ON}$$
$$\cong \frac{h_{ie} + (h_{oe}h_{ie} - h_{fe}h_{re})R_L}{1 + h_{oe}R_L} \, i_B \text{ ON}$$

The ON condition of the transistor is taken to be the intersection point of the load line and the 300-μa characteristic. This point is very close

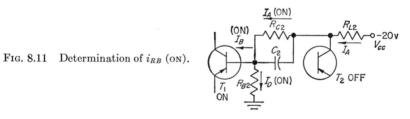

Fig. 8.11   Determination of i_{RB} (ON).

to the saturation region of the transistor. Substituting values, and solving,

$$e_{B\text{-}E} \text{ ON} \cong 0.4 \text{ volt}$$

e_{RE} can be computed from the following:

$$e_{RE} \cong \frac{R_E V_{cc}}{R_L + R_E} = \frac{70 \times 20}{1.670 \times 10^3} \cong 0.84 \text{ volt}$$

and

$$e_{RB} \text{ ON} = 0.84 + 0.40 = 1.33 \text{ volts}$$

In order to determine i_{RB} of the ON transistor, let us analyze the voltage-divider arrangement, which is illustrated in Fig. 8.11.

When T_1 is conducting, electron current flows as illustrated in Fig. 8.11. Recall that T_2 is OFF when T_1 is ON. e_{RB2} ON was found to be 0.49 volt, and the following relationship between the currents exists:

$$I_A = I_B + I_D$$

I_B is i_{B1} ON and equals 300 μa, as determined from the collector characteristics. I_A can be computed from

$$I_A = \frac{V_{cc} - e_{RB2} \text{ ON}}{R_{L2} + R_{C2}} = \frac{20 - 1.33}{34.6 \times 10^3}$$
$$\cong 540 \ \mu\text{a}$$

Note that I_A could be made larger by reducing R_{C2} and smaller by increasing R_{C2}. This is a convenient method of varying I_A without alternating the load-line work done up to this point.

i_{RB2} ON, which is I_D in Fig. 8.11, can now be calculated:

$$I_D = I_A - I_B$$
$$= 540 \ \mu\text{a} - 300 \ \mu\text{a} = 240 \ \mu\text{a}$$

Since I_D is the current through R_{B2}, and e_{RB2} is known, R_{B2} can be found:

$$R_{B2} = \frac{e_{RB2} \text{ ON}}{i_{RB2} \text{ ON}} = \frac{1.33}{240 \times 10^{-4}}$$
$$\cong 5.5 \text{ kilohms}$$

Step 4: *Circuit Construction.* The circuit is now constructed. Operation is checked with a voltmeter and oscilloscope. The calculated component values are:

$$R_{B1}, R_{B2} \cong 5.5 \text{ kilohms}$$
$$R_{C1}, R_{C2} = 33 \text{ kilohms}$$
$$C_1, C_2 = 200 \ \mu\text{f}$$
$$R_{L1}, R_{L2} = 1.6 \text{ kilohms}$$
$$R_E = 70 \text{ ohms}$$

Asymmetrical Flip-Flop. Figure 8.12 illustrates an asymmetrical flip-flop circuit. T_2 may be considered as in the common-collector configuration and T_1 as in the common-base configuration. This type of circuit is designed so that the base-to-common voltage of the ON transistor is greater than the base-to-common voltage of the OFF transistor. Specifically, e_{RB3} should be greater than e_{RB2} when T_2 is ON and T_1 is OFF. Base bias for T_1 is obtained from the voltage-divider arrangement of R_{B1} and R_{B2}. The voltage-divider network of R_{L1}, R_C, and R_{B3} determines the base bias of T_2. R_E provides temperature stabilization for the circuit. The maximum frequency is determined by C. As in the preceding flip-flop, a reduction in C increases the maximum repetition rate of the circuit.   The trigger input pulse may be applied to either base with the base of T_1 being most suitable.

*All-*ON*-All-*OFF *Bistable Multivibrator.*   Figure 8.13 illustrates a PNP-NPN bistable multivibrator which has provisions for using either base for the application of the input trigger pulse.   A positive pulse at the base of the PNP will turn it OFF, whereas the same pulse applied to the base of the NPN will turn it ON.   Notice the placement of the diodes in the emitter circuit of each transistor, which are to provide sufficient reverse bias to the OFF transistors to ensure cutoff.

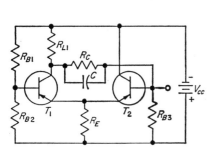

FIG. 8.12   Asymmetrical flip-flop.

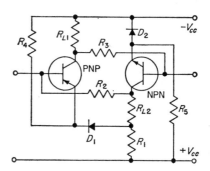

FIG. 8.13   PNP-NPN bistable multivibrator.

Consider the chain of events when a negative pulse is applied to the base of the PNP transistor: This drives the PNP into saturation, thereby reducing its collector voltage to a low negative value.   Note that this results in an increase in e_{RL1} and e_{R3}, which is across the emitter-base circuit of the NPN.   The base to $-V_{cc}$ voltage of the NPN is made positive, which drives the NPN into saturation.   Hence both the PNP and NPN are ON together.

A positive pulse applied to the base of the PNP transistor triggers the opposite type of action.   The PNP is turned OFF, resulting in $e_{RL1} + e_{R3}$ decreasing.   The NPN base is thereby made less positive and is turned OFF by this action.

The application of a trigger pulse to the base of the NPN produces the same chain of events, but with a pulse of opposite polarity.   The repetition rate is determined by the trigger pulse rate.

PROBLEMS

8.35   Refer to Fig. 8.10.   Explain how the bias change occurs when T_1 is turned OFF.

8.36   In the saturated flip-flop circuits of Fig. 8.9, why is it desirable to have both the collector-base and base-emitter junctions of the OFF transistor reverse-biased?

8.37 In Fig. 8.9b, what components determine the frequency?

8.38 Refer to the symmetrical saturated flip-flop circuit of Fig. 8.9b. The 2N465 transistors are to be used (see Fig. 8.8). $V_{cc} = 15$ volts; $R_{L1} = R_{L2} = 1.9$ kilohms; $R_{E1} = 100$ ohms; $C_1 = C_2 = 0.5$ μf; $R_{C1} = R_{C2} = 50$ kilohms. Construct the load line for transistors 1 and 2.

8.39 In the circuit of Prob. 8.38, assume the ON operating point of each transistor is at the intersection of the load line and the 150-μa characteristic. Find:

(a) R_i (b) i_B ON

(c) e_{B-E} ON   (d) e_{RE}

(e) e_{RB} ON

8.40 In the saturated flip-flop of Probs. 8.38 and 8.39 (refer to Fig. 8.11 for notation), find (a) I_A; (b) I_B; (c) I_D; (d) R_{B2}, R_{B1}.

8.41 Refer to the symmetrical saturated flip-flop circuit of Fig. 8.9b. The 2N337 is to be used (see Fig. 8.26). $V_{cc} = 20$ volts; $R_{L1} = R_{L2} = 1.9$ kilohms; $R_E = 100$ ohms; $C_1 = C_2 = 1.0$ μf; $R_{C1} = R_{C2} = 30$ kilohms. Construct the load line for transistors 1 and 2.

8.42 In the circuit of Prob. 8.41, assume the ON operating point of each transistor is at the intersection of the load line and the 250-μa characteristic. Find:

(a) e_{B-E} ON (by use of the i_C versus e_{B-E} characteristic) (b) i_B ON

(c) e_{RE} (d) e_{RB} ON

8.43 In the saturated flip-flop of Probs. 8.41 and 8.42 (refer to Fig. 8.11 for notation), find (a) I_A; (b) I_B; (c) I_D; (d) R_{B2}, R_{B1}.

8.8 The Unijunction Transistor. The unijunction transistor, although a three-terminal semiconductor device, behaves in a manner which is quite different from the conventional two-junction transistor. A bar of N-type silicon is used, upon which the emitter junction (PN) is formed.

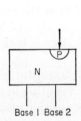

FIG. 8.14 The unijunction transistor.

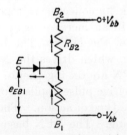

FIG. 8.15 Voltage-divider representation of a unijunction transistor.

The two base leads have ohmic contact with the silicon bar, but no junction is made. Hence the name unijunction transistor. Notice that the emitter junction is formed closer to one of the base leads (base 2). The resistance between base 1 and base 2, called the *interbase resistance* R_{BB}, ranges from 5 to 10 kilohms. The unijunction-transistor circuit can be compared to a voltage divider. With no input signal and zero emitter current,

$$e_{E\text{-}B1} = \frac{R_{B1}}{R_{B1} + R_{B2}} V_{bb}$$

where $e_{E\text{-}B1}$ = emitter to base 1 voltage

$\quad R_{B1}$ = resistance between the emitter and base 1 terminal

$\quad R_{B2}$ = resistance between the emitter and base 2 terminal

Base 1 is customarily placed at circuit common, and a positive voltage is applied at base 2. The emitter is under zero bias condition when no signal is applied. At this condition, a small leakage emitter current flows, and the transistor is essentially at cutoff. When $e_{E\text{-}B1}$ becomes less than the zero bias value, the emitter junction is driven further into its cutoff region, resulting in the flow of leakage emitter current only. The emitter junction is under forward bias when the emitter to base 1 resistance is reduced in a manner which is described in a following paragraph.

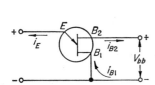

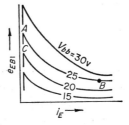

FIG. 8.16   Current flow in the unijunction transistor.

FIG. 8.17   Emitter characteristics.

The application of forward bias to the emitter increases the electron-hole pairs in the silicon bar region between the emitter contact and B_1, which causes an increase in the emitter current and a decrease in $e_{E\text{-}B1}$. Increasing the number of electron-hole pairs in the area between the emitter and base 1 reduces the resistance of this region, causing the increase in emitter current and the reduction in $e_{E\text{-}B1}$.

R_{B1} decreases as a function of i_E in a nonlinear manner. Of special interest is that the emitter voltage *decreases* as the emitter current *increases*. The characteristic of such a relationship is of *negative resistance*. The emitter characteristics of a unijunction transistor are similar to those shown in Fig. 8.17.

Point A, called the *peak point*, is also the cutoff point. Notice the vertical slope of the characteristic from A to C. The voltage at point A is the demarcation between reverse and forward bias on the emitter junction. At the point A value of $e_{E\text{-}B1}$, electron-hole pairs are generated in the emitter–base 1 region, resulting in a rise in i_E and a fall in $e_{E\text{-}B1}$, and the transistor is in its *negative resistance region*. The increase of emitter current and decrease in emitter voltage continues in a nonlinear manner up to point B. Point B, called the *valley point*, is the end of the negative resistance region, since $e_{E\text{-}B1}$ no longer continues its decrease, and a positive resistance region is encountered to the right of point B. Positive dynamic resistance values of 5 to 20 ohms are common in this portion of the characteristic, which is called the *saturation region*. The characteristics of the GE 2N492 are shown in Fig. 8.27.

Definition of Several Unijunction Transistor Parameters. The use of unijunction transistors involves the use of several new terms, which may be defined as follows.

Interbase Resistance R_{BB}. This is the resistance between base 1 and base 2 with the emitter open-circuited. It is a conventional ohmmeter measurement. R_{BB} increases at about 0.8 per cent per °C.

Intrinsic Standoff Ratio n. This is the ratio of R_{B1} to $R_{B1} + R_{B2}$ when the emitter is open-circuited. The intrinsic standoff ratio can be computed from

$$n = \frac{E_p - E_D}{V_{bb}}$$

where E_p = peak-point emitter voltage
    E_D = voltage across the emitter-base junction
    n = 0.7 volt at 25°C, and decreases about 3 mv/°C

Peak-point Current I_p. i_E is at the peak point of the characteristic. This is the minimum current for transistor operation.

Peak-point Emitter Voltage E_p. This is the maximum emitter voltage and is dependent upon E_D and V_{bb}. Transposing the intrinsic standoff ratio equation,

$$E_p = nV_{bb} + E_D$$

E_D, as stated earlier, decreases at an approximate rate of 3 mv per degree rise in temperature, resulting in E_p decreasing at the same rate. The use of a small resistor in series with B_2 tends to stabilize this condition.

Emitter Saturation Voltage $e_{E,\text{sat}}$. This is the forward voltage drop from the emitter to base 1 at the time maximum emitter current flows.

Interbase Modulated Current $i_{B2,\text{mod}}$. This is the base 2 current which flows at $e_{E,\text{sat}}$ condition.

Emitter Reverse Current $I_{E,o}$. This current is comparable to I_{co} of the conventional junction transistor in that it flows when the emitter is reverse-biased. $I_{E,o}$ also varies with temperature.

Valley Voltage e_v. The emitter voltage at the valley point of the characteristic is the valley voltage. e_v increases as V_{bb} increases, and decreases with the use of larger resistances in series with base 1 or base 2.

8.9 The Unijunction Relaxation Oscillator. The negative resistance characteristics of the unijunction transistor can be utilized in a number of circuits, several of which are derived from the basic relaxation oscillator. In Fig. 8.18 the emitter is at a reverse bias value at the beginning of the operating cycle, at which time it is at cutoff. C_E charges through R_E as shown by the solid arrows. As this action continues, $e_{E\text{-}B1}$ exponentially rises in value toward the magnitude of V_{bb}. This eventually brings $e_{E\text{-}B1}$ up to its peak-point value, which is the threshold of the negative resistance region. At the peak-point value, $e_{E\text{-}B1}$ is under forward bias and the emitter conducts. The resistance between the emitter and base 1 decreases to a low value when emitter current flows, as pointed out in the preceding section.

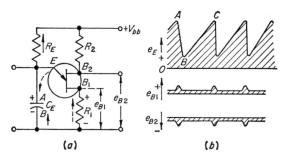

FIG. 8.18   Basic unijunction relaxation oscillator: (a) circuit, (b) waveforms. [*Taken with permission from "Transistor Manual" (4th Edition), published by General Electric Co., Charles Building, Liverpool, N.Y.*]

Because of the sharp reduction in the emitter to base 1 resistance, $e_{E\text{-}B1}$ begins to decrease in magnitude. Plate A of C_E takes on electrons via the emitter to base 1 circuit, as shown by the dotted arrows in Fig. 8.18.

As plate A takes on electrons, $e_{E\text{-}B1}$ exponentially decreases (becomes less positive). Referring to the emitter characteristics of Fig. 8.17, the operating point is moving from point A (peak point) toward point B. When $e_{E\text{-}B1}$ reduces to about 2 volts, the emitter stops conducting and plate A of C_E ceases to take on electrons. At this time, plate A of C_E discharges electrons through R_E, which is the beginning of the next cycle of operation. The period of oscillation can be determined by

$$T = R_E C_E \log \frac{V_{bb} - e_{E\text{-}B1,min}}{V_{bb} - n e_{B2\text{-}B1} - E_D} + 2t_f$$

where V_{bb} = supply voltage

    $e_{E-B1,min}$ = minimum emitter to base 1 voltage (about 2 volts)

       n = intrinsic standoff ratio, which is the ratio of $R_{B1}/(R_{B1} + R_{B2})$ when the emitter is open-circuited

     E_D = emitter junction voltage (0.7 volt at 25°C and decreases about 3 mv/°C)

      t_f = emitter voltage fall time (the time it takes the emitter voltage to fall from 90 to 10 per cent of its waveform)

  e_{B2-B1} = instantaneous interbase voltage

When R_1 and R_2 are small, the period of oscillation is approximated by

$$T \cong R_E C_E \log \frac{1}{1 - n}$$

The resistor in series with base 2 is used for temperature stabilization and can be found by

$$R_2 \cong \frac{0.65 R_{BB}}{n V_{bb}}$$

The maximum and minimum emitter–base 1 voltages (points A and B of the emitter waveforms in Fig. 8.18b) can be calculated from

$$e_{E-B1,max} = E_p \cong n V_{bb} + 0.7 \text{ volt}$$

and $e_{E-B,min} \cong 0.5 e_{E-B1,sat}$

8.10 Criteria for Designing a Unijunction Transistor Relaxation Oscillator

Step 1. The load line is determined by V_{bb} and R_E. The slope of the load line is equal to the reciprocal of R_E. Point C of Fig. 8.19 is equal

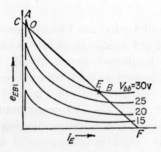

Fig. 8.19 Unijunction transistor characteristics with load line.

to V_{bb}. The characteristic of concern is determined by the V_{bb} to be used. The load line must intersect the characteristic curve to the right of the peak point (point O is to the right of point A) and to the left of the valley point (point E is to the left of point B).

In order to ensure that the load line intersects the characteristic to the right of the peak point, the following condition must be met:

$$\frac{V_{bb} - E_p}{R_E} > I_p$$

where E_p = peak-point voltage
   I_p = peak-point current

In order to ensure that the load line intersects the characteristic to the left of the valley point, the following condition must be met:

$$\frac{V_{bb} - E_v}{R_E} < I_v$$

where E_v = valley-point voltage
   I_v = valley-point current

The first condition ensures that sufficient current is supplied by R_E to turn the unijunction transistor on, and the second condition ensures that the unijunction transistor will turn off after it is turned on.   If

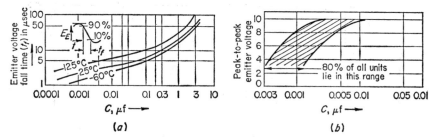

Fig. 8.20   Relaxation oscillator characteristics: (a) emitter voltage fall time versus capacitance, (b) peak-to-peak emitter voltage versus capacitance.   (*General Electric Co.*)

the load line should intersect the characteristic to the right of the valley point, the unijunction transistor will not turn off, and this factor must be taken into consideration in the design of the unijunction-transistor relaxation oscillator.

Step 2: Determination of C_E.   The use of a large C_E lengthens the fall time of the emitter voltage.   Reducing the size of C_E, on the other hand, results in a reduction of e_{E-B1} waveform, for values of C less than 0.01 μf. Figure 8.20a illustrates the relationship between emitter fall time t_f and C_E (with no resistor in the base 1 lead), while 8.20b shows the effect of capacitance on the magnitude of e_{E-B1}.

The fall time t_f can be approximated from the following relationship when there is no resistor in the base 1 lead:

$$t_f \cong (2 + 5C_E)e_{E-B1,\text{sat}}$$

where $C_E = \mu$f.

When a resistor is used in the base 1 lead, t_f increases in proportion to the time constant of R_1 and C_E.

8.11 The Unijunction Transistor as a Pulse and Saw-tooth Generator.

The relaxation oscillator of Fig. 8.18 can be conveniently used as a pulse generator since a pulse of current flows in each terminal (emitter, base 1, base 2) when the unit is fired. The pulse associated with each terminal for this relaxation oscillator is shown in Fig. 8.18b.

Possible Pulse Generator Configurations. Several possible configurations of the relaxation oscillator for use as a pulse generator are shown in Fig. 8.21. The first three configurations possess low output impedance because the output pulse is generated by the capacitor discharge current. The base 2 current is utilized in the pulse generation of Fig. 8.21d; therefore it possesses a higher output impedance than the other three. Also, the fourth configuration is capable of a higher output voltage.

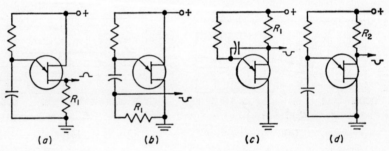

Fig. 8.21 Configurations of the relaxation oscillator for use as pulse generators. (*General Electric Co.*)

The pulse width of the emitter current (between 90 per cent and 10 per cent) is about equal to twice the fall time t_f. The maximum emitter current, with small values of series base 1 resistance, is approximated by

$$i_{E,\max} \cong \frac{E_p C_E}{t_f}$$

and

$$i_{B2,\max} \cong \frac{i_{B2,\mathrm{mod}} (i_{E,m})^{1/2}}{7}$$

where all currents are in milliamperes.

As pointed out in the introductory section of the unijunction transistor, the emitter-diode voltage decreases with increases in temperature. Temperature compensation is best provided by use of a resistor in series with base 2, which can be approximated by

$$R_2 \cong \frac{0.65 R_{BB}}{n V_{bb}}$$

The base 2 series resistance also serves as a protective device for the unijunction transistor when large values of C_E and V_{bb} are used. A general rule of thumb is to use a resistor in series with C_E when C_E is 10 μf or more and E_p is 30 volts or more. This series resistance should possess at least 1 ohm of resistance per 1 μf of C_E.

The Saw-tooth Generator. The voltage output of the unijunction transistor emitter, as illustrated in Fig. 8.18b, is a reasonably good saw-tooth waveform. The emitter of the unijunction transistor may be directly coupled to the base of a junction transistor, as shown in Fig. 8.22. Recall that the minimum emitter voltage of the unijunction transistor is between 1 and 2 volts, which is sufficiently large to keep the junction transistor in its conduction region. Notice that the junction transistor is in the common-collector configuration, which is called an *emitter follower*. The emitter follower stage tends to load down the circuit, which alters the frequency of the saw-tooth oscillations. Furthermore, if beta and

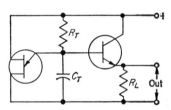

FIG. 8.22 Unijunction saw-tooth generator with an NPN emitter-follower output stage. *(General Electric Co.)*

R_L of the junction transistor are too small, oscillations will not take place. The following conditions must be met if the circuit is to oscillate:

$$\frac{(\beta + 1)R_L}{R_T + (\beta + 1)R_L} > n \text{ (max)}$$

Since beta of the emitter follower decreases with temperature rises, the corresponding change in loading effect changes the oscillator frequency. This effect is reduced to a minimum by making $(\beta + 1)R_L$ much greater than R_T. I_{co} of the junction transistor can increase the frequency with increases in temperature. With the use of the NPN silicon type, this effect is negligible up to 100°C. A PNP silicon junction transistor results in even better circuit operation.

8.12 Unijunction Multivibrator Circuits. With the advent of the unijunction transistor, a new improved multivibrator circuit has been devised. The leading disadvantages of the conventional transistor multivibrators are centered around the cross-coupling capacitors, which result in:

1. Limitation of accuracy and stability of the time periods
2. Collector voltage waveform distortion, making it difficult to obtain reasonably good rectangular waveforms

The improved multivibrator circuits utilize 2 PNP or NPN transistors and one unijunction transistor. The advantages of this type of circuit are nondistorted rectangular output waveforms, increased timing stability, and the use of a single timing capacitor of relatively small value. The unijunction transistor serves as a timing and triggering source.

The Symmetrical Multivibrator. Figure 8.23 illustrates a saturated flip-flop, which can be designed in the manner described in a previous section. The circuit is switched from one state to the other by means of the negative pulses of the unijunction transistor, which are developed across R_A. When the unijunction transistor is conducting, electron flow occurs from B_1 to the right plate of C_T. The left plate of C_T discharges electrons through R_A, the voltage supply, and back to B_1. This action develops an abrupt negative pulse across R_A, which cuts off the ON

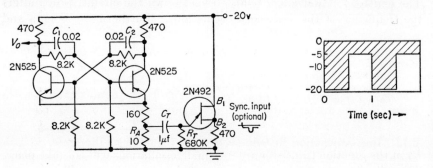

FIG. 8.23 Unijunction multivibrator circuit with collector-voltage waveforms. (*General Electric Co.*)

transistor and fires the OFF transistor. e_{E-B2} increases until e_{E-B1} is brought to its minimum conducting value, at which time the unijunction transistor stops conducting. When the unijunction transistor is cut off, the right plate of C_T discharges through R_T. As the discharge continues, e_{RT} decreases, thereby bringing the emitter back to its peak-point value, which is the start of the next cycle, since the unijunction transistor then fires ON and the right plate of C_T takes on electrons.

The time of the negative pulse across R_A is largely determined by C_T and R_T; therefore the timing of the switching from one state to the other is controlled by these two components.

One-shot Unijunction Multivibrator Circuit. Figure 8.24 illustrates a one-shot multivibrator arrangement in which Q_1 is normally OFF and Q_2 is ON. When the unijunction transistor Q_3 is off, a positive voltage is applied to the base of Q_2, which turns Q_2 OFF. The collector voltage of Q_1 rises to about 20 volts, causing C_T to charge through R_T. When the emitter to base 1 voltage is at its peak-point value, the unijunction transistor fires. This applies a negative pulse at the base of Q_2, thereby causing that transistor to turn ON again. C_T discharges until the emitter–base 1 voltage cuts off the unijunction transistor, which is the beginning of the next trigger cycle. R_1 biases the emitter of the unijunction

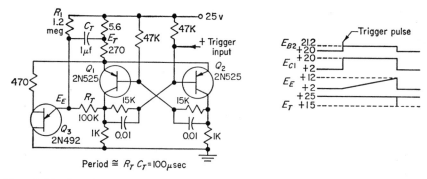

Period $\cong R_T C_T = 100 \mu\text{sec}$

Fig. 8.24 One-shot unijunction multivibrator circuit with typical waveforms. (*General Electric Co.*)

transistor to about 2 volts in its OFF condition, thereby ensuring the return of the emitter voltage to its beginning value.

8.13 The Unijunction Bistable Circuit. Figure 8.25 illustrates a basic unijunction transistor bistable circuit with a typical emitter-characteristic curve and load line. The emitter characteristic selected is

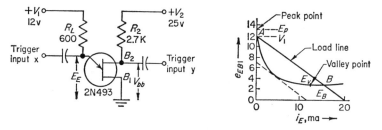

Fig. 8.25 Basic unijunction transistor bistable circuit with typical emitter-characteristic curve. (*General Electric Co.*)

determined by the V_{bb} to be used, since each emitter-characteristic curve is a function of i_E versus $e_{E\text{-}B1}$ with V_{bb} fixed. The load line determines the value of R_L to be used in the emitter circuit.

In the bistable circuit, the load line should intersect the emitter characteristics to the *left* of the peak point and to the *right* of the valley point. Choosing the emitter supply voltage such that it is smaller than the peak-point voltage will ensure the first condition, and the second condition is met by the selection of a sufficiently small R_L. Notice that

$$R_L = \frac{V_1}{i_{E,m}}$$

Point A is in the cutoff region of the characteristic, with only a very small cutoff emitter current flowing. Point A can also be considered the OFF condition of the circuit. Notice that point B is well into the saturation region of the characteristic, since it is to the right of the valley point, and may be considered as the ON condition. The unijunction transistor presents 10 megohms or more during its OFF time and 49 ohms or less during its ON time. The OFF power is negligible whereas the ON power may be up to 1.5 watts by the proper selection of R_L and supply voltages.

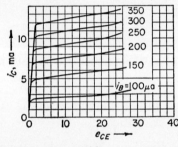 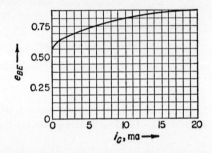

Absolute values (25°C).

$$e_{c\text{-}E} = 45 \text{ volts}$$
$$i_c = 20 \text{ ma}$$
$$\text{PD} = 125 \text{ mw (derate 1 mw/°C)}$$
$$h_{fe} = 65 \qquad h_{ob} = 0.2 \ \mu\text{mho}$$
$$h_{ib} = 50 \qquad h_{rb} = 2 \times 10^{-4}$$

FIG. 8.26   Characteristics of the 2N337(NPN).   (*Texas Instruments, Inc.*)

The circuit is normally OFF. A positive trigger voltage at the emitter, with a magnitude greater than $(E_p - V_1)$, or a negative trigger voltage at base 2, with a value larger than $E_p - V_1/n$, will turn the circuit ON. Since point B is also a stable point, the unijunction transistor will not turn OFF without the application of a negative trigger at the emitter. The turn OFF trigger magnitude is equal to $E_B - E_V$. With point B to the right of the valley point, the unijunction transistor cannot be turned OFF by a trigger voltage at base 2.

With the application of the ON trigger voltage, the emitter potential is raised to slightly greater than its peak-point value. This fires the unijunction transistor, resulting in a reduced base 1 to emitter resistance. The emitter current increases, and the emitter voltage decreases, because of the conductivity of the emitter to base 1 region. The unijunction transistor is in its negative resistance region, which is unstable. This action continues up to an emitter current value determined by the magnitude of V_1 and R_L, which is point B. If the point is to the right of the valley point, the unijunction transistor is out of its negative resistance region, and hence at a stable point. The unijunction transistor will remain at point B until the emitter voltage is reduced by the application of the next trigger voltage. Since the emitter to base 1 voltage is positive, the application of a small negative pulse will bring the transistor

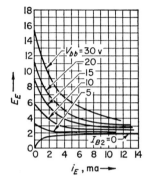 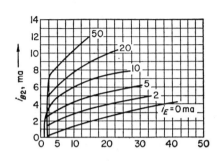

FIG. 8.27 Typical characteristics of the GE 2N492 unijunction transistor. [*Taken with permission from "Transistor Manual" (4th Edition), published by General Electric Co.; Charles Building, Liverpool, N.Y.*]

back to its valley point, which is at the right edge of the negative resistance (unstable) region. The conductivity of the emitter-base region is decreased; i_E becomes smaller and $e_{E\text{-}B1}$ increases until the negative resistance region has been completely traversed. This brings the unijunction transistor to cutoff and point A, where it remains until another ON trigger voltage is applied.

PROBLEMS

8.44 What is the interbase resistance of a unijunction transistor?

8.45 Describe the effect of applying forward bias to the emitter of a unijunction transistor.

8.46 What is the relationship between R_{B1} and i_E in a unijunction transistor?

8.47 Refer to Fig. 8.17. Describe the behavior of the unijunction transistor in its negative resistance region.

8.48 Refer to the unijunction relaxation oscillator of Fig. 8.18. If E_D is 0.7 volt at 25°C and decreases about 3 mv per °C, find E_D at the following temperatures:
(*a*) 30°C
(*b*) 40°C
(*c*) 100°C

8.49 In the unijunction relaxation oscillator of Fig. 8.18, $R_{BB} = 7$ kilohms; $V_{bb} = 30$ volts; $n = 0.7$ volt. Calculate the value of the temperature compensation resistor R_2.

8.50 Refer to Fig. 8.27. State the relationship between V_{bb} and the peak-point potential.

CHAPTER 9

DIODE APPLICATIONS (SEMICONDUCTOR AND VACUUM-TUBE)

The chief characteristics of the semiconductor diode are investigated in Chap. 3. The fundamental characteristics of the vacuum diode are analyzed in the following section. This chapter reveals many of the important functions of the diode in several types of power supplies, demodulation, and other types of circuits. Design information consistent with the background of the technician is presented for many of the applications, along with actual working circuits.

The all important new semiconductor diode developments are also studied, namely, the silicon-controlled rectifier, the zener diode, and the tunnel diode. Typical applications of these devices are analyzed.

9.1 The Vacuum Diode. In Chap. 3, it was found that a mobile electron is one which has sufficient energy to permit that electron to remain independent of any nuclear influence. The mobile electron was more or less free to move about the confines of the material in accordance with the conditions it found at the time. If no electric field is present, the free electron would wander about in a haphazard, directionless manner. The introduction of an external field imposes a directional characteristic upon the motion of the mobile electron, resulting in drift or current flow.

Electron Emission. The free electrons within a material are restricted to the bulk of the material unless they acquire additional packages of energy. Electrons can actually break through the surface of a material into the surrounding space if they acquire the required increment of energy. The process of their leaving the confines of the material to go into the surrounding space is called *electron emission*. The additional energy required for emission may be obtained from:

1. Heat
2. Impact
3. Radioactive substances
4. Light
5. Electrostatic fields

Electron emission by the application of heat is called *thermionic emission* and is the technique utilized in vacuum tubes. The emitted electrons per square centimeter of surface area can be determined from the following equation:

$$I = AT^2\epsilon - \frac{b}{T}$$

where A = a constant for the given material

T = absolute temperature = C° + 273

$\epsilon \cong 2.72$

b = a constant for the given material

$\quad = \frac{e}{k} W = 11,600W$

k = Boltzmann's constant

W = work function in volts

e = electron charge

The Cathode. Many materials are poorly suited for use as emissive surfaces because of the great amount of additional energy required to produce thermionic emission. Materials which are suited for thermionic emission can be placed in three categories:

1. Certain pure metals (such as tungsten, platinum, molybdenum, and tantalum)

2. Certain metals with a surface layer of another metal (such as tungsten coated with calcium)

3. Nonmetallic coatings (such as barium oxide, strontium oxide, and calcium oxide)

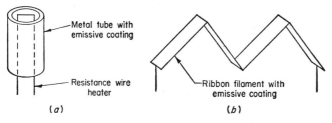

Metal tube with emissive coating

Resistance wire heater

Ribbon filament with emissive coating

(*a*)   (*b*)

FIG. 9.1   Cathode heating techniques: (*a*) indirect, (*b*) direct.

An electron emissive surface is called a *cathode*, without regard to the type of material from which it is constructed. The type of cathode hinges upon the intended operating temperature and other considerations. The cathode may receive its thermal energy directly or indirectly.

Refer to Fig. 9.1*a*, which illustrates one way in which the cathode can be indirectly heated. The emissive material is coated on a metal tube. A resistance wire is placed within the tube but carefully insulated from it. The application of a predetermined amount of voltage across the resistance wire terminals results in the current flow which develops the required amount of heat.   This results in thermionic emission from the emissive coating on the metal tube.

The direct-heating technique is illustrated in Fig. 9.1b.   The cathode is in the form of a ribbon.   When a predetermined potential is applied across the ribbon, current flows, the ribbon heats to the correct temperature, and thermionic emission occurs.   The direct-heating type is also called the *filament* type.

The filament type is incorporated where very high cathode temperatures are required, whereas the indirectly heated type suffices in those cases where moderate to low temperatures are needed.

The Anode (Plate).   Now that the manner in which thermionic emission can be achieved has been analyzed, let us place a positive electrode in the immediate vicinity of the cathode.   This electrode is positive with respect to the cathode and is called the *anode* or *plate*.

The placement of an anode in the vicinity of the cathode forms a two-electrode device, called the *diode*.   Both electrodes are placed in an airtight container or shell.   The gases within the shell are exhausted, resulting in the formation of the *vacuum-tube diode*.   Figure 9.2a is the schematic representation of the indirectly heated diode, and the directly heated type is depicted in Fig. 9.2b.

Space Charge.   The cathode-emitted electrons form a cloud surrounding the cathode called a *space charge*.   If the anode is not positive, the space-charge electrons will eventually lose their newly acquired energy and reenter the cathode surface.   A positive anode attracts electrons within the space charge to itself, resulting in an electron flow from the cathode to the plate via the space charge.

The number of electrons involved in the diode current is determined by many factors, several of which are:

1. Cathode material
2. Cathode temperature
3. Distance between plate and cathode
4. Plate potential

As pointed out earlier, certain materials at a given temperature are more emissive than others.   Higher cathode temperatures, for any given material, result in larger values of electron emission and therefore greater diode current.   Placing the anode physically closer to the cathode, for a given anode potential, cathode material, and temperature, produces an increase in diode current.   With all other factors constant, increasing the positive anode potential increases the diode current up to a point.

Figure 9.3 illustrates the relationship between plate voltage and plate current as a function of cathode temperature. As the plate voltage is increased from 0 to B, there is only a small plate-current increase because of the space charge. The plate current in this region increases approximately as the square of the plate voltage. During the time these low-voltage values occur, the anode is not as successful in drawing electrons. Furthermore, the space charge surrounds the cathode and is largest at low anode voltages, since the cathode temperature is constant. This space charge tends to hinder and oppose further emission, since a negative space charge opposes negative approaching electrons. Therefore anode current between zero volts and B is limited by the space charge.

Point B is that plate potential where the relationship between anode current and voltage becomes approximately linear. The linear portion of the characteristic lies between points B and A. As the anode voltage

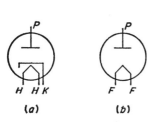

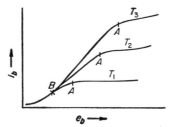

FIG. 9.2   Representation of the vacuum-tube diode: (a) indirectly heated, (b) directly heated.

FIG. 9.3 e_b versus i_b as a function of cathode temperature.

is increased, the anode receives a greater percentage of the cathode-emitted electrons. At point A, called the *saturation point*, the anode is receiving approximately all the electrons thermally emitted by the cathode. Notice that the saturation point has a direct relationship to the temperature. If the cathode temperature in a given diode is increased, its emission increases and the saturation point occurs at a high anode potential (see Fig. 9.3).

Note, however, that the anode current continues to increase, but more slowly, beyond the saturation point or knee of the curve. This is due to the electrostatic effect of the highly positive anode upon cathode surface electrons. That is, with high plate voltages, some electrons are actually pulled off the cathode, which is called the *Schottky effect*. It should be noted that the Schottky effect produces relatively small plate-current increases for large plate-voltage increases. Vacuum diodes customarily are restricted in their normal use to that region of the characteristic which lies to the left of point A in Fig. 9.3.

9.2 Volt-Ampere Characteristics of Diodes.

The volt-ampere characteristics of semiconductors and vacuum-tube diodes display certain similarities and differences which are brought out in this section. Figure 9.4 illustrates both types of volt-ampere characteristics.

Volt-Ampere Characteristics of the Vacuum Diode. Figure 9.4a depicts the volt-ampere characteristic of a vacuum diode. The space-charge effect upon the plate current becomes almost negligible at such a low plate voltage that it does not appear in the characteristic. Also, the maximum plate-voltage rating of most diodes is below the saturation point, thereby avoiding the need to use the curve in that region. The over-all *E-I* characteristic is relatively linear, and is arbitrarily considered linear in some design problems. Notice that reverse electron flow cannot

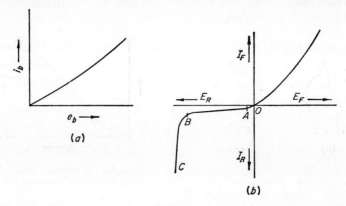

Fig. 9.4 *E-I* characteristics of diodes: (a) vacuum, (b) semiconductor.

occur without physically damaging the diode. The negative plate potential at which electrons will arc from the plate to the cathode is called the *peak inverse plate voltage.* This potential is to be avoided in actual operation, since it may lead to destruction of the unit.

Vacuum diodes have found use in applications where less than 1 volt and more than 1,000 volts are encountered. Because of this wide divergence in usage, a large variety of vacuum diodes have been made available.

Volt-Ampere Characteristics of the Semiconductor Diode. The *E-I* characteristics of the semiconductor diode are shown in Fig. 9.4b. The forward *E-I* characteristic of the unit resides in the first quadrant. This portion of the characteristic is only slightly nonlinear for voltage values within the limitations stated by the manufacturer. Unlike its vacuum-tube counterpart, current does flow in the opposite direction with the application of reverse potential. This is called the reverse current of the diode and is generally 10^{-3} times the magnitude of typical forward currents. For example, if the forward current of a particular unit is stated in milliamperes, the reverse current is expressed in microamperes.

The reverse current area between points A and B of Fig. 9.4*b* is called the *saturation region*. The reverse current undergoes a very small increase for large changes in reverse voltage within this region because the involved electrons are generated by deathnium center action, which is relatively independent of external potential. At some reverse potential, an avalanche action sets in, whereby the reverse current undergoes an abrupt increase. This region, between B and C of Fig. 9.4*b*, is called the *zener region* of the characteristic.

In conclusion, it is seen that the semiconductor diode characteristic can be divided into three distinct regions:

1. Forward
2. Reverse—saturation
3. Reverse—zener

All these regions are useful. Some applications utilize the forward and saturation regions (rectification), others use only the saturation region (constant-current regulation), and still others incorporate the zener region (constant-voltage regulation).

PROBLEMS

9.1 What is electron emission?

9.2 State five ways in which the required additional energy for emission may be obtained.

9.3 What is thermionic emission?

9.4 State the three basic types of materials which are suited for thermionic emission.

9.5 What is a cathode?

9.6 State the basic differences between the indirect and directly heated cathodes.

9.7 What is the chief purpose of the anode?

9.8 Describe the vacuum-tube diode.

9.9 What effect does the space charge have on vacuum-tube diode operation?

9.10 State four factors which determine the plate current in a vacuum-tube diode.

9.11 What is the saturation point of a vacuum-tube diode?

9.12 Describe the Schottky effect.

9.13 What is meant by the peak inverse voltage of a vacuum-tube diode?

9.14 Why is the volt-ampere characteristic of a vacuum-tube diode restricted to the first quadrant?

9.15 Why does the semiconductor diode have a portion of its volt-ampere characteristic in the third quadrant?

9.3 Principles of Rectification. The perfect rectifier is one which offers zero resistance to current flowing in one direction and infinite resistance to current flowing in the opposite direction. It may be thought of as a synchronously operated switch which passes current of only one polarity and opens when current in the opposite direction attempts to flow. It is the first step in converting a bidirectional voltage into unidirectional potential. In actual practice, a rectifier does possess some forward resistance and a relatively high resistance to reverse current.

Half-wave Vacuum Rectifier. Figure 9.5 illustrates the fundamental half-wave rectifier circuit with the pure resistance load condition. Assume e_g is a sinusoidal source voltage for this analysis. When e_g undergoes its positive excursion, the diode conducts and the voltage distribution in the circuit is

$$e_g = e_b + e_{RL} = e_b + i_b R_L$$

where e_g = instantaneous generator voltage
e_b = diode plate to cathode voltage
i_b = diode plate current

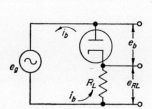

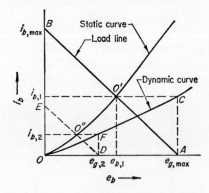

Fig. 9.5 Basic half-wave vacuum-rectifier circuit.

Fig. 9.6 Load line and dynamic-curve analysis.

Examination of this relationship reveals that e_{RL} increases as i_b increases, since R_L is of fixed ohmic value, and e_b decreases by the same amount:

$$\Delta e_{RL} = -\Delta e_b$$

When e_g undergoes its negative excursion, e_b is negative and the diode does not conduct. The source voltage "sees" the rectifier as an open circuit, resulting in the following voltage relationship:

$$e_b = e_g$$
and
$$e_{RL} = 0$$

Load-line and Dynamic-curve Analysis. The load line, as shown in Fig. 9.6, is constructed in the following manner:

1. The x-axis intercept is the value of $e_{g,\max}$ (point A).
2. The y-axis intercept is determined by $e_{g,\max}/R_L$ (point B).

The intersection of the load line and the static characteristic determines the i_{b1} and e_{b1} values at this condition, and $e_{RL} = e_g - e_{b1}$ at this time. Recall that static-characteristic curves are determined when the device has no load. A plot of e_g versus i_b results in a *dynamic*-characteristic curve. When e_g is at its maximum value, i_{b1} flows, resulting in the establishment of point C. When e_g decreases to some value, such as e_{g2}, i_b decreases. Let us determine the manner in which this value of i_b is found: Since R_L is fixed, the slope of the load lines does not change. The x-axis intercept of the load line is now at the e_{g2} value (point D). The load line is redrawn DE parallel to the first load line AB. The intersection of this second load line and the static curve O'' yields the current which flows at this condition i_{b2}. Thus point F is determined. This process may be repeated for additional e_g values if desired. The

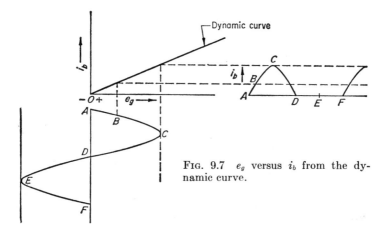

Fig. 9.7 e_g versus i_b from the dynamic curve.

line joining points C and F to the origin forms the dynamic curve of the rectifier for this load.

There is a dynamic curve for each value of R_L used, just as there is a new load line for each R_L. The dynamic tube resistance r_p is determined by

$$r_p = \frac{e_b}{i_b}$$

Both the static and dynamic curves are relatively linear and may be replaced with straight lines for most practical purposes. Therefore r_p may be considered constant for positive values of plate voltage.

Output Current Waveform Analysis for the Half-wave Rectifier. The relationship between the input potential e_g and the output current i_b is conveniently determined by a graphical analysis of e_g versus i_b from the dynamic-characteristic curve. Assume the rectifier of Fig. 9.5 has the dynamic curve shown in Fig. 9.7. When e_g is at point A (zero volts), the diode is not conducting, resulting in point A (zero amperes) in the i_b waveform. As e_g swings positive, i_b increases in the same manner, assuming the dynamic curve is linear. The positive pulse of e_g is reproduced by i_b, and i_b is zero for the negative e_g excursion. The most faithful reproduction occurs with the most linear dynamic curve. This technique is of value in determining the magnitude of e_g required for a desired i_b.

Recall that r_p may be assumed constant for positive anode potentials if the static curve is assumed linear, resulting in the following relationship:

$$e_b = i_b r_p$$

and

$$e_g = e_b + i_b R_L$$

Substituting the value of e_b,

$$e_g = i_b r_p + i_b R_L = i_b(r_p + R_L)$$

Determination of Average, RMS, and Power Values. The following discussion assumes a sinusoidal input voltage waveform. The plate

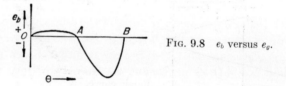

FIG. 9.8 e_b versus e_g.

current flows for half of each cycle, and the average plate current is found by

$$I_{av} = I_{b,s} = \frac{i_{b,m}}{\pi}$$

This is the reading which would be obtained from a d-c ammeter.

The rms value of the half-sinusoidal waveform is

$$I_p = \frac{i_{b,m}}{2}$$

The average plate voltage involves the small voltage drop across the tube while it is conducting and the entire supply voltage across the diode when its plate is negative. Figure 9.8 illustrates the preceding statement.

The diode resistance equals r_p when in its forward condition, and

$$e_b = i_b r_p$$

When the tube is not conducting,

$$e_b = e_g$$

The average plate voltage, which averages all these changes, is

$$E_{b,s} = - \frac{i_{b,m} R_L}{\pi}$$

The power delivered to the entire circuit by the input generator is

$$P_i = (I_p)^2 (R_L + r_p)$$

And the average power delivered to the load is

$$P_{RL,\text{av}} = (I_p)^2 R_L = \left(\frac{i_{b,m}}{2}\right)^2 R_L$$

Efficiency of Rectification. The efficiency of rectification is the ratio of $P_{RL,\text{av}}$ to the power dissipated by the plate circuit. This ratio becomes important in applications where the presence of alternating components changes the amount of power dissipated in a circuit. This ratio may also be called the conversion or theoretical efficiency:

$$n_r = \frac{40.6}{1 + r_p/R_L} \qquad \text{per cent}$$

Percentage of Regulation. The variation of the average output voltage as a function of the average output current is termed the regulation of the circuit. The regulation is most conveniently expressed as a percentage and may be determined from the following relationship:

$$\text{Per cent regulation} = \frac{E_{\text{no load}} - E_{\text{full load}}}{E_{\text{full load}}} \times 100$$

where

$$E_{\text{no load}} = \frac{E_m}{\pi}$$

and

$$E_{\text{full load}} = \frac{E_m}{\pi} - I_{\text{av}} r_p$$

Examination of the preceding equation reveals that zero regulation is achieved when the output voltage does not vary with the load. The circuit of Fig. 9.5 has very poor regulation because of the large change in output voltage between no load and load conditions. Regulation is improved somewhat by use of a filter in conjunction with the load, as is studied subsequently.

Ripple Factor. Another important concept pertaining to rectifier circuits is the ripple factor, which indicates the magnitude of the output current fluctuations:

$$rf = \frac{I_{\text{ac,rms}}}{I_{\text{av}}} = \frac{E_{\text{ac,rms}}}{E_{\text{av}}}$$

and

$$rf = \left[\left(\frac{I_{\text{ac,rms}}}{I_{\text{av}}}\right)^2 - 1\right]^{1/2}$$

The factors introduced in the preceding paragraphs are applied in conjunction with other rectifier circuits which follow.

The Full-wave Vacuum Rectifier. Figure 9.9 illustrates a full-wave rectifier, which is much more desirable than the half-wave variety in many cases. There are usually two or more secondary windings within the input transformer. The center-tapped secondary is commonly a step-up coil, thereby making it possible to apply voltages to each anode which exceeds the input voltage. The full-wave rectifier is actually two half-wave rectifiers connected in such a manner that each diode conducts on alternate half cycles. When the top of the step-up secondary is negative, D_1 fails to conduct since its anode is negative during this half cycle. But the plate of D_2 is positive at this time and it conducts. An electron current flows from the plate of D_2 up the lower half of the secondary to common, up R_L, back to the cathode of D_2 to complete the series loop.

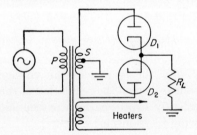

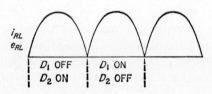

FIG. 9.9 The full-wave vacuum rectifier.   FIG. 9.10 Output waveforms of a full-wave vacuum rectifier.

During the next half cycle, D_2 is cut off since its anode is negative. The plate of D_1 is positive, however, resulting in electrons flowing from the plate of D_1, down the upper half of the step-up secondary to common, up R_L to the cathode of D_1. The resultant current and e_{RL} waveform are shown in Fig. 9.10.

Assuming a sinusoidal input potential, the output waveform is made up of two half sinusoids per one cycle of input voltage.

The average load current I_{RL} is twice that found in the previously discussed half-wave rectifier:

$$I_{RL} = \frac{2I_m}{\pi}$$

Notice that the second pulse in Fig. 9.10 is identical to the first pulse and has the same polarity. In a pure, complete sinusoid, the second pulse is identical to the first but opposite in polarity. Recall that the direction of current through a resistive load has no bearing on the power dissipation. Therefore the "heating effect" or rms value for the two pulses of the Fig. 9.10 waveform is identical with that of a pure sinusoid:

$$I_{rms} = \frac{I_m}{1.414} = 0.707 I_m$$

The r_p values of the diodes must be taken into consideration when computing the average power input. This is most easily understood by examining Fig. 9.11, which illustrates the equivalent circuit for the full-wave rectifier being considered. Since each diode possesses some forward resistance, they are imperfect components. A convenient technique for the analysis of imperfect components is to consider the device as a combination of perfect components. Using this approach, each half-wave rectifier is considered as a perfect diode in series with its forward resistance r_p. Assuming that D_1 and D_2 are matched and that a center-tapped secondary coil is used, which leads to the following equalities:

$$r_{p1} = r_{p2}$$

and

$$e_{g1} = e_{g2}$$

Since D_1 and D_2 alternately conduct, r_{p1} and r_{p2} are alternately in series with R_L. Note that the total circuit resistance is always $R_L + r_{p1}$

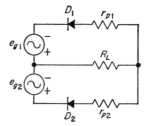

FIG. 9.11   Full-wave rectifier equivalent circuit.

or $R_L + r_{p2}$, which are equalities. Therefore the power input is determined by

$$P_i = (I_{rms})^2(r_p + R_L)$$

Considering R_L as the load resistance, then the average power output is readily found from

$$P_o = P_{RL} = (I_{rm})^2 R_L$$

The efficiency of rectification is twice the value obtainable from the half-wave rectifier:

$$n_r = \frac{81.2}{1 + r_p/R_L} \qquad \text{per cent}$$

The ripple factor is computed by the same equation as stated for the half-wave rectifier. The following example illustrates the fact that a much lower ripple factor is obtained with full-wave rectification.

Example: Determine the ripple factor for the (a) half-wave rectifier; (b) full-wave rectifier.

Solution:

(a) Determine the ripple factor for the half-wave rectifier:

$$rf = \left[\left(\frac{I_{ac \cdot rms}}{I_{av}}\right)^2 - 1\right]^{\frac{1}{2}}$$

The rms value of a half-wave rectifier is $0.5I_m$, and the average value is I_m/π. Substituting these values into the ratio within the ripple-factor equation,

$$\frac{I_m/2}{I_m/\pi} = \frac{\pi}{2} = 1.57$$

Now solving,

$$rf = [(1.57)^2 - 1]^{\frac{1}{2}} = 1.21$$

(b) Determine the ripple factor of the full-wave rectifier:
The rms current value in the full-wave rectifier circuit is $0.707I_m$, and the average value is $2I_m/\pi$. Substituting these values and solving,

$$rf = \left[\left(\frac{0.707I_m}{2I_m/\pi}\right)^2 - 1\right]^{\frac{1}{2}}$$
$$= [(1.11)^2 - 1]^{\frac{1}{2}} = 0.482$$

The average output voltage is twice that possible with the half-wave circuit, with all other conditions being identical; hence

$$E_{RL} = \frac{2E_m}{\pi} - I_{av}r_p = I_{av}R_L$$

The preceding solutions reveal that the ripple voltage is greater than the average output voltage in the half-wave arrangement whereas in the full-wave circuit, the ripple voltage is a much smaller portion of the average output voltage. The advantages of full-wave rectification are further analyzed in a following section.

PROBLEMS

9.16 A half-wave rectifier circuit has the following parameters: $r_p = 400$ ohms; $R_L = 2,500$ ohms; $e_g(rms) = 250$ volts. Calculate:

(a) maximum load current   (b) average load current
(c) rms load current   (d) average tube voltage
(e) P_i (f) $P_{RL,av}$
(g) efficiency of rectification   (h) percentage of regulation (no load to full load)

9.17 Refer to Fig. 9.12 for the static volt-ampere characteristics of the 6BL4. Construct the dynamic characteristics for $R_L = 300$ ohms; e_g(rms) $= 40$ volts.

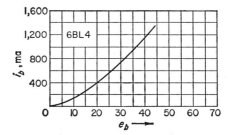

FIG. 9.12 Average plate characteristics. (RCA.)

9.18 Repeat Prob. 9.17, with $R_L = 60$ ohms; e_g(rms) $= 40$ volts.

9.19 Calculate the following for Prob. 9.18 (assume $r_p = 50$ ohms):
(a) maximum load current   (b) average load current
(c) rms load current   (d) average tube voltage
(e) P_i (f) $P_{RL,av}$
(g) efficiency of rectification (h) percentage of regulation (no load to full load)

9.20 The duo-diode 6AX5-GT is to be used in a full-wave rectifier circuit. Assume r_p of each diode $= 50$ ohms; $R_L = 250$ ohms; e_g(rms) for each diode $= 300$ volts. Calculate:
(a) average load current   (b) average current of each diode
(c) rms load current   (d) $P_{RL,av}$
(e) efficiency of rectification (f) percentage of regulation (no load to full load)

9.21 Repeat Prob. 9.20 with $R_L = 100$ ohms.

9.22 Calculate the ripple factor in Probs. 9.16, 19.19, 9.20, and 9.21.

9.23 The duo-diode 6AX5-GT is to be used in a full-wave rectifier circuit. r_p of each diode $= 50$ ohms; $R_L = 400$ ohms; e_g(rms) for each diode $= 155$ volts. Find:
(a) average load current   (b) average current of each diode
(c) rms load current   (d) $P_{RL,av}$
(e) efficiency of rectification (f) percentage of regulation (no load to full load)
(g) ripple factor

9.24 Repeat Prob. 9.23, with $R_L = 800$ ohms.

9.25 State the relationship (as seen in Probs. 9.23 and 9.24) between R_L and (a) percentage of regulation; (b) ripple factor.

9.26 State the relationship between ripple factor and the type of rectification (half-wave versus full-wave).

9.4 Basic Rectifier-filter Circuits.

Four basic rectifier-filter circuits are discussed in this section. One of the chief purposes of a rectifier filter is to "smooth" the current pulsations so as to provide a relatively constant output current and voltage. Reasonably constant voltages and currents are often necessary for the operation of many control device circuits, the chief application of rectifier-filter circuits.

The Inductor Filter. Figure 9.13 illustrates both the half-wave and full-wave rectifier circuits with an inductor or "choke" filter. Recall that an inductor opposes changes in current. Since the current attempts

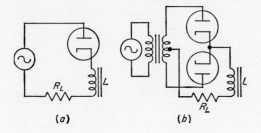

FIG. 9.13 Vacuum rectifier with an inductor filter: (*a*) half-wave, (*b*) full-wave.

(*a*)   (*b*)

to change as a function of the plate voltage, the inductor tends to smooth out these changes, thereby reducing the variations between maximum and minimum current.

The smoothing out or filter action is increasingly more effective with larger ratios of L/R. The choke input type is rarely used with the half-wave rectifier circuit, since the current is still zero for a portion of the time. The pi-section type, discussed in the following paragraphs, is used in such cases. In the case of the full-wave rectifier, the inductor filter results in a reduction of output voltage variations, as shown in Fig. 9.14.

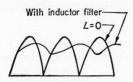

FIG. 9.14 Full-wave rectifier output voltage with and without an inductor filter.

FIG. 9.15 Equivalent circuit of FIG. 9.13*b*.

A substantial ripple content is still present, even with large values of inductors. Most filters incorporate the use of a more elaborate network.

An equivalent circuit, which separates the ripple and unidirectional components, may be drawn as shown in Fig. 9.15. The alternating current generator potential is determined by

$$- \frac{4E_m}{3\pi} \cos 4\pi \; ft$$

and the unidirectional current generator potential is

$$\frac{2E_m}{\pi}$$

When the ratio of L/R is large, the ripple factor is

$$rf = \frac{R_L}{4.242(2\pi fL)} = \frac{R_L}{4.242X_L}$$

and the output voltage is

$$E_0 = \frac{2E_m}{\pi} - I_{\mathrm{av}}R$$

Note that R is the total circuit resistance (including that of the transformer secondary and tube) but excluding R_L.

The Capacitor Filter. The characteristics of capacitance may be utilized in smoothing out the output voltage. While the rectifier is conducting, the capacitor is allowed to charge, thereby building up a cemf. When the rectifier is not conducting, the capacitor discharges through the load resistance, resulting in a substantial reduction in output voltage variations. The capacitor is connected in parallel with R_L.

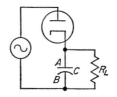

Fig. 9.16 Half-wave rectifier with a capacitor filter.

Refer to Fig. 9.16 for the following discussion. Let us first consider the operation of the half-wave rectifier with a capacitor filter at the condition where R_L is infinite. When the generator is turned on, the tube conducts, plate B of the capacitor charges, and plate A discharges through the rectifier. This action continues only until the input generator achieves its maximum potential. Assuming the tube and generator resistance are negligible, $e_c = e_{g,\max}$ at this time. When the generator potential decreases, plate B of C cannot discharge, since R_L is infinite and the rectifier will not pass reverse current. This results in e_c remaining at the $e_{g,\max}$ value. Therefore the cathode of the rectifier is more positive than the anode, and tube conduction cannot occur. When the input generator reverses polarity, the cathode becomes positive by the sum of e_g and e_c, which is $2e_{g,\max}$ at the time e_g is at its maximum value. As long as R_L is infinite, plate B of the capacitor cannot discharge, and e_c remains fixed at the value of $e_{g,\max}$. Since R_L is in parallel with e_c, the output voltage remains fixed at that value. Thus perfect filtering is achieved when R_L is infinite ohms.

Since an R_L of infinite ohms is not the usual condition, let us further analyze this circuit with R_L equal to a finite value, which is most conveniently done with the assistance of the waveforms shown in Fig. 9.17.

When e_g is first turned on (point O), the diode begins to conduct, and plate B of the capacitor begins to charge. Assuming negligible values of r_p and generator resistance, e_c will increase in step with e_g up to point A. At point A, the tube is cut off, since e_g begins to decrease, and the cathode is more positive than the anode. Since plate B must discharge through R_L, e_c decreases more slowly than e_g (long discharge time constant), as shown by AC in Fig. 9.17. Notice that the plate-B discharge provides R_L with a current flow which is in the same direction as the original tube current. This current flows to plate A of the capacitor, to complete the discharge path of plate B.

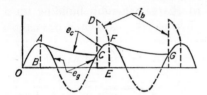

FIG. 9.17 e_g, e_c, and i_b of a half-wave rectifier with capacitor filter (circuit of Fig. 9.16).

While this is going on, e_g continues with its sinusoidal variations. If R_L is comparatively large, e_c has time to decrease by only a small amount (see AC) by the time e_g is positive and equal to e_c. When $e_g = e_c$, the tube conducts, which is called the *cut-in point* (point C). The rectifier conducts only when e_g is greater than e_c, since this is the only time the anode is positive with respect to the cathode. Since the capacitor has a short charging time constant (r_p and r_{gen} are small), e_c increases as e_g increases, causing the tube to stop conducting shortly after the generator voltage is maximum F, which is called the *cut-out point*. $e_c = e_g$ (point F), and the capacitor begins to discharge again in the manner previously described.

The variations of e_c are most smooth with larger values of R_L, since its discharge time constant is lengthened. This filter action is decidedly more effective with full-wave rectifiers, since the capacitor is charged twice as often and allowed to discharge only half as long between charging actions. Figure 9.18 illustrates the output voltage waveforms for both the half-wave and full-wave rectifiers with capacitor filters.

In viewing Fig. 9.18, notice that the ripple voltage resembles a triangular wave. In many practical cases, it may be treated as such. This approximation does not seriously affect the accuracy in circuit computations. Redrawing the waveform with this approximation results in the waveform shown in Fig. 9.19.

Notice that E_c denotes the capacitor discharge voltage and is approximated by a straight line. The average output voltage is obtained from

$$E_{av} = E_m - \frac{E_{c(pp)}}{2}$$

The rms value of the ripple (triangular) voltage is

$$E_{c,rms} = \frac{E_{c(pp)}}{3.464}$$

In Fig. 9.19, the nonconducting time occurs between points A and B, and may be found by

$$T_{nc} = \frac{E_{av}T_p}{E_m}$$

where T_{nc} = nonconducting time
T_p = total period time (conducting plus nonconducting)

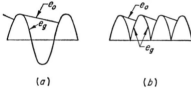

(a)   (b)

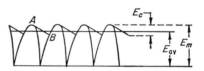

Fig. 9.18   Output voltage waveforms for capacitor filter rectifiers: (a) half-wave, (b) full-wave.

Fig. 9.19   Approximation of e_o.

The total change in capacitor voltage may be found from

$$E_{c(pp)} = \frac{I_{av}T_{nc}}{C} = \frac{I_{av}E_{av}}{2fCE_m}$$

The average output voltage for the full-wave rectifier may also be computed from the following:

$$E_{av} = \frac{E_m}{1 + I_{av}/4fCE_m}$$

The ripple factor may be determined from

$$rf = \frac{E_{c(pp)}}{3.464E_{av}} = \frac{I_{av}}{6.928fCE_m}$$

The following equation can be used for determining the average voltage for the half-wave rectifier with the capacitor filter:

$$E_{av} = \frac{E_m - I_{av}/4fC}{1 + I_{av}/4fCE_m}$$

The capacitor type of filter is most effective with light loads (large R_L)

and large values of capacitors. Electrolytic capacitors are frequently used in such circuits.

The average output voltage equation is useful in determining the *operation characteristics* of the rectifier, as indicated by the following example.

Example 1: $f = 60$ cps; $C = 4$ μf; $E_m = 450$ volts per plate. This is a full-wave rectifier. Find the average output voltage for the following load currents: (*a*) 50 ma; (*b*) 100 ma; (*c*) 150 ma.

Solution:

(*a*) When $I_{av} = 50$ ma:

$$E_{av} = \frac{450}{1 + [50 \times 10^{-3}/(4 \times 60 \times 4 \times 10^{-6} \times 450)]}$$
$$\cong 404 \text{ volts}$$

(*b*) When $I_{av} = 100$ ma:

$$E_{av} = \frac{450}{1.231} \cong 365 \text{ volts}$$

(*c*) When $I_{av} = 150$ ma:

$$E_{av} = \frac{450}{1.344} \cong 334 \text{ volts}$$

A plot of average load current versus average output voltage of a specified plate rms voltage may be called the *operation characteristic* of the rectifier.

Example 2: Draw the operation characteristic of the rectifier in example 1.

Such curves are very useful in determining the average output voltage for a given plate voltage and load current.

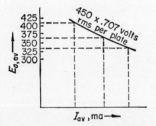

FIG. 9.20 I_{av} versus $E_{o,av}$.

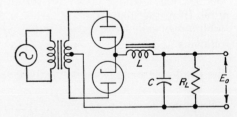

FIG. 9.21 Full-wave rectifier with the *L*-section filter.

The L-section Filter. The choke and capacitor filters can be combined to form the *L*-section filter. A full-wave rectifier with the *L*-section filter is shown in Fig. 9.21. The inductor presents a high series impedance to the ripple component while the capacitor provides a low parallel impedance to the current variations.   In other words, the smoothing out action of both the inductance and capacitance are combined in this filter, resulting in an output current which is considerably smoother than would be the case with either component alone.

The cathode to center-tap voltage in Fig. 9.21 may be computed from

$$e_{k,ct} = \frac{2E_m}{\pi} - \frac{4E_m}{3\pi} \cos 4\pi ft$$

The average output voltage is

$$E_{o,av} = \frac{2E_m}{\pi} - I_{av}R$$

where R is the sum of all the resistances excluding R_L but including half the transformer secondary and the tube.

In order to maintain a low-level ripple component, the series reactance X_L should be large in comparison to the parallel impedance of the capacitor X_C and load resistance R_L.   The parallel impedance of X_C and R_L is easily made small by the selection of a sufficiently large capacitor, which ensures an X_C value much smaller than R_L.   Also, X_L is increased by use of a larger inductance.   It can be seen that the L-section filter most effectively reduces the ripple component with large values of L and C.

The ripple factor may be found from the following:

$$rf = \frac{1.414X_C}{3X_L} \cong \frac{X_C}{2X_L}$$

and

$$rf = \frac{E_{ac,rms}}{E_{av}}$$

In the design of the L-section filter, the minimum inductance required to permit each rectifier to conduct for its portion of the input cycle is of great importance if good filtering is to be achieved.   This minimum inductance, called the *critical inductance* L_c, may be calculated from the following:

$$L_c \cong \frac{R_L}{6\pi f} \times 1.25$$

Examination of the preceding equation reveals that the critical inductance hinges upon the value of R_L.   The load resistance may vary over considerable limits in actual practice, requiring some technique for stabilizing the condition.   There are a couple of ways in which this problem may be handled:

1. Use of a *bleeder resistor*
2. Use of a *swinging choke*

When a bleeder resistor is used, the actual load resistance is in parallel with the bleeder resistance. The bleeder resistor is selected such that the parallel total of R_{bleeder} and the maximum R_L meets the requirements for the inductor value selected. In such cases, the equivalent value of R_L and R_{bleeder} is used in the equation for the determination of critical inductance; thus,

$$L_c = \frac{R_L R_B/(R_L + R_B)}{6\pi f} \times 1.25$$

where R_L is the maximum load resistance to be encountered.

The second method of stabilizing the change of critical inductance with a change in load is by utilization of a swinging choke, which is a coil with variable inductance. The inductance of an iron core hinges on the magnitude of the unidirectional current flowing through its windings. Recall that the B-H curve of the iron-core coil has its steepest slope at minimum current. The slope of the B-H curve decreases with larger

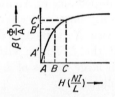

FIG. 9.22 B-H curve of an iron-core coil.

FIG. 9.23 I versus L of a "swinging choke."

values of current because the saturation point is being approached. The greater the flux density, originally, the smaller the flux density increase for a given NI (ampere-turn) increase, as shown in Fig. 9.22.

Refer to Fig. 9.22, which shows a B-H curve of an iron-core coil, for the following discussion. When $\Delta H = AB$, then $\Delta\beta = A'B'$; and when $\Delta H = B$ to C, then $\Delta\beta = B'$ to C', which is a smaller increase in flux density for the same amount of change in H. Hence the flux-cutting action is less pronounced with increasing values of H. Recall that the inductance is determined by the *amount of flux cutting;* therefore, the inductance of the coil decreases with larger values of H, which is usually varied by changes in the winding current. The relationship between the inductance and unidirectional winding current is depicted in Fig. 9.23.

The use of the swinging choke results in a large inductance when the current is small and a decreasing inductance as the load current becomes larger. This technique is very effective in maintaining good filtering action for decreased load currents, and less effective for increased load currents.

Design of the L-section Filter. The following parameters are desired:

100 ma at 300 volts; ripple voltage is to be 5 volts or less, frequency is 60 cps.   Determine the required C and L.

Step 1: *Determination of* R_L

$$R_L = \frac{300}{0.1} = 3 \text{ kilohms}$$

Step 2: *Determination of* L_c

$$L_c = \frac{3 \times 10^3 \times 1.25}{6 \times 3.14 \times 60} \cong 3.31 \text{ henrys}$$

Hence any commercial inductor of this value or greater is satisfactory, such as 4 henrys.

Step 3: *Determination of the Ripple Factor*

$$rf = \tfrac{5}{300} = 0.0167$$

Step 4: *Determination of* C

Since

$$rf = \frac{1.414 X_C}{3 X_L}$$

then,

$$X_C = \frac{rf 3 X_L}{1.414}$$

Using 4 henrys for the coil value, its reactance is

$$X_L = 6.28 \times 60 \times 4 \cong 1.5 \text{ kilohms}$$

Now solving for X_C,

$$X_C = \frac{1.67 \times 10^{-2} \times 3 \times 1.5 \times 10^3}{1.414} \cong 53 \text{ ohms}$$

and finding C,

$$C = \frac{1}{6.28 \times 60 \times 53} = 50 \ \mu\text{f}$$

Hence a commercial unit of 50 μf or larger may be used.

NOTE: If a swinging choke is to be used, the minimum value of that choke must be that value computed in step 2.

Pi-section Filter.   A full-wave rectifier with a pi-section filter is shown in Fig. 9.24.   This filter can be considered an L-section (L, C_2) working

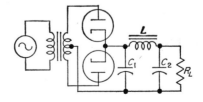

FIG. 9.24 Full-wave rectifier with a pi-section filter.

upon the output of a capacitor filter C_1. Recall that the output voltage of the capacitor filter has a ripple which is triangular. The L-section portion of the filter serves to further reduce and smooth out these variations.

The ripple factor may be found by

$$rf = \frac{X_{C1}X_{C2} \times 1.414}{R_L X_L}$$

where X_{C1}, X_{C2}, and X_L are computed with $f = $ 2nd harmonic of the input generator frequency for the full-wave rectifier and $f = $ generator fundamental for the half-wave rectifier.

The average output voltage is about equal to that of the capacitor filter minus the voltage drop across the inductor:

$$E_{o,\text{av}} = \frac{E_m}{1 + I_{\text{av}}/4fCE_m} - I_{\text{av}}R_{\text{ind}}$$

where R_{ind} is the winding resistance of the inductor.

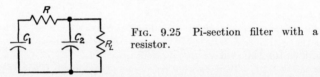

FIG. 9.25 Pi-section filter with a resistor.

In cases where only low current drains are imposed, it is advantageous to replace the inductor with a resistor, as shown in Fig. 9.25. The ripple factor is then

$$rf = \frac{1.414 \times X_{C1}X_{C2}}{R_L R}$$

and the output voltage (average) is found by

$$E_{o,\text{av}} = \frac{E_m}{1 + I_{\text{av}}/4fCE_m} - I_{\text{av}}R$$

PROBLEMS

NOTE: f in the following problems is the ripple frequency.

9.27 Refer to the equivalent circuit of the full-wave rectifier with an inductor filter as shown in Fig. 9.15. $E_m = 220$ volts; $f = 120$ cps; $R_L = 500$ ohms; $L = 5$ henrys; $r_p = 100$ ohms. Find (a) unidirectional current generator voltage; (b) ripple factor; (c) output voltage (average).

9.28 Refer to Fig. 9.15. $E_m = 110$ volts; $f = 120$ cps; $R_L = 250$ ohms; $r_p = 50$ ohms. Find (a) unidirectional current generator voltage; (b) ripple factor; (c) output voltage.

9.29 In the circuit of Fig. 9.16, what is meant by "cut-in point"?

9.30   In the circuit of Fig. 9.16, what is meant by "cut-out point"?

9.31   A rectifier circuit has a capacitor filter (see Fig. 9.19 for notation): $E_m = 150$ volts; $E_{c(pp)} = 10$ volts; $C = 5$ μf; $f = 120$ cps.   Find:
(a) E_{av} (b) $E_{c,rms}$
(c) T_{nc} (d) ripple factor

9.32   A full-wave rectifier circuit has a capacitor filter; $f = 120$ cps; $C = 6$ μf; $E_m = 400$ volts per plate.   Find the average output voltage for the following currents:
(a) 25 ma
(b) 50 ma
(c) 100 ma

9.33   Draw the operation characteristic of the rectifier in Prob. 9.32.

9.34   A full-wave rectifier circuit has a capacitor filter: $f = 120$ cps; $C = 4$ μf; $E_m = 500$ volts per plate.   Find the average output voltage for the following load currents:
(a) 50 ma
(b) 100 ma
(c) 150 ma

9.35   Draw the operation characteristic of the rectifier in Prob. 9.34.

9.36   Refer to Fig. 9.21.   What is meant by critical inductance?

9.37   Refer to Fig. 9.21. $f = 120$ cps; $C = 5$ μf; $L = 10$ henrys; $R_L = 500$ ohms.   Find (a) rf; (b) L_c.

9.38   In Prob. 9.37, assume a bleeder resistance of 200 ohms is used.   Find L_c.

9.39   State the relationship between L_c and the circuit resistance.

9.40   What is a swinging choke?

9.41   Refer to Fig. 9.21.   The following parameters are desired: 100 ma at 400 volts; ripple to be 10 volts or less; $f = 120$ cps.   Find:
(a) R_L (b) L_c
(c) rf   (d) C

9.42   Refer to Fig. 9.21.   The following parameters are desired: 150 ma at 250 volts; ripple to be 5 volts or less; $f = 120$ cps.   Find:
(a) R_L (b) L_c
(c) rf   (d) C

9.43   Refer to the full-wave rectifier with a pi-section filter of Fig. 9.24. $C_1 = 10$ μf; $C_2 = 10$ μf; $L = 5$ henrys; $R_L = 2$ kilohms; $f = 120$ cps.   Find the ripple factor.

9.44   In Prob. 9.43, let $C_1 = C_2 = 5$ μf.   Find the ripple factor.

9.45   In comparing the ripple factor of Probs. 9.43 and 9.44, what is the relationship between the size of C_1 and C_2 and the ripple factor?   Explain.

9.46   Refer to Fig. 9.24. $C_1 = C_2 = 8$ μf; $L = 10$ henrys; $R_L = 500$ ohms; $f = 120$ cps.   Find the ripple factor.

9.47   In Prob. 9.46, find the ripple factor if R_L is (a) 100 ohms; (b) 1 kilohm.

9.48   Refer to Probs. 9.46 and 9.47.   What is the relationship between the size of R_L and the ripple factor?   Explain.

9.49   Refer to the pi-section filter with a resistor as shown in Fig. 9.25. $R = 100$ ohms; $R_L = 2$ kilohms; $C_1 = C_2 = 8$ μf; $f = 60$ cps.   Find the ripple factor.

9.50   In the circuit of Prob. 9.49, let $R = 500$ ohms.   Find the ripple factor.

9.51   Refer to Probs. 9.49 and 9.50.   What is the relationship between the size of R and the ripple factor?

9.52   In Probs. 9.49 and 9.50, what is the relationship between the size of R and $E_{o,av}$?   Explain.

9.5 Design of a Full-wave Vacuum-rectifier Power Supply. Using the relationships and equations revealed in the preceding sections, a full-wave rectifier circuit can be designed. In most practical cases, such a power supply is designed around three or four basic specifications, namely:

1. $E_{o,\text{av}}$
2. I_{av}
3. Ripple factor
4. Voltage regulation

Let us state a typical set of specifications and actually design a full-wave power supply to meet the requirements.

Specifications. $E_{o,\text{av}} = 275$ volts; $I_{\text{av}} = 100$ ma; ripple to be no greater than 5 volts; voltage regulation to be as high as possible; the source potential is 117 volts (rms).

Solution:

Step 1: Selection of a Transformer. The transformer must have a center-tapped secondary with such a turns ratio that the rms voltage per plate is about 350 volts or 700 volts (rms) plate-to-plate. This makes allowances for the transformer, tube, and inductor voltage drops. Hence the primary to center-tapped secondary turns ratio is

$$\frac{N_p}{N_s} = \frac{E_p}{E_s} = \frac{117}{700} \cong \frac{1}{6}$$

An additional secondary winding is required to furnish the required filament voltage to the rectifier. This is to be determined by the full-wave rectifier tube to be used, which is selected in one of the following steps.

Step 2: Selection of the Filter Components. The L-section filter is arbitrarily selected, since it is effective in maintaining a smooth output, has good regulation, and is reasonably economical. Let us first find R_L:

$$R_L = \frac{E_{o,\text{av}}}{I_{\text{av}}} = \frac{275}{100 \times 10^{-3}} = 2.75 \text{ kilohms}$$

And the critical inductance,

$$L_c = \frac{R_L}{6\pi f} \times 1.25 = \frac{2.75 \times 10^3 \times 1.25}{6 \times 3.14 \times 120}$$
$$\cong 1.52 \text{ henrys}$$

NOTE: The ripple frequency is twice the input voltage frequency in the full-wave rectifier circuit.

A commercial coil of 4 henrys or greater may be used. The ripple factor is

$$rf = \frac{5}{275} \cong 0.0182$$

We shall next find C:

$$X_C = \frac{rf 3 X_L}{1.414}$$

where $X_L = 6.28 \times 120 \times 4 = 3{,}014$ ohms

Solving for X_C,

$$X_C = \frac{0.0182 \times 3(3,014)}{1.414} \cong 116 \text{ ohms}$$

Now determining C,

$$C = \frac{1}{6.28 \times 120 \times 116} \cong 11.42 \ \mu\text{f}$$

A commercial capacitor of 15 μf or larger may be used.

Step 3: *Comparison of $E_{o,\text{av}}$ Desired to $E_{o,\text{av}}$ Obtained by Design*

$$E_{o,\text{av}} = \frac{2E_m}{\pi} - I_{\text{av}}R$$

R in the preceding equation includes the resistance of the filter coil, r_p of the conducting tube, and the resistance of one-half of the secondary coil (since only one-half of the secondary passes current at any one time). Typical values which will be used here are:

$$r_p = 100 \text{ ohms}$$
$$R_{\text{ind}} = 200 \text{ ohms}$$
$$R_{x,\text{former}} = 60 \text{ ohms (for one-half of the step-up secondary)}$$

Therefore, $\qquad R = 100 + 200 + 60 = 360$ ohms

and $\qquad E_m = 350 \times 1.414 \cong 495$ volts

Now, finding $E_{o,\text{av}}$,

$$E_{o,\text{av}} = \frac{2(495)}{3.14} - 0.1(360) \cong 280 \text{ volts (full load)}$$

Note that the average output voltage at full load (100 ma) is 5 volts above the desired value, which is usually considered well within the permissible tolerance.

Step 4: *Determination of Voltage Regulation.* In the calculation of voltage regulation, the "no-load" condition is usually taken to be the condition of *minimum* load. Let us assume the minimum load to be half of the full load, or 50 ma. The $I_{\text{av}}R$ portion of the average output voltage equation then reduces to

$$I_{\text{av}}R = 50 \times 10^{-3} \times 360 = 18 \text{ volts}$$

and $\qquad E_{o,\text{av}}$ (half load) $= 316 - 18 = 298$ volts

Using the full-load and half-load average output voltage values, the regulation is now determined:

$$\text{Per cent regulation} = \frac{E \text{ half load} - E \text{ full load}}{E \text{ full load}} \times 100$$

Substituting values,

$$\text{Per cent regulation} = \frac{298 - 280}{280} \times 100 \cong 6.4\%$$

In terms of potential, the average output voltage varies by 18 volts between half- and full-load conditions.

Step 5: *Selection of the Tube.* A full-wave rectifier tube is to be selected which has an inverse peak voltage rating at least equal to the peak value of the entire secondary, which is about 1,000 volts in this case. The 5Y3GT meets this requirement and may be used. Since the filament of the 5Y3GT requires 5 volts at 2 amp, a suitable step-down secondary is also required.

NOTE: In those cases where an input filter capacitor is used (the pi-section filter), the maximum current rating of the tube must be great enough to exceed the value:

$$I_m = \frac{E_m}{R_{x,\text{former}} + r_p}$$

The reason for such a high current rating is that this current could flow in the event the input voltage passes through its maximum value at the instant the circuit is turned on, at which time the input filter capacitor appears as a short circuit.

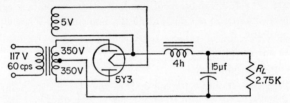

FIG. 9.26 The designed full-wave vacuum-rectifier power supply.

Step 6: *Circuit Construction.* Figure 9.26 illustrates the designed full-wave power supply. The insertion of a 2.75-kilohm resistor of proper wattage to serve as a "dummy load" will enable the designer to check out the circuit without the use of the actual load.

9.6 Design of a Full-wave Silicon-rectifier Power Supply.

Many modern power supplies utilize a semiconductor type of rectifier in place of the vacuum tube. The silicon rectifier is one of the most notable semiconductor rectifiers used for this purpose. Some of its chief advantages are:

1. Economy of space and weight
2. No filament power required
3. Outstanding dependability and long life
4. Very small voltage drop across the rectifier (2 volts or less in many cases)

Although the silicon rectifier displays a flow of reverse current, its magnitude is generally small enough to be ignored. The earlier introduced design equations are equally applicable to the silicon-rectifier circuit, as is shown in the following design problem.

Specifications. The input source voltage = 117 volts (rms); $f = 60$ cps; $E_{o,\text{av}} = 30$ volts; ripple voltage is not to exceed 1 volt; full load = 100 ma; minimum load = 30 ma; L-section filter is to be used.

Solution:

Step 1: Selection of a Transformer. Assume the silicon-rectifier voltage drop is about 2 volts, which is typical. Also assume the filter-choke resistance is about 200 ohms, and its voltage drop is maximum at full-load condition, causing $E_{o,av}$ to be lowest at that time. The maximum voltage drop across the filter choke is

$$E_{R,ind} = I_{av} \text{ (full load)} R_{ind}$$
$$= 100 \times 10^{-3} \times 2 \times 10^2 = 20 \text{ volts}$$

Therefore, the secondary potential of the transformer should be large enough to furnish this potential, the drop across its own resistance (half of the secondary), the rectifier, and still enable $E_{o,av}$ at full-load condition to be about 30 volts. Hence

$$E_s \cong 25 + 30 \cong 55 \text{ volts}$$

A transformer with a unity transformation ratio and a center-tapped secondary will serve this purpose. The transformer should be capable of handling at least 100 ma without being damaged.

Step 2: Selection of the Filter Components. The input capacitor C_1 can be the same value (or slightly smaller) than the value of C_2 when the pi-section filter is to be used. A more specific method for the determination of C_1 is discussed in a preceding paragraph.

C_2 and L can be found by the same techniques used in the vacuum-tube rectifier designed in the preceding section.

$$R_L = \frac{E_{o,av}}{I_{av}} = \frac{30}{100 \times 10^{-3}} = 300 \text{ ohms}$$

And the critical inductance,

$$L_c = \frac{R_L \times 1.25}{6\pi f} = \frac{300 \times 1.25}{6 \times 3.14 \times 120}$$
$$\cong 0.166 \text{ henry}$$

A commercial coil of 0.2 henry or greater can be used. For convenience, we shall use a 4-henry, 200-ohm coil.

The ripple factor is

$$rf = \frac{1}{30} = 0.0333$$

Determination of C_2:

$$X_{C2} = \frac{rf 3 X_L}{1.414}$$

where
$$X_L = 6.28 \times 120 \times 4 = 3{,}014 \text{ ohms}$$

and
$$X_{C2} \cong \frac{0.0333 \times 9{,}042}{1.414} \cong 215 \text{ ohms}$$

then
$$C_2 = \frac{1}{6.28 \times 120 \times 215} \cong 6.16 \ \mu f$$

A commercial capacitor of 6 μf or larger may be used. As stated earlier, C_1 can be equal or slightly smaller than C_2 when the pi-section filter is used. Hence

$$C_1 \cong C_2 \cong 6 \ \mu f$$

Step 3: *Determination of Voltage Regulation*

$$\text{Per cent regulation} = \frac{E_{o,\text{av}} \text{ min load} - E_{o,\text{av}} \text{ full load}}{E_{o,\text{av}} \text{ full load}} \times 100$$

Recall that

$$E_{o,\text{av}} = \frac{2E_m}{\pi} - I_{\text{av}}R$$

where
$$R = R_{\text{ind}} + r_{\text{diode}} + R_{x,\text{former}} \ (\text{one-half of secondary})$$
$$R_{\text{ind}} \cong 200 \text{ ohms}$$
$$r_{\text{diode}} \cong 5 \text{ ohms (dynamic resistance of the rectifier)}$$
$$R_{x,\text{former}} \cong 20 \text{ ohms}$$

hence
$$R = 200 + 5 + 20 = 225 \text{ ohms}$$
and
$$E_m = 55 \times 1.414 = 78 \text{ volts}$$

Now solving,

$$E_{o,\text{av}} \ (\text{min load}) = \frac{2E_m}{\pi} - I_{\text{av,min}}R$$

$$= \frac{2(78)}{3.14} - 0.03(225) = 42.85 \text{ volts}$$

and
$$E_{o,\text{av}} \ (\text{full load}) = \frac{2E_m}{\pi} - I_{\text{av,full load}}R$$

$$= 49.6 - 0.1(225) = 27.1 \text{ volts}$$

then
$$\text{Per cent regulation} = \frac{42.85 - 27.1}{27.1} \times 100 \cong 58\%$$

Step 4: *Selection of the Silicon Rectifier.* A full-wave silicon rectifier is to be selected which has an inverse peak voltage rating equal to the peak value of the entire secondary, which is about 156 volts in this case. The 1N254 silicon rectifier can be used in the choke input circuit designed. The maximum average current of this rectifier is 1.2 amp.

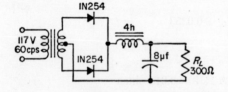

Fig. 9.27 The designed full-wave silicon-rectifier power supply.

Step 5: *Circuit Construction.* Figure 9.27 illustrates the designed power supply. The insertion of a 300-ohm resistor to serve as a dummy load will enable the designer to thoroughly check this circuit without connecting the actual load.

Note that the regulation of this circuit would be markedly improved by reducing the internal resistance of the power supply. The use of a lower resistance inductor and transformer, which is frequently done in actuality, would enhance the voltage regulation of the circuit substantially.

Pi-section Filter Silicon-rectifier Circuit Considerations. When designing a pi-section filter for silicon rectifiers, a number of factors must be taken into consideration.

As in its vacuum-tube counterpart, the peak surge current must be determined:

$$I_m = \frac{E_m}{R_s}$$

where $R_s = R_{x,\text{former}} + r_{\text{diode}}$.

Perhaps the reader wonders why such a high maximum current must be tolerated by the rectifier when the capacitor input (pi-section) filter is used. If the input capacitor is completely discharged at the time the supply is turned on and the input voltage is at its peak value at the same instant, the transformer winding resistance and r_p of the rectifier are the only resistances offered to the initial current. This is so, since the uncharged input capacitor appears as a short circuit in parallel with the load resistance and filter choke at this initial condition. It is imperative that the rectifier be capable of tolerating this instantaneous peak current. A wise practice is to select a rectifier whose permissible *peak surge current* is about twice that predicted for the designed circuit. In many cases, a small surge resistor is placed in series with each rectifier to reduce this momentary surge current.

In the previously designed power supply,

$$I_m = \frac{E_m}{R_s} = \frac{78}{25} = 3.12 \text{ amp}$$

Therefore, a silicon rectifier which is capable of withstanding a peak surge of about 6 amp should be selected if the pi-section filter is to be used.

The peak surge current exponentially decreases to the normal surge values as the input capacitor accumulates its charge in accordance with the time constant $R_s C_1$ (where C_1 is the input filter capacitor), which is called the *surge time constant*.

The input capacitance can be determined from

$$C_1 = \frac{800}{R_s} \text{ μf} \qquad \text{for 60 cps}$$

and
$$C_1 = \frac{120}{R_s} \text{ μf} \qquad \text{for 400 cps}$$

In the previously designed power supply, the input capacitance (if a pi-section filter was selected) would be

$$C_1 = {}^{800}\!/_{25} = 32 \ \mu\mathrm{f} \qquad \text{for 60 cps input}$$

or $\qquad C_1 = {}^{120}\!/_{25} = 4.8 \ \mu\mathrm{f} \qquad \text{for 400 cps input}$

It has been determined that *optimum voltage regulation* is achieved when R_sC_1 is between 0.5 and 1 msec for 60-cps input frequency. In the case of the condition just analyzed,

$$R_sC_1 = 25 \times 32 \times 10^{-6} = 0.8 \text{ msec}$$

Hence optimum voltage regulation would be achieved.

The ripple content is reduced by increasing the surge time constant. But this also increases the time it takes the current surges to return to their normal values, thereby introducing the possibility of creating additional heat within the rectifier. Furthermore, increasing this time constant also results in degrading the voltage regulation. It is deemed advisable to select a silicon rectifier which can withstand the peak surge current for $2(R_sC_1)$ sec or more. In the previously described case, the selected rectifier should be capable of withstanding about 6 amp for $2(25 \times 32 \times 10^{-6}) = 1.6$ msec.

The reverse voltage that the rectifier is expected to encounter should not be more than 75 per cent of its rating, thereby allowing for switching transients. Again referring to the previous case, 78 volts should be no more than 75 per cent of the peak inverse voltage of the selected rectifier. Hence a rectifier possessing a reverse voltage rating of at least 100 volts is required.

9.7 The Voltage Doubler. In the types of power supplies analyzed up to this point, the average output voltage was never in excess of the input

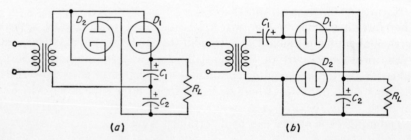

Fig. 9.28   Voltage-doubler circuits: (a) full-wave, (b) half-wave.

voltage. It is possible to obtain an output potential greater than the input voltage by circuits such as those shown in Fig. 9.28.

Figure 9.28a illustrates the *full-wave voltage-doubler* circuit. This is actually the combination of two half-wave rectifiers connected in such a way that D_1 charges up C_1 when it conducts, and D_2 charges up C_2 when it conducts. With no load connected, the potential across C_1 and C_2 in series is equal to twice the peak value of the input potential. With R_L connected, $E_{o,\text{av}}$ is less than this value but still greater than the peak value of the input voltage if the circuit is lightly loaded.

The chief disadvantage of the full-wave doubler is that the common side of the load is not at the same potential point as the common end of the input transformer secondary. The danger of electric shock exists because of the difference of potential between these two points.

Figure 9.28*b* illustrates the *half-wave voltage doubler*, which has a common "ground" for the transformer and the load resistance. C_1 charges up to the peak input voltage when D_2 conducts and D_1 is at cutoff. When the input potential polarity undergoes its reversal, D_1 conducts while D_2 is cut off. During the time D_1 conducts, C_2 charges up to the peak input voltage. Therefore the output voltage is equal to the sum of

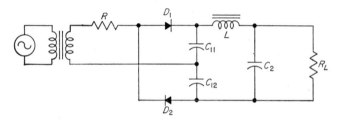

Fig. 9.29 Silicon full-wave voltage doubler with a pi-section filter.

E_{C1} and E_{C2}. With R_L connected, the peak values are not obtained, but $E_{o,\text{av}}$ is still considerably greater than the peak input potential for light loads.

Figure 9.29 illustrates a silicon-rectifier full-wave voltage doubler with a pi-section filter. The transformer is selected in the same manner as used in designing the full-wave L-section power supply. C_{11} serves as the input capacitor for D_1 and C_{12} serves the same purpose for D_2. Their values can be determined by

$$C_{11} = C_{12} = \frac{800}{R_s} \qquad \text{for 60 cps input}$$

or
$$C_{11} = C_{12} = \frac{120}{R_s} \qquad \text{for 400 cps input}$$

where $R_s = R_{x,\text{former}} + R + r_{\text{diode}}$.

The resistor R in series with the rectifiers is a *surge limiting resistor* and is usually small in value (from one to several ohms).

The filter choke and output capacitor C_2 are determined in the same way as the previously designed power supply.

Since the voltage doubler utilizes the capacitor input type of filter, special attention must be given to the expected peak surge current, as analyzed in the preceding paragraphs.

9.8   The Bridge Rectifier.   Figure 9.30 illustrates the full-wave bridge rectifier with both the pi-section and L-section types of filters.   In both circuits, D_1 and D_2 are in series, as are D_3 and D_4. D_1 and D_3 conduct during one portion of the input voltage cycle while D_2 and D_4 are under reverse voltage at that time.   Upon reversal of the input voltage, D_2 and D_4 are under forward bias condition and D_1 and D_3 are cut off.   Therefore the over-all rectification is basically that of the full-wave variety.   Since two of the rectifiers are in series, the inverse voltage to which each rectifier is subjected is only one half of E_m.   This permits the use of rectifiers with smaller reverse breakdown voltage values.   Using the safety factor stated earlier, the reverse breakdown voltage of each rectifier

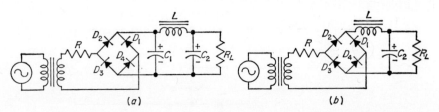

FIG. 9.30   Full-wave bridge rectifiers: (a) pi-section filter, (b) L-section filter.

need only be $1.33(E_m/2)$.   Although four rectifiers are used instead of two, the lower reverse breakdown voltage requirements may result in the use of lower-cost rectifiers.

The values of filter components, surge current, optimum voltage regulation, transformer, rectifiers, and output voltage are determined in accordance with the procedures specified in earlier paragraphs.   The techniques for the pi-section filter are applied to the full-wave pi-section type, and the L-section principles are applicable to the L-section type. The chief difference in design is the reduction of reverse potential encountered by each rectifier.

9.9   The Vibrator-type Power Supply.   Figure 9.31 illustrates a common type of vibrator power supply.   The vibrator unit essentially consists of two contacts (A and B) and a magnetically vibrating reed C.   When the switch is closed, current flows from $-E_B$, the vibrator coil D, the upper portion of the transformer primary, back to $+E_B$.   The vibrator coil is magnetized by this current, which attracts the reed to point A.   This results in shorting out the coil, allowing it to become demagnetized.

Since the reed is no longer attracted to contact A, it leaves A and swings past its rest position and hits contact B.   Current now flows from $-E_B$ through B, the bottom half of the transformer primary back to $+E_B$.   In the meantime, the vibrator coil again becomes magnetized, and the reed is drawn to contact A.   This mechanical action of the vibrator reed occurs at a repetition rate of about 115 cps in the common auto vacuum tube radio type of vibrator.

The making and breaking of contact by the vibrator reed results in hash and sparking.   Practical circuits incorporate filter arrangements to reduce this problem. R_1 and R_2 are used for this purpose. C_1, commonly called the "buffer" capacitor, absorbs the high voltage surges which occur because of the contact action just described.

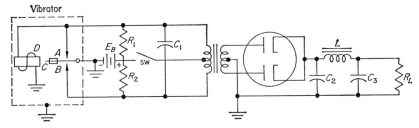

FIG. 9.31   The vibrator-type power supply.

The remainder of the power supply is essentially identical to the full-wave pi-section circuit described earlier.   The chief difference is that the input frequency is about 115 cps.   The design information previously discussed is applicable.

PROBLEMS

9.53   A full-wave vacuum-tube rectifier power supply is to be designed.   The specifications are: $E_{o,av} = 275$ volts; $I_{av} = 75$ ma; ripple to be no greater than 5 volts; voltage regulation of 10 per cent or less is required; L-section filter is to be used; 5Y3GT is to be used.   Find:

(a) required transformer turns ratio   (b) R_L
(c) L_c (d) rf
(e) C (f) $E_{o,av}$ (full load)
(g) $E_{o,av}$ (half load)   (h) per cent regulation

9.54   A full-wave vacuum-tube rectifier power supply is to be designed.   The specifications are: $E_{o,av} = 325$ volts; $I_{av} = 125$ ma; ripple to be no greater than 5 volts; voltage regulation of 10 per cent or less is required; L-section filter is to be used.   Find:

(a) required transformer turns ratio   (b) R_L
(c) L_c (d) rf
(e) C (f) $E_{o,av}$ (full load)
(g) $E_{o,av}$ (half load)   (h) per cent regulation

9.55 In Prob. 9.53, let $I_{av} = 150$ ma; all other specifications are the same. Find:
(a) R_L (b) L_c
(c) rf (d) C
(e) $E_{o,av}$ (full load)   (f) $E_{o,av}$ (half load)
(g) per cent regulation

9.56 In Prob. 9.54, let $I_{av} = 50$ ma; all other specifications are the same. Find:
(a) R_L (b) L_c
(c) rf (d) C
(e) $E_{o,av}$ (full load)   (f) $E_{o,av}$ (half load)
(g) per cent regulation

9.57 State the effect a change of I_{av} has upon the following in a full-wave vacuum-tube power supply with the L-section filter:
(a) L_c
(b) C
(c) per cent regulation

9.58 A full-wave silicon-rectifier power supply is to be designed. The specifications are: input source voltage = 117 volts(rms); $f = 60$ cps; $E_{o,av} = 22$ volts; ripple voltage not to exceed 1 volt; full load = 100 ma; minimum load = 20 ma; L-section filter is to be used; filter choke $R = 10$ ohms; silicon-rectifier voltage drop = 2 volts; $r_{diode} = 5$ ohms; one-half of transformer secondary resistance = 10 ohms. Find:
(a) $E_{R,ind}$ (maximum)   (b) E_s
(c) required transformer turns ratio   (d) R_L
(e) L_c (f) rf
(g) C_2 (h) $E_{o,av}$ (min load)
(i) $E_{o,av}$ (full load)   (j) per cent regulation

9.59 In the power supply of Prob. 9.58, a pi-section filter is to be used instead of the L-section filter. Find:
(a) I_m
(b) C_1
(c) surge time constant

9.60 A full-wave silicon-rectifier power supply is to be designed. The specifications are: input source voltage = 117 volts(rms), 60 cps; $E_{o,av} = 40$ volts; ripple voltage not to exceed 1 volt; full load = 150 ma; minimum load = 30 ma; L-section filter is to be used; filter choke $R = 20$ ohms; silicon-rectifier voltage drop = 2 volts; $r_{diode} = 5$ ohms; one half of transformer secondary resistance = 15 ohms. Find:
(a) $E_{R,ind,m}$ (b) E_s
(c) required transformer turns ratio   (d) R_L
(e) L_c (f) rf
(g) C_2 (h) $E_{o,av}$ (min load)
(i) $E_{o,av}$ (full load)   (j) per cent regulation

9.61 In the power supply of Prob. 9.60, a pi-section filter is to be used instead of the L-section filter. Find:
(a) I_m
(b) C_1
(c) surge time constant

9.62 In Prob. 9.59, was optimum voltage regulation achieved?

9.63 In Prob. 9.61, was optimum voltage regulation achieved?

9.64 Refer to Fig. 9.29. $R = 2$ ohms; R of half the transformer secondary = 10 ohms; r_p of each rectifier = 5 ohms. Find
(a) C_{11} (b) C_{12}
(c) L (d) C_2

9.10 Voltage and Current-regulator Circuits. In many applications, voltage regulation which is superior to that inherently possible in any of the power supplies analyzed is required. In such cases, special circuits are added between the filter and the actual load. There are also many cases where a constant output current is needed. This section studies some of the fundamental semiconductor circuits for:

1. Shunt voltage regulators
2. Series voltage regulators
3. Constant-current regulators

Voltage regulation by use of gas tubes is analyzed in Chap. 14.

Shunt Voltage Regulators. The most simple shunt voltage regulator circuit is shown in Fig. 9.32a. D_1 is a breakdown or *zener diode*, which

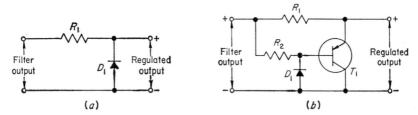

(a)   (b)

FIG. 9.32 Basic shunt voltage-regulator circuits.

is connected in its reverse direction across the output. The reverse characteristics of the zener diode are illustrated in Fig. 9.33.

The characteristic between points O and A, called the *saturation region*, is that portion of the characteristic where a relatively constant reverse current flows for large changes in reverse voltage. The *avalanche effect* occurs at point A, which is in the *zener region*. Notice that large values

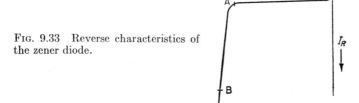

FIG. 9.33 Reverse characteristics of the zener diode.

of reverse current can be permitted with virtually no change in the diode reverse potential. Using D_1 in Fig. 9.32a in its zener region thereby allows the output voltage (which is the zener potential of the diode) to remain relatively constant for large changes in output current.

If such a circuit is used, R_1 and D_1 should be so selected that the D_1 maximum power dissipation is not exceeded. Zener diodes are commercially available with potentials up to 200 volts and power-dissipation ratings of 50 watts.

The current through R_1 is the zener current plus the load current, and the voltage across R_1 is equal to the filter output voltage minus the zener potential; hence

$$R_1 = \frac{E_{o,\min} - E_{D1,\max}}{I_{o,\max} + I_{D1,\min}}$$

where $E_{o,\min}$ = minimum filter output voltage (including ripple)

$\quad E_{D1,\max}$ = highest zener voltage to be used

$\quad I_{o,\max}$ = maximum load current

$\quad I_{D1,\min}$ = minimum zener current

The maximum power dissipation of the zener diode can be predicted from the following:

$$P_{D1,\max} = E_{D1,\max} I_{D1,\max}$$

The shunt voltage regulator of Fig. 9.32b may be used in those cases where the impedance of the zener diode is too low for the particular application. The zener-diode voltage is relatively constant, as in the previous circuit, and serves as the input potential for transistor T_1. Notice that T_1 is in the common-collector configuration. T_1 should be near cutoff when maximum output current flows. D_1 should be capable of handling $I_{R2,\max}$ plus $I_{B,\max}$ of T_1 which flows when T_1 is conducting most heavily (no-load condition).

R_1 is determined as in the first circuit, and

$$R_2 = \frac{E_{R1} + E_{C\text{-}B}}{I_{D1,\min} + I_{B,CO}}$$

where E_{R1} = as determined in the previous circuit

$\quad E_{C\text{-}B}$ = collector-base voltage of T_1

$\quad I_{D1,\min}$ = minimum zener-diode current

$\quad I_{B,CO}$ = base cutoff current of the transistor

The shunt voltage regulators are inherently more safe than the types to be discussed shortly, since an overload or shorted output does not endanger any of the regulator circuit components. The efficiency of the shunt type is low and is most commonly used where fixed-load (fixed-potential) conditions are encountered.

Series Voltage Regulators. Figure 9.34 illustrates one possible series voltage regulator. When the output current increases, the voltage across the transistor decreases accordingly, thereby allowing the regulated output voltage to remain relatively constant. The transistor is in the common-collector configuration, and the potential across the zener diode is its input voltage. R_1 serves as the path for the zener current and the transistor base current, and can be determined by

$$R_1 = \frac{E_{B\text{-}C}}{I_{D1} + I_B}$$

where $E_{B\text{-}C}$ = base-to-collector voltage of T_1
I_{D1} = zener-diode current
$I_B = T_1$ base current

E_{D1} changes with the unregulated input, as in the shunt circuits, resulting in a change in the regulated output voltage. This circuit is more efficient than the shunt type, since there is no current path which shunts the output. The chief drawback of the circuit is that a short circuit in the output can destroy T_1.

Both the shunt and series voltage regulators discussed result in a relatively low output impedance.

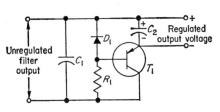   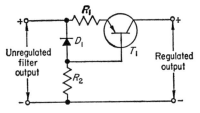

Fig. 9.34   The series voltage regulator.   Fig. 9.35   A constant-current regulator.

Constant-current Regulators.   One possible constant-current regulator is illustrated in Fig. 9.35. E_{R1} is felt by the emitter of T_1, and E_{D1} is felt by the base of T_1. The transistor will conduct such that E_{R1} and E_{D1} are nearly equal. The output resistance is the collector-base resistance of T_1. This circuit is most effective with large zener voltages and large values of R_1.

Notice that R_1 tends to offset the changes in the load current. As the load current tends to increase, i_E increases, causing E_{R1} to increase. This is an increase in reverse bias, causing i_C to decrease. E_{D1} provides forward bias to the base of T_1, and E_{R1} (which is reverse bias) tends to remain close to that value. The magnitude of the output current hinges on the difference between these two voltages.

9.11   D-C to D-C Converters.   The vibrator-type power supply discussed in Sec. 9.9 is actually a type of d-c to d-c converter. Recall that the input voltage is a unidirectional potential which is mechanically interrupted with a vibrator. The interrupted voltage is utilized as the transformer primary potential, which is stepped up to the desired magnitude. The larger secondary potential is then rectified and filtered in the usual way. This process is relatively inefficient, and vibrator failure is commonplace. High efficiency and reliability are achieved in the transistorized d-c to d-c power converters.

The successful operation of the d-c to d-c power converter hinges upon the use of a transformer whose core displays a square B-H curve, as shown in Fig. 9.36.

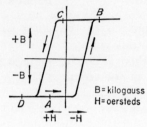

FIG. 9.36 B-H curve of the transformer core.

B = kilogauss
H = oersteds

A typical d-c to d-c power converter is illustrated in Fig. 9.37. The circuit operation commences upon sensing an unbalance. Assume such an unbalance exists, causing T_1 to conduct, driving current through N_2, as shown by the arrows. The transformer core is saturated at this condition, such as point A in Fig. 9.36. The core flux begins to change from point A toward point D, which is the saturation region. This change induces a potential in N_1 which drives T_1 far into saturation, while

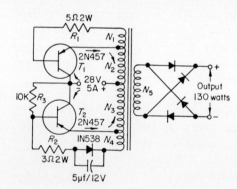

$N_2, N_3 = 88$ turns #18
$N_2, N_4 = 12$ turns #22

$$N_5 = \frac{N_2(E_s)}{28}$$

Core: Toroidal ribbon-wound core with maximum flux density = 14.8 kilogauss at a cross-sectional area of 1.37 sq cm

FIG. 9.37. A typical d-c to d-c power converter. (*Texas Instruments, Inc.*)

the N_4 potential causes T_2 to reach cutoff. Because of the heavy current through N_2, the majority of the input voltage appears across it.

When the core actually reaches saturation (such as point D in Fig. 9.36), the induced voltage abruptly decreases. This cuts off T_1. Meanwhile the core condition reverts to point A, which fires T_2 into saturation. The majority of the input voltage appears across N_3. The T_2 and N_3 current flow is shown by arrows in Fig. 9.37. T_1 is cut off and T_2 is conducting heavily when the core flux is changing from the negative saturated region A to the positive saturated region B. T_1 is heavily conducting and T_2 is cut off when the core flux is changing from the positive saturation region C to the negative saturation region D.

Because of the abruptness at which the B-H curve changes, the transistors are fired on and cut off in an equally abrupt fashion. This results in the N_5 voltage being an alternating square wave. The alternating square wave is rectified by the bridge circuit and may be filtered by a capacitor if so desired. A single capacitor is often sufficient for good filtering because of the linear nature of the square wave.

In some applications, the desired output voltage is an alternating voltage. In such cases, no rectification or filtering is necessary, with the output taken across N_5. Such an arrangement is called a *d-c to a-c power converter*.

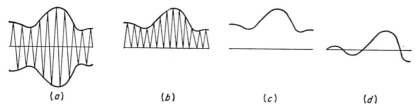

(a) (b) (c) (d)

FIG. 9.38 Diode demodulation: (a) input, (b) diode output, (c) filter output, (d) output after blocking C.

9.12 A-M Demodulation.

Demodulation (or detection) may be defined as the process by which the modulation (intelligence) of a carrier is removed from the waveform and retained. A-m demodulation can be accomplished by means of a diode (semiconductor or vacuum-tube), followed by a suitable filter arrangement.

Refer to Fig. 9.38. Diagram (a) illustrates the amplitude-modulated wave which is applied to the diode. The diode output is the rectified version of the input, as shown in diagram (b). The diode output is then applied to a filter circuit. The capacitor of the filter offers low impedance to the carrier frequency but high impedance to the signal variations. This filtering action results in the retention of the signal variations only, as shown in diagram (c). The filtered output is then applied to a blocking capacitor, resulting in the bidirectional signal variations of diagram (d).

Figure 9.39 illustrates a detector circuit which utilizes either a semiconductor or vacuum-tube diode. A modulated carrier (such as 455 kc) is applied to the diode by the secondary of the transformer. The resulting diode current contains the variations of both the carrier and modulation, and it is rectified. C_1 is so selected that it is a low impedance path for the rectified carrier, thereby allowing only the intelligence (audio frequencies in this case) to appear across R_1 and R_v, which is then coupled by C_2 to the audio amplifier, which magnifies the desired signal.

Let us further examine the filtering action of C_1 and R_1 so as to understand how the carrier frequency variations are bypassed. Refer to Fig. 9.40 for the following analysis.

The rectified portion of the carrier variations, which is the diode output, is shown in Fig. 9.40. Notice that the varying amplitude of the carrier variations is the desired modulation. C_1 is so selected that its reactance

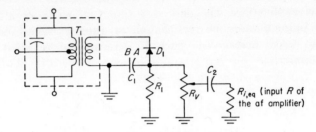

FIG. 9.39 Diode a-m detector circuit.

at the carrier frequency is small as compared to that of R_1. When the carrier voltage increases from O to A (see Fig. 9.40a), plate A of C_1 takes on electrons and plate B discharges electrons. Let us call this the charging action of C_1. The complete charging path of C_1 is from plate B through the secondary coil of the transformer (see Fig. 9.39), through the diode, and then to plate A. The transformer secondary resistance and the forward resistance of the diode are both low in value. Therefore the charging time constant of C_1 is very short. This enables e_{C1} to

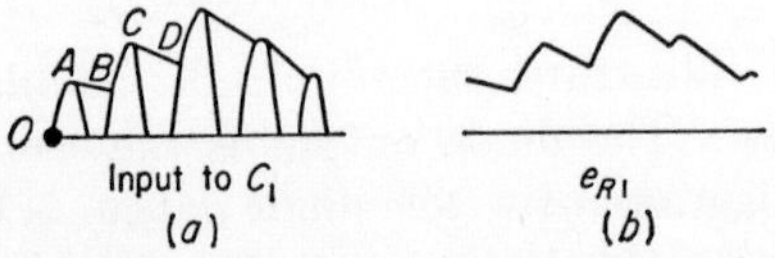

Input to C_1

(a)

e_{R1}

(b)

FIG. 9.40 Action of C_1 and R_1.

follow the variations of the increasing carrier voltage (such as O to A in Fig. 9.40a).

When the carrier voltage begins to decrease, as it does at point A of Fig. 9.40a, C_1 begins to discharge. Plate A discharges electrons through R_1, to common, and plate B of C_1 takes on electrons from common. R_1 is many times larger than the resistance of the charging path, thereby creating a long discharge time constant. Since C_1 has a long discharging time constant, e_{C1} decreases much more slowly than the carrier voltage, thereby cutting off the diode. With a sufficiently large R_1, e_{C1} undergoes only a small decrease during the time the carrier voltage decreases to zero and then increases again. e_{C1} continues to discharge at its relatively slow rate until the carrier voltage has increased to the value of e_{C1}, such as point B of Fig. 9.40a. At this time the diode again conducts. C_1 now begins to charge. Since its charging time constant is very short, it will follow the carrier voltage increase up to point C. Then C_1 begins its discharge action at point C, since the carrier voltage has again decreased to a value less than e_{C1}, and the diode is again cut off. The charging action is resumed at point D, where the carrier voltage is again equal to e_{C1}. Thus it can be seen that C_1 can respond only to the rises in the carrier voltage that exceed e_{C1}.

The preceding paragraphs have shown that C_1 can respond only to the rises in the carrier voltage that exceed e_{C1} because of the short charge–long discharge time-constant relationships of C_1. Because of the action of C_1, e_{R1} appears as shown in Fig. 9.40b.

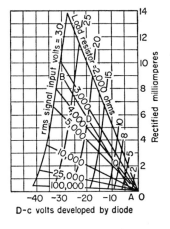

Fig. 9.41 E-I characteristics of the 6AL5. (*Characteristics from RCA.*)

Distortion by Diode Clipping. One possible cause of distortion in the diode detector is the *clipping* of voltages by the diode when a high percentage of modulation is present. An analysis of the rectification characteristics in conjunction with the circuit of Fig. 9.39 will reveal how this clipping action may develop. The average rectification characteristics of a typical diode are illustrated in Fig. 9.41. Notice that the operation characteristics, which are a plot of average load current versus average output voltage, are given in Fig. 9.41. Recall that the technique by which they can be computed is illustrated in Fig. 9.20. Also notice that load lines for R_L values from as small as 2 kilohms to as large as 100 kilohms are included in Fig. 9.41.

The zero signal diode load resistance is equal to the equivalent value of R_1 and R_v. Assume R_1 and R_v have an equivalent value of 5 kilohms.

In the absence of a detected signal, C_2 appears as an open circuit, and $R_{i,eq}$ of the a-f amplifier is isolated from the diode circuit. This zero signal load line of 4 kilohms is shown in Fig. 9.41.

With the application of a signal to the diode detector circuit, C_2 drops out and the signal load resistance of the diode becomes the equivalent value of R_1, R_v, and $R_{i,eq}$. This results in a reduced load resistance.

For the sake of an example, assume the signal load resistance is 3 kilohms. Let us consider the effect of the change in load resistance when the carrier signal is 10 volts (rms). The zero signal Q point is located at the intersection of the 4-kilohm load line and the 10-volt operation characteristic. The signal load line (3 kilohms) is drawn through the same operating point Q, shown as line AB in Fig. 9.41. When the carrier level is 10 volts (rms) or 14.14 volts peak, 100 per cent modulation would cause the carrier level to vary from $14.14 \times 2 = 28.28$ volts as a maximum to 0 volts as a minimum. This excursion could have been handled by the zero signal load line. But let us see what happens as a result of a reduction in load resistance with the new load line AQB.

As the carrier level increases from 14.14 volts peak at Q to 28.28 volts at D (20 volts rms), it is seen that the diode can respond to the complete change. When the carrier decreases from 14.14 volts Q toward zero, the diode cuts off at point A, which is only an excursion of 14.14 to 4.5 volts. In other words, the diode clipped a portion of the modulated signal.

The problem of diode clipping is minimized by making $R_{i,eq}$ of the succeeding amplifier as large as possible. If the maximum percentage of modulation is known, a diode signal load resistance may be chosen to reduce the problem. In the preceding example, larger values of R_1 and R_v might have been chosen so that $R_{i,eq}$ would not reduce the equivalent value (of R_1, R_v, and $R_{i,eq}$) to such an extent. The point at which the lower portion of the modulated carrier is clipped can be determined from the following:

$$\text{Clipping point} = \frac{E_{\max} - E_{\min}}{E_{\max} + E_{\min}} \times 100$$

where $E_{\max}$ = maximum peak voltage developed by diode B of Fig. 9.41

$E_{\min}$ = minimum peak voltage developed by diode A of Fig. 9.41

Taking the values from Fig. 9.41:

$$\text{Clipping point} = \frac{32 - 4.5}{32 + 4.5} \times 100$$

$$= \frac{27.5}{36.5} \times 100 \cong 75\%$$

For the given equivalent value of R_1, R_v, and $R_{i,eq}$, the detector diode of the preceding example clipped when the modulation was 75 per cent or greater at a carrier level of 10 volts (rms).

Distortion by diode clipping can be further complicated by the fact that R_v is a potentiometer in many cases. This is quite often the case in a radio receiver, where it serves as the volume control. Reducing R_v decreases the diode output voltage, which is the input signal to the first audio amplifier, thereby reducing the volume of the ultimate speaker output. A change in R_v establishes a new load resistance for the diode. At higher settings of R_v, the signal load resistance is highest, and vice versa. Therefore it is conceivable to find diode clipping occur at 75 per cent modulation at one R_v setting and 60 per cent (or less) at a lower resistance setting.

Distortion by Filter Clipping. Refer to Fig. 9.42 for the following discussion. It will be recalled that the detector filter capacitor (C_1 in Fig. 9.39) has a short charging time constant and a long discharging

Fig. 9.42   Filter clipping.

(a)   (b)

time constant. Because of its short charging time constant, the capacitor voltage easily follows the rises in the carrier voltage. If the signal should undergo a relatively large decrease, as shown in Fig. 9.42a, the capacitor voltage cannot follow it, resulting in distortion (see Fig. 9.42b).

At first glance it seems that this type of distortion can be minimized by reducing the discharging time constant of the capacitor. But this would result in an increase in the saw-toothlike variations, which is also undesirable. Therefore, a compromise between these two undesirable features is to have a discharge time constant sufficiently long to minimize the saw-toothlike variations, but short enough not to produce an excessive amount of filter clipping.

Distortion by Square-law Detection. The characteristics of a diode are most nonlinear in the region closest to the origin. In this low forward voltage region, the increases of diode current are approximately a direct function of the square of the diode voltage. When the diode voltage is considerably greater than zero, the characteristic becomes more linear and the square-law relationship is no longer true. Therefore a diode detector also distorts the signal at the lower forward voltage values. This type of distortion can be minimized by placing the diode at a small forward potential at zero signal level.

9.13 Diode Mixers. In a mixer circuit, two signal frequencies are applied to the input. The principal function of this circuit is to obtain from the output that frequency which is equal to the *difference* of the two input frequencies. A semiconductor or vacuum diode can be used for this purpose.

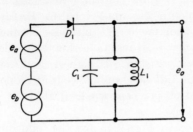

FIG. 9.43 Diode mixing circuit.   FIG. 9.44 Waveforms of the diode mixer.

Figure 9.43 illustrates a diode mixing circuit. C_1 and L_1 form a tank circuit, which is made to resonate at the difference frequency $(f_a - f_b)$. The diode reacts to the instantaneous composite value of $e_a + e_b$.

Figure 9.44 illustrates the waveforms associated with the diode mixer circuit. The diode responds to $e_a + e_b$. The frequency of the envelope of $e_a + e_b$ is equal to $f_a - f_b$. The diode rectifies the envelope, and the tank circuit resonates at $f_a - f_b$, which is e_o. With a reasonably high tank circuit, Qe_o will be sinusoidal.

9.14 Diode Clamps. Diode clamps have a large number of applications, several of which are discussed in Chap. 12, which is concerned with transistor applications. One of the common uses for the diode clamp

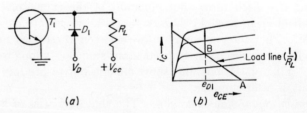

FIG. 9.45 Collector-voltage clamp: (a) circuit, (b) output characteristics.

is to clamp the collector voltage of a transistor to a sufficiently high value to prevent the transistor from going into its saturation region.

Figure 9.45 illustrates a collector-voltage clamp circuit and the effect of the diode clamp upon the operating points. With increases of forward bias at the base of the transistor, i_C increases, causing e_{RL} to increase and e_{C-E} to decrease. If e_{C-E} is allowed to become too small, T_1 will be in its saturation region. Furthermore, the collector voltage, without the clamp, might actually become smaller than the base voltage, which would result in forward bias applied to the collector junction.

With D_1 connected as shown and V_D equal to the lowest desirable collector voltage, $e_{C\text{-}E}$ cannot fall below that value (point B in Fig. 9.45), thereby preventing transistor saturation and its associated problems.

Figure 9.46 shows another diode clamp arrangement. In this circuit, i_B is controlled so that i_C cannot reach its saturation magnitude. When

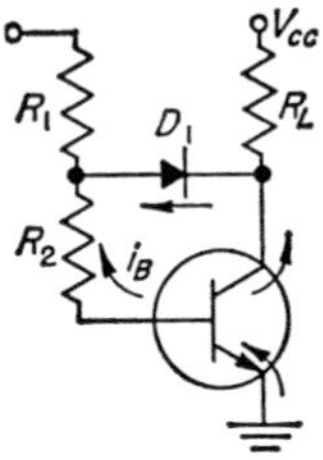

Fig. 9.46   Collector-current clamp.

$e_{C\text{-}E}$ becomes less than $e_{R2} + e_{B\text{-}E}$, D_1 conducts. This permits a portion of the collector terminal current to be diverted into the base circuit, instead of going into the collector, thereby counteracting the actual base current. The result is a decrease in the base current, which prevents an increase in i_C.

9.15   Diode Dampers. The purpose of the diode in Fig. 9.47 is to absorb the transient energy surges of the inductor. In this way, the transient potentials will not be so great as to damage the transistor. The diode would be of the type which can absorb the voltage and current transients expected from the inductor without being damaged or destroyed.

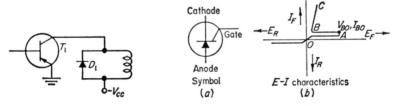

Fig. 9.47   A diode damper.

Fig. 9.48   The silicon-controlled rectifier (scr): (a) symbol, (b) E-I characteristics.

9.16   The Silicon-controlled Rectifier. The symbol and volt-ampere characteristics of the silicon-controlled rectifier are illustrated in Fig. 9.48. This rectifier is a PNPN type structure. A positive cathode is the reverse voltage condition while a negative cathode is the forward voltage condition. Very small values of forward current flow (O to A)

until the forward breakover voltage V_{BO} is attained. When V_{BO} is achieved, which is often 6 volts at 300 ma, the scr is turned ON, resulting in a high current (A to B to C) which is limited only by the supply potential and external circuit resistance. The scr potential is about 1 to 2 volts during its ON condition.

A small potential (typically 1.5 volts at 30 ma) applied across the gate to cathode circuit can switch the scr into high conduction when the anode-to-cathode voltage is already at some low forward value. The scr will remain in its high conduction state even after the removal of the gate pulse. The anode current must be diverted or interrupted for about 20 μsec to turn the scr OFF.

The unijunction relaxation oscillator (analyzed in Chap. 8) is a convenient method for the application of pulses to the scr gate, as shown in Fig. 9.49. E_{B1} can be used as the gate pulses, and these pulses will be about $0.2R_E$ μsec apart. This circuit is commonly used for firing scr's

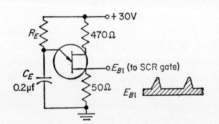

FIG. 9.49 The unijunction relaxation oscillator used for the scr gate pulse. [*Taken with permission from "Transistor Manual" (4th Edition), published by General Electric Co., Charles Building, Liverpool, N.Y.*]

in d-c to a-c inverters and many types of equipment which function from d-c supplies. Since the temperature and supply voltage do not alter the time interval between pulses, the timing is determined by the designer in his choice of R_E and C_E.

Figure 9.50 illustrates a circuit in which the scr firing pulses occur with a fixed relationship to the time the supply voltage is at its zero crossing.

Let us first consider the supply voltage of the unijunction transistor. As shown by the waveforms, a full-wave rectified sine wave is clipped by D_1 (which could be a zener diode), so that a maximum of 30 volts is applied to the ujt. Upon closing the circuit, C_E begins charging through R_E. When e_{RE} becomes small enough, the emitter will be sufficiently positive to fire the ujt, creating a pulse across R_{B1}, which fires the scr. The actual timing of this pulse is determined by the value of C_E and R_E. When the ujt supply voltage drops to zero (end of half cycle), C_E will discharge. When the next half sine wave appears, the action is repeated. Notice that the firing of the ujt is timed to its supply voltage.

The NPN transistor serves as a time delay for the firing of the ujt and the scr. When the capacitor charges, a portion of the current passes through the NPN, thereby delaying the time when the ujt will fire. This results in a reduction in the average load current.

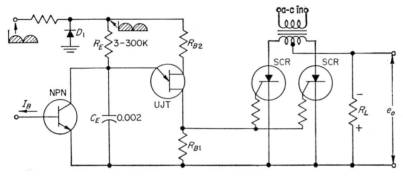

FIG. 9.50   Basic circuit for a regulated power supply. [*Taken with permission from "Transistor Manual" (4th Edition), published by General Electric Co., Charles Building, Liverpool, N.Y.*]

It should be pointed out that the scr has many more possible applications, which are too numerous to mention in this text.

9.17   The Tunnel Diode.   The behavior of the tunnel diode is best described by examining its volt-ampere characteristics, which are shown in Fig. 9.51b.   The tunnel diode is a narrow PN junction through which the majority carriers tunnel through at the speed of light at those voltages

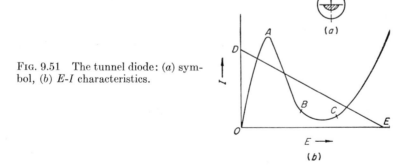

FIG. 9.51   The tunnel diode: (a) symbol, (b) E-I characteristics.

between points O to C.   The normal diode-current flow, due to minority-carrier flow, occurs at those diode forward voltage values greater than point C.   The rapid carrier transit time in the tunnel region of the characteristics enables the use of the tunnel diode in high-frequency applications.

Germanium tunnel diodes with peak currents between 100 μa and 10 amp are commercially available.

Tunnel diodes have been used in a wide variety of circuits, including oscillators, amplifiers, and switching circuits. A simple bistable circuit can be designed with a tunnel diode in series with a voltage supply and one resistor. The load line of the resistor should be such that it intersects the characteristic at points on its positive slope (such as load line DE in Fig. 9.51). A current pulse which reduces the current to a value less than the peak-point current will trigger the circuit from OFF to ON. An opposite current pulse, which places the current in excess of the peak value, will drive the diode from ON to OFF.

PROBLEMS

9.65 State three kinds of regulators.

9.66 Refer to Fig. 9.33. What portion of the characteristic of the zener diode can be utilized in maintaining a constant output voltage for large changes in the output current? Explain.

9.67 Refer to Fig. 9.32. $E_{o,min} = 25$ volts; $E_{D1,max} = 20$ volts; $I_{o,max} = 100$ ma; $I_{D1,min} = 5$ ma; $I_{D1,max} = 200$ ma. Find:
(a) R_1
(b) P_{R1} (100 per cent safety factor)
(c) $P_{D1,max}$

9.68 Refer to Fig. 9.32a. $E_{o,min} = 38$ volts; $E_{D1,max} = 30$ volts; $I_{o,max} = 200$ ma; $I_{D1,min} = 5$ ma; $I_{D1,max} = 400$ ma. Find:
(a) R_1
(b) P_{R1} (100 per cent safety factor)
(c) $P_{D1,max}$

9.69 Refer to Fig. 9.32b. $E_{o,min} = 22$ volts; $E_{D1,max} = 20$ volts; $I_{o,max} = 100$ ma; $I_{D1,min} = 5$ ma; $E_{C-B} = 20$ volts; $I_{B,co} = 10$ μa. Find:
(a) E_{R1} (b) R_1
(c) $P_{D1,max}$ (d) R_2

9.70 In the shunt voltage regulator circuit of Prob. 9.69, $E_{o,min} = 40$ volts; $E_{D1,max} = 30$ volts; all other parameters are the same. Find:
(a) E_{R1} (b) R_1
(c) $P_{D1,max}$ (d) R_2

9.71 Refer to Fig. 9.32b. $E_{o,min} = 30$ volts; $E_{D1,max} = 25$ volts; $I_{o,max} = 200$ ma; $I_{D1,min} = 5$ ma; $E_{C-B} = 25$ volts; $I_{B,co} = 5$ μa. Find:
(a) E_{R1} (b) R_1
(c) $P_{D1,max}$ (d) R_2

9.72 In the shunt voltage regulator of Prob. 9.71, $I_{o,max} = 500$ ma; all other factors are the same. Find:
(a) E_{R1} (b) R_1
(c) $P_{D1,max}$ (d) R_2

9.73 Refer to the series voltage regulator of Fig. 9.34. $E_{B-C} = 20$ volts; $I_{D1} = 200$ ma; $I_B = 10$ ma. Find R_1.

9.74 In the series voltage regulator of Prob. 9.73, $E_{B-C} = 25$ volts; all other parameters are the same. Find R_1.

9.75 Refer to the series voltage regulator of Fig. 9.34. $E_{B\text{-}C} = 10$ volts; $I_{D1} = 500$ ma; $I_B = 30$ ma. Find R_1.

9.76 In the series voltage regulator of Prob. 9.75, $I_B = 5$ ma; all other parameters are the same. Find R_1.

9.77 Refer to the d-c to d-c power converter of Fig. 9.37. What fires the circuit into operation? Explain.

9.78 What is the difference between a d-c to d-c power converter and a d-c to a-c converter?

9.79 Refer to the diode a-m detector circuit of Fig. 9.39. $R_{i,eq} = 2$ kilohms; R_v is set at 1.5 kilohms; $R_1 = 1.8$ kilohms; $r_{d1} = 10$ ohms; $r_{z,former,sec} = 3$ ohms. Find (a) resistance of C_1 charging path; (b) resistance of C_1 discharging path.

9.80 In the detector circuit of Prob. 9.79, carrier frequency = 455 kc; 0.1 tc (or less) charging = T_p of carrier frequency; 10 tc (or more) discharging = T_p of carrier frequency. Find a suitable value of C_1.

9.81 In the detector circuit of Prob. 9.79, find (a) zero signal load resistance; (b) signal load resistance.

9.82 Refer to the diode a-m detector circuit of Fig. 9.39. $R_{i,eq} = 1.3$ kilohms; R_v is set at 2 kilohms; signal load resistance = 500 ohms. Find R_1.

9.83 In Prob. 9.82, signal load resistance = 200 ohms. Find R_1.

9.84 Refer to the diode a-m detector circuit of Fig. 9.39. $R_{i,eq} =$ kilohms; R_v is set at 2 kilohms; $R_1 = 2.2$ kilohms; $r_{d1} = 5$ ohms; $r_{v,former,sec} = 5$ ohms. Find (a) resistance of C_1 charging path; (b) resistance of C_1 discharging path.

9.85 In Prob. 9.84, carrier frequency = 455 kc; 0.1 tc (or less) of charging = T_p of carrier frequency; 10 tc (or more) of discharging = T_p of carrier frequency. Find a suitable value of C_1.

9.86 In Prob. 9.84 find (a) zero signal load resistance; (b) signal load resistance.

9.87 In the circuit of Fig. 9.39, how can diode clipping be minimized? Explain.

9.88 In the circuit of Fig. 9.39, how can filter clipping be minimized? Explain.

9.89 How can square-law detection distortion be minimized in a circuit such as 9.39? Explain.

9.90 Assume a diode a-m detector circuit has the following parameters: $E_m = 30$ volts; and E_{min} is 2 volts because of diode clipping. Find the clipping point (percentage of modulation at which clipping occurs).

CHAPTER 10

THE AMPLIFYING VACUUM TUBE

The vacuum diode, as stated in the preceding chapter, is capable of reproducing only that portion of the input potential which places the device in the forward region of its volt-ampere characteristic. This fact makes the diode an excellent device for the performance of rectification and those applications where rectification is desired. Amplification cannot be achieved with the diode.

There are several types of vacuum tubes which are capable of voltage, current, and/or power amplification. These tubes incorporate additional electrodes within the evacuated region between the anode and cathode. The characteristics and possibilities of each type are analyzed in this chapter.

10.1 Triodes. The triode is a three-electrode vacuum tube; the electrodes are the cathode, control grid, and the anode. The *control grid* is a wire mesh which is physically near the cathode. In most cases, the control grid surrounds the cathode and is often called the "grid" in much of the literature on the subject. The purpose of the grid is to control

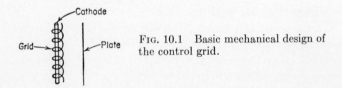

FIG. 10.1 Basic mechanical design of the control grid.

the flow of electrons from the cathode to the anode. The basic mechanical design of the grid is shown in Fig. 10.1.

A negative potential, called the *grid bias*, is applied to the control grid. This voltage acts upon the cathode-emitted electrons such that it regulates the cathode-to-anode electron drift (which is called the plate current). In conjunction with the grid bias, the anode has a positive potential which tends to attract the thermionic-emitted electrons by its

electrostatic field.   The electrostatic field is changed by variations in
grid voltage, which is an effective method for controlling the plate
current.

Figure 10.2 illustrates the electrostatic field of the triode for a conducting and a nonconducting condition.   Such an illustration may be
called an *electrostatic map*.   All the points having the same potential
with respect to the cathode are joined together, forming an *equipotential
line*.   If the tube had no grid, all the equipotential lines would be more
or less parallel to the surface of the anode.   When the plate is sufficiently
positive, the positive equipotential lines are close enough to the cathode
to accelerate the space-charge electrons on toward the anode, resulting
in the flow of plate current.

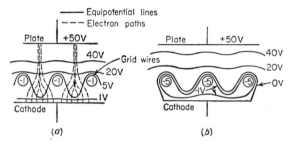

FIG. 10.2   Vacuum-tube electrostatic field: (*a*) conducting, (*b*) nonconducting.

Let us now consider the effect of introducing a control grid between
the plate and the cathode, thereby forming a triode.   Assume a condition
such as shown in Fig. 10.2*a*, where a small negative potential is applied
to the grid.   Notice the manner in which the equipotential lines are
distorted because of the negative grid wires.   Also note that the positive,
1-volt equipotential line is in the immediate vicinity of the cathode surface.   A number of the space-charge electrons are accelerated toward
the plate.   If the plate is made more positive with respect to the cathode
or the grid is made less negative with respect to the cathode, a higher-
value equipotential line (such as +5 volts) would be very close to the
cathode surface area.   This provides a greater accelerating force for the
space-charge electrons, which enables more of them to travel to the anode
(the plate current magnitude increases).   Since the grid is physically
closer to the cathode than the anode, a relatively small change in grid
potential can produce a variation in plate current which would require a
relatively large change in plate voltage (with the grid voltage fixed).
This feature is analyzed in greater detail in the section concerned with
tube constants.

Consider Fig. 10.2*b*, where there is no electron drift from cathode to anode. Using the same plate voltage, the grid potential is made more negative. Notice that the equipotential line at the cathode surface is a negative value with respect to the cathode. This provides a repelling effect upon the space-charge electrons, resulting in no electrons being accelerated toward the anode. This condition of zero plate current is called *cutoff*, brought about by a sufficiently negative grid. A later section reexamines this action in greater detail.

10.2 Volt-ampere Characteristics of the Triode. Vacuum tubes are voltage-controlled devices. There are two potentials applied to electrodes in the triode which have some degree of control over the flow of cathode-to-anode current. Increasing the negative grid potential decreases the tube current, and an increase in the positive plate voltage results in a larger plate current.

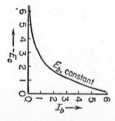

Fig. 10.3   Triode transfer characteristic.

The graphical relationship of negative grid volts versus the plate current with a constant plate voltage, sometimes called the *transfer characteristic*, illustrates the effect of the grid voltage upon the plate current, as shown in Fig. 10.3. In this figure, the magnitude of the plate current varies inversely with negative grid voltage. Perhaps the reader wonders why a negative potential is applied to the grid. The grid is designed to regulate the flow of electrons from the cathode to the anode. If the grid wires were made positive with respect to the cathode, a number of the cathode-emitted electrons would be attracted to the wire meshwork of the grid, thereby depriving the anode of some of the otherwise plate-destined electrons. The development of grid current in this fashion results in power dissipation in the grid circuit, which is a highly undesirable situation in many applications. Maintaining the grid at a negative potential discourages the development of grid current and minimizes grid circuit power dissipation.

Later sections reveal that changing the negative grid potential is the most effective method for the control of anode current and is the most common technique in practice.

Also important to the designer is the effect of the plate voltage upon the plate current. In order to most accurately evaluate this relationship, the grid voltage is maintained constant. In this way, any change in the plate current is caused by the plate-voltage variation. The relationship of the plate current as a function of the plate voltage at several fixed grid potentials is called the family of *plate volt-ampere characteristics*. Figure 10.4 illustrates a typical family.

A third family of characteristics, called the *constant-current characteristics*, are also of use. The constant-current characteristic is E_c versus E_b with I_b constant.

Figure 10.5 illustrates the constant-current characteristics of a triode. Notice that the characteristics are quite linear for negative grid-voltage values. Given any one of the three sets of characteristics, the remaining two sets can be determined.

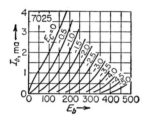

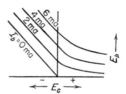

FIG. 10.4 Plate volt-ampere characteristics of the 7025. (*RCA.*)

FIG. 10.5 Triode constant-current characteristics.

The majority of design work for triode amplifiers can be accomplished by use of one or more of the preceding characteristics.

10.3 Tube Constants.

The volt-ampere characteristics studied in the preceding section are useful in obtaining the static characteristics of a particular vacuum tube. The chief dynamic characteristics, often referred to as tube constants, are:

1. Amplification factor μ
2. Plate resistance r_p
3. Control grid–plate transconductance g_m

Amplification Factor. The amplification factor μ is defined as the ratio of a plate voltage change to control grid voltage change in the opposite direction with the plate current held constant. In equation form,

$$\mu = \frac{\Delta E_b}{\Delta E_c} \quad \text{with } I_b \text{ constant}$$

This ratio actually determines the effectiveness of the control grid in creating changes in the plate voltage.

Example 1: Refer to Fig. 10.4. I_b is fixed at 1.5 ma; E_c is changed from -1.5 to -2 volts. Find (*a*) E_b change required to maintain I_b constant at 1.5 ma; (*b*) μ.

Solution:

(a) From the characteristics: when $E_c = -1.5$ volts, $E_b = 218$ volts; when $E_c = -2.0$ volts, $E_b = 270$ volts. Therefore $\Delta E_b = 270 - 218 = 52$ volts.

(b)
$$\mu = \frac{52 \text{ volts}}{0.5 \text{ volts}} = 104$$

In the preceding example, the control grid voltage is 104 times more effective than the plate voltage in producing a change in plate current.

Plate Resistance. The plate resistance r_p is defined as the opposition offered by the tube to the changing cathode-to-plate electron flow; i.e.,

$$r_p = \frac{\Delta E_b}{\Delta I_b} \quad \text{with } E_c \text{ constant}$$

Example 2: Refer to the characteristics of Fig. 10.4. E_c is fixed at -1.5 volts; I_b is changed from 1 to 1.5 ma.

Find (a) corresponding plate voltages to maintain $E_c = -1.5$ volts; (b) r_p.

Solution:

(a) From the characteristics: when $I_b = 1$ ma, $E_b = 180$ volts, $E_c = -1.5$ volts; when $I_b = 1.5$ ma, $E_b = 220$ volts.

(b)
$$r_p = \frac{220 - 180}{1.5 - 1.0} = \frac{40}{0.5 \times 10^{-3}} = 80 \text{ kilohms}$$

Transconductance. The control grid–plate transconductance g_m, often called transconductance, is defined as the ratio of a small plate current change to the small control grid voltage change which caused it, with the plate voltage held constant; i.e.,

$$g_m = \frac{\Delta I_b}{\Delta E_c} \quad \text{with } E_b \text{ constant}$$

Example 3: Refer to the characteristics of Fig. 10.4. E_b is fixed at 200 volts; I_b is changed from 1 to 1.5 ma.

Find (a) corresponding E_c changes required to maintain E_b constant at 200 volts; (b) g_m.

Solution:

(a) From the characteristics of Fig. 10.4, with E_b fixed at 200 volts: when $I_b = 1$ ma, $E_c = -1.5$ volts plus $\frac{4}{13}$ of a half volt; i.e., the point formed by $I_b = 1$ ma, $E_b = 200$ volts, lies $\frac{4}{13}$ of the distance between the -1.5- and -2-volt characteristics.

Hence,
$$\frac{4}{13} \times \frac{1}{2} = \frac{4}{26} = -0.15 \text{ volt}$$
$$E_c = -1.5 + (-0.15) = -1.65 \text{ volts}$$

When $I_b = 1.5$ ma,

$$E_c = -1 + 1\tfrac{2}{8}(-0.5) = -1 + (-0.33) = -1.33 \text{ volts}$$

(b) $$g_m = \frac{(1.5 - 1.0) \times 10^3}{1.65 - 1.33} = \frac{5 \times 10^{-4}}{0.32} \cong 15.6 \times 10^{-4} \text{ mho}$$

or 1,560 μmho.

Note: Because of the small conductance values, the unit of *micromho* is in common usage.

Of importance in design work is the relationship of the tube constants as a function of the grid voltage. These relationships for the 7025

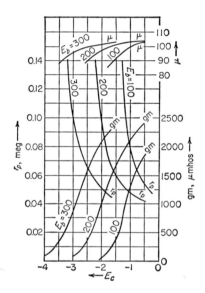

Fig. 10.6 μ, r_p, g_m versus E_c of the 7025. (*RCA.*)

triode are illustrated in Fig. 10.6. Notice that r_p is largest when E_c is most negative because a more negative control grid presents more opposition to the cathode-emitted electrons. The amplification factor is greatest at lower grid potentials, and the transconductance is also maximum at lower grid potentials. Also of significance is the effect of E_b upon the tube constants. For a given value of E_c, a higher plate voltage results in a larger g_m, smaller r_p, and a larger μ.

10.4 Waveform Analysis of the Common-cathode Configuration.

Figure 10.7 illustrates the common-cathode connection, which is the most popular configuration. The input signal is applied between the control grid and the cathode. The output is obtained across plate to cathode in Fig. 10.7, but could also be taken across R_L by use of a suitable coupling arrangement. Let us proceed to the theory of operation, with a waveform analysis of this circuit.

Assume the input waveform is sinusoidal, as shown in waveform a of Fig. 10.8. When the input signal swings positive, the control grid becomes less negative, thereby causing the plate current to increase (see waveforms b and c). Since electrons are leaving the tube by way of the anode, the plate current is assigned a positive value. This increase in the plate current results in a larger voltage drop across R_L, a fixed plate

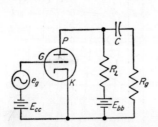

FIG. 10.7 Common-cathode configuration.

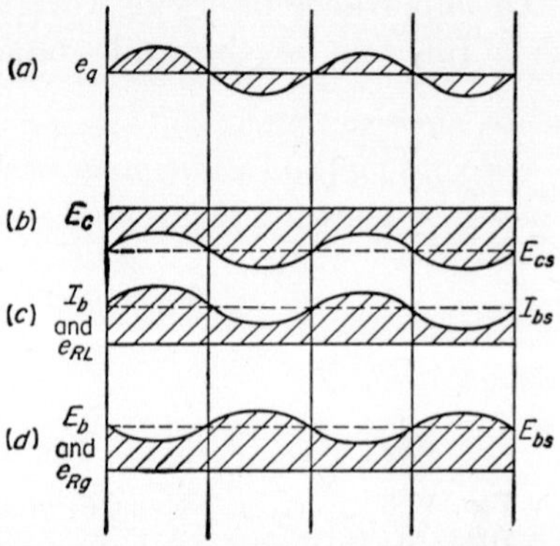

FIG. 10.8 Waveforms for the common-cathode configuration.

load resistor. For all practical purposes, the I_b and e_{RL} waveforms are identical in shape. E_{bb}, the plate supply voltage, is constant, and

therefore
$$e_{RL} + E_b = E_{bb}$$
$$\Delta e_{RL} = -\Delta E_b$$

Assume, for example, that e_{RL} increased by 10 volts, then E_b must decrease by 10 volts, since E_{bb} is equal to their sum and is fixed. The plate-to-cathode voltage variations are identical in shape and 180° out of phase with the variations of e_{RL} (see waveform d).

If the output voltage is taken across the plate-to-cathode circuit, as is the case with RC coupling, the output signal variations are inverted with respect to the input signal. On the other hand, no signal inversion occurs if the variations are taken across R_L.

10.5 Load-line Analysis. The load-line construction techniques for the amplifying vacuum tube are essentially the same as those studied in the transistor amplifier analysis of earlier chapters. The static load line is constructed upon the E_b versus I_b characteristics. The x-axis intercept is determined by the magnitude of the supply voltage E_{bb} at the condition when the tube appears as an open circuit in Fig. 10.9,

Point $B = E_{bb}$ when the tube is open-circuited

and the y-axis intercept is determined by the magnitude of the plate current at the condition when the tube appears to be a short circuit.   In Fig. 10.9,

$$\text{Point } A = \frac{E_{bb}}{R_L}$$

This results in the load line having a slope that is equal to $1/R_L$.   Since R_L is assumed to be purely resistive in nature, the load line is linear. Since the load line in such cases is linear, only two points need be known for its construction.

A second method which may be utilized for static load-line construction is based on knowing the desired zero signal parameters of the tube and the

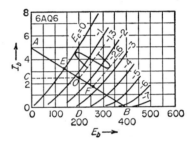

Fig. 10.9   Triode load line.   (*Characteristics from RCA.*)

value of E_{bb}.   Assume, for example, that an amplifier is to be designed at which the zero signal plate voltage is to be $0.5E_{bb}$ or 200 volts, and I_b is to be 2.5 ma at this time.   The zero signal operating point is determined by the intersection of the line drawn from 2.5 ma on the y axis parallel to the x axis (CO of Fig. 10.9) and the line constructed from $0.5E_{bb}$ on the x axis parallel to the y axis (line DO of Fig. 10.9).   The required zero signal grid voltage (bias) is determined from the location of point O. When the operating point is between two characteristics, the required grid bias can be easily interpolated, which is about -1.3 volts in our example.   Drawing a straight line between points O and B and extending it to the y axis results in the construction of the load line.

As in previous load-line analysis work, the operating point should be in the center of the linear range of the characteristic to allow for the largest possible input signal in linear work.

In the load line of Fig. 10.9, the maximum input signal voltage is 1.3 volts or $1.3 \times 0.707 \cong 0.92$ volts rms, if grid current is to be avoided. The maximum output voltage, for the stated conditions, is about 65 volts or $65 \times 0.707 \cong 46$ volts rms.   The voltage gain A_v may be determined graphically:

$$A_v = \frac{\Delta E_b}{\Delta E_c}$$

Using the peak-to-peak variation of plate voltage and peak-to-peak variation of E_c,

$$A_v = \frac{E_{b,\max} - E_{b,\min}}{E_{c,\max} - E_{c,\min}}$$

Since in class A work, the zero signal E_c value is midway between the peak values, $E_{c,\max} - E_{c,\min} = 2E_{cc}$,

$$A_v = \frac{E_{b,\max} - E_{b,\min}}{2E_{cc}}$$

From the values in Fig. 10.9,

$$A_v = \frac{265 - 135}{2(1.3)} = \frac{130}{2.6} = 50$$

The maximum signal plate current $i_{p,\max}$ is equal to the difference between the zero signal plate current I_b and the maximum composite plate current $I_{b,\max}$; i.e.,

$$i_{p,\max} = I_{b,\max} - I_b$$

Substituting values from Fig. 10.9,

$$i_{p,\max} = 3.3 \text{ ma} - 2.5 \text{ ma} = 0.8 \text{ ma}$$

And the rms plate signal current I_p, with sinusoidal variations, is

$$I_p = 0.707(I_{b,\max} - I_b)$$
$$= 0.707(0.8 \text{ ma}) \cong 0.57 \text{ ma}$$

The power delivered by the power supply is $E_{bb}I_b$ watts. The power input P_i of the tube is E_bI_b watts. The difference between these two powers is dissipated by R_L, that is, $(E_{bb}I_b - E_bI_b)$ watts. The plate circuit power dissipation is equal to the input power of the tube. A portion of this power is dissipated simply to operate the tube, while the remainder is the actual signal power output. The *signal power output* $P_{o,\text{sig}}$ may be determined by

$$P_{o,\text{sig}} = I_pE_p$$
$$= \frac{(I_{b,\max} - I_{b,\min})(E_{b,\max} - E_{b,\min})}{8}$$

The portion of the input power converted into output signal power is called the *plate circuit efficiency* (plate eff) of the circuit:

$$\text{Plate eff} = \frac{P_{o,\text{sig}}}{P_i}$$
$$= \frac{(I_{b,\text{max}} - I_{b,\text{min}})(E_{b,\text{max}} - E_{b,\text{min}})/8}{E_b I_b}$$

The *circuit efficiency* is determined by the fraction of the power delivered by the power supply to the power that is converted into output signal power:

$$\text{Circuit eff} = \frac{P_{o,\text{sig}}}{E_{bb} I_b}$$
$$= \frac{(I_{b,\text{max}} - I_{b,\text{min}})(E_{b,\text{max}} - E_{b,\text{min}})/8}{E_{bb} I_b}$$

The chief source of distortion in the triode tube is the presence of second harmonics. The percentage of second harmonic distortion may be determined by

$$\text{Per cent distortion} = \frac{0.5(I_{b,\text{max}} + I_{b,\text{min}}) - I_b}{I_{b,\text{max}} - I_{b,\text{min}}} \times 100$$

where I_b is the zero signal plate current. In most practical applications, the second harmonic distortion should not exceed 5 per cent.

Example: Calculate the following for the amplifier whose load line is shown in Fig. 10.9: (a) $P_{o,\text{sig}}$; (b) plate eff; (c) circuit eff; (d) per cent distortion.

Solution:

(a) Determine $P_{o,\text{sig}}$:
$$P_{o,\text{sig}} = I_p E_p = 5.7 \times 10^{-4} \times 46$$
$$\cong 26.2 \text{ mw}$$

(b) Find plate efficiency:
$$\text{Plate eff} = \frac{P_{o,\text{sig}}}{E_b I_b} = \frac{26.2 \times 10^{-3}}{200 \times 2.5 \times 10^{-3}}$$
$$= 0.0524 = 5.24\%$$

(c) Calculate circuit efficiency:
$$\text{Circuit eff} = \frac{P_{o,\text{sig}}}{E_{bb} I_b} = \frac{26.2 \times 10^{-3}}{400 \times 2.5 \times 10^{-3}}$$
$$= 0.0262 = 2.62\%$$

(d) Find per cent distortion:
$$\text{Per cent distortion} = \frac{0.5(3.3 \times 10^{-3} + 1.8 \times 10^{-3}) - 2.5 \times 10^{-3}}{3.3 \times 10^{-3} - 1.8 \times 10^{-3}} \times 100$$
$$= 3.3\%$$

10.6 Equivalent Constant-voltage Circuit. Figure 10.10 is the equivalent constant-voltage circuit for the amplifier illustrated in Fig. 10.7. Recall that the grid-voltage change is mu times more effective than a plate-voltage change in producing a plate-current variation. Hence the circuit voltage appears to be $-\mu e_g$.

Assuming the characteristic of the tube in question is linear, then by Ohm's law

$$i_p = \frac{-\mu e_g}{r_p + R_L}$$

where r_p = plate resistance of the tube

μ = amplification factor of the tube

The vacuum tube is customarily coupled into some impedance (such as R and C in RC coupling). This impedance Z_i must be considered in

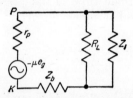

Fig. 10.10 Equivalent constant-voltage circuit for Fig. 11.7.

the preceding equation as being in parallel with R_L. Also, the internal impedance of the plate supply Z_b may have to be included if it is sufficiently large. The equation is then modified to

$$i_p = \frac{-\mu e_g}{Z_b + r_p + R_{L,\text{eq}}}$$

where

Z_b = plate supply impedance

$R_{L,\text{eq}} = R_L Z_i / (R_L + Z_i)$

and

$$e_{rl} = \frac{-\mu e_g R_{L,\text{eq}}}{Z_b + r_p + R_{L,\text{eq}}}$$

and the voltage gain

$$A_v = \frac{e_{rl}}{e_g} = \frac{-\mu R_{L,\text{eq}}}{Z_b + r_p + R_{L,\text{eq}}}$$

With r_p and μ constant, A_v increases as R_L is made larger. It should be noted that the largest possible value of A_v is μ. The voltage gain is equal to half the value of μ when $R_L = r_p$. In order to maintain a constant r_p, the plate current I_b must be fixed, since r_p and I_b are inversely related to each other. In order to maintain a fixed r_p with larger values of R_L, a larger E_{bb} value must be used. It is impractical to use E_{bb} magnitudes beyond a certain value in actual amplifier design, however. Therefore a compromise between these two factors is chosen, so that the largest possible R_L for the largest practical E_{bb} is used. In this way, a voltage gain less than μ, but the maximum practical value, is achieved in actual practice.

10.7 Equivalent Constant-current Circuit. The equivalent constant-current circuit of the amplifier illustrated in Fig. 10.7 is shown in Fig. 10.11. This circuit is equivalent to the constant-voltage circuit of Fig. 10.10.

FIG. 10.11 Equivalent constant-current circuit for Fig. 11.7.

In Fig. 10.10, the plate current is fixed when e_g is constant. g_m is also constant under these conditions, even if r_p is changed; and ignoring Z_b,

$$e_p = e_{rl} = -g_m e_g \frac{r_p R_{L,eq}}{r_p + R_{L,eq}}$$

Since $g_m r_p = \mu$, then

$$e_p = e_{rl} = \frac{-\mu e_g R_{L,eq}}{r_p + R_{L,eq}}$$

Thus the equivalent constant-voltage and constant-current circuits are equivalent to each other.

In Fig. 10.11, notice that R_L and r_p are in parallel; and ignoring Z_i and Z_b,

$$A_v = -g_m \frac{r_p R_L}{r_p + R_L}$$

When r_p is 10 or more times greater than R_L, the preceding equation may be approximated to

$$A_v = -g_m R_L$$

Z_b, the internal impedance of E_{bb}, can be ignored in most practical cases. On the other hand, Z_i (the input impedance of the following stage) must be considered in many applications. In such cases

$$A_v = -g_m \frac{1}{1/r_p + 1/R_L + 1/Z_i}$$

since Z_i appears as a third component in parallel with r_p and R_L.

Since the equivalent constant-voltage circuit of Fig. 10.10 and the equivalent constant-current circuit of 10.11 are equivalent to each other, either one may be used in the analysis of any given amplifier. In actual practice, the equivalent constant-voltage circuit is commonly used in the analysis of triode amplifiers, and the constant-current circuit is frequently used in the analysis of certain pentode amplifiers. The reason for this lies in the fact that r_p is generally not too large as compared to R_L in triode amplifiers while r_p may be very large as compared to R_L in pentode amplifiers.

The equivalent circuits are applicable for the operation of the tube where the characteristics are linear. Since this is generally the case, they are useful in most applications.

PROBLEMS

10.1 State the purpose of the control grid in the amplifying vacuum tube.

10.2 What is the purpose of grid bias?

10.3 By use of the electrostatic map of Fig. 10.2, explain the action of the control grid.

10.4 State two ways in which zero plate current can result.

10.5 Refer to the triode transfer characteristic curve of Fig. 10.3. Assume E_b is fixed at a larger value than shown in the characteristic. What effect would this have on the E_c cutoff value? Explain.

10.6 Refer to Fig. 10.3. Assume E_b is fixed at a smaller value than shown in the characteristic. What effect would this have on I_b for a given E_c value? Explain.

10.7 Refer to the plate volt-ampere characteristics of Fig. 10.4. $E_{bb} = 400$ volts; $R_L = 100$ kilohms. Construct the load line.

10.8 Refer to Fig. 10.4. $E_{bb} = 400$ volts; $R_L = 200$ kilohms. Construct the load line.

10.9 In Probs. 10.7 and 10.8, what relationship exists between the slope of the load line and the value of R_L? Explain.

10.10 Why is control grid current generally undesirable?

10.11 Refer to Fig. 10.4 for the plate volt-ampere characteristics of the 7025. Construct the transfer characteristic of this tube with E_b fixed at:

(a) 100 volts　　　　　　　　(b) 150 volts
(c) 200 volts　　　　　　　　(d) 250 volts
(e) 300 volts

10.12 Refer to Fig. 10.4 for the plate volt-ampere characteristics of the 7025. Construct the constant-current characteristics of the tube with I_b fixed at:

(a) 0 ma　　　　　　　　(b) 0.5 ma
(c) 1.0 ma　　　　　　　　(d) 1.5 ma
(e) 2 ma　　　　　　　　(f) 2.5 ma
(g) 3 ma　　　　　　　　(h) 3.5 ma

10.13 Refer to Fig. 10.9 for the plate volt-ampere characteristics of the 6AQ6. Construct the transfer characteristics of this tube with E_b fixed at:

(a) 100 volts　　　　　　　　(b) 150 volts
(c) 200 volts　　　　　　　　(d) 250 volts
(e) 300 volts　　　　　　　　(f) 350 volts
(g) 400 volts　　　　　　　　(h) 450 volts

10.14 Refer to Fig. 10.9 for the plate volt-ampere characteristics of the 6AQ6. Construct the constant-current characteristics of this tube with I_b fixed at:
(a) 0 ma (b) 1 ma
(c) 2 ma (d) 3 ma
(e) 4 ma (f) 5 ma
(g) 6 ma

10.15 Refer to Fig. 10.20 for the plate volt-ampere characteristics of the pentode 7199. Construct the transfer characteristics of this tube with E_b fixed at:
(a) 50 volts (b) 100 volts
(c) 150 volts (d) 200 volts
(e) 250 volts (f) 300 volts

10.16 Refer to Fig. 10.20 for the plate volt-ampere characteristics of the pentode 7199. Construct the constant-current characteristics for this tube with I_b fixed at:
(a) 2.5 ma (b) 5 ma
(c) 7.5 ma (d) 10 ma
(e) 12.5 ma (f) 15 ma
(g) 20 ma

10.17 Refer to Fig. 10.4. $I_b = 1$ ma; E_c is changed from -0.5 to -1.0 volt. Find (a) E_b change required to maintain I_b constant at 1 ma; (b) μ.

10.18 Refer to Fig. 10.4. $I_b = 1$ ma; E_c is changed from -3.5 to -4.0 volts. Find (a) E_b change required to maintain I_b constant at 1 ma; (b) μ.

10.19 In Probs. 10.17 and 10.18, state the relationship between E_c (more negative or less negative) and μ. Explain.

10.20 Refer to Fig. 10.4. $I_b = 2$ ma; E_c is changed from 0 to -0.5 volt. Find (a) E_b change required to maintain I_b constant at 2 ma; (b) μ.

10.21 Refer to Fig. 10.4. $I_b = 2$ ma; E_c is changed from -2 to -2.5 volts. Find (a) E_b change required to maintain I_b constant at 2 ma; (b) μ.

10.22 In Probs. 10.20 and 10.21, state the relationship between E_c (more negative or less negative) and μ. Explain.

10.23 Refer to the pentode characteristics of Fig. 10.20. $I_b = 12.5$ ma; E_c is changed from 0 to -1 volt. Find (a) E_b change required to maintain I_b constant at 12.5 ma; (b) mu.

10.24 Refer to Fig. 10.20. $I_b = 6$ ma; E_c is changed from -2 to -2.5 volts. Find (a) E_b change required to maintain I_b constant; (b) μ.

10.25 In Probs. 10.23 and 10.24, state the relationship between E_c (more negative or less negative) and μ. Explain.

10.26 Refer to Fig. 10.6. State the relationship between μ and E_b for a given E_c. Explain.

10.27 Refer to Fig. 10.4. E_c is fixed at 0 volts; I_b is changed from 2 to 2.5 ma. Find (a) corresponding plate voltages to maintain $E_c = 0$ volts; (b) r_p.

10.28 Refer to Fig. 10.4. E_c is fixed at -3 volts; I_b is changed from 0.5 to 1 ma. Find (a) corresponding plate voltages to maintain $E_c = -3$ volts; (b) r_p.

10.29 In Probs. 10.27 and 10.28, state the relationship between the fixed value of E_c (more negative or less negative) and r_p.

10.30 Refer to Fig. 10.9. E_c is fixed at -1 volt; I_b is changed from 2 to 3 ma. Find (a) corresponding plate voltages to maintain $E_c = -1$ volt; (b) r_p.

10.31 Refer to Fig. 10.9. E_c is fixed at -4 volts; I_b is changed from 2 to 3 ma. Find (a) corresponding plate voltages to maintain $E_c = -4$ volts; (b) r_p.

10.32 In Probs. 10.30 and 10.31, state the relationship between the fixed value of E_c (more negative or less negative) and r_p.

10.33 Refer to Fig. 10.20. E_c is fixed at -1 volt; I_b is changed from 11.25 to 12.5 ma. Find (a) corresponding plate voltages to maintain $E_c = -1$ volt; (b) r_p.

10.34 Refer to Fig. 10.20. E_c is fixed at -3 volts; E_b is changed from 50 to 275 volts. Find (a) coresponding plate currents to maintain $E_c = -3$ volts; (b) r_p.

10.35 In Probs. 10.33 and 10.34, state the relationship between the fixed value of E_c (more negative or less negative) and r_p.

10.36 Refer to Fig. 10.4. E_b is fixed at 100 volts; I_b is changed from 1 to 2 ma. Find (a) corresponding E_c changes required to maintain E_b constant at 100 volts; (b) g_m.

10.37 Refer to Fig. 10.4. E_b is fixed at 300 volts; I_b is changed from 1 to 2 ma. Find (a) corresponding E_c changes required to maintain E_b constant at 300 volts; (b) g_m.

10.38 In Probs. 10.36 and 10.37, state the relationship between the fixed value of E_b and g_m.

10.39 Refer to Fig. 10.9. E_b is fixed at 100 volts; I_b is changed from 2 to 3 ma. Find (a) corresponding E_c changes required to maintain E_b constant at 100 volts; (b) g_m.

10.40 Refer to Fig. 10.9. E_b is constant at 300 volts; I_b is changed from 2 to 3 ma. Find (a) corresponding E_c changes required to maintain E_b constant at 300 volts; (b) g_m.

10.41 In Probs. 10.39 and 10.40, state the relationship between the fixed value of E_b and g_m.

10.42 Refer to Fig. 10.20. E_b is fixed at 100 volts; I_b is changed from 5 to 10 ma. Find (a) corresponding E_c changes required to maintain E_b constant at 100 volts; (b) g_m.

10.43 Refer to Fig. 10.20. E_b is fixed at 200 volts; I_b is changed from 5 to 10 ma. Find (a) corresponding E_c changes required to maintain E_b constant at 200 volts; (b) g_m.

10.44 In Probs. 10.42 and 10.43, state the relationship between the fixed value of E_b and g_m.

10.45 Refer to Figs. 10.7 and 10.8. State the relationship between (a) e_{RL} and E_c; (b) E_b and E_c; (c) e_{RL} and E_b.

10.46 Refer to Fig. 10.4. $E_{bb} = 300$ volts; zero signal plate voltage $= 0.5E_{bb}$; $I_{b,s} = 2$ ma.
(a) Construct the load line.
(b) Determine R_L.
(c) Determine E_{cc}.

10.47 In the amplifier of Prob. 10.46, maximum input signal voltage $= 0.5$ volts. Find:
(a) rms input signal voltage   (b) $E_{b,mxa}$
(c) $E_{b,min}$ (d) $E_{c,min}$
(e) $E_{c,max}$ (f) A_v

10.48 In the amplifier of Probs. 10.46 and 10.47, find:
(a) $I_{b,max}$ (b) I_b
(c) $i_{p,max}$ (d) I_p

10.49 In the amplifier of Probs. 10.46 to 10.48, find:
(a) power delivered by the power supply   (b) tube input power
(c) P_{RL} (d) $P_{o,sig}$
(e) plate eff   (f) circuit eff
(g) per cent distortion

10.50 Refer to Fig. 10.9. $E_{bb} = 400$ volts; zero signal plate voltage $= 0.5E_{bb}$, $I_{b,s} = 3$ ma.

(a) Construct the load line.   (b) Determine R_L.

(c) Determine E_{cc}.

10.51 In the amplifier of Prob. 10.50, maximum input signal voltage is 1 volt. Find

(a) rms input signal voltage   (b) $E_{b,max}$

(c) $E_{b,min}$ (d) $E_{c,max}$

(e) $E_{c,min}$ (f) A_v

10.52 In the amplifier of Probs. 10.50 and 10.51, find:

(a) $I_{b,max}$ (b) I_b

(c) $i_{p,max}$ (d) I_p

10.53 In the amplifier of Probs. 10.50 to 10.52, find:

(a) power delivered by the power supply   (b) tube input power

(c) P_{RL} (d) $P_{o,sig}$

(e) plate eff   (f) circuit eff

(g) per cent distortion

10.54 Refer to Fig. 10.20. $E_{bb} = 250$ volts; $I_{b,s} = 6.25$ ma; zero signal plate voltage $= 0.5E_{bb}$.

(a) Construct the load line.

(b) Determine R_L.

(c) Determine E_{cc}.

10.55 In the amplifier of Prob. 10.54, the maximum input signal is 1 volt. Find:

(a) rms input signal voltage   (b) $E_{b,max}$

(c) $E_{b,min}$ (d) $E_{c,max}$

(e) $E_{c,min}$ (f) A_v

10.56 In the amplifier of Probs. 10.54 and 10.55, find:

(a) $I_{b,max}$ (b) I_b

(c) $i_{p,max}$ (d) I_p

10.57 In the amplifier of Probs. 10.54 to 10.56, find:

(a) power delivered by the power supply   (b) tube input power

(c) P_{RL} (d) $P_{o,sig}$

(e) plate efficiency   (f) circuit efficiency

(g) per cent distortion

10.58 Refer to the equivalent constant-voltage circuit of Fig. 10.10. $\mu = 17$; $e_g = 10$ volts; $r_p = 8$ kilohms; $R_L = 15$ kilohms; Z_b is negligible, $Z_i = 50$ kilohms. Find:

(a) $R_{L,eq}$ (b) i_p

(c) e_{rl} (d) A_v

10.59 In Prob. 10.58, $R_L = 8$ kilohms; all other parameters are the same. Find:

(a) $R_{L,eq}$ (b) i_p

(c) e_{rl} (d) A_v

10.60 In Probs. 10.58 and 10.59, at what value of R_L did Z_i produce the greatest effect? Explain.

10.61 In Probs. 10.58 and 10.59, how does the change in R_L affect A_v? Explain.

10.62 Refer to Fig. 10.11. $r_p = 400$ kilohms; $g_m = 7,000$ μmho; $R_L = 200$ kilohms; $e_g = 5$ volts; $Z_i = 200$ kilohms. Find (a) $R_{L,eq}$; (b) e_p; (c) A_v.

10.63 Refer to Fig. 10.11. $r_p = 1$ megohm; $g_m = 2,000$ μmho; $R_L = 25$ kilohms; $Z_i = 500$ kilohms; $e_g = 3$ volts. Find (a) $R_{L,eq}$; (b) e_p; (c) A_v.

10.64 Which equivalent circuit (Figs. 10.10 and 10.11) is most commonly used for (a) triodes; (b) pentodes.

10.8 Relationship of Plate Current and Input Signal for All Classes of Operation.

Vacuum-tube amplifiers are classified in much the same manner as their transistor counterparts.

Class A. The input signal voltage and grid bias are of such values that the plate current flows for the entire signal cycle. In most class *A* vacuum-tube amplifiers, the zero signal grid bias is about 60 per cent of its

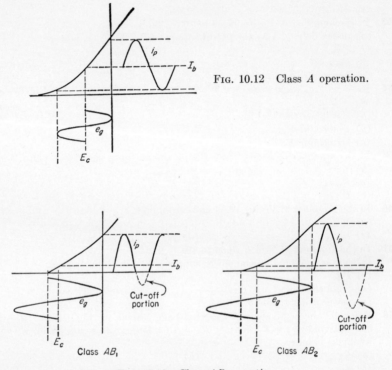

FIG. 10.12   Class *A* operation.

FIG. 10.13   Class *AB* operation.

cutoff value. As with transistors, most vacuum-tube voltage amplifiers are operated in class *A*. This class of operation is illustrated in Fig. 10.12.

Class AB. The input signal and grid bias are of such value that the plate current flows for more than half but less than the entire signal cycle. This class of operation can be further broken down. Class AB_1 meets these requirements and the grid is never driven positive, whereas AB_2 fulfills the conditions but the grid becomes positive for a portion of the signal cycle. Figure 10.13 illustrates both class *AB* conditions.

Class B. The signal and grid bias are such that the plate current flows for only one-half of the input signal cycle. Figure 10.14 illustrates this mode of operation.

The grid bias is adjusted to a value approximately equal to E_b/μ. The grid voltage may or may not be driven positive (class B_1 or B_2).

Class C. A class C amplifier is one in which the input signal and grid bias are such that plate current flows for less than half of the signal cycle. E_c is at least twice the grid cutoff voltage value in most practical cases. The grid is usually driven positive for a portion of the input signal cycle. Figure 10.15 illustrates class C operation.

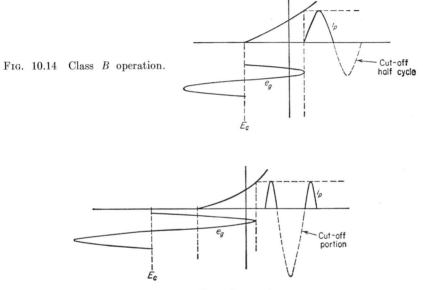

Fig. 10.14 Class B operation.

Fig. 10.15 Class C operation.

Conclusions. Class A has the lowest efficiency but has the least distortion, and is commonly used in voltage and low-power amplifiers. Class AB is more efficient than class A but less efficient than class B, and is frequently used in push-pull circuitry. Class B is more efficient than A or AB, produces more distortion, and is most generally used in push-pull work. Class C is most efficient, has the most distortion, and is useful in certain narrow-band amplifiers and oscillators.

10.9 Interelectrode Capacitance. Recall that the electrodes within the triode are metallic, and they are separated by a vacuum. The metallic electrodes act as capacitor plates, and the vacuum serves as the dielectric, thereby causing three capacitances to be present with the triode, as shown in Fig. 10.16.   These capacitances are generally small, in the order of several micromicrofarads.   They present few problems when the tube is used in a low-frequency circuit.   But with higher frequencies, these resistances become small enough to adversely affect the operation of the tube.

C_{pg} is particularly troublesome, since it produces positive feedback from the anode back to the grid when its reactance becomes small (at high frequencies), which is called the Miller effect.   This capacitance changes the input impedance of the tube. C_{gk}, which is larger than C_{pg}, is

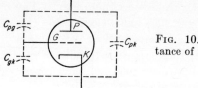

FIG. 10.16 Interelectrode capacitance of a triode.

less troublesome but must also be considered in designing tuned circuits. C_{pk} creates the least disturbance of the three and is generally smallest in value.

10.10   The Tetrode.   One of the first attempts to minimize the effect of C_{pg} was to place a fourth electrode, called the *screen grid*, between the control grid and the anode.   This electrode is placed at a positive potential, so as not to seriously retard electron flow to the plate.   By use of the proper bypass capacitor in the external circuit, the screen is placed at ground potential as far as the signal is concerned.   The use of the screen grid reduced C_{pg} to a negligible value but introduced a new type of problem.   Since the screen is placed at a positive potential, it further accelerates the anode-going electrons to such high velocities that upon striking the plate, many of them bounce off and fall prey to the screen itself.   Thus the plate is deprived of many electrons by this phenomenon, called *secondary emission*.

The problems associated with secondary emission are particularly pronounced when the plate voltage is equal or less than the screen voltage.   In more severe cases, an increase in plate voltage actually results in reduced plate current, since the secondary emission becomes more severe.   This effect is considerably reduced when the plate voltage is not allowed to fall below the value of the screen potential.   The secondary emission severely limits the usefulness of the tetrode type of vacuum tube.

10.11   The Pentode.   The pentode retains the good points of the tetrode (particularly in reducing C_{pg} to a negligible value) and overcomes the detrimental effect of secondary emission. This is accomplished by the introduction of a fifth electrode, called the *suppressor grid*. The suppressor grid is placed between the screen grid and anode, and its potential is maintained at the cathode level. Figure 10.17 illustrates the pentode.

Fig. 10.17 The pentode.

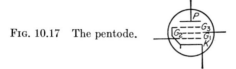

The suppressor grid, although actually at zero potential with respect to the cathode, is negative with respect to the anode. The electrons en route to the plate are traveling at such high velocity that this relatively negative electrode does not seriously impede their drift action. On the other hand, the secondary electrons (those that bounce off the plate) are moving back toward the suppressor grid at much lower velocities. These secondary electrons, when confronted with the repelling effect of the suppressor, are encouraged to return to the plate. In this way, the undesirable features of secondary emission are overcome.

Pentodes can be divided into two types, based upon the effect of the control grid on the plate current. These two types are the *sharp-cutoff* and *remote-cutoff* pentodes.

Figure 10.18*a* illustrates the basic grid construction of the sharp-cutoff pentode. Notice that the grid meshwork is closely wound and of uniform

Fig. 10.18 Features of the sharp-cutoff pentode: (*a*) grid construction, (*b*) transfer characteristics.

spacing throughout. As shown in Fig. 10.18*b*, plate current cutoff occurs at relatively small values of negative control grid voltage.

The grid construction of the remote cutoff type of pentode is shown in Fig. 10.19*a*. The grid meshwork is closely woven at the ends with wider

Fig. 10.19 Features of the remote-cutoff pentode: (*a*) grid construction, (*b*) transfer characteristics.

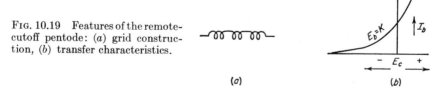

spacings in its center region. The effect of this type of grid construction is illustrated by the transfer characteristics (Fig. 10.19*b*). As the control grid is made more negative, the plate current decreases but less abruptly than in the previous type of construction.   The reason for this is due to the reduced effect of the negative grid voltage upon the plate current because of the wider spacing between the center portion of the meshwork. Electron flow through the meshwork at the two ends is stopped at relatively low potentials, whereas a much larger bias value is required to prohibit current flow through the larger spacings.

The remote-cutoff pentode is widely used in circuits where automatic types of bias which function in accordance with the signal strength are utilized, such as avc (discussed in Chap. 13).

Because of the addition of two more electrodes between the anode and cathode, the amplification factor and plate resistance of the pentode are generally higher than found in the triode (typical $\mu = 2,000$ and typical $r_p = 1$ megohm or more).

10.12   Volt-Ampere Characteristics of the Pentode.   Figure 10.20 illustrates the plate volt-ampere characteristics of the pentode.   The plate

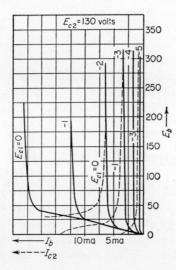

FIG. 10.20   Characteristics of the pentode 7199.   (*RCA.*)

current undergoes large changes for small changes of plate voltage in the lower plate-voltage region.   For larger plate-voltage values, the plate voltage has very little effect upon the plate current.   This is due to the relative isolation of the plate from the cathode because of the screen grid, resulting in most of the electrons going to the screen.   The broken-line characteristics of Fig. 10.20 show the relationship between screen (grid 2) current and plate voltage.   Notice that the screen current is highest at low plate voltages and sharply decreases as the plate voltage is increased.   At a relatively low plate potential, the plate current is almost at its maximum magnitude, which is the time the screen current is just about at its minimum value. For example, from the characteristics of Fig. 10.20, consider the E_b versus I_b and E_b versus I_{c2} characteristics when $E_c = -1$ volt: $I_b = 11.5$ ma when $E_b = 50$ volts, and $I_{c2} = 4.5$ ma at the same E_b value. When $E_b = 250$ volts, $I_b = 12.5$ ma, and $I_{c2} = 3.75$ ma. I_b increased by only 1 ma for the plate increase of 200 volts, and I_{c2} decreased by 0.75 ma. Therefore the 50-volt region is the area of the characteristics where the plate current levels off to an almost constant value. The small changes in I_b from this point to the maximum plate voltage are due to the slight decrease in I_{c2} and the Schottky effect of the higher E_b values upon the cathode. The "plateau" region of the characteristics is relatively linear and is the region most commonly used in linear amplifier work.

Because of the relatively small I_b change for a large change in E_b with fixed E_c, pentodes have r_p and mu values that are considerably larger than found in triodes. The tube constants (r_p, μ, and g_m) can be graphically determined by the same techniques used in the triode analysis.

10.13 The Beam-power Tube. In the beam-power tube, the cathode-emitted electrons are directed into beams, which flow to the plate. In

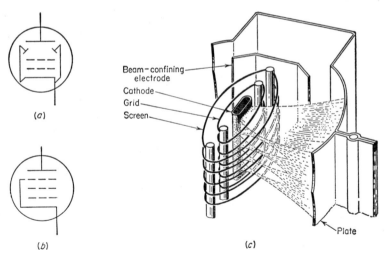

Beam–confining electrode
Cathode
Grid
Screen

(a)

(b)   (c)

Plate

Fig. 10.21 The beam power tube: (a) symbol (tetrode), (b) symbol (pentode), (c) structure (tetrode). [(c) *from RCA.*]

this way, the power capabilities of the tube are substantially improved. Some beam-power tubes are pentodes, as shown in Fig. 11.21b. The suppressor grid serves the same function as in the ordinary pentode, to reduce the effect of secondary emission.

The tetrode type of beam-power tube is in common usage (see Fig. 10.21a). The control grid and screen grid are spaced such that the electrons are focused into beams, as they are in the pentode type. The structure of the tetrode type is illustrated in Fig. 10.21c. The plate is at a lower potential than the screen in this type. As the beam of electrons

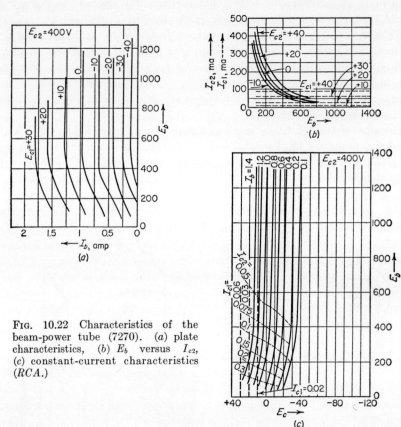

Fig. 10.22 Characteristics of the beam-power tube (7270). (a) plate characteristics, (b) E_b versus I_{c2}, (c) constant-current characteristics (RCA.)

passes through the screen grid region toward the plate, the electrons undergo a decrease in velocity since they are leaving the more positive screen area. The concentrated, slower moving beam of electrons repels secondary emission electrons, thereby forcing them back to the plate. In this manner, the concentrated electron beams serve the same purpose as the suppressor grid in the conventional pentode.

The screen current is kept at relatively low magnitudes by special construction techniques, resulting in high output power and high efficiency. Typical characteristics of the type 7270 beam-power tube are shown in Fig. 10.22.

10.14 Special-purpose and Multiunit Tubes. Multipurpose tubes may be classified in three general categories:

1. Special-purpose
2. Multiunit
3. Special-purpose multiunit

Special-purpose tubes are designed to render optimum service for a particular type of application and are generally multielectrode types. The 6BY6, which has five grids, a cathode, plate, and heater is an example. Since it has five grids, it is also called a *pentagrid* tube. This tube is primarily designed as a combined sync-separator and sync-clipper in some television receivers. The 6CB6, which is a sharp-cutoff pentode, is another example of this type. The suppressor grid is taken to a separate pin connection rather than internally tied to the cathode. Its chief use is as an intermediate-frequency amplifier up to 45 Mc and as a radio-frequency amplifier in vhf television tuners.

There are many multiunit tubes available. The 5Y3-GT is an example of the multiunit type, as are all full-wave rectifiers, since two diodes are included in the same evacuated envelope. Twin-diode triodes are available, such as the 6BF6 and 6AV6. The 6U8-A and 6X8 are triode pentodes. Twin triodes with similar or dissimilar triode units are also available. The 6CG7 and 12AX7 are twin triodes which are similar, while the 6CM7 has two dissimilar triode units.

The special-purpose multiunit tubes contain seven electrodes, five grids, a cathode, and a plate. The pentagrid converter type is a typical example—1R5, 6BE6, and 6SA7. This type serves as an oscillator and mixer-amplifier in superheterodyne receivers.

10.15 The Cathode Follower (Common-plate Connection). A vacuum tube connected in the common-anode configuration is called the cathode follower. The signal input is applied between the control grid and plate by selecting C such that it displays low reactance for all signal frequencies. The output is taken from cathode to plate, as shown in Fig. 10.23. The

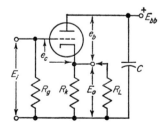

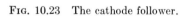

FIG. 10.23   The cathode follower.

input impedance of the circuit is high while the output impedance is low. The cathode follower is useful when the output is to be connected to a low impedance load and a transformer is undesirable.

In Fig. 10.23, the voltage gain of the amplifier is determined by

$$A_{v,0} = \frac{A_v}{1 - \beta A_v} = \frac{R_k}{[(\mu + 1)/\mu]R_k + r_p/\mu}$$

or

$$A_{v,0} = \frac{R_k}{R_k[(\mu + 1)/\mu] + 1/g_m}$$

where $A_{v,0}$ = voltage gain with feedback

A_v = voltage gain without feedback

$= \mu[R_k/(r_p + R_k)]$

βA_v = feedback factor

When a cathode follower, such as shown in Fig. 10.23, utilizes a pentode, where μ is customarily very large, the voltage gain can be approximated by

$$A_{v,0} \cong \frac{R_k}{R_k + 1/g_m}$$

In some cases, R_L is of such a value that it must be considered in determining the voltage gain of the cathode follower. It is seen that R_L is in parallel with R_k; therefore their equivalent value is used in the preceding equations; i.e.,

$$A_{v,0} = \frac{R_k R_L/(R_k + R_L)}{[(\mu + 1)/\mu][R_k R_L/(R_k + R_L)] + r_p/\mu}$$

or

$$A_{v,0} = \frac{R_k R_L/(R_k + R_L)}{[R_k R_L/(R_k + R_L)][(\mu + 1)/\mu] + 1/g_m}$$

For pentodes,

$$A_{v,0} \cong \frac{R_k R_L/(R_k + R_L)}{R_k R_L/(R_k + R_L) + 1/g_m}$$

The load impedance, as seen by R_k in Fig. 10.23, is determined by

$$Z_L = \frac{r_p R_k}{r_p + (1 + \mu)R_k}$$

The output voltage is found from

$$E_o = \frac{R_k}{R_k + r_p/(\mu + 1)} \frac{E_i \mu}{\mu + 1}$$

When R_L is connected and its effect must be considered, R_k in the preceding equation is replaced with

$$\frac{R_k R_L}{R_k + R_L}$$

Maximum power transfer occurs when

$$R_k = \frac{r_p}{\mu + 1}$$

Dynamic Response Curve of a Cathode Follower. The dynamic response curve of the cathode follower is a plot of e_i versus e_o. This curve is of value in that it provides the designer with a means for determining the most suitable grid bias E_{cc}, maximum voltage that can be accepted by the amplifier, and the voltage gain. The dynamic response curve is easily constructed in conjunction with the conventional load line, as shown in the following example.

Example: Refer to Fig. 10.24 for the average plate characteristics of the 6SN7. $R_k = 20$ kilohms; $E_{bb} = 400$ volts.
(*a*) Construct the load line.
(*b*) Construct the dynamic response curve.
(*c*) Determine E_{cc}.
(*d*) Find $E_{i,\text{max}}$.
(*e*) Find $A_{v,o}$.

Solution:

(*a*) Construct the load line (refer to Fig. 10.24):

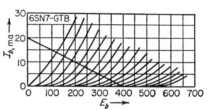

Fig. 10.24 Characteristics of the 6SN7-GTB (E_c from 0 to 36 volts in 2-volt steps). (*RCA.*)

Point A is determined by the selected value of E_{bb}. Point B is found by

$$I_{b(SC)} = \frac{E_{bb}}{R_k} = \frac{400}{20 \times 10^3} = 20 \text{ ma}$$

Joining points A and B with a straight line completes the construction of the load line.
(*b*) Construct the dynamic response curve:
From the cathode-follower circuit of Fig. 10.23, it is seen that $e_o = e_{Rk} = i_b R_k$; and the input voltage $e_i = e_c + e_{Rk} = e_c + E_o$, where e_c (grid voltage) is taken off the plate characteristics. Find the e_i and e_o values for a number of i_b values in the next step, as listed below.

i_b, ma	$i_b R_k = e_o$, volts	$e_c + e_o = e_i$
0	0	$-26 + 0 = -26$
2.5	$2.5 \times 10^{-3} \times 20 \times 10^3 = 50$	$-17 + 50 = 33$
5.0	$5 \times 10^{-3} \times 20 \times 10^3 = 100$	$-12 + 100 = 88$
10.0	$10 \times 10^{-3} \times 20 \times 10^3 = 200$	$-5 + 200 = 195$
13.5	$13.5 \times 10^{-3} \times 20 \times 10^3 = 270$	$0 + 270 = 270$

The relationship of e_i versus e_o is then plotted, as shown in Fig. 10.25.

Point A is the cutoff point of the tube, where $e_c = -26$ volts. Point E, the opposite extreme, is that point where $e_c = 0$ volts (which is the condition when grid current begins to flow).

(c) Determine E_{cc}:

E_{cc}, the zero signal grid bias, should be located midway between points A and E if the largest possible input signal is to be considered. The midpoint of the

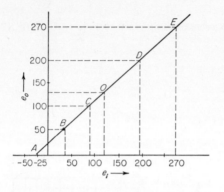

Fig. 10.25 Dynamic response curve.

response curve AE is at point O, where $e_i = 122$ volts and $e_o = 135$ volts. Since $e_i = e_c + e_o$, then

$$e_c = E_i - E_o = 122 - 135 = -13 \text{ volts}$$

Therefore $E_{cc} = -13$ volts.

(d) Find $e_{i,\max}$:

$e_{i,\max}$ is the input signal swing from point O to either extreme of the dynamic response curve of Fig. 10.25. This swing from point O to point $E(270 - 122)$ is equal to 148 volts. If the input signal is sinusoidal, the rms input signal is $148 \times 0.707 = 105$ volts.

(e) Find $A_{v,o}$:

The voltage gain of the circuit can be graphically determined:

$$A_{v,o} = \frac{\Delta e_o}{\Delta e_i} = \frac{270 - 0}{270 - (-26)} = 0.912$$

In terms of the input signal, the input is applied between grid and plate, thereby resulting in a high input impedance. R_g, the grid bias resistor, functions independently of the load resistor R_k, which makes it possible to match R_g to the output impedance of the device ahead of the cathode follower. The output is taken from between cathode and plate, which results in a relatively low impedance output. R_k may be the actual load resistance, or may be matched to the load resistance, whichever is most desirable for the particular application.

The cathode follower is capable of handling low frequencies down to about zero cps. Its high input impedance enhances its possibilities as a direct-coupled amplifier, and its low output impedance makes it very suitable as a driver for class AB and B power amplifiers.   As pointed out in the preceding example, the voltage gain is slightly less than unity.

PROBLEMS

10.65 The cutoff grid voltage for a certain tube $= -5$ volts; it is to be used as a class A amplifier.   Find the zero signal grid bias required so that the maximum input signal can be utilized.

10.66 In Prob. 10.65, what is the maximum input signal which the tube can utilize in class A service?

10.67 State the difference between AB_1 and AB_2.

10.68 In Prob. 10.67, which type of service results in the development of greatest input circuit power?   Explain.

10.69 State the difference between class B_1 and B_2.

10.70 In Prob. 10.69, which type of service results in the development of greatest input circuit power?   Explain.

10.71 What is the minimum E_{cc} value in class C service in most practical cases? Explain.

10.72 Refer to Fig. 10.9 with the load line AB.   Maximum input signal $= 1$ volt Find the smallest acceptable E_c for operation in (a) class A_1; (b) class AB_1; (c) class B_1; (d) class C.

10.73 At what frequencies is C_{pg} most troublesome?   Explain.

10.74 Describe the effect of the screen grid upon the behavior of the (a) tetrode; (b) pentode.

10.75 What causes secondary emission?

10.76 How is the problem of secondary emission reduced in (a) tetrode beam-power tube; (b) conventional pentode.

10.77 State the difference between the sharp cutoff and remote cutoff pentodes.

10.78 Why is μ and r_p of a pentode higher than in a triode?

10.79 State the difference between the tetrode beam-power tube and the pentode beam-power tube.

10.80 Refer to the plate volt-ampere characteristics of the 6SN7-GTB in Fig. 10.24, and the cathode follower of Fig. 10.23. $E_{bb} = 300$ volts; $R_K = 20$ kilohms; $R_L = 20$ kilohms; $e_{i,\mathrm{max}} = 40$ volts.   Draw the load line.

10.81 In the cathode follower considered in Prob. 10.80, $\mu = 20$; $r_p = 7.7$ kilohms: $g_m = 2600$ μmho.   Find (a) A_v; (b) $A_{v,o}$; (c) β.

10.82 In the cathode follower of Probs. 10.80 and 10.81, find (a) Z_L; (b) e_o.

10.83 In the circuit of Probs. 10.80 to 10.82, find E_o for i_b values of:

(a) 0 ma (b) 5 ma
(c) 10 ma (d) 12 ma
(e) 14 ma (f) 16 ma

10.84 Find the corresponding E_i values for the E_o values determined in Prob. 10.83.

10.85 Plot the dynamic response curve with the E_o and E_i values determined in Probs. 10.83 and 10.84.

10.86 In the cathode follower of Probs. 10.80 to 10.85, find:

(a) E_{cc} (b) $e_{i,\mathrm{max}}$
(c) $E_{i,\mathrm{rms}}$ (d) A_v

10.87 Why is the voltage gain of a cathode follower no greater than unity?

10.88 State the relative values of Z_i and Z_o of the cathode follower as compared to the common-cathode connection.   Explain.

10.89 What transistor configuration is similar to the common-plate connection?

10.90 What is the low-frequency response of the cathode follower?

CHAPTER 11

VACUUM-TUBE AMPLIFIERS AND OSCILLATORS

This chapter is concerned with the fundamental aspects of vacuum-tube amplifiers in all classes of operation. The design techniques are presented in a manner which is consistent with the background of the technician, so as to lend more meaning to the material. The most significant bias networks used in vacuum-tube circuitry are first analyzed. The various coupling techniques used in cascaded amplifiers, along with the equivalent circuits at low, middle, and high frequencies, are then considered. Upon completing the study of the bias networks and coupling possibilities, the class A amplifier is examined, and a class A RC-coupled amplifier is designed in a step-by-step fashion. The class B push-pull amplifier, which has found popular use, is studied and then designed in detail.

The fundamentals of oscillator action are reviewed prior to the analysis of various sine-wave (LC) oscillators, the three basic multivibrators, and sine-wave (RC) oscillators.

11.1 Bias Networks. It will be recalled from the load-line analysis in the preceding chapter that the control grid of the tube is placed at a negative potential E_{cc} at zero signal condition. The magnitude of E_{cc}, along with the magnitude of the input signal voltage, determines the amplifier's class of operation. Let us consider the more common bias networks used in vacuum-tube circuitry.

Cathode Bias (Self-bias). Refer to Fig. 11.1a, which illustrates a triode amplifier with cathode bias. When the circuit is connected, electrons flow from $-E_{bb}$ to common, to the cathode, through the tube to the plate, through R_L to the positive terminal of E_{bb}. This current develops the voltage drop e_{RK}. Notice that e_{RK} is of such a polarity that the cathode becomes *positive* with respect to common. The control grid is at the common potential. Therefore e_{RK} effectively makes the control grid *negative* with respect to the cathode, and may be used as a bias voltage for the amplifier.

Recall that under usual circumstances, control-grid current does not flow, resulting in the cathode current and plate current being equal in the triode amplifier. The zero signal plate current and grid bias can be readily determined from the load line drawn upon the plate volt-ampere

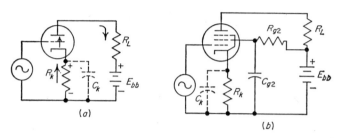

FIG. 11.1 Cathode bias: (a) triode, (b) pentode.

characteristics. Knowing these values, the required R_K can be computed:

$$R_K = \frac{e_{G\text{-}K,S}}{i_K} = \frac{E_{c,s}}{I_{b,s}} \quad \text{in the triode}$$

where $e_{G\text{-}K,S} = E_{c,s} =$ zero signal grid-to-cathode potential
$i_K = I_{b,s} =$ zero signal plate current

Example 1: Refer to Fig. 10.9. The operating point O parameters are: $E_c = -1.3$ volts; $I_b = 2.5$ ma. Find R_K.

Solution:

$$R_K = \frac{E_{c,s}}{I_{b,s}} = \frac{1.3}{2.5 \times 10^{-3}} = 520 \text{ ohms}$$

e_{RK} varies directly with the magnitude of the current passing through R_K. With higher values of plate current, e_{RK} is larger for a fixed R_K, which places the control grid at a larger negative potential with respect to the cathode. The increase in negative grid voltage results in increased opposition to current flow through the tube, and I_b decreases. In other words, R_K creates degenerative feedback. As pointed out in the earlier study of transistors, negative feedback produces a reduction in gain and also a reduction in distortion. While the sacrifice in gain may be undesirable, the reduction in distortion may be considered necessary. If the distortion problem takes precedence, then R_K is left unbypassed. When the gain considerations are most important, R_K may be bypassed by a capacitor of suitable value, thereby reducing the degeneration to a negligible value. A general rule of thumb is to select C_K such that X_{CK} at the *lowest signal frequency* is 10 per cent or less of R_K.

Example 2: Consider the amplifier of example 1. Assume the signal frequency range is 40 cps to 15 kc; degeneration is to be reduced to a negligible value. Find C_K.

Solution:

$$X_{CK} \text{ at } 40 \text{ cps} = 0.10R_K = 0.10(520) = 52 \text{ ohms}$$

Solving for C_K,

$$C_K = \frac{1}{2\pi f_{min}X_{CK}} = \frac{1}{6.28 \times 40 \times 52}$$
$$= 76 \ \mu\text{f or larger}$$

Cathode bias can also be used with pentode amplifiers, as shown in Fig. 11.1b. Note that i_{RK} is the sum of the plate and screen currents. Both currents must be considered in the computation of R_K. The zero signal value of the screen and plate currents, as well as the grid-to-cathode potential, can be determined from the plate volt-ampere characteristics. R_K can be found by

$$R_K = \frac{E_{c,s}}{I_{c2,s} + I_{b,s}}$$

where $I_{c2,s}$ = zero signal screen current
$I_{b,s}$ = zero signal plate current
$E_{c,s}$ = zero signal grid-to-cathode potential

Example 3: Refer to Fig. 10.20. Assume the operating point is at $E_b = 200$ volts; $E_{c2} = 130$ volts; $E_c = -1$ volt. Find the required R_k.
Solution: From the characteristics of Fig. 10.20, we find $I_{b,s} = 12.5$ ma. From the E_b versus I_{c2} characteristic with $E_{c1} = -1$ volt, we find $I_{c2} = 3.75$ ma with $E_b = 200$ volts. R_K may now be computed:

$$R_K = \frac{1}{(3.75 + 12.5) \times 10^{-3}} \cong 62 \text{ ohms}$$

The reduction in gain and distortion by use of an unbypassed R_K is also present in the pentode arrangement. Bypass provisions may be made, if so desired.

Example 4: Assume the frequency range of the pentode amplifier of example 3 is 5 to 50 kc; degeneration due to R_K is to be reduced to a negligible value. Find a suitable C_K.

Solution:

$$X_{CK} \text{ at } 5 \text{ kc} = 0.1R_K = 6.2 \text{ ohms}$$

Solving for C_K,

$$C_K = \frac{1}{6.28 \times 5,000 \times 5.3} = 5.1 \ \mu\text{f or larger}$$

Contact Bias. Refer to Fig. 11.2 for the following discussion. A phenomenon occurs in most vacuum tubes which is utilized in the development of contact bias. Earlier sections stated that maintaining the control grid at a negative potential prohibits the flow of control-grid current, thereby keeping the input power dissipation down to a minimum. Positive-grid current, i.e., electron flow out of the control grid, does occur at small values of negative control-grid voltages. Point A in Fig. 11.2b is the point where a positive I_c begins to flow and is called the contact potential point of the tube. When the grid is made more negative than point A, no positive I_c flows. Therefore the region to the right of point

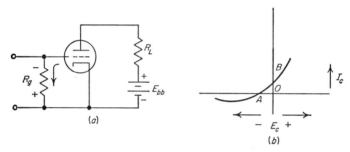

FIG. 11.2   Analysis of contact bias: (*a*) circuit, (*b*) E_c versus I_c characteristic.

A is called the contact-current region. The magnitude of E_c at point A may be called the *contact potential* of the tube. It is this point which is utilized in a contact-bias circuit.

Contact potential is caused by several factors. The velocity of the cathode-emitted electrons approaching and passing through the control-grid meshwork helps to create positive-grid current at the contact potential. A number of the electrons en route from the cathode to the plate strike the meshwork of the control grid. Some of the electrons involved in this collision process bounce off and continue their journey to the plate, but a certain number of them remain on the grid because of the small emf which exists between grid and cathode. Recall that the control grid and cathode are made of different materials. The energy level of mobile electrons at the surface of the cathode is higher than that found at the surface of the control grid, resulting in a small emf between the two electrodes; i.e., the control grid is slightly positive with respect to the cathode. This slight difference of potential enables the control grid to hold on to some of the electrons which collide with its structure. Also contributing to the grid-to-cathode emf (contact potential) is the fact that the control grid is at a lower temperature than the cathode. When the grid is made sufficiently negative (greater than point A of Fig. 11.2b), the contact potential has a less pronounced effect upon the electrons impinging upon the control grid.

The positive-grid current caused by the point contact potential is limited by the presence of R_g. In order to prevent the input resistance of the tube from becoming too low, R_g must be large in value so as to keep I_c at a small value. A large magnitude of positive I_c would result in loading down the preceding stage. The contact potential at point A of Fig. 11.2b can be as great as 1.5 volts. In most vacuum-tube amplifiers, the bias is made greater than this value, so as to reduce the possibility of positive-grid current and its associated disadvantages, i.e., grid circuit

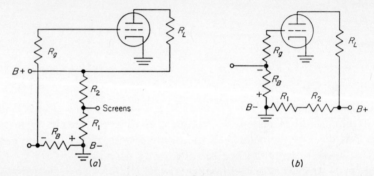

FIG. 11.3   Fixed grid bias.

power dissipation, reduction of the dynamic input resistance of the tube, and loading the output of the preceding stage.

There are certain amplifiers where the positive-grid current flowing through R_g is used as the bias for the tube. Such an arrangement may be called *contact bias*, and its use is most commonly restricted to small signal voltage amplifiers whose zero signal tube dissipation is very low. In such cases, only 1 volt or so of bias is required, and R_g is in the order of 2 to 15 megohms.

Fixed Grid Bias. A third basic method of maintaining the control grid at a negative potential with respect to the cathode is illustrated in Fig. 11.3 and is called *fixed grid bias*. In this type of bias, a fixed negative potential is applied to the grid, which may be obtained by a separate voltage supply or from a resistor placed in a voltage divider, which is across the filter output of the power supply. Notice that R_B, the bias resistor, is connected between the negative side of the filter and common. The total supply current flows through R_B to common, in the direction indicated by the polarity of E_{RB}.

The total current consists of all the load currents plus the bleeder current I_{R1}. In the circuit illustrated in Fig. 11.3a, notice that the load current flows from common $B-$, to the cathode of the tube, to its anode, back to $B+$, thereby bypassing R_1 and R_2. If other tubes are in the circuit, they may obtain their plate voltages in the same manner. A lower value of positive voltage is available between $B+$ and $B-$ for the screen grids of pentodes, or lower plate potentials for other tubes. Recall that the bleeder current is generally 10 to 25 per cent of the total load current. The desired grid bias is obtained from the load-line analysis, and R_B can be computed from

$$R_B = \frac{\text{desired } E_c}{I_{\text{load}} + I_{\text{bleeder}}}$$

Example 5: Assume the desired $E_c = -4$ volts; $I_{\text{load}} = 100$ ma; $I_{\text{bleeder}} = 10$ ma. Find R_B.

Solution:

$$R_B = \frac{4}{(100 + 10) \times 10^{-3}} \cong 36 \text{ ohms}$$

11.2 Direct Coupling. Direct coupling in vacuum-tube circuits is similar to that found in transistor circuitry in that the output terminal of stage 1 is directly connected to the input terminal of the second stage. A direct-coupled amplifier is also called a *d-c amplifier*.

Figure 11.4 illustrates a two-stage d-c amplifier. The bias for stage 1 is obtained by the potential E_{K1}, and the plate-supply voltage for T_1 is E_{bb1}. The potential applied to the grid of T_2 is $E_{bb1} - E_{RL1}$ and is positive. In order to make E_{G-K2} negative by the desired amount, the cathode of T_2 must be made more positive than its control grid by the required magnitude of grid bias; i.e.,

$$E_{K2} = E_{G\text{-COMMON}} + \text{desired grid bias}$$

The plate-supply voltage of T_2 is E_{bb2}. It will be noticed that this circuit requires four voltage values, which can be obtained by using a voltage divider with the required tap potentials. This complexity is a serious disadvantage, particularly when the d-c amplifier has more than two stages. In such circuits, E_{K1} is smaller than E_{K2}, and E_{bb1} is smaller than E_{bb2}. Furthermore, a power supply with a high degree of regulation is required, since any variation of the power supply potential would change all the voltages.

A more suitable type of direct coupling is illustrated in Fig. 11.5. This type of direct coupling requires the use of an additional bias supply E_{cc} as well as the usual plate supply E_{bb}. As pointed out in the preceding section, this can be obtained from the same rectifier system. R_{g1} and R_{g2} form a voltage divider between the plate of T_1 and the negative terminal of E_{cc}. Notice the polarity of e_{Rg1} and e_{Rg2}. Point B is positive with respect to point A, and point D is positive with respect to point C. The voltage across R_{g2} is determined by

$$e_{Rg2} = \frac{R_{g2}}{R_{g1} + R_{g2}} (E_b T_1 + E_{cc})$$

E_{cc} is generally large, but not as large as E_{bb}. The ratio of R_{g2} to $R_{g1} + R_{g2}$ determines the voltage applied to the control grid of T_2. Although point B is positive with respect to point A, point B can be made negative with respect to the cathode of T_2 by selecting the correct ratio

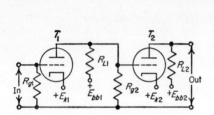

FIG. 11.4   A two-stage d-c amplifier.   FIG. 11.5   D-C coupling with voltage divider.

of R_{g2} to $R_{g1} + R_{g2}$. In this way, the plate of T_1 can be directly coupled to the control grid of T_2.

The variations of the T_1 plate voltage are the signal that is coupled to the control grid of T_2. The amplitude of the T_1 plate-voltage variations is reduced in accordance with the ratio of R_{g2} to $R_{g1} + R_{g2}$.

Example 1: Assume Δe_p of $T_1 = 140 - 150$ volts, and $R_{g1} = R_{g2}$; then $\Delta e_p = 10$ volts and $\Delta e_{Rg1} = \Delta e_{Rg2} = 5$ volts (each). The required value of E_{cc} for class A operation is determined by

$$E_{cc} = E_{b1,s} + 2E_{cT2,s}$$

where $E_{b1,s}$ = zero signal plate voltage of T_1
$E_{cT2,s}$ = zero signal control-grid voltage of T_2

A numerical example will help to clarify the preceding equations.

Example 2: In Fig. 11.5, let $E_{b1,s} = 200$ volts; $E_{cT2,s} = -5$ volts. Determine the required (a) E_{cc}; (b) e_{Rg2}.
NOTE: Let $R_{g1} = R_{g2}$.

Solution:

(a)

$$E_{cc} = E_{b1,s} + 2E_{cT2,s}$$
$$= 200 + 10 = 210 \text{ volts}$$

(b)

$$e_{Rg2} = \frac{R_{g2}}{R_{g1} + R_{g2}} (E_{b1,s} + E_{cc})$$

Since $R_{g2} = R_{g1}$,

$$e_{Rg2} = \tfrac{1}{2}(E_{b1,s} + E_{cc})$$
$$= \tfrac{1}{2}(200 + 210) = 205 \text{ volts}$$

Going around the loop from the cathode to point C results in the value of $E_{cT2,s}$, which is a good check:

$$E_{cT2,s} = -210 + 205 = -5 \text{ volts}$$

The value of $E_{b1,s}$ is determined from the conventional load-line analysis of T_1, and $E_{c,T2}$ is found from the same analysis of T_2. It is suggested that R_{g1} and R_{g2} be large in value (1 megohm or so), thereby imposing only a small drain upon E_{cc}. By combining the load-line analysis of both tubes with the preceding equations, the two-stage d-c amplifier of Fig. 12.5 can be designed with little or no complications. The same type of arrangement can be extended to more than two stages, by providing a voltage-divider network between the plate of the preceding stage and control grid of the following stage.

Upon observation of the d-c amplifiers of Figs. 11.4 and 11.5, it is seen that a change in the grid voltage results in the plate current changing to a new fixed value. In effect, a small change in d-c grid voltage can produce a correspondingly large change in d-c plate voltage, and it can be said that it amplifies d-c voltages, resulting in the name of d-c amplifier. If the control-grid voltage is continuously changed at a rapid rate, the plate-voltage changes will be larger and at the new rapid rate; i.e., the d-c amplifier is also capable of functioning as an a-c amplifier. Direct coupling is customarily used only in those circuits where a frequency response down to zero cps is required, which is the chief advantage of direct coupling. The high-frequency limit is determined by C_{pk} of T_1, C_i of T_2, and the stray-wiring capacitance of the circuit. These capacitances all act in shunt and establish the high-frequency limit of the amplifier in a manner similar to that studied in RC coupling in the next section.

11.3 RC Coupling. In the preceding section, several methods are presented by which the output of one stage (d-c and a-c components) can be coupled to a second stage. It was seen that this technique, called direct coupling, presents a number of problems because of the relatively large positive potential applied to the control grid of the second stage. Because of this drawback and its associated problems, direct coupling is usually restricted in application to those circuits where a frequency response down to zero cps is a prerequisite. The more common amplifier applications do not require such a low-frequency response, thereby permitting the use of other coupling techniques.

Perhaps the most economical coupling network is the RC coupling variety, which is illustrated in Fig. 11.6. The RC coupling network consists of C_c and R_{g2}. The coupling capacitor C_c serves a dual function:

1. Blocks the d-c component of the first-stage plate from the second-stage grid

2. Couples the a-c component of the first-stage plate to the second-stage grid

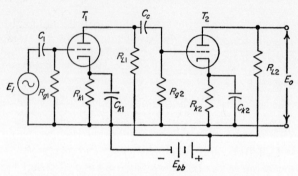

FIG. 11.6 A two-stage RC-coupled amplifier.

The use of C_c eliminates the problem of placing the control grid of the second stage at a large positive potential, but allows the signal component to be felt by the control grid.

R_{g2} also serves several purposes:

1. Places the control grid at d-c "ground" potential

2. Develops the signal variations of plate 1 between the control grid of stage 2 and common

The control grid of stage 2 must have a d-c path to ground for stable operation. In the section which analyzed contact bias, it was found that the control grid collected a fraction of the electrons which are en route from the cathode to the plate. With R_{g2} connected between the control grid and common, the captured electrons can be repelled out of the control grid to common via R_{g2}. With no external d-c return between these two points, a *floating-grid* condition exists. Under this condition, the electrons captured by the grid would repell electrons attempting to pass through the grid meshwork and on toward the plate. Unpredictable tube behavior and even cutoff can result from a floating grid.

Notice that C_c and R_{g2} are effectively across the output voltage of stage 1; that portion of the circuit can be redrawn as shown in Fig. 11.7.

X_{cc} varies inversely with frequency; i.e., X_{cc} is largest at the lowest signal frequency and minimum at the highest frequency. E_c is a lagging quadrature voltage, and E_{i2} is an in-phase potential, with E_{o1} the resultant of the two (see Fig. 11.8).

Because X_{cc} varies inversely with frequency, E_{i2} is proportional to frequency. By the proper selection of C_c, E_{i2} can be made to be equal to $0.707E_{o1}$ at the lowest desirable frequency. E_{o1} is larger than this value for all higher frequencies. In effect, the RC coupling circuit is a *low-*

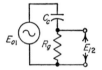

Fig. 11.7 RC coupling network.

frequency discriminator, since the output is taken across R_g. It should also be pointed out that the RC coupling network is an *attenuator*, and the output voltage can never be greater than the input potential.

In order to evaluate RC coupling to a greater extent, let us consider the equivalent circuit of the entire two-stage amplifier at low, middle, and high frequencies.

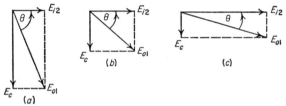

Fig. 11.8   Voltage relationship in the RC network: (a) low frequency, (b) $f_{co,\mathrm{min}}$, (c) high frequency.

Refer to Fig. 11.9. Notice that four capacitances appear in the equivalent circuit. C_{pk1} is the plate-to-cathode interelectrode capacitance of the first stage tube. C_w is the stray capacitance of the wiring and is generally negligible when the two stages are physically close together. C_i is the input capacitance of the stage-2 tube. Since C_{pk1}, C_w, and C_i are

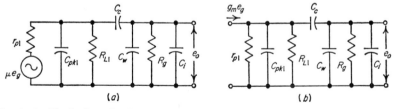

Fig. 11.9   Equivalent circuits of Fig. 11.6: (a) constant voltage, (b) constant current.

parallel capacitances, it is most desirable that they be small in value, resulting in high reactive values. When this is the case, a minimum of signal energy is shunted to common before the input of stage 2. C_c, on the other hand, is a series capacitance and is most noticeable at the low frequencies where X_{cc} is largest.

Low-frequency Analysis. Figure 11.10 illustrates the equivalent circuit of the RC-coupled amplifier at low frequencies. C_{pk1}, C_w, and C_i do not appear in the circuit because their reactances are very high and do not shunt very much signal energy away from the second stage input. C_c, on the other hand, does remain in the circuit, since its reactance is high and reduces e_o. The lower the signal frequency, the larger X_{cc} and e_{cc}, resulting in a smaller e_o.

The equivalent circuit of Fig. 11.10a may be further simplified to that of Fig. 11.10b. In Fig. 11.10a, consider the condition when C_c is not

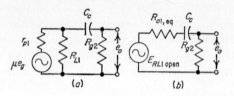

FIG. 11.10   Equivalent circuits at low frequencies.

connected to R_{L1}, resulting in a simple series circuit of generator μe_g and resistances r_p and R_{L1}. This current i_{open} is

$$i_{\text{open}} = \frac{\mu e_g}{r_p + R_{L1}}$$

i_{open} develops a voltage across R_{L1}, which may be called $e_{RL1,\text{open}}$:

$$e_{RL1,\text{open}} = i_{\text{open}} R_{L1}$$

where $e_{RL1,\text{open}}$ is the voltage across R_{L1} when C_c and R_g are not connected. Substituting into the equation for i_{open},

$$e_{RL1,\text{open}} = \frac{\mu e_g}{r_p + R_{L1}} R_{L1}$$
$$= \frac{\mu e_g R_{L1}}{r_p + R_{L1}}$$

Notice that $e_{RL1,\text{open}}$ is the generator in the equivalent circuit of Fig. 11.10b.

In Fig. 11.10a, the resistance to the left of C_c is the equivalent value of r_p and R_{L1} and is called $R_{o1,\text{eq}}$ in Fig. 11.10b; i.e.,

$$R_{o1,\text{eq}} = \frac{r_p R_{L1}}{r_p + R_{L1}}$$

With the new generator $E_{RL1,\text{open}}$ and $R_{o1,\text{eq}}$, the circuit of Fig. 11.10b is a series-RC circuit. The total resistance is the sum of R_{g2} and $R_{o1,\text{eq}}$, which may be called R_a:

$$R_a = \frac{r_p R_{L1}}{r_p + R_{L1}} + R_{g2}$$
$$= R_{o1,\text{eq}} + R_{g2}$$

The current of this circuit may be determined by

$$i = \frac{e_{RL1,\text{open}}}{R_a - jX_{cc}}$$

And the output voltage e_o is

$$e_o = iR_{g2} = \frac{e_{RL1,\text{open}}R_{g2}}{R_a - jX_{cc}}$$

The low-frequency gain may now be found:

$$A_{v(\text{l-f})} = \frac{e_o}{e_g}$$

where e_g is the input signal voltage to stage 1.

This equation may be expanded to

$$A_{v(\text{l-f})} = \frac{e_{RL1,\text{open}}R_{g2}}{R_a - jX_{cc}}$$
$$= \frac{[\mu e_g R_{L1}/(r_p + R_{L1})]R_{g2}}{(R_{o1,\text{eq}} + R_{g2}) - jX_{cc}}$$

Low-frequency cutoff may be defined as that frequency where the gain is 3 db down from some mid-frequency value. The mid-frequency reference is generally at a frequency where X_{cc} is considered negligible as compared to the resistance $(R_{o1,\text{eq}} + R_{g2})$. Using this as a criterion, low-frequency cutoff occurs when $X_{cc} = R_{o1,\text{eq}} + R_{g2}$; i.e.,

$$f_{co,\text{min}} = \frac{1}{2\pi C_c(R_{o1,\text{eq}} + R_{g2})}$$

Middle-frequency Analysis. At the middle frequencies, X_{cc} is small enough to be negligible, yet X_{Cpk1}, X_{Cw}, and X_{Ci} are still large enough not to create a shunting problem. The middle-frequency equivalent circuit,

Fig. 11.11   Mid-frequency equivalent circuit.

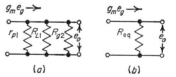

(a)   (b)

therefore, consists of three parallel resistances (see Fig. 11.11*a*), which can be simplified to their equivalent value (Fig. 11.11*b*):

$$R_{\text{eq}} = \frac{1}{1/r_p + 1/R_{L1} + 1/R_{g2}}$$

The output voltage may be found by

$$e_o = (g_m e_g)R_{\text{eq}}$$

The voltage gain is determined by

$$A_{v(\text{m-f})} = \frac{e_o}{e_g} = \frac{(g_m e_g)R_{\text{eq}}}{e_g}$$

and simplifying,

$$A_{v(\text{m-f})} = g_m R_{\text{eq}}$$

High-frequency Analysis. The shunt capacitances discussed in an earlier paragraph all appear at high frequencies (see Fig. 11.12a). The equivalent value of the shunt capacitances C_{st} is found by

$$C_{st} = C_{pk1} + C_w + C_i$$

Figure 11.12b illustrates the same equivalent circuit, but the individual capacitances have been replaced with their equivalent value.

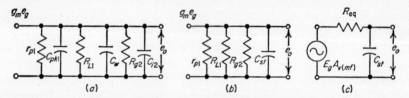

FIG. 11.12   High-frequency equivalent circuit.

In looking back into the circuit from the output terminals, with the output terminals open, the internal resistance appears as R_{eq} (see Fig. 11.12c), which may be determined by

$$R_{\text{eq}} = \frac{1}{1/r_p + 1/R_{L1} + 1/R_{g2}}$$

The equivalent circuit has now been reduced to a series circuit which consists of R_{eq} and C_{st}. The generator is the mid-frequency output voltage (which is $e_g A_{v(\text{m-f})}$), or $e_g g_m R_{\text{eq}}$.

In Fig. 11.12c, the current i_{eq} is

$$i_{\text{eq}} = \frac{e_g A_{v(\text{m-f})}}{R_{\text{eq}} - jX_{Cst}}$$

and the output voltage is

$$e_o = i_{\text{eq}}(-jX_{Cst})$$

Expanding,

$$e_o = \frac{e_g A_{v(\text{m-f})}}{R_{\text{eq}} - X_{Cst}}(-jX_{Cst})$$

The voltage gain may be determined by

$$A_{v(\text{h-f})} = \frac{e_o}{e_g} = \frac{e_g A_{v(\text{m-f})}(-jX_{Cst})}{(R_{\text{eq}} - X_{Cst})e_g}$$

and simplifying,

$$A_{v(\text{h-f})} = \frac{A_{v(\text{m-f})}(-jX_{Cst})}{(R_{eq} - X_{Cst})}$$

High-frequency cutoff may be defined as that frequency where e_o is 3 db down from some predetermined frequency value. When the mid-frequency is such that X_{Cst} may be neglected, then high-frequency cutoff occurs when $X_{Cst} = R_{eq}$, and

$$f_{co,\max} = \frac{1}{2\pi R_{eq} C_{st}}$$

The Universal Amplification Curve. In the previous analysis of the RC coupling network for the various frequency ranges, it was found that the relationship between the reactance and resistance determines the gain of the network. Recall that the RC coupling network is an attenuator circuit, with an optimum voltage gain of unity. In other words, the amplifier gain is reduced when the coupling network voltage gain is less than unity. In the low-frequency equivalent circuit, the optimum voltage gain is achieved when the ratio of R_a to X_{cc} is greater than 10. In the middle-frequency equivalent circuit, where X_{cc} is negligible, the voltage gain is unity (the optimum value). In the high-frequency equivalent circuit, where X_{Cst} behaves as a shunt, the voltage gain is determined by the ratio of X_{Cst} to R_{eq} (see Fig. 11.12). Let us verify these relationships with a typical example.

Example 1: Refer to the low-frequency equivalent circuit of Fig. 11.10a. $R_{L1} =$ 10 kilohms; $R_{g2} = 50$ kilohms; $r_p = 7$ kilohms; $g_m = 5,300$ μmho; $\mu = 38$; $e_g = 1$ volt. Find $A_{v(\text{l-f})}$ when X_{cc} is (a) $0.01R_a$; (b) $0.1R_a$; (c) R_a; (d) $10R_a$; (e) $100R_a$.

Solution:

Recall that

$$R_a = \frac{r_p R_{L1}}{r_p + R_{L1}} + R_{g2}$$

Substituting values and solving,

$$R_a = \frac{7,000 \times 10,000}{7,000 + 10,000} + 50,000$$
$$\cong 54 \text{ kilohms}$$

(a) Find $A_{v(\text{l-f})}$ when $X_{cc} = 0.01R_a = 540$ ohms:

$$A_{v(\text{l-f})} = \frac{[\mu e_g R_{L1}/(r_p + R_{L1})]R_{g2}}{(R_{o1,eq} + R_{g2})(-jX_{cc})}$$
$$= \frac{[38 \times 1 \times 10 \times 10^3/(7 \times 10^3 + 10 \times 10^3)] \times 50 \times 10^3}{\{[7 \times 10^3 \times 10 \times 10^3/(7 \times 10^3 + 10 \times 10^3)] + 50 \times 10^3\}(-j540)}$$
$$= \frac{1.118 \times 10^6}{(54 \times 10^3) - j540}$$

Converting the denominator into polar form,

$$A_{v(l\text{-}f)} = \frac{1.118 \times 10^6}{54 \times 10^3 \underline{/-0.6°}}$$
$$\cong 20.7$$

(b) Find $A_{v(l\text{-}f)}$ when $X_{cc} = 0.1R_a = 5.4$ kilohms:
The only change in the preceding equation is the value of X_{cc}:

$$A_{v(l\text{-}f)} = \frac{1.118 \times 10^6}{54 \times 10^3 - j5,400}$$
$$= \frac{1.118 \times 10^6}{54 \times 10^3 \underline{/-5.7°}}$$
$$\cong 20.7$$

(c) Find $A_{v(l\text{-}f)}$ when $X_{cc} = R_a = 54$ kilohms:

$$A_{v(l\text{-}f)} = \frac{1.118 \times 10^6}{54 \times 10^3 - j54 \times 10^3}$$
$$= \frac{1.118 \times 10^6}{76.4\text{K} \underline{/-45°}}$$
$$= 14.6$$

(d) Find $A_{v(l\text{-}f)}$ when $X_{cc} = 10R_a = 540$ kilohms:

$$A_{v(l\text{-}f)} = \frac{1.118 \times 10^6}{54 \times 10^3 - j540\text{K}}$$
$$= \frac{1.118 \times 10^6}{543\text{K} \underline{/-84.3°}}$$
$$\cong 2.06$$

(e) Find $A_{v(l\text{-}f)}$ when $X_{cc} = 100R_a = 5,400$ kilohms:

$$A_{v(l\text{-}f)} = \frac{1.118 \times 10^6}{54 \times 10^3 - j5,400 \times 10^3}$$
$$= \frac{1.118 \times 10^6}{5,400\text{K} \underline{/-89.4°}}$$
$$\cong 0.207$$

The relationship of R_a/X_{cc} versus relative gain indicates the frequency response of any RC coupling network, where R of the network is R_a. As revealed in the preceding relationship, the gain is maximum when X_{cc} is one-tenth or less of R_a. A universal amplification curve illustrates the following:

1. Amplifier and RC coupling network gain versus R_a/X_{cc}
2. RC coupling network gain versus R_a/X_{cc}

When R_a is 10 or more times greater than X_{cc}, the RC coupling network voltage gain is unity, and the maximum amplifier gain is utilized with virtually no attenuation from the RC coupling network.

The relative voltage gain of the RC coupling network can be computed by

$$A_{v,RC} = \frac{A_{v(\text{l-f})}}{A_{v(\text{l-f,max})}}$$

where $A_{v(\text{l-f,max})}$ = maximum low-frequency voltage gain
$A_{v(\text{l-f})}$ = low-frequency voltage gain for the R_a/X_{cc} ratio under consideration

Example 2: Calculate the relative voltage gain of the RC coupling network in example 1.

Solution:

$$A_{v,RC} = \frac{20.7}{20.7} = 1 \qquad \text{when } R_a/X_{cc} = {}^{100}\!/_1$$

$$A_{v,RC} = \frac{20.7}{20.7} = 1 \qquad \text{when } R_a/X_{cc} = {}^{10}\!/_1$$

$$A_{v,RC} = \frac{14.6}{20.7} \cong 0.707 \qquad \text{when } R_a/X_{cc} = {}^{1}\!/_1$$

$$A_{v,RC} = \frac{2.06}{20.7} \cong 0.1 \qquad \text{when } R_a/X_{cc} = {}^{1}\!/_{10}$$

$$A_{v,RC} = \frac{0.207}{20.7} \cong 0.01 \qquad \text{when } R_a/X_{cc} = {}^{1}\!/_{100}$$

The universal amplification curve can now be plotted, as shown in Fig. 11.13.

Notice that $f_{co,\min}$ occurs where $R_a/X_{cc} = 1$; $A_{v,RC} = 0.707$; and $A_{v(\text{l-f})} = 0.707 A_{v(\text{l-f,max})}$. In most practical applications, this is the lowest acceptable frequency of the RC-coupled amplifier.

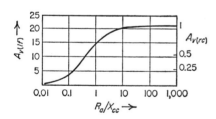

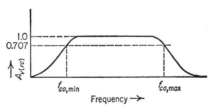

FIG. 11.13 Universal amplification curve.   FIG. 11.14 RC-coupled amplifier response curve.

It should be pointed out that the same amplification curve can be constructed by using the $A_{v(\text{h-f})}$ relationship in conjunction with the $X_{Cst/R_{eq}}$ ratio.

The over-all response curve of the RC-coupled amplifier is illustrated in Fig. 11.14. $f_{co,\min}$ occurs where $R_a/X_{cc} = 1$, the lower-frequency limit of the amplifier. $f_{co,\max}$ occurs where $R_{eq}/X_{Cst} = 1$, the upper-frequency limit of the amplifier. $f_{co,\min}$ is kept to the desired value by use of large values of R_g, R_{L1}, and C_c. $f_{co,\max}$, on the other hand, is greatly determined by the interelectrode capacitances of the first- and second-stage tubes. In the high-frequency equivalent circuit of Fig. 11.12, it is seen that

$$C_{st} = C_{pk1} + C_w + C_i$$

where C_{pk1} = plate-to-cathode capacitance of the first-stage tube
C_w = stray-wiring capacitance
C_i = input capacitance of the second-stage tube

The input capacitance of the second-stage tube is a function of C_{gk2}, C_{gp2}, and A_{v2}. Recall that C_{gp} creates positive feedback from the plate to the control grid, which can result in uncontrollable oscillations when $X_{C_{gp}}$ is small. At low frequencies, $X_{C_{gp}}$ is relatively high, and the regenerative effects of C_{gp} are not particularly troublesome. In amplifiers to be used in high-frequency work, the second-stage tube is generally a pentode because C_{gp} of the pentode is much smaller than found in triodes. C_{gp} is often 1 to 3 $\mu\mu$f for triodes, while values of 0.005 $\mu\mu$f and less are obtainable in pentodes.

The input capacitance of the second-stage tube, when its plate load is resistive, can be found by

$$C_i = C_{gp}(1 + A_{v2}) + C_{gk}$$

where A_{v2} is the voltage gain of stage 2.

Notice that the effect of C_{gp} is most pronounced because of the positive feedback property created by it. In order to extend the upper portion of the response curve (Fig. 11.14), the C_{pk} of the first tube and the C_i of the second tube must be kept to a minimum. Because of the relatively low C_{gp} values found in pentodes, high-frequency amplifiers generally incorporate pentodes.

Example 3: Refer to the two-stage RC-coupled amplifier of Fig. 11.6. Assume both tubes are the triode 6BC4: let $C_{pk} = 0.26$ $\mu\mu$f; $C_{gk} = 2.9$ $\mu\mu$f; $C_{gp} = 1.6$ $\mu\mu$f; $A_{v2} = 40$; $C_w = 2$ $\mu\mu$f. Find (a) C_i; (b) C_{st}.

Solution:

(a)
$$\begin{aligned} C_i &= C_{gp}(1 + A_{v2}) + C_{gk} \\ &= 1.6(1 + 40) + 2.9 \\ &= 68.5 \ \mu\mu\text{f} \end{aligned}$$

(b)
$$\begin{aligned} C_{st} &= C_{pk1} + C_w + C_i \\ &= 0.26 + 2.0 + 68.5 \\ &= 70.76 \ \mu\mu\text{f} \end{aligned}$$

Example 4: Refer to the two-stage RC-coupled amplifier of Fig. 11.6. Assume both tubes are the pentode 6AU6; let $C_{pk} = 5$ $\mu\mu$f; $C_{gk} = 5.5$ $\mu\mu$f; $C_{gp} = 0.0035$ $\mu\mu$f; $A_{v2} = 25$; $C_w = 2$ $\mu\mu$f. Find (a) C_i; (b) C_{st}.

Solution:

(a)
$$C_i = C_{gp}(1 + A_{v2}) + C_{gk}$$
$$= 0.0035(1 + 25) + 5.5 \ \mu\mu\text{f}$$
$$= 5.591 \ \mu\mu\text{f}$$

(b)
$$C_{st} = C_{pk1} + C_w + C_i$$
$$= 5 + 2 + 5.591$$
$$= 12.591 \ \mu\mu\text{f}$$

In comparing C_{st} of examples 3 and 4, it is seen that the two-stage RC-coupled amplifier shunt capacitance C_{st} is substantially reduced by use of pentodes. This, in turn, increases $f_{co,\max}$ of the amplifier.

11.4 Transformer Coupling. A third general coupling technique is by means of a transformer, as shown in Fig. 11.15, to which the reader can

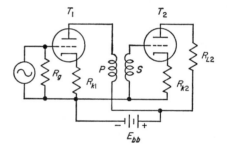

FIG. 11.15 Two-stage transformer-coupled amplifier.

refer for the following analysis. The primary coil contains both the unidirectional and bidirectional components of the plate current. The unidirectional voltage appearing across the primary is

$$E_{\text{pr}} = \frac{R_{\text{pr}}}{R_{\text{pr}} + r_p} E_{bb}$$

In many practical cases, the resistance of the primary R_{pr} is small, resulting in a very low d-c voltage drop across the primary. This is an advantage in that smaller values of E_{bb} can be used than would be the case with RC coupling. The bidirectional component of the plate current is of particular interest, since it contains the signal variations and is that portion of the plate current which generates the flux lines that cut across the secondary. The rms value of i_p may be determined by

$$I_p = \frac{\mu E_g}{[(r_p)^2 + (X_L)^2]^{1/2}}$$

where μ = amplification factor of T_1

$\quad E_g$ = rms input signal voltage of T_1

$\quad r_{p1}$ = plate resistance of T_1

$\quad X_L$ = inductive reactance of the primary

The rms value of e_{pr} may be found by

$$E_{\mathrm{pr}} = \mu E_g \frac{X_L}{[(r_{p1})^2 + (X_L)^2]^{1/2}}$$

And the voltage gain of stage 1 is

$$A_{v1} = \frac{\mu X_L}{[(r_{p1})^2 + (X_L)^2]^{1/2}}$$

Upon examination of the preceding equation for the determination of A_{v1}, it is seen that a large X_L is required for large voltage gains. Recalling the fundamentals of inductance, it is apparent that a large number of turns in the primary (large primary inductance) is necessary to fully utilize this possibility. The use of a primary with many turns results in an expensive, bulky transformer and tends to increase the capacitance of the primary. This capacitance becomes noticeable at higher frequencies when it is at a low value but could become troublesome at lower frequencies if it is large in value. Therefore the use of a primary inductance of large value is not practical or economical.

It would seem that the use of a large step-up ratio would result in an over-all amplifier gain that would be even greater than the amplification factor of the tube. In order to realize this possible gain, however, the primary inductance must be large. Because of the problems introduced, as mentioned in the preceding paragraph, a large primary inductance is not practical. Furthermore, the secondary would have to be even more bulky than the primary, since it would have a greater number of turns. The capacitive effects of the secondary would be even more pronounced than found in the primary, resulting in a reduced value of high-frequency cutoff.

Because of these drawbacks, a maximum step-up ratio of 1 to 3 is utilized. The major applications of transformer coupling are generally restricted to impedance matching, tuned-transformer coupling, and push-pull power work.

The over-all frequency response of a well-designed transformer is illustrated in Fig. 11.16. $f_{co,\mathrm{min}}$ occurs at that frequency where the transformer response is 70.7 per cent of its mid-frequency value. At reduced frequencies the primary inductive reactance decreases, creating this shunting effect at lower frequencies. The capacitance associated with the primary, and also that capacitance associated with the secondary, creates a shunting effect at higher frequencies. At some high frequency ($f_{co,max}$ of Fig. 11.16), the primary and secondary shunt capacitances cause the transformer output voltage to be 70.7 per cent of its mid-frequency value; this point is called $f_{co,max}$. Generally, better frequency response can be obtained with a carefully designed RC network when impedance matching is not a major problem. When the output impedance of stage 1 is considerably different from its ultimate load, then

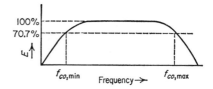

FIG. 11.16 Transformer frequency response.

transformer coupling may have a decided advantage over the other coupling techniques.

Recall that the impedance matching possibilities of the transformer are analyzed in Chap. 5. It was found that

$$R_L = (a)^2 R_s$$

where R_L = reflected primary resistance
a = transformer turns ratio
R_s = ultimate (secondary) load resistance

In designing a transformer-coupled amplifier, the signal load line is determined by the reflected primary resistance. Selection of the proper turns ratio enables the designer to match any two resistances. For further details, refer to the transformer coupling section of Chap. 5.

11.5 Design of a Class A Audio Amplifier. Figure 11.17 illustrates the two-stage class A audio amplifier that is designed in this section. Let us

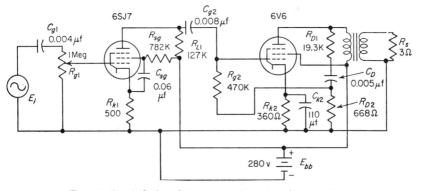

FIG. 11.17 A designed two-stage class A audio amplifier.

proceed to design this amplifier in a step-by-step fashion. The selection of tubes and power supply is determined by practical considerations and the intended use of the amplifier. Assume that an audio amplifier, whose ultimate load is a speaker with a 3-ohm voice coil, is the type to be designed. The 6SJ7 is selected as the first-stage tube and the 6V6 for the second stage. The amplifier is to have a volume control, which can be located in the grid circuit of stage 1. The maximum input signal is to be 0.10 volt, with a frequency range of 40 cps to 10 kc. E_{bb} is to be 280 volts.

Step 1: *Selection of* R_{L1}. A suitable value of R_{L1} may be determined by examination of the average plate characteristics of the 6SJ7 (see Fig. 11.18). Point A is one end of the load line, which is the selected

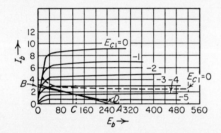

Fig. 11.18 Average plate characteristics of the 6SJ7. (*Characteristics from RCA*)

E_{bb} value of 280 volts. Using point A as a pivot, several load lines are drawn and examined. Load line AB is found to be suitable. The value of R_{L1} which coincides with this load line is determined by

$$R_{L1} = \frac{x \text{ intercept } (A)}{y \text{ intercept } (B)} = \frac{E_{bb}}{I_{b,sc}}$$

where $I_{b,sc}$ is the plate current when the tube is short-circuited and $R_{L1,eq}$ is in the circuit

Solving, $$R_{L1} = \frac{280}{2.8 \times 10^{-3}} = 100 \text{ kilohms}$$

Step 2: *Determination of Stage 1 Zero Signal Parameters.* Since this is to be a class A amplifier, the zero signal operating point should be such that:

1. E_{c1} never becomes positive.
2. I_b always flows.

The maximum input signal is 0.10 volt; therefore $E_{c1,s}$ of -4 volts will meet the requirements (see point C in Fig. 11.18). Reading from the y axis, $I_{b,s}$ is found to be about 1.8 ma; and $E_{b,s}$ is determined to be about 120 volts. I_{c2} must be estimated since the I_{c2} versus E_b characteristic with $E_{c1} = -4$ volts is not included. When $E_{c1} = 0$ volts, $I_{c2} \cong 2.8$ ma at the selected operating point (see Fig. 11.18). From the manufacturer's data, it is found that $I_{c2} \cong 0.8$ ma when $E_{c1} = -3$ volts. As a first approximation, we may interpolate:

When $E_{c1} = 0$, $I_{c2} \cong 2.8$ ma.

When $E_{c1} = -3$ volts, $I_{c2} \cong 0.8$ ma.

When $\Delta E_{c1} = 3$ volts, $\Delta I_{c2} \cong 2.0$ ma; therefore, when $\Delta E_{c1} = 4$ volts, $\Delta I_{c2} \cong 2.0/3 = 0.67$ ma per 1 volt of E_{c1}.

Since $I_{c2} \cong 0.8$ ma when $E_{c1} = -3$ volts, then $I_{c2} \cong (0.8 - 0.67)$ ma when $E_{c1} = -4$ volts; that is, $I_{c2,s} \cong 0.23$ ma.

NOTE: This is only a first approximation, which is sufficient for the initial selection of several components to be considered in the following steps.

Step 3: Calculation of R_{k1}. R_{k1} is the self-bias resistor for stage 1, and is determined by

$$R_{k1} = \frac{\text{desired zero signal bias}}{\text{total zero signal cathode current}}$$

$$= \frac{E_{c1,s}}{I_{b,s} + I_{c2,s}}$$

Substituting values and solving,

$$R_{k1} = \frac{1}{(1.8 + 0.23) \times 10^{-3}} \cong 500 \text{ ohms}$$

Step 4: Determination of R_{sg} and C_{sg}. The purpose of R_{sg} is to reduce the supply voltage felt by the screen to 100 volts, and its value may be determined by

$$R_{sg} = \frac{E_{bb} - E_{c2,s}}{I_{c2,s}}$$

Substituting values and solving,

$$R_{sg} = \frac{280 - 100}{0.23 \times 10^{-3}} \cong 782 \text{ kilohms}$$

C_{sg} is to place the screen grid at zero potential in terms of the signal variations. Therefore C_{sg} should offer a low impedance path to common for the signal variations. Since X_{Csg} is maximum at the lowest signal frequency, then

$$X_{Csg} \text{ at 40 cps} = 0.1 R_{sg} \cong 78 \text{ kilohms}$$

Solving,
$$C_{sg} = \frac{1}{6.28 \times 40 \times 78 \times 10^3}$$
$$\cong 0.06 \ \mu\text{f or larger}$$

NOTE: The reduced value of screen voltage may be obtained by use of a voltage divider across the filtered output of the power supply, when such an arrangement is available.

Step 5: *Selection of* R_{g1} *and* C_{g1}. R_{g1} is to serve the dual purpose of controlling volume and providing a d-c path between the control grid of stage 1 and common. In order to prevent the variable R_{g1} from shunting the input signal at the lower values of R_{g1}, the movable wiper is connected to the control grid, and the entire value of R_{g1} appears in series with C_{g1} at all times. The value of R_{g1} is not too critical: a 1-megohm or 500-kilohm potentiometer will serve well. If the 1-megohm potentiometer creates hum, the d-c resistance between the control grid and common is too high, and it should be replaced with a lower-value potentiometer, such as 500 kilohms.

The purpose of C_{g1} is to couple the signal to the control grid of the first stage and block any d-c component of the signal from the control grid. At $f_{co,min}$,

$$\frac{X_{cg1}}{R_a} = \frac{X_{cg1}}{R_i + R_{g1}} = 1$$

where R_i is the internal resistance of the generator.

The value of R_i hinges on the type of signal source. Let us assume it is such a small portion of 1 megohm (the value of R_{g1}) that $R_a = R_{g1}$; therefore, at $f_{co,min}$

$$\frac{X_{cg1}}{R_{g1}} \cong 1$$

i.e., $\qquad X_{cg1}$ at 40 cps = 1 megohm

Solving for C_{g1},

$$C_{g1} = \frac{1}{6.28 \times 1 \times 10^6 \times 40}$$
$$\cong 0.004 \ \mu\text{f or larger}$$

Step 6: *Determination of* R_{g2} *and* $R_{L1,eq}$. The purpose of R_{g2} is to provide a d-c path from the control grid of the second stage to common.

R_{L1} was selected to be 100 kilohms in step 1. Let $R_{g2} = 470$ kilohms, which is a typical value for this resistor. Solving for $R_{L1,eq}$,

$$R_{L1,eq} = \frac{1}{1/(100 \times 10^3) + 1/(470 \times 10^3)}$$
$$\cong 83.3 \text{ kilohms}$$

The signal load line is then determined (this technique is described in Sec. 5.16).

$$\tan \alpha = \frac{1 \times 10^3}{R_{L1,eq}} = \frac{1}{83.3}$$
$$\cong 0.012$$

The distance between points C and D in Fig. 11.18 is found by

$$D - C = \frac{I_{b,s} \times 10^3}{\tan \alpha} = \frac{1.8}{0.012} \cong 150 \text{ volts}$$

The x-axis intercept of the signal load line of T_1 is next determined:

$$x\text{-axis intercept} = E_{b,s} + (D\text{-}C) = 120 + 150 = 270 \text{ volts}$$

The signal load line is now constructed (see Fig. 11.18). It should be noticed that the signal load line is not too far removed from the d-c load line because of the small loading effect that R_{g2} imposes upon the output of T_1.

Step 7: Determination of C_{g2}

At $f_{co,\min}$

$$\frac{X_{C_{g2}}}{R_a} = 1$$

where

$$R_a = \frac{r_{p1}R_{L1}}{r_{p1} + R_{L1}} + R_{g2}$$

r_{p1} is determined by

$$r_{p1} = \frac{\Delta E_b}{\Delta I_b} \qquad \text{with } E_c = k$$

Let $\Delta E_b = 80$ to 160 volts $= 80$ volts; from Fig. 11.18, with E_c fixed at -1 volt, $\Delta I_b = 6.7$ to 6.8 ma $= 0.1$ ma. Solving for r_{p1},

$$r_{p1} = \frac{80}{1 \times 10^{-4}} = 800 \text{ kilohms}$$

R_a can now be calculated,

$$R_a = \frac{8 \times 10^5 \times 100 \times 10^3}{8 \times 10^5 + 100 \times 10^3} + 4.7 \times 10^5$$
$$= 558 \text{ kilohms}$$

Therefore $X_{C_{g2}}$ at 40 cps $= 558$ kilohms.
Solving for C_{g2},

$$C_{g2} = \frac{1}{6.28 \times 5.6 \times 10^5 \times 4 \times 10^1}$$
$$\cong 0.007 \ \mu f \text{ or larger}$$

Step 8: Construction of the V_2 Load Line. Refer to Fig. 11.19. A load line consistent with the selected E_{bb} and maximum tube values is constructed in the conventional manner. AB in Fig. 11.19 is such a load line, and

$$R_L = \frac{x\text{-axis intercept}}{y\text{-axis intercept}} = \frac{280}{140 \times 10^{-3}}$$
$$= 2 \text{ kilohms}$$

Step 9: *Calculation of the Output Transformer Turns Ratio.* The ultimate load is the 3-ohm voice coil and is to be "seen" as 2 kilohms by the plate of the second-stage tube.

$$R_L = (a)^2 R_s$$

and $$a = \left(\frac{R_L}{R_s}\right)^{\frac{1}{2}}$$

Solving for the required turns ratio,

$$a = \left(\frac{2{,}000}{3}\right)^{\frac{1}{2}} = \frac{25.8}{1}$$

Step 10: *Determination of Zero Signal Parameters of* V_2. The maximum input signal to the second stage occurs when the input signal to the first

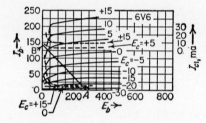

FIG. 11.19   Plate characteristics of the 6V6. (*Characteristics from RCA*)

stage is maximum (0.1 volt). By use of the load line of stage 1 (Fig. 11.18), the maximum output signal of stage 1 can be determined.

The 0.1 input volt excursion moves the operating point from C to maximum, back to C, to minimum, and back to C to complete one cycle. The E_b values at these points are found in Fig. 11.18 to be:

Maximum: $E_b = 107$ volts
Point C: $E_b = 120$ volts
Minimum: $E_b = 133$ volts

Therefore $\Delta E_b(\text{pp}) = 133 - 107 = 26$ volts and $\Delta E_{b,\max} = 13$ volts.

In the middle frequency range, $X_{C_{g2}}$ is negligible, and the maximum input signal to stage 2 is 13 volts. To ensure class A operation, the grid bias of stage 2 should be at least 13 volts and yet not so large as to drive the tube to cutoff when the signal swings negative.

We next turn to the plate characteristics and load line of the second-stage tube (Fig. 11.19) to select a suitable operating point. Point O will meet the predetermined requirements, where $E_{c1,s} = -15$ volts. From the load line $I_{b,s}$ is found to be about 37 ma, and from the manufacturer's data, the value of $I_{c2,s}$ is found to be about 4.5 ma.

Step 11: *Calculation of R_{k2} and C_{k2}.* The required -15 grid-bias potential is to be obtained by the cathode-bias method, i.e.

$$R_{k2} = \frac{E_{c1,s}}{I_{b,s} + I_{c2,s}}$$

Substituting values and solving,

$$R_{k2} = \frac{15}{(37 + 4.5) \times 10^{-3}} \cong 360 \text{ ohms}$$

NOTE: Since the values of $I_{b,s}$ and $I_{c2,s}$ used in the preceding relationship are approximate, the exact value of R_{k2} is determined upon construction of the circuit. R_{k2} is varied until $E_{Rk2} = 15$ volts.

C_{k2} is to serve as a bypass capacitor for R_{k2}, in order to reduce its degenerative effects. Based on the previously used rule of thumb,

$$X_{Ck2} \text{ at } f_{co,\min} = 0.1R_{k2}$$

which is $\qquad X_{Ck2} = 36$ ohms at 40 cps

Solving for C_{k2},

$$C_{k2} = \frac{1}{6.28 \times 36 \times 40} \cong 110 \ \mu\text{f or larger}$$

Step 12: *Determination of A_v of the Amplifier at Various Frequencies without Second-stage Degeneration.* It is seen that the primary of the output transformer is connected between the plate of the second-stage tube and $+E_{bb}$. Since X_L varies directly with frequency, the primary voltage tends to be larger at higher frequencies. This results in a non-uniform output in terms of frequency and is called *frequency distortion.* A logical approach toward the reduction of frequency distortion is to incorporate a negative feedback network which varies in step with the gain of the amplifier; i.e., when the voltage gain of the amplifier increases, the negative feedback is increased, thereby reducing the gain to a predetermined level. The first step in the design of such a degenerative network is to determine the voltage gain of the amplifier at a frequency where A_v is at a normal value and at a frequency where A_v is at or near its maximum value.

The frequency response curve of a typical audio output transformer indicates that A_v tends to rise with frequency increases. The over-all response of the amplifier is affected by this action, and it should be properly evaluated if good amplifier performance is to be expected.

Following is a simple procedure for determining the frequency response of the amplifier without provisions for second-stage degeneration:

1. Construct the designed second stage, and connect the ultimate load. Connect R_{g2} and one end of the ultimate load resistance to common for this test.

2. Apply an input signal at the allowed maximum value determined by the previous design work. Connect a good voltmeter (such as a VTVM) from the control grid to common of the second stage. Connect an oscilloscope across the ultimate load. See Fig. 11.20.

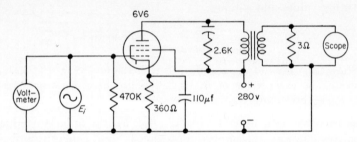

FIG. 11.20   Test circuit for determination of frequency response for stage 2.

3. Calibrate the oscilloscope at some convenient vertical deflection with the input signal at the designed maximum value to the second-stage control grid at a frequency of 1 kc. This deflection will be the reference gain of the second stage for the various frequencies.

4. Determine the relative gain for a number of frequencies throughout the signal range of the amplifier. Figure 11.20 illustrates the test circuit.

With each new input frequency, see that the input signal voltage remains at its predetermined maximum value (check with the voltmeter in Fig. 11.20). Record the vertical deflection of the oscilloscope (as the number of squares) with each change in frequency. Do not adjust the vertical gain of the oscilloscope after the reference setting.

Table 11.1 lists the relative gains of the designed amplifier without provisions for negative feedback.

Table 11.1   Relative Gain of Stage 2 without Degeneration

f	A_v (relative)	f	A_v (relative)
50	14	6 kc	21.3
100	17	8 kc	22
500	19.5	10 kc	22.5
1 kc	20	12 kc	24
2 kc	20.5	14 kc	24.5
3 kc	20.8	61 kc	26
4 kc	21	17 kc	27
5 kc	21		

In order to eliminate frequency distortion, the voltage gain of the amplifier must be made constant for the signal frequency range. One method to approach this ideal condition is to provide a degenerative network that will reduce A_v toward a predetermined value, which is the lowest A_v found by running the response curve of the second stage. With the particular components used in the design of this circuit, the relative A_v was found to be a minimum of 14 at 50 cps. The response curve of the amplifier without provisions for negative feedback is shown in Fig. 11.21

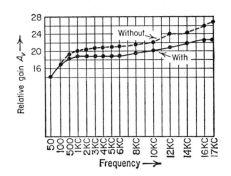

FIG. 11.21 Second-stage frequency response with and without degeneration.

(dotted line). A_v was found to be at a maximum of 27 at 17 kc. Therefore maximum degeneration should be present at 17 kc.

Step 13: *Determination of the Second-stage Degenerative Network.* The second-stage degenerative network is to be comprised of R_{D1}, C_D, and R_{D2}. The purpose of this network is to maintain the voltage gain down to as near a constant value for the frequency range of the amplifier as possible. Recall that

$$A_{v,o} = \frac{A_v}{1 - \beta A_v}$$

where in this application,

$A_{v,o}$ = voltage gain with feedback (which is $A_{v,\min}$ from the response curve)

A_v = voltage gain without feedback (which is $A_{v,\max}$ from the response curve)

= desired amount of feedback

Transposing, we may find β,

$$-\beta = \frac{A_v - A_{v,o}}{(A_v)(A_{v,o})}$$

Using the values found in the response curve of Fig. 11.20,

$$-\beta = \frac{27 - 14}{27(14)} = -0.0344$$

Let us now examine the proposed degenerative network in greater detail; it is redrawn in Fig. 11.22. R_{D1}, C_D, and R_{D2} form a voltage divider from the second-stage plate to common. In terms of the signal,

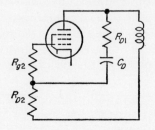

Fig. 11.22 The second-stage degenerative network.

they are in parallel with the primary of the output transformer. In order to avoid loading down the primary to a great extent, let

$$R_{D1} + R_{D2} = 10R_L$$

The fraction of the output voltage across R_{D2} is returned to the second-stage control grid. Since this feedback signal is the inverse phase (180°) of the input signal, the composite input signal is reduced by the magnitude of the feedback voltage.

When X_{CD} is considered negligible, the portion of the output voltage that appears across R_{D2} is

$$E_{\text{feedback}} = \frac{R_{D2}}{R_{D1} + R_{D2}} E_o$$

This portion of the output voltage is to be beta, determined in the preceding paragraph,

$$-\beta = \frac{R_{D2}}{R_{D1} + R_{D2}}$$

therefore, $$R_{D2} = -\beta(R_{D1} + R_{D2})$$

and $$R_{D1} = \frac{R_{D2}(1 - \beta)}{\beta}$$

and $$R_{D1} + R_{D2} = 10R_L = 20 \text{ kilohms}$$

Using the values of R_L and β found earlier,

$$R_{D2} = \beta(R_{D1} + R_{D2}) = 0.0344(20 \times 10^3)$$
$$= 688 \text{ ohms}$$

and $$R_{D1} = \frac{R_{D2}(1 - \beta)}{\beta} = \frac{688(1 - 0.0344)}{0.0344}$$
$$\cong 19.3 \text{ kilohms}$$

In order to ensure that beta be equal to the desired value when A_v is maximum, X_{CD} should be $0.1(R_{D1} + R_{D2})$ at the frequency when $A_{v,\max}$ occurs.

$$X_{CD} \text{ at } 17 \text{ kc} = 0.1(R_{D1} + R_{D2})$$
$$X_{CD} = 2 \text{ kilohms}$$

Solving for C_D,

$$C_D = \frac{1}{6.28 \times 2 \times 10^3 \times 17 \times 10^3}$$
$$\cong 0.005 \ \mu f$$

Placing the computed values of R_{D1}, R_{D2}, and C_D into the test circuit affects the relative gain as shown in Table 11.2.

Table 11.2 f versus $A_{v,o}$ (with $-\beta = 0.0344$)

f	$A_{v,o}$ (with $\beta = -0.0344$)	f	$A_{v,o}$ (with $\beta = -0.0344$)
50	14	6 kc	19
100	17	8 kc	20
500	19	10 kc	20.5
1 kc	19	12 kc	21
2 kc	19	14 kc	22
3 kc	19	16 kc	23
4 kc	19	17 kc	23
5 kc	19		

Table 11.2 shows the effect of the degenerative circuit in flattening the response of the amplifier. The relative $A_{v,o}$ versus frequency response is shown in Fig. 11.21 (solid line). Notice that the relative gain is virtually unaffected at the lower frequencies but reduced substantially at the higher frequencies, resulting in a more uniform gain over the entire frequency range.

At those frequencies below that frequency where A_v is maximum, X_{CD} is larger and must be considered in the degenerative. voltage-divider network.

$$\beta = \frac{R_{D2}}{R_{D2} + (R_{D1} - jX_{CD})}$$

As the frequency is reduced, X_{CD} becomes larger and beta becomes smaller. In other words, the negative feedback at lower frequencies is reduced because of two factors:

1. Since A_v is less at lower frequencies, the amount of feedback is reduced.

2. Beta is reduced at lower frequencies because of the increase of X_{CD} in the voltage-divider arrangement in accordance with the relationship stated in the preceding paragraph.

In observing the movement of the operating point on the load line of Fig. 11.19, it is apparent that the voltage swing below rest is not equal in amplitude to the voltage swing above rest, resulting in *harmonic distortion*.

Refer to Fig. 11.23. Because of the unequal distances between the characteristics, ΔE_b without feedback would appear as shown in diagram A when an input voltage like that of diagram B is applied. Notice that the output voltage is largest during the second half cycle. At this time, E_{RD2} is larger, reducing ΔE_{c1} to that shown in diagram C, which caused ΔE_b to resemble diagram D. Therefore the degenerative network also effectively reduces harmonic distortion as well, since the amount of feedback varies directly with the output voltage.

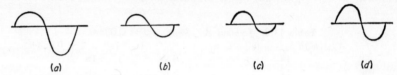

(a)   (b)   (c)   (d)

Fig. 11.23   Harmonic distortion reduction by negative feedback: (a) ΔE_b without feedback, (b) ΔE_{c1} without feedback, (c) ΔE_{c1} with feedback, (d) ΔE_b with feedback.

Step 14: Construction of the Amplifier. After construction of the amplifier, the frequency response should be checked. The output waveform can be examined over the frequency range with an oscilloscope, and the voltage gain as a function of frequency can be checked with the oscilloscope or voltmeter. An improvement of the output waveform and more constant A_v values for the frequency range can be brought about by increasing the negative feedback (increasing R_{D2} or decreasing R_{D1} or C_D). The improved response is accompanied by a reduction in A_v, however. A compromise value is to be determined by the designer.

11.6 Special Amplifier Filter Circuits. There are several corrective filters which can be incorporated in audio amplifiers which reduce distortion or improve the frequency response of the amplifier. The degenerative network utilized in the output stage of the preceding amplifier was used for the purpose of reducing frequency distortion by regulating the negative feedback in direct proportion to the voltage gain. Several other networks which improve the response of the amplifier are examined in this section.

Constant Load-impedance Filter for a Power Output Stage. The preceding section pointed out that the plate load impedance of a transformer-coupled amplifier tends to increase with frequency because of the inductance of the transformer primary. This effect is most pronounced at the higher portion of the amplifier's frequency range. In cases where the primary inductance is relatively large, such as the output transformer

of a power stage, the resultant frequency distortion is sufficiently severe
to warrant the use of a special network for its reduction.   The degenerative network designed in the preceding section is one such circuit.   Negative feedback is not desirable in some applications, however, in which case
the network illustrated in Fig. 11.24 may be incorporated.

R_L in Fig. 11.24b designates the reflected primary impedance of the
amplifier.   Because of the inductance value of the primary, the reflected

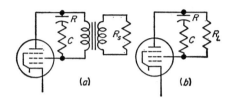

FIG. 11.24   Constant load impedance
filter.

primary impedance increases with frequency.   The series-RC circuit
connected across the primary possesses an impedance which decreases
with frequency increases.   It is desirable to select a value of R and C
such that the impedance of this leg decreases at about the same rate at
which R_L increases.   By the correct selection of R and C, the load
impedance of the tube can be maintained at a relatively constant value.
At the same time, R and C must be of such values that the plate load
impedance is not reduced below the recommended R_L value determined
by the load-line analysis.

A common rule of thumb is to select a value of R which is equal to
$1.3R_L$.   The value of C can be quickly determined on an empirical basis
with the circuit set up as shown in Fig. 11.25.

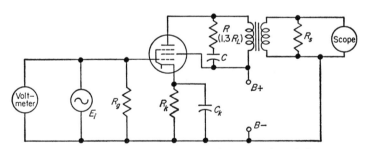

FIG. 11.25   Test circuit for the determination of C.

Refer to Fig. 11.25.   As a first approximation in audio work, select C
so that the time constant $(R \times C)$ is about $6.5 \times 10^{-8}R_L$ seconds; i.e.,

$$tc = 1.3R_L C = 6.5 \times 10^{-8}R_L \text{ sec}$$

and
$$C = \frac{tc}{1.3R_L} = \frac{6.5 \times 10^{-8}R_L}{1.3R_L} = 0.05 \ \mu f$$

After the construction of the test circuit as shown in Fig. 11.25, apply an input signal of 500 cps, and record the relative output (with the scope or some other suitable means). Using the same magnitude of input voltage, apply an input signal of 1 kc and record the output voltage again. If the selected C is of the correct value, the output voltage will be the same magnitude for both frequencies. A 0.05 μf is generally quite close to the correct value for the power stage of audio amplifiers.

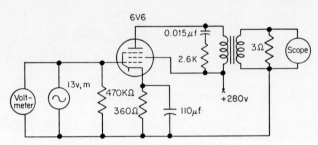

FIG. 11.26 Test circuit for determination of R and C for second stage of designed amplifier.

Using the power stage designed in the preceding section, the circuit was constructed without the degenerative network, and the values are shown in Fig. 11.26.

Since $R_L = 2$ kilohms, then

$$R = 1.3 \times 2 \times 10^3 = 2.6 \text{ kilohms}$$

and, as a first approximation,

$$C = 0.05 \ \mu\text{f}$$

In the particular circuit under consideration,

When $E_i = 10$ volts at 500 cps: $E_o = 16$ squares
When $E_i = 10$ volts at 1 kc: $E_o = 22$ squares

Since the output is not constant at these two frequencies, another value of capacitance should be considered. Reducing the value of C results in a larger X_C value, increasing the impedance of the series-RC circuit, which is in parallel with R_L. This increases E_o at the lower frequencies. It was found that E_o is nearly identical at 500 cps and 1 kc with $C = 0.015 \ \mu$f for this circuit. Table 11.3 lists the gain for various frequencies for the test circuit of Fig. 11.26.

Notice that the relative gain increased from 50 to about 500 cps. In this region, the impedance of the RC branch is relatively high and does not shunt R_L sufficiently to completely offset the increase of X_L with frequency.   At frequencies above 500 cps, X_C reduces the impedance of the RC branch to the extent that the increases of X_L are offset.   This results in the relative gain leveling off at the higher frequencies.

Table 11.3   Relative Gain of Fig. 11.26 ($c = 0.015$ μf)

f	$A_{v,o}$ (relative)	f	$A_{v,o}$ (relative)
50	13	6 kc	14.5
100	17.5	8 kc	13.5
500	21	10 kc	13
1 kc	20.5	12 kc	13
2 kc	18.5	14 kc	12.5
3 kc	17	16 kc	12.5
4 kc	16	17 kc	12.5
5 kc	15.5		

A comparison of the frequency response curve with and without the constant output impedance filter is shown in Fig. 11.27.   This method of dealing with frequency distortion is recommended only when the degenerative network of the preceding stage cannot be utilized.

Tone-control Circuits.   The purpose of a tone-control circuit is to provide a means for adjusting the low- and/or high-frequency response of an amplifier.   Many tone-control circuits, both simple and elaborate,

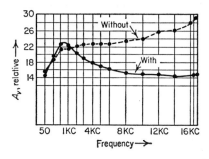

FIG. 11.27   Comparison of f versus A_v with and without constant output impedance filter.

have been designed to fulfill this objective to varying degrees.   A simple tone-control circuit is shown in Fig. 11.28.

C_T is selected so that it offers negligible reactance to the high frequencies. R_2 is a potentiometer whose full resistance should be at least $10R_L$ so as to minimize its shunting effect upon the output load impedance. Since X_{CT} is negligible at high frequencies, the extent to which the high frequencies are bypassed from the output is determined by the setting of R_2.   When R_2 is at position B, virtually no high frequencies are bypassed, and this can be called the *treble* setting of R_2.   When R_2 is at position A, a substantial portion of the high frequencies are bypassed and may be termed *bass*.   The purpose of R_1 is to maintain a small resistance in series with C_T at position A, thereby protecting R_2 from overheating.   To

Table 11.4   Frequency Response for Treble and Bass Positions of Fig. 11.28

f	A_v (treble)	A_v (bass)
50	7.5	7.5
100	10.5	10.5
500	16	11.5
1 kc	20	12.5
2 kc	23	14
3 kc	24	15
4 kc	24.5	15.5
5 kc	24.5	15.5
6 kc	25	16
8 kc	25.5	15.5
10 kc	26.5	15
12 kc	27.5	15
14 kc	29.5	14.5
16 kc	32.5	14.5
17 kc	33	14.5

enable R_2 to be most effective, R_1 should be no greater than 10 per cent the value of R_2.

The frequency response of the designed tone-control circuit of Fig. 11.28 with R_2 at positions A and B is shown in Table 11.4 and Fig. 11.29.

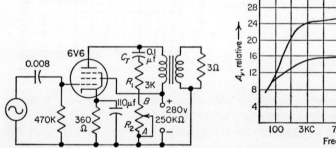

FIG. 11.28 A designed tone-control circuit.

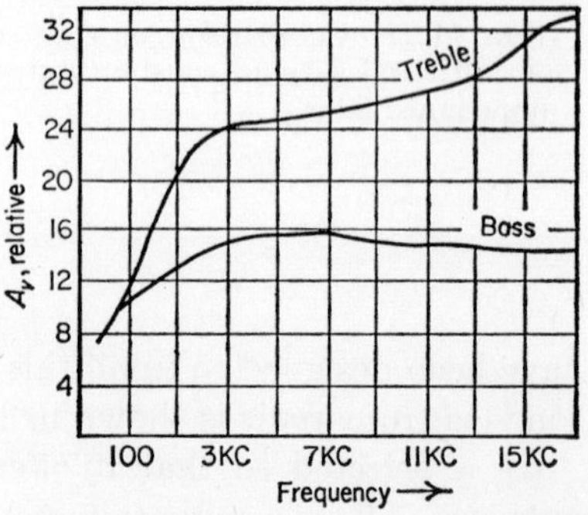

FIG. 11.29 Frequency response of the designed tone-control circuit at bass and treble positions.

Upon examination of the frequency response illustrated in Fig. 11.29, it is seen that the bass position actually reduces the over-all gain of the amplifier. In this way, the low-frequency gains are *relatively higher*, since the high-frequency gains are reduced. Because of the manner in which the low frequencies are accentuated, this type of circuit can be called a *false bass*.

PROBLEMS

11.1 A triode is to have the following zero signal parameters: $E_{c,s} = -2$ volts; $I_{b,s} = 4$ ma. Calculate R_k.

11.2 Assume the triode of Prob. 11.1 is to have the following zero signal parameters: $E_{c,s} = -4$ volts; $I_{b,s} = 3$ ma. Calculate R_k.

11.3 A pentode is to have the following zero signal parameters: $E_{c,s} = -5$ volts; $I_{b,s} = 10$ ma; $I_{c2,s} = 2$ ma. Calculate R_k.

11.4 A pentode is to have the following zero signal parameters: $E_{c,s} = -10$ volts; $I_{b,s} = 40$ ma; $I_{c2,s} = 6$ ma. Calculate R_k.

11.5 Calculate the required wattage values of R_k in Probs. 11.1 to 11.4.

11.6 Assume the frequency range of the triode in Probs. 11.1 and 11.2 is 50 cps to 20 kc. Find the minimum C_k to reduce the degeneration of R_k to a negligible value in Probs. 11.1 and 11.2.

11.7 The frequency range of the pentodes in Probs. 11.3 and 11.4 is 20 cps to 15 kc. Find the minimum C_k to reduce the degeneration of R_k to a negligible value in Probs. 11.3 and 11.4.

11.8 Assume a triode class A amplifier has the following parameters: $R_g = 10$ megohms; $E_{c1,s} = -1.0$ volts; contact bias is utilized. Calculate the grid contact current.

11.9 A fixed bias resistor R_B is to be inserted in a power supply. Load $I = 100$ ma; bleeder $I = 12$ ma; required bias $= 12$ volts. Calculate R_B.

11.10 The fixed-bias method is to be used in an amplifier. Load $I = 80$ ma; bleeder $I = 10$ ma; $E_{c1} = -4$ volts, and E_{c1} for a second tube $= -15$ volts. Calculate (a) R_B, total; (b) R_{B1}; (c) R_{B2}.

11.11 Refer to the direct coupling circuit of Fig. 11.5. $E_{b1,s} = 240$ volts; $E_{cT2,s} = -6$ volts; $R_{g1} = R_{g2}$. Find (a) E_{cc}; (b) e_{Rg2}.

11.12 Refer to the direct coupling circuit of Fig. 11.5. $E_{b1,s} = 150$ volts; $E_{cT2,s} = -2$ volts; $R_{g1} = R_{g2}$. Find (a) E_{cc}; (b) e_{Rg2}.

11.13 Refer to the low-frequency equivalent circuit of Fig. 11.10. $\mu = 30$; $e_g = 2$ volts; $r_p = 100$ kilohms; $R_{L1} = 200$ kilohms; $R_g = 500$ kilohms; $C_c = 0.05$ μf. Calculate:

(a) i_{open} (b) $e_{RL1,open}$
(c) $R_{o1,eq}$ (d) R_a
(e) e_o (f) $A_{v(1-f)}$
(g) $f_{co,min}$

11.14 Refer to the low-frequency equivalent circuit of Fig. 11.10. $\mu = 100$; $e_g = 1.5$ volts; $r_p = 80$ kilohms; $R_{L1} = 250$ kilohms; $C_c = 0.02$ μf; $R_g = 470$ kilohms. Calculate:

(a) i_{open} (b) $e_{RL1,open}$
(c) $R_{o1,eq}$ (d) R_a
(e) e_o (f) $A_{v(1-f)}$
(g) $f_{co,min}$

11.15 Refer to Fig. 11.11 for the equivalent middle-frequency circuit. $r_p = 70$ kilohms; $R_{L1} = 150$ kilohms; $R_g = 470$ kilohms; $g_m = 1,000$ μmho; $e_g = 1.5$ volts. Calculate (a) R_{eq}; (b) e_o; (c) $A_{v(m-f)}$.

11.16 Refer to Fig. 11.11 for the equivalent middle-frequency circuit. $r_p = 7$ kilohms; $R_{L1} = 30$ kilohms; $R_g = 270$ kilohms; $g_m = 3,000$ μmho; $e_g = 3$ volts. Calculate (a) R_{eq}; (b) e_o; (c) $A_{v(m-f)}$.

11.17 Refer to the high-frequency equivalent circuit of Fig. 11.12. $C_{pk1} = 4$ $\mu\mu$f; $C_w = 2$ $\mu\mu$f; $C_{gp2} = 0.3$ $\mu\mu$f; $C_{gk3} = 10$ $\mu\mu$f; $A_{v2} = 20$; $r_{p1} = 7.5$ kilohms; $R_{L1} = 25$ kilohms; $R_{g2} = 250$ kilohms; $e_g = 0.25$ volt. Calculate:
(a) C_i (b) C_{st}
(c) R_{eq} (d) i_{eq}
(e) e_o (f) $A_{v(\text{h-f})}$
(g) $f_{co,\max}$

11.18 Refer to the high-frequency equivalent circuit of Fig. 11.12. $C_{pk1} = 2.8$ $\mu\mu$f; $C_w = 2$ $\mu\mu$f; $C_{gp2} = 0.34$ $\mu\mu$f; $C_{gk2} = 8.5$ $\mu\mu$f; $A_{v2} = 32$; $r_{p1} = 45$ kilohms; $R_{L1} = 150$ kilohms; $R_{g2} = 470$ kilohms; $e_g = 0.5$ volt. Calculate:
(a) C_i (b) C_{st}
(c) R_{eq} (d) i_{eq}
(e) e_o (f) $A_{v(\text{h-f})}$
(g) $f_{co,\max}$

11.19 Refer to the low-frequency equivalent circuit of Fig. 11.10a; $R_{L1} = 83$ kilohms; $R_{g2} = 470$ kilohms; $r_{p1} = 800$ kilohms; $g_{m1} = 1,600$ μmho; $\mu = 400$; $e_g = 0.1$ volt; $C_c = 0.008$ μf. Find X_{cc} at the following frequencies:
(a) 20 cps   (b) 50 cps
(c) 100 cps   (d) 500 cps
(e) 1 kc   (f) 5 kc
(g) 10 kc   (h) 15 kc

11.20 In Prob. 11.19, calculate R_a.

11.21 In the amplifier of Probs. 11.19 and 11.20, calculate $A_{v(\text{l-f})}$ at the stated frequencies.

11.22 Plot the frequency response curve (f versus $A_{v(\text{l-f})}$) for the amplifier of Probs. 11.19 to 11.21.

11.23 Refer to the low-frequency equivalent circuit of Fig. 11.10a. $R_{L1} = 250$ kilohms; $R_{g2} = 470$ kilohms; $r_{p1} = 75$ kilohms; $g_{m1} = 1,100$ μmho; $\mu = 100$; $e_g = 0.5$ volt; $C_c = 0.004$ $\mu\mu$f. Find X_{cc} at the following frequencies:
(a) 20 cps   (b) 50 cps
(c) 100 cps   (d) 500 cps
(e) 1 kc   (f) 5 kc
(g) 10 kc   (h) 15 kc

11.24 In Prob. 11.23, calculate R_a.

11.25 In the amplifier of Probs. 11.23 and 11.24, calculate $A_{v(\text{l-f})}$ at the stated frequencies.

11.26 Plot the frequency response curve (f versus $A_{v(\text{l-f})}$) for the amplifier of Probs. 11.23 to 11.25.

11.27 In the amplifier of Probs. 11.19 to 11.22, calculate the following at the stated frequencies:
(a) $A_{v,RC}$
(b) $A_{v,RC}$ in decibels

11.28 In the amplifier of Probs. 11.23 to 11.26, calculate the following at the stated frequencies:
(a) $A_{v,RC}$
(b) $A_{v,RC}$ in decibels

11.29 Refer to the transformer-coupled amplifier of Fig. 11.15. $R_{pr} = 50$ ohms; $r_{p1} = 80$ kilohms; $E_{bb} = 200$ volts; $\mu = 100$; $L_p = 0.4$ henry; $E_g = 0.5$ volt. Calculate E_{pr}.

11.30 In the transformer-coupled amplifier of Prob. 11.29, calculate I_p for the following frequencies:
(a) 20 cps   (b) 100 cps
(c) 1 kc   (d) 10 kc

11.31 Why is I_p a different value for each frequency in Prob. 11.30?

11.32 In the transformer-coupled amplifier of Probs. 11.29 and 11.30, calculate A_{v1} for the stated frequencies.

11.33 Refer to Prob. 11.32. Plot the frequency response curve (f versus A_{v1}).

11.34 Assume a transformer-coupled amplifier is to have the following parameters: $R_{L1} = 50$ kilohms; $R_{g2} = 700$ kilohms. Calculate the required turns ratio for properly matching the two impedances.

11.35 What determines the signal load line for an RC-coupled amplifier?

11.36 Assume an RC-coupled amplifier has the following parameters: $r_{p1} = 1$ megohm; $R_{L1} = 100$ kilohms; $R_{g2} = 500$ kilohms; $f_{co,min} = 40$ cps. Calculate (a) $R_{o1,eq}$; (b) R_a; (c) C_c.

11.37 Assume a pentode amplifier has the following parameters: $E_{bb} = 300$ volts; $E_{c2,s} = 100$ volts; signal frequency range is 50 cps to 20 kc. Find (a) R_{sg}; (b) C_{sg}.

11.38 Assume an amplifier has the following parameters: $A_{v,max} = 26$; $R_L = 5$ kilohms; $A_{v,min} = 14$; frequency range is 40 cps to 20 kc; the degenerative network of Fig. 11.22 is to be designed. Find:

(a) beta   (b) R_{D2}

(c) R_{D1} (d) C_D

11.39 Assume the amplifier of Prob. 11.38 is to have the same frequency range and the following parameters: $R_L = 10$ kilohms; $A_{v,min} = 16$; $A_{v,max} = 35$. The degenerative network of Fig. 11.22 is to be designed. Find:

(a) beta   (b) R_{D2}

(c) R_{D1} (d) C_D

11.40 Assume the constant load impedance filter of Fig. 11.24 is to be utilized. $R_L = 4$ kilohms. Find the following as a first approximation:

(a) R

(b) C

11.7 The Class AB Push-Pull Amplifier.

The basic concepts of the class A push-pull amplifier are analyzed in Chap. 6. These principles are applicable to vacuum-tube circuitry as well. The load-line analysis of the class AB amplifier is given in Sec. 7.1 and illustrated in Fig. 7.1. The advantages of class AB amplifiers (vacuum tube) are fundamentally the same.

As in transistor circuitry, class AB push-pull amplifiers are more commonly used than class B in vacuum-tube work. Larger values of grid bias are required, since the tube is to be at cutoff for a portion of the input signal cycle. The plate current is restricted to lower magnitudes because of the larger grid-bias values utilized, permitting the use of high plate and screen voltages. The power output is made greater because of the larger plate and screen potentials.

Class AB amplifiers may be divided into two types:

1. AB_1
2. AB_2

The class AB_1 amplifier is biased such that the control grid is always negative, and no appreciable grid current flows. In the class AB_2 amplifier, the grid is permitted to become positive for a portion of the input signal peak value, resulting in the flow of grid current. Therefore the class AB_2 amplifier requires a larger driving power than the class AB_1.

The push-pull amplifier of Fig. 11.30 may be operated in class A or class AB, which is determined by the zero signal potential developed by R_k. Notice that R_k serves as the self-bias resistor for both tubes. In some cases, an individual R_k and C_k may be provided for each push-pull tube.

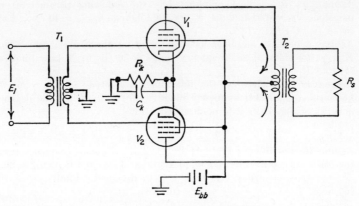

FIG. 11.30 A push-pull amplifier.

Recall that one of the chief purposes of T_1 (the input transformer) is to provide V_1 and V_2 with signals of equal magnitude but exactly opposite in polarity. This is achieved by use of the center-tap secondary. T_1 serves as an input signal phase inverter. A later section examines the vacuum-tube phase inverter, which may be more desirable in some instances for economy and space reasons.

T_2 is the output transformer and matches R_s to the output of the push-pull stage. The over-all action of T_2 is identical to that discussed in conjunction with transistor push-pull amplifiers in Chaps. 6 and 7.

Although the circuit of Fig. 11.30 illustrates pentode-type tubes, triodes can also be used. The beam-power tube (such as the 6V6) has found great popularity in this type of application.

Circuit efficiency, like that of the class AB transistor circuit, is between 50 per cent (class A) and 78 per cent (class B). The plate-to-plate load resistance is determined by

$$R_{LL} = \frac{4E_{bb}}{I_{\max,y}}$$

where R_{LL} = plate-to-plate load resistance

E_{bb} = supply voltage

$I_{\max,y}$ = maximum plate current at the intersection of the load line and the y axis

The power output of the class AB amplifier may be found from

$$P_o = \left(\frac{I_{\max}}{1.414}\right)^2 = \frac{R_{LL}}{4} = \frac{(I_{\max})^2 R_L}{8}$$

where $I_{\max}$ = peak-plate current at $E_c = 0$ volts

R_{LL} = plate-to-plate load resistance

The average plate current at maximum signal condition is

$$I_{\text{av(sig,max)}} = \frac{2I_{\max}}{\pi} = 0.636 I_{\max}$$

where $I_{\max}$ is the peak-plate current at $E_c = 0$ volts.

The average power input at maximum signal condition is

$$P_i(\text{sig,max}) = \frac{0.636 I_{\max}}{E_{bb}}$$

As a first approximation in the design of a class AB amplifier, a load line may be drawn such that it intersects the $E_c = 0$-volt characteristic where $E_b = 0.6E_{bb}$. The plate dissipation which would result from the use of this load line should then be compared to the maximum rated value of the tube; i.e.,

$$P_{i,\text{av}} \text{ for both tubes} = 0.636 I_{\max} E_{bb}$$

and
$$P_o = \left(\frac{I_{\max}}{1.414}\right)^2 \frac{R_{LL}}{4}$$

and $\quad PD$ (both tubes) $= P_{i,\text{av}}$ (both tubes) $- P_o$ (both tubes)

and $\quad PD$ (one tube) $= \dfrac{PD \text{ (both tubes)}}{2}$

If the power dissipation exceeds the maximum rated value of the tube, a larger load resistance (smaller load-line slope) is to be selected.

The zero signal bias value and plate current can be determined by the graphic technique illustrated in Fig. 11.31. In diagram B, E_c is plotted on the x axis and I_b is on the y axis at the same scale as on the plate characteristics. The intersection point of the load line with each characteristic is carried to diagram B (see points C and C', also D and D', E and E'). The points are joined by a curve (C'-E'). A line tangent to this curve is next drawn, as shown in diagram B. The zero signal bias is determined by the intersection of the tangent line and the x axis (-32 volts in Fig. 11.31b). The zero signal plate current can now be read from the plate characteristics (point O' in diagram A), which is about 8 ma in this illustration. The operating point for each tube is thereby determined (point O). The maximum input signal voltage for class AB_1 operation is easily determined by E_c at the operating point, which is 32 volts in this example.

The same procedures can be used in the design of the class AB_2 amplifier except in the determination of zero signal bias. The zero signal bias is

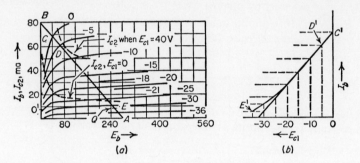

FIG. 11.31 Graphical technique for the determination of zero signal parameters: (a) average plate characteristics of the 6K6, (b) E_{c1} versus I_b on the load line. [(a) *characteristics from RCA*]

selected so that the input signal is permitted to drive the control grid positive for a portion of the time. This requires a greater amount of input driving power because of the grid circuit power dissipation. A well-regulated power supply should be used in class AB_1 and AB_2 work.

11.8 Design of a Class AB_1 Push-Pull Amplifier. Let us design the class AB_1 push-pull amplifier of Fig. 11.32. A correction filter (R and C) will be used across the primary of the output transformer to improve the frequency response. Cathode bias will be used for the tubes. The 6V6 will be used and E_{bb} is 250 volts. Signal frequency range is 50 cps to 20 kc.

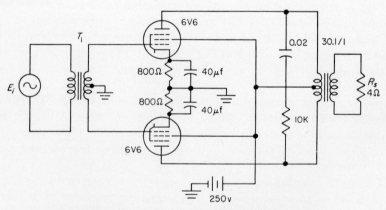

FIG. 11.32 The designed class AB_1 push-pull amplifier.

Step 1: *Construction of the Load Line.* The approximation technique discussed in the preceding section is to be utilized. See Fig. 11.33a for the 6V6 plate characteristics. The x-axis intercept is the selected E_{bb} of 250 volts (point A). The second point is to be the point located by $0.6E_{bb}$ and $E_c = 0$ volts, i.e., where $E_b = 0.6(250) = 150$ volts and $E_c = 0$ volts (point B).

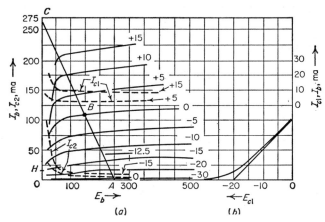

FIG. 11.33   Characteristics of the 6V6.   (*Characteristics from RCA*)

Let us compute the plate dissipation and compare it with the rated value of the tube:

$$PD \text{ (both tubes)} = 0.636 I_{max} E_{bb} - \frac{I_{max}^2 R_{LL}}{8}$$

where I_{max} = plate current where $E_c = 0$ volts
$\quad R_{LL}$ = plate-to-plate load resistance
$\quad\quad = 4 E_{bb}/I_{max,y}$
The plate-to-plate load resistance is

$$\frac{4(250)}{275 \times 10^{-3}} \cong 3.65 \text{ kilohms}$$

and from Fig. 11.33,

$$I_{max} = 110 \text{ ma}$$

Substituting values and solving,

PD (both tubes)

$$= 0.636(110 \times 10^{-3}) \times 250 - \frac{(110 \times 10^{-3})^2 \times 3.65 \times 10^3}{8}$$

$$= 17.50 - 5.52 \cong 12 \text{ watts}$$

and $\quad\quad\quad PD \text{ (one tube)} \cong \dfrac{12 \text{ watts}}{2} = 6 \text{ watts}$

From the manufacturer's data, the maximum plate dissipation of the 6V6 is 14 watts; therefore the selected load line is acceptable.

Step 2: Determination of Zero Signal Parameters. Refer to Fig. 11.33*a* and *b*. The corresponding I_b for each intersection of the load line with the characteristics of E_c of 0 volts to E_c of -30 volts is plotted in Fig. 11.33*b*. These points are joined to form the curve. A line tangent to this curve is then drawn (see diagram *b*). The zero signal grid bias (for each tube) is found to be -20 volts. Returning to the characteristics in Fig. 11.33*a*, $I_{b,s}$ is found to be 20 ma (point *H*). In conclusion, it is determined that

$$I_{b,s} \cong 20 \text{ ma}$$
$$E_{c,s} \cong -20 \text{ volts}$$

Step 3: Calculation of R_k and C_k for Each Tube. Both tubes are to have the same bias potential; therefore

$$R_{k1} = R_{k2}$$

and

$$R_{k1} = \frac{E_{c1,s}}{I_{b1,s} + I_{c2(1),s}}$$

From the manufacturer's data, $I_{c2(1),s} \cong 5$ ma; substituting values and solving,

$$R_{k1} = R_{k2} = \frac{20}{(20 + 5) \times 10^{-3}} = 800 \text{ ohms}$$

$X_{Ck} = 0.1R_k$ at the lowest signal frequency; i.e.,

$$X_{Ck} = 80 \text{ ohms at 50 cps}$$

and

$$C_{k1} = C_{k2} = \frac{1}{6.28 \times 50 \times 80} \cong 40 \text{ } \mu\text{f or larger}$$

Step 4: Determination of the Output Transformer Turns Ratio. The ultimate load is 4 ohms. The reflected primary impedance is to be equal to the plate-to-plate load impedance, which was earlier found to be 3.65 kilohms. Solving for the required turns ratio,

$$a = \left(\frac{R_{LL}}{R_s}\right)^{1/2}$$
$$= \left(\frac{3.65 \times 10^3}{4}\right)^{1/2} = \frac{30.1}{1}$$

Step 5: Selection of an Input Transformer. The input transformer is selected so that the source voltage generator impedance is matched to the grid circuits of the push-pull tubes.

Step 6: *Determination of Maximum Input Voltage.* The maximum input voltage is that excursion from $E_c = -20$ volts to $E_c = 0$ volts, or 20 volts.

Step 7: *Determination of the Corrective Filter (R and C).* Using the previously stated rule of thumb, let $R = 1.3R_{LL}$, as a first approximation; i.e.,

$$R = 1.3(3.65\text{K}) \cong 4.75 \text{ kilohms}$$

As a first approximation, let $C = 0.05$ μf. Check the frequency response at 400 cps and 1 kc for uniformity; decrease C if the output at 400 cps exceeds that at 1 kc, and vice versa. In this circuit, the final values chosen were 0.02 μf for C and 10 kilohms for R. R was increased in order to increase A_v at the high frequencies; and C was decreased in order to increase A_v at the low frequencies, resulting in a fairly uniform gain throughout the frequency range, as shown in Table 11.5.

Table 11.5 f Versus A_v (relative) for the Circuit of Fig. 11.32

f	A_v (relative)
50	26.5
100	32
1 kc	28.5
5 kc	26
10 kc	24
14 kc	24
20 kc	24

Step 8: *Circuit Construction.* Upon construction of the circuit, the exact value of the components are selected for best performance.

NOTE: The input signal of the designed amplifier is usually the output of a driver stage, which is a conventional class A amplifier designed to furnish the power required for the input of the push-pull stage.

11.9 The Class B Push-Pull Amplifier.

The class B push-pull amplifier has the advantage of possessing an efficiency which approaches 78 per cent whereas Class AB efficiency lies between 50 and 78 per cent. Each tube is at or near cutoff for approximately half of the signal cycle, thereby making it possible to deliver a large power output without exceeding the dissipation value of the tube.

Triodes, tetrodes, or pentodes may be used in class B work. Triodes which are best suited for class B applications are designed to operate at bias values which are close to zero, thereby permitting their utilization without the use of any bias. When triodes are used in this manner, class B_2 operation results; i.e., grid current flows for nearly half the signal cycle, since the grid is positive for that portion of the time. This results in a considerable amount of grid power dissipation, which must be considered when selecting a suitable driver and input transformer. The input transformer is generally of the stepdown variety in this application.

When the tetrode or pentode is to be utilized in class B work, the tube is biased at or near cutoff at zero signal condition. The fixed-bias method is the most common type of biasing technique used in these cases. The input signal may or may not be allowed to drive the grid positive, depending upon whether B_1 or B_2 operation is desired. The advantage of class B_2 is that a larger input signal excursion can be taken by the tube. On the other hand, the permitted grid current develops power dissipation in the grid circuit, imposing a larger load upon the driver stage. In order to keep distortion at a minimum, the driver power output capabilities should be considerably greater than the load imposed upon it.

11.10 The Phase Inverter. In the vacuum-tube push-pull circuits analyzed in the preceding sections, the required phase inversion for the grids of the push-pull tubes was achieved by use of an input transformer with a center-tapped secondary. This technique, as is the case with transistor push-pull circuits, may be undesirable from the standpoint of economy or space. Phase inversion may be achieved by a number of circuits. One of the more popular phase-inversion circuits is illustrated in Fig. 11.34.

In Fig. 11.34, V_1 and V_2 are to serve the same function as the input transformer of Fig. 11.32. The outputs of V_1 and V_2 are to be identical but inverted in phase. In this way, when V_1 is delivering a positive going signal voltage excursion to the input of V_3, V_2 is delivering a negative going signal voltage of equal magnitude to the input of V_4.

V_1 and V_2 are generally the same type of tube and may even be contained in one envelope. A common-cathode bias resistor can be used for both tubes, since they possess identical parameters. R_{L1} and R_{L2} are also the same value. In order to enable V_2 to provide V_4 with a signal which is inverted with respect to the signal provided to V_3 by V_1, the input of V_2 must be inverted with respect to the input of V_1. Recall that the output voltage of V_1 is inverted with respect to its input voltage, which appears across the grid-to-common circuit of V_3, i.e., across R_{g3} and R_{g2} in series. Therefore, a portion of this voltage e_{Rg2} may be used as the input of V_2, thereby providing the required inverted input signal for V_2.

Since V_1 and V_2 are identical tubes and possess the same zero signal parameters, their voltage gains are also identical for an input signal of identical magnitudes. In order to provide V_2 with an input signal whose magnitude is equal to the input of V_1, the following relationship must exist:

$$E_{i1} = E_{i2} = \frac{R_{g2}}{R_{g2} + R_{g3}} = \frac{1}{A_{v1}}$$

Since $A_{v1} = A_{v2}$, the output of V_2, which appears across R_{g4}, is equal to the output of V_1. Final balance adjustments are made by relatively small changes in R_{g2}. Increasing R_{g2} increases the input and output signal voltages of V_2, while reducing R_{g2} produces the opposite type of change.

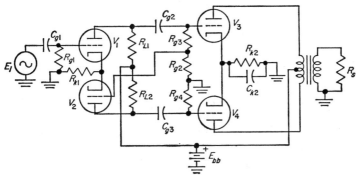

FIG. 11.34   Push-pull amplifier with a vacuum-tube phase inverter circuit.

In order to ensure the proper balance between the grid-to-cathode circuits of the push-pull tubes (V_3 and V_4), the following relationship should exist:

$$R_{g3} + R_{g2} = R_{g4}$$

Criteria for the Design of a Vacuum-tube Phase Inverter

1. Select identical tubes (preferably in one envelope) for V_1 and V_2.
2. Determine C_{g1}, R_{g1}, R_{k1}, R_{L1}, and R_{L2} in the conventional manner used for class A amplifiers.   Let $R_{L1} = R_{L2}$ and $A_{v1} = A_{v2}$.
3. Determine R_{g3}, R_{g2}, R_{g4} such that

$$\frac{R_{g2}}{R_{g2} + R_{g3}} = \frac{1}{A_{v1}}$$

and

$$R_{g2} + R_{g3} = R_{g4}$$

4. Determine the parameters and components of the push-pull portion of the circuit in the manner described in Sec. 11.8.

PROBLEMS

11.41   Refer to the push-pull amplifier of Fig. 11.30. $I_{b1,s} = I_{b2,s} = 20$ ma; required $E_{c,s}$ for each tube $= -25$ volts; signal frequency range $= 40$ cps to 20 kc. Find (a) R_k; (b) C_k.

11.42 Refer to Fig. 11.30. $I_{b1,s} = I_{b2,s} = 5$ ma; required $E_{c,s}$ for each tube $= -10$ volts; signal frequency range $= 60$ cps to 20 kc. Find (a) R_k; (b) C_k.

11.43 Refer to the push-pull amplifier of Fig. 11.30. Assume the 6K6 is to be used (see Fig. 11.31a for the plate characteristics). $E_{bb} = 320$ volts. Class AB_1 operation is to be obtained. Draw a load line such that it intersects the $E_c = 0$ characteristic where $E_b = 0.6E_{bb}$.

11.44 In the push-pull amplifier of Prob. 11.43, find:

(a) R_{LL} (b) P_o

(c) $I_{av(\text{sig,max})}$ (d) $P_{i(\text{sig,max})}$

(e) PD for both tubes (f) PD for one tube

11.45 In the amplifier of Probs. 11.43 and 11.44, is the maximum plate dissipation of each tube exceeded? If so, what should be done?

11.46 In the amplifier of Probs. 11.43 and 11.44, construct the E_c versus I_b curve on the load-line characteristic (in the same manner as seen in Fig. 11.31b).

11.47 From the E_c versus I_b characteristic constructed in Prob. 11.46, find the required E_{cc}.

11.48 Using the E_{cc} value found in Prob. 11.47, determine (a) R_k; (b) C_k (assume the frequency range is 60 cps to 15 kc).

11.49 Assume the amplifier of Probs. 11.43 to 11.48 has an ultimate load resistance R_s of 5 ohms. Calculate the required turns ratio of the output transformer.

11.50 Refer to the push-pull amplifier of Fig. 11.30. Assume the 6V6 is to be used (see Fig. 11.33a for the plate characteristics); $E_{bb} = 300$ volts. Class AB_1 operation is to be obtained. Draw a load line such that it intersects $E_c = 0$ characteristic where $E_b = 0.6E_{bb}$.

11.51 In the push-pull amplifier of Prob. 11.50, find:

(a) R_{LL} (b) P_o

(c) $I_{av(\text{sig,max})}$ (d) $P_{i(\text{sig,max})}$

(e) PD for both tubes (f) PD for one tube

11.52 In the amplifier of Probs. 11.50 to 11.51, is the maximum plate dissipation of each tube exceeded?

11.53 Refer to the amplifier of Prob. 11.50 and 11.51. Construct the E_c versus I_b characteristic on the load-line characteristic (in the same manner illustrated in Fig. 11.33b).

11.54 From the E_c versus I_b characteristic constructed in Prob. 11.53, find the required E_{cc}.

11.55 Using the E_{cc} value found in Prob. 11.54, determine (a) R_k; (b) C_k (assume the frequency range is 50 cps to 20 kc).

11.56 Assume the amplifier of Probs. 11.50 to 11.55 has an ultimate load resistance R_s of 8 ohms. Calculate the required turns ratio of the output transformer.

11.57 A constant plate impedance network is to be used in the amplifier of Probs. 11.43 to 11.49. Let $C = 0.04$ μf. Determine the first approximation value of R.

11.58 The amplifier of Probs. 11.50 to 11.56 is to utilize a constant plate impedance network. Let $C = 0.025$ μf. Determine the first approximation value of R.

11.59 Refer to the phase inverter circuit of Fig. 11.34. $A_{v1} = 50$; $R_{g4} = 470$ kilohms. Find (a) R_{g2}; (b) R_{g3}.

11.60 In the phase-inverter circuit of Fig. 11.34, $A_{v1} = 15$; $R_{g4} = 270$ kilohms. Find (a) R_{g2}; (b) R_{g3}.

11.11 The Class C Amplifier.

Vacuum-tube class C operation, like its counterpart in transistor circuitry, is most generally used for oscillator work. The principles of feedback as analyzed in Sec. 7.9 are applicable to vacuum-tube circuitry as well. Positive feedback is incorporated in the same manner so that class C operation can be achieved. Class C efficiency in vacuum-tube work can approach 90 per cent with lower angles of plate current. The fundamentals of vacuum-tube oscillator action are the same as those principles developed in Sec. 7.10.

The general types of oscillators treated with transistors in Chap. 7 are treated with vacuum tubes in the following sections. Before studying them, it is recommended that the reader review Secs. 7.9 and 7.10.

11.12 The Armstrong Oscillator. One of the most basic oscillator circuits is the Armstrong oscillator, which is illustrated in Fig. 11.35.

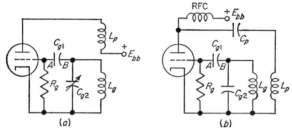

Fig. 11.35 Tuned grid–untuned plate Armstrong oscillators: (a) series-fed, (b) shunt-fed.

The series-feed type of tuned grid–untuned plate Armstrong oscillator is shown in Fig. 11.35a. Notice that the plate current flows through the L_p portion of the oscillator coil. In the shunt-feed type, the unidirectional component of the plate current is blocked from the oscillator coil by the plate capacitor C_p. The L_p portion of the oscillator coil is the feedback or tickler coil, since signal energy is magnetically coupled to the tank circuit (C_{g2} and L_g) by L_p.

The self-biasing technique analyzed in Sec. 7.11 is provided by C_{g1} and R_g. During the small portion of the cycle when the tube conducts, plate A of C_{g1} takes on electrons until charged, then plate A discharges electrons through R_g to common. This places the grid at sufficient negative potential to cut off tube conduction. The discharging time constant should be long enough to maintain i_{R_g} sufficiently large to enable e_{R_g} to be greater than the cutoff value of the tube for the majority of the signal cycle. The discharging time constant of C_{g1} is determined by

$$\text{tc discharge} = R_g C_{g1}$$

and, since a long discharging time constant is desired,

$$\text{tc discharge} = 10 T_{p(f_o)}$$

This type of bias, when used in vacuum-tube circuitry, is called *grid-leak bias*.

The frequency of oscillations is determined by the same relationships in both the series-feed and shunt-feed types. Notice that R_g is in parallel with the tank circuit. When R_g is large, it produces no significant loading effect on the tank circuit, and

$$f_o = \frac{1}{2\pi(L_g C_{g2})^{1/2}}$$

In many practical cases, R_g is sufficiently small to cause the loaded-Q value of the tank circuit to be substantially smaller than its unloaded-Q value. In such instances, the frequency of oscillation is determined by

$$f_o = \frac{1}{2\pi(L_g C_{g2})^{1/2}[1 + (R_g/r_p)(L_p/L_g)]^{1/2}}$$

and the stable oscillation condition is found by

$$g_m = \frac{\mu R_g C_{g2} L_g}{M(\mu L_p - M)}$$

where M = mutual inductance of L_p and L_g
$= k(L_g L_p)^{1/2}$

Another variation of the Armstrong oscillator is the tuned plate–untuned grid, both series and shunt feed. Their operation is similar to the tuned grid–untuned plate types, with the chief difference being that the tank circuit is placed in the plate circuit.

11.13 The Hartley Oscillator. The series-fed and shunt-fed Hartley oscillators are shown in Fig. 11.36. The oscillator coil is tapped and has three terminals. The series-fed has found great popularity because of its relative simplicity; therefore it shall be examined in greater detail.

The amount of positive feedback may be determined from

$$\beta = \frac{X_{Lp}}{X_{Lp} + X_{Cg2}}$$

NOTE: Additions are vectorial (X_{Cg2} is negative, X_{Lp} is positive).

Notice that the amount of feedback is determined by the position of the coil tap with a position about one-third from the bottom commonly used. The plate load impedance of the tube can be found by

$$Z_L = \frac{-X_{Lp}X_{Lg} - X_{Cg2}X_{Lg}}{jX_{Lp} + jX_{Lg} + jX_{Cg2}}$$

The gain of the tube, with feedback, is

$$A_v = \mu \frac{X_{Lp}}{\beta X_{Lg}}$$

The frequency of oscillations can be found by

$$f_o = \frac{[1 + R_p/r_p]^{\frac{1}{2}}}{2\pi[(L_g + L_p + 2M)C_{g2}]^{\frac{1}{2}}}$$

where M = mutual inductance of L_g and L_p

R_p = resistance of L_p

When the resistance of L_p is very small as compared to r_p, the numerator of the equation approaches unity, thereby simplifying the relationship. This is quite often the case in actual practice.

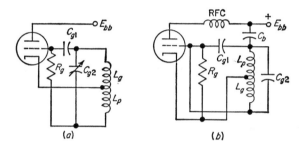

FIG. 11.36 The Hartley oscillator: (a) series-fed, (b) shunt-fed.

Very often the frequency of oscillations is predetermined, and the task of the designer is to select the tube and components for the oscillator. Following is a method for designing a series-fed Hartley oscillator.

Criteria for the Design of a Series-fed Hartley Oscillator

1. Select a tube of medium-to-high mu which is suitable for the frequency of oscillations.

2. Select an oscillator coil with a variable tap. The inductance of the coil should be lower with higher oscillation frequencies.

3. As a first approximation, select C_{g2} by use of the following relationship:

$$C_{g2} = \frac{1}{(2\pi)^{\frac{1}{2}}(L_g + L_p + 2M)f_o}$$

4. Select C_{g1} and R_g so that the grid-leak action analyzed in the preceding section is obtained. It is suggested that $C_{g1} \cong C_{g2}$ for a first approximation. Then determine R_g so that

$$R_g C_{g1} \cong 10 T_{p(f_o)}$$

5. Since $\beta A_v = 1$ to maintain oscillations, L_p and L_g should be in such a ratio that

$$A_v\beta = \mu \frac{X_{Lp}}{X_{Lg}} = 1$$

Adjust the position of the tap to achieve this relationship.

6. Construct the circuit. If the circuit fails to oscillate, move the coil so that L_p is increased. If this doesn't trigger the oscillations, then reduce R_g. Oscillations should then result. Check the oscillator frequency (by use of an oscilloscope), and adjust for the exact frequency by varying C_{g2}.

11.14 The Colpitts Oscillator. One version of the Colpitts oscillator is illustrated in Fig. 11.37. R_g and C_g establish the grid-leak bias required

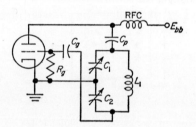

Fig. 11.37 A Colpitts oscillator.

for the operation of the oscillator. Notice that the capacitors C_1 and C_2 divide the tank voltage, with e_{C2} being applied to the grid of the tube. The tank circuit capacitance is

$$C_T = \frac{C_1 C_2}{C_1 + C_2}$$

The feedback is determined by

$$\beta = \frac{X_{C2}}{X_{C2} + X_{L1}}$$

The plate load impedance of the tube can be found by

$$Z_L = \frac{-X_{C2}X_{C1} - X_{C1}X_{L1}}{jX_{C1} + jX_{C2} + X_{L1}}$$

and the gain with feedback is

$$A_v = \mu \frac{X_{C2}}{\beta X_{C1}}$$

The frequency of oscillation can be determined by

$$f_o = \frac{1}{2\pi \{L_1[C_1C_2/(C_1 + C_2)]\}^{1/2}} \left[1 + \left(\frac{R_{L1}}{r_p} \frac{C_1}{C_1 + C_2} \right) \right]^{1/2}$$

where R_{L1} is the resistance of L_1.

The stable oscillation condition is achieved when

$$g_m = \frac{\mu R_{L1}(C_1 + C_2)}{L_1(\mu - C_2/C_1)}$$

When the resistance of the tank coil is very small as compared to rp of the tube, the approximate frequency of oscillation can be determined by

$$f_o = \frac{1}{2\pi \{L_1[C_1C_2/(C_1 + C_2)]\}^{1/2}}$$

As stated in the preceding section, the more practical cases predetermine the frequency of oscillations, and the designer is to determine the required components. A practical approach is as follows.

Criteria for the Design of a Colpitts Oscillator

1. Select a tube of medium-to-high mu which is suitable for the frequency of oscillations.
2. Select an oscillator coil of suitable inductance and high Q for the frequency to be handled (higher frequencies require lower inductance values).
3. Select C_1 and C_2 such that their series total is the required capacitance for the desired frequency. C_T, which is the series sum of C_1 and C_2, can be found by

$$C_T = \frac{1}{4(\pi)^2(f_o)^2L_1}$$

Since $\beta A_v = 1$ to sustain oscillations, C_1 and C_2 should be such that

$$\beta A_v = \mu \frac{X_{C2}}{X_{C1}} = 1$$

4. Determine R_g and C_g, the grid-leak bias network, in the manner previously analyzed.
5. Construct the circuit. If the circuit fails to oscillate, decrease C_2 (which increases X_{C2}) in order to increase the amount of feedback. If oscillations still do not occur, reduce R_g. Oscillations should then result. Check the oscillator frequency (by use of an oscilloscope), adjust C_2 or C_1 for the exact frequency.

11.15 The Wien Bridge Oscillator. The fundamental Wien bridge vacuum-tube oscillator is illustrated in Fig. 11.38. Prior to entering into the analysis of this circuit, the basic concepts of the Wien bridge in Sec. 7.19 should be reviewed by the student.

Consider the Wien bridge oscillator of Fig. 11.38 for the special case where

$$R_3 = R_4$$

and

$$C_a = C_b$$

The input voltage of V_1 is e_{R4}, called E_{ob} in the circuit. Let us determine

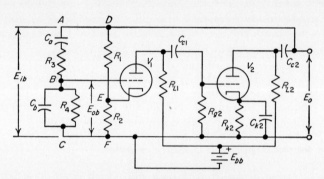

FIG. 11.38   The Wien bridge oscillator.

the impedance and current equations for branch ABC, from which E_{ob} can be found:

$$Z_{ABC} = R_3 - jX_{Ca} + \frac{R_4(-jX_{Cb})}{R_4 - jX_{Cb}}$$

Let $R_3 = R_4 = R$, and $C_a = C_b = C$, thereby simplifying the equation to

$$Z_{ABC} = R - jX_C + \frac{R(-jX_C)}{R - jX_C}$$

The total current in the left branch ABC may be determined by

$$i_{ABC} = \frac{e_{ABC}}{Z_{ABC}} = \frac{e_{ABC}}{R - jX_C + [R(-jX_C)/(R - jX_C)]}$$

e_{ob} can be computed from

$$e_{ob} = i_{ABC}Z_{BC} = i_{ABC}\frac{R_4(-jX_{Cb})}{R_4 - jX_{Cb}}$$

Again using R for R_4 and C for C_b,

$$e_{ob} = i_{ABC}\frac{R(-jX_C)}{R - jX_C}$$

Using the familiar voltage-divider relationship,

$$e_{ob} = \frac{Z_{BC}}{Z_{ABC}} e_{ib}$$

where $\qquad\qquad e_{ib} = e_{AC}$

Substituting values,

$$e_{ob} = \frac{R(-jX_C)/(R - jX_C)}{(R - jX_C) + [R(-jX_C)/(R - jX_C)]} e_{ib}$$

and simplifying,

$$e_{ob} = \frac{-jRX_C}{-j3RX_C} e_{ib} = \frac{e_{ib}}{3}$$

Therefore the input voltage to the grid of V_1 (see Fig. 11.38) is $e_{ib}/3$ at the balance frequency, which is $E_{ob,\max}$. Notice that

$$e_{ib} = E_{b,s}(T2)$$

and $\qquad\qquad E_{c1,s}(T_1) = e_{ob} = \frac{e_{ib}}{3} = \frac{E_{b,s}(T2)}{3}$

The circuit will oscillate at the condition where $e_{ob} = e_{ib}/3$.

Let us consider the phase angle of branch ABC at the frequency of oscillations.

Example 1: Assume that $R_3 = R_4 = X_{ca} = X_{cb} = 10$ kilohms

Solution:

$$Z_{ABC} = (10\mathrm{K} - j10\mathrm{K}) + \frac{10\mathrm{K}(-j10\mathrm{K})}{10\mathrm{K} - j10\mathrm{K}}$$

Solving for the parallel combination first,

$$Z_{ABC} = (10\mathrm{K} - j10\mathrm{K}) + \frac{100 \times 10^6 \underline{/-90^\circ}}{7.07\mathrm{K} \underline{/-45^\circ}}$$

$$= (10\mathrm{K} - j10\mathrm{K}) + 14.14\mathrm{K} \underline{/-45^\circ}$$

Expanding to rectangular form,

$$Z_{ABC} = (10\mathrm{K} - j10\mathrm{K}) + (10\mathrm{K} - j10\mathrm{K})$$
$$= (20\mathrm{K} - j20\mathrm{K}) \text{ ohms}$$

Notice that the phase angle of the parallel RC combination (R_4 and C_b) and series RC (R_3 and C_a) are the same. Therefore the phase angle of the output voltage is zero degrees at the frequency of oscillations.

At the frequency of oscillations, the following relationships exist in the Wien bridge of Fig. 11.38:

$$R_3 = R_4 = X_{Ca} = X_{Cb}$$
$$R_1 = 2R_2 \text{ (minimum ratio)}$$
$$f_b = f_o = \frac{0.159}{R_3 C_a}$$

Since the oscillations occur only at the balance frequency, the oscillator frequency can be altered by changing R_3 and R_4, or C_a and C_b.

Notice that R_2 is the cathode-bias resistor of V_1 and must be selected to meet the class A operating requirements of V_1. R_2 is frequently a light bulb or other device which has a positive temperature coefficient of resistance. A feedback loop is provided between points D and A. Each stage shifts the signal by 180°, resulting in a 360° shift from both tubes, thereby providing positive feedback from the second-stage plate circuit to the Wien bridge. C_{C2} is often made very large, so as to minimize the possibility of introducing additional phase shift in the feedback loop. The amount of feedback is sufficiently strong to maintain oscillations only at the balance frequency, since this is the only signal which is being sufficiently amplified by the two stages. E_{ob} is not large enough at other frequencies to allow the two stages to build up an output signal voltage of the minimum magnitude required to provide the required amount of positive feedback.

The parameters of V_1 and V_2 are determined by the conventional class A design techniques, since both tubes are operated in the class A mode. The components are carefully selected so that the oscillation frequency lies in the middle-frequency range of the amplifier.

The Wien bridge may be designed after the over-all gain of the two-stage amplifier has been determined. Recall that

$$\frac{R_1}{R_2} = 2 \text{ (at balance)}$$

This ratio must be slightly larger, as determined by $1/A_{v,t}$ of the amplifier, in order to maintain oscillations. For example, assume the over-all voltage gain of an amplifier is 100; then

$$\frac{R_1}{R_2} = \frac{2 + \frac{1}{100}}{1} = \frac{2.01}{1} \text{ (the minimum ratio)}$$

In actual practice R_2 is often the resistance of a lamp bulb, or some similar device, at the temperature when the zero signal cathode current is passing through it. This resistance serves as the cathode-bias resistance of V_1, thereby establishing the zero signal operating point on the load line as determined by E_{bb} and R_{L1}. R_1 is then selected so that at least the minimum ratio is achieved. It is permissible to select R_1 such that the minimum ratio is exceeded. In the preceding example, where the minimum ratio was found to be 2.01/1, R_1 could be greater than $2.01R_1$. When the minimum ratio is exceeded, it takes the circuit a little longer to develop a sinusoidal output. The initial output closely resembles a square wave; with each succeeding signal cycle, the square wave becomes more rounded until it finally becomes a sinusoidal waveform.

The components in the RC branch of the Wien bridge may be selected last in the design of the bridge circuit. Notice that R_4 and C_b are in the grid circuit of V_1, and the total impedance of this parallel combination should not be too low for the V_1 input circuit. The most simple approach is to utilize the special case mentioned in the preceding paragraphs.

In such cases, the frequency of oscillations is determined by

$$f_o = \frac{0.159}{RC}$$

The capacitors may be arbitrarily selected, and then the required R may be determined by transposing the preceding equation; i.e.,

$$R = \frac{0.159}{f_o C}$$

Or, if so desired, the value of resistance may be arbitrarily selected, and the capacitance values determined; i.e.,

$$C = \frac{0.159}{f_o R}$$

Example 2: Refer to Fig. 11.38. Assume $A_{v,t} = 80$; $C_a = C_b = 500\ \mu\mu\mathrm{f}$; $R_2 = 1$ kilohm; $f_o = 10$ kc. Find (a) R_1; (b) R_3 and R_4.

Solution:

(a) Find R_1:

$$\frac{R_1}{R_2} = \frac{2 + (1/A_{v.t})}{R_2} = \frac{2 + \frac{1}{80}}{1 \times 10^3}$$

and

$$R_1 = 2.013 \text{ kilohms (minimum)}$$

(b) Find R_3 and R_4:

$$R_3 = R_4 = R = \frac{0.159}{f_o C} = \frac{0.159}{1 \times 10^4 \times 5 \times 10^{-10}}$$

$$R \cong 31.8 \text{ kilohms}$$

One of the common applications of the Wien bridge oscillator is for use in signal generators. The frequency is made variable within a given range by use of variable capacitors for C_a and C_b. More than one range is possible by using tapped resistors for R_3 and R_4. The required resistance value is switched from one tap to another when switching the range. The following example helps to clarify the manner in which a variable frequency Wien bridge oscillator can be designed.

Example 3: A signal generator has the following frequency ranges: 20 to 200 cps; 200 cps to 2 kc; 2 to 20 kc. C_a and C_b are ganged variable capacitors. R_3 and

R_4 are three-tapped and ganged. $R = 150$ kilohms on the highest range, 1.5 megohms on the next range, and 15 megohms on the lowest range. Find (a) $C_{a,\min}$; (b) $C_{a,\max}$.

Solution:

(a) Find $C_{a,\min}$:

$$C_{a,\min} = C_{b,\min} = C = \frac{0.159}{f_{o,\min} R_{\max}} = \frac{0.159}{f_{o,\max} R_{\min}}$$

C is at its minimum value at the highest frequency. Solving for range 1 (20 to 200 cps), where $R = 1.5$ megohms,

$$C_{\min} = \frac{0.159}{2 \times 10^2 \times 1.5 \times 10^7} = 53 \ \mu\mu\text{f}$$

Note that this is also the required minimum C for the other two ranges. Since the upper frequency of the remaining two ranges is in multiples of ten (200 cps, 2 kc, 20 kc), and the resistance decreases in multiples of ten (15 megohms, 1.5 megohms, 150 kilohms), the same minimum C will serve for the three ranges. The same type of reasoning is applicable to the maximum value of C.

(b) Find $C_{a,\max}$:

$$C_{a,\max} = C_{b,\max} = C = \frac{0.159}{f_{o,\min} R_{\text{range}}}$$

Using the values of range 1,

$$C = \frac{0.159}{2 \times 10^1 \times 1.5 \times 10^7} = 530 \ \mu\mu\text{f}$$

Therefore, the two variable capacitors are to have a range of 53 to 530 $\mu\mu$f. When the two ganged capacitors are at their minimum setting, the generator is at the maximum frequency of the range, while the maximum setting (capacitors in mesh) establishes the minimum frequency of that range. The range is determined by the position of the ganged-tapped resistors (R_3 and R_4). When the resistors are set at their maximum value, the generator is at its lowest-frequency range; when R_3 and R_4 are at their minimum value the generator is at its maximum-frequency setting.

11.16 Design of a Wien Bridge Oscillator.

Let us design a Wien bridge vacuum-tube oscillator with a frequency range of 2 to 20 kc (see Fig. 11.39). C_a and C_b have a range of 10 to 365 $\mu\mu$f; $E_{bb} = 250$ volts; the 12AU7 (twin triode) is to be used.

Step 1: Determination of Stage 2 Parameters. E_{bb} is given as 250 volts. The plate characteristics of the 12AU7 are illustrated in Fig. 11.40. Let us select the input voltage of stage 2 to be a maximum of 2 volts. Let us arbitrarily select the operating point of the second-stage amplifier so that it is at $E_{c1,s} = -2$ volts and $E_{b,s} = 0.5E_{bb} = 125$ volts. The load line can then be constructed; the zero signal parameters of the second stage are:

$$I_{b,s} = 10 \text{ ma}$$
$$E_{b,s} = 125 \text{ volts}$$
$$E_{c1,s} = -2 \text{ volts}$$

The required R_{L2} is then found:

$$R_{L2} = \frac{E_{bb}}{I_{b,SC}} = \frac{250}{20 \times 10^{-3}} = 12.5 \text{ kilohms}$$

R_{L2} can be determined after the design of the Wien bridge. Since the zero signal plate current and grid bias are known, R_{k2} is next computed:

$$R_{k2} = \frac{E_{c1,s}}{I_{b,s}} = \frac{2}{10 \times 10^{-3}} = 200 \text{ ohms}$$

NOTE: Use a 5-kilohm potentiometer for R_{k2} at initial construction, and adjust for the most sinusoidal waveform.

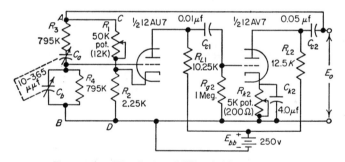

FIG. 11.39   The designed Wien bridge oscillator.

Now determining the cathode bypass capacitor,

$$X_{Ck2} = 0.1R_{k2} \text{ at 2 kc}$$

i.e.,
$$X_{Ck2} = \frac{0.159}{C_{k2} \times 2 \times 10^3}$$

Transposing, and solving for C_{k2},

$$C_{k2} = \frac{0.159}{2 \times 10^1 \times 2 \times 10^3}$$
$$\cong 4.0 \ \mu\text{f or larger}$$

The feedback coupling capacitor C_{C2} should be sufficiently large so as not to introduce any significant phase shift in the feedback loop; that is,

X_{CC2} must be very small as compared to R_a.   Recall that,

$$R_a = \frac{1}{1/r_p + 1/R_L} + R_g$$

where, in this case:

$r_p =$ dynamic plate resistance of T_2

$R_L = R_{L2}$

$R_g =$ the resistance at the grid of T_1

But the resistance at the grid of T_1 is determined after the design of the

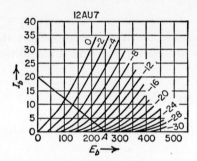

FIG. 11.40   Characteristics and load line for the designed Wien bridge oscillator.   (*Characteristics from RCA*)

Wien bridge.   Therefore the determination of C_{C2} will be one of the final steps in the design.

R_{g2} must be within the limits stated by the manufacturer, which can be up to 2.2 megohms for the 12AU7.   Therefore, R_{g2} is arbitrarily selected as 1 megohm.

Step 2: Design of Stage 1 Parameters.   Since *RC* coupling is utilized between stages 1 and 2, then the signal load line is determined by the equivalent value of R_{L1} and R_{g2}; i.e.,

$$R_{L1,eq} = \frac{1}{1/R_{L1} + 1/R_{g2}}$$

Using the same load line for stage 1 as used in stage 2, then

$$R_{L1,eq} = R_{L2,eq} = 12.5 \text{ kilohms}$$

R_{g2} was determined in step 1 as 1 megohm.   By transposing, R_{L1} can be found:

$$R_{L1} = \frac{1}{1/R_{L1,eq} - 1/R_{g2}} = \frac{1}{1/(12.5 \times 10^3) - 1/(1 \times 10^6)}$$
$$\cong 12.5 \text{ kilohms}$$

therefore $R_{L1} \cong R_{L1,eq}$

NOTE: R_2 serves as an unbypassed cathode resistor for T_1 and appears as a portion of the load resistance in the load-line considerations.   Therefore the actual value of R_{L1} is to be 12.5 kilohms minus the calculated value of R_2 (found in the next step).

C_{C1}, the coupling capacitor between stages 1 and 2, must be such a value that its reactance at the lowest frequency is negligible. At the lowest frequency,

$$\frac{X_{CC1}}{R_{a1}} = 1$$

where
$$R_{a1} = \frac{1}{1/r_{p1} + 1/R_{L1}} + R_{g2}$$

Recall that the dynamic plate resistance can be found by

$$r_{p1} = \frac{\Delta e_b}{\Delta i_b} \qquad \text{with } E_c \text{ constant}$$

Calculating r_{p1} for the i_b swing from 10 to 15 ma,

$$r_{p1} = \frac{160 - 125}{(15 - 10) \times 10^{-3}} = \frac{35}{5 \times 10^{-3}}$$
$$= 7 \text{ kilohms}$$

R_{a1} may now be computed:

$$R_{a1} = \frac{1}{1/(7 \times 10^3) + 1/(12.5 \times 10^3)} + 1 \times 10^6$$
$$\cong 1.005 \text{ megohms}$$

Therefore, $X_{CC1} = 1.005$ megohms at 2 kc. Solving for C_{C1},

$$C_{C1} = \frac{0.159}{1.005 \times 10^6 \times 2 \times 10^3}$$
$$\cong 80 \ \mu\mu\text{f or larger}$$

Let $C_{C1} = 0.01 \ \mu\text{f}$.

Step 3: Determination of R_2. Consider the selected case when the input signal voltage of T_2 is a maximum of 2 volts. From the plate characteristics of Fig. 11.40, it is found that this swing in the operating point is accompanied by a swing in E_b of 25 volts (125 to 100). This voltage is fed back to the Wien bridge. Recall from a previous analysis of the Wien bridge that it was found that one-third of the signal potential across the bridge is applied to the grid of stage 1 at the frequency of oscillations. The maximum input signal to the grid of T_1 is therefore one-third of 25 volts, or 8.3 volts.

Since the maximum input voltage of stage 1 is to be 8.3 volts and class A operation is desired, the operating point is selected at the intersection of the load line and the $E_{c1} = -9$-volt characteristic. $I_{b1,s}$ is found to be about 4 ma. The cathode-bias resistor of T_1, which is R_2, can now be computed:

$$R_2 = \frac{E_{c1,s}}{I_{b1,s}} = \frac{9}{4 \times 10^{-3}}$$
$$= 2.25 \text{ kilohms}$$

Step 4: *Determination of the Over-all Two-stage Gain* $A_{v,t}$. The over-all two-stage gain can be found by

$$A_{v,t} = \frac{E_{o,T2,\max}}{E_{i,T1,\max}} = \frac{25}{8.3} = 3$$

Step 5: *Determination of* R_1. Recall that

$$\frac{R_1}{R_2} = \frac{2 + 1/A_{v,t}}{R_2}$$

Substituting and solving for R_1,

$$R_1 = 2.33(R_2) = 2.33(2.25 \times 10^3)$$
$$= 5.25 \text{ kilohms (minimum)}$$

Because the Wien bridge shunts the output of T_2, R_1 should be made larger so that $R_1 + R_2$ is greater than the selected $R_{L2,eq}$ of 12.5 kilohms. Let $R_1 = 12$ kilohms, thereby making $R_1 + R_2 = 14.25$ kilohms.

NOTE: Use a 50-kilohm potentiometer for R_1 when first constructing the circuit. Adjust R_1 for the required amount of feedback.

Step 6: *Determination of* R_3 *and* R_4. When C_a and C_b are at their minimum setting, the oscillator is at its maximum frequency; i.e.,

$$C_a = C_b = C = 10 \ \mu\mu\text{f when } f = 20 \text{ kc}$$

Solving for R_3 and R_4,

$$R_3 = R_4 = R = \frac{0.159}{f_{o,\max} C_{\min}}$$
$$R = \frac{0.159}{2 \times 10^4 \times 10 \times 10^{-12}} = 795 \text{ kilohms}$$

Step 7: *Determination of* $R_{g1,eq}$, R_{L2}, *and* C_{C2}. $R_{L2,eq}$ was found in step 1 as 12.5 kilohms. The R_g in parallel with R_{L2} can be called $R_{g,eq}$ and is the Wien bridge circuit. Referring to the notation in Fig. 11.39,

$$R_{g,eq} = \frac{Z_{AB}Z_{AC}}{Z_{AB} + Z_{AC}}$$

Expanding,

$$R_{g,eq} = \frac{\{R_3 - jX_{Ca} + [R_4(-jX_{Cb})/(R_4 - jX_{Cb})]\}(R_1 + R_2)}{\{R_3 - jX_{Ca} + [R_4(-jX_{Cb})/(R_4 - jX_{Cb})]\} + (R_1 + R_2)}$$

This impedance is seen across the output of stage 2. Since $C_a = C_b$, then $X_{Ca} = X_{Cb}$. Also, this is the special condition where

$$R_3 = R_4 = X_{Ca} = X_{Cb}$$

at the oscillation frequency. Therefore we need only determine one of them. R_3 and R_4 were found to be 795 kilohms in step 6; therefore,

$$X_{Ca} = X_{Cb} = 795 \text{ kilohms}$$

Substituting values and solving the preceding equation results in

$$R_{g,\text{eq}} \cong 14.5\text{K}\underline{/0°} \text{ ohms}$$

Notice that $R_{g,\text{eq}}$ is the total impedance of the Wien bridge and is purely resistive, since the phase angle is zero degrees.

$R_{L2,\text{eq}}$ can now be determined, since both $R_{g,\text{eq}}$ and $R_{L2,\text{eq}}$ are known:

$$R_{L2,\text{eq}} = \frac{1}{1/R_{L2} + 1/R_{g,\text{eq}}}$$

Substituting values and solving,

$$R_{L2,\text{eq}} = \frac{1}{1/(12.5 \times 10^3) + 1/(14.5 \times 10^3)}$$
$$= 6.7 \text{ kilohms}$$

C_{C2} is the coupling capacitor for stage 2, and

$$\frac{X_{CC2}}{R_{a2}} = 1$$

We must first determine R_{a2}, which in this case is

$$R_{a2} = \frac{1}{1/r_{p2} + 1/R_{L2}} + R_{g,\text{eq}}$$

r_{p2} can be taken as equal to r_{p1}, since the same tube and load line is being used. Substituting values and solving,

$$R_{a2} = \frac{1}{1/(7 \times 10^3) + 1/(12.5 \times 10^3)} + 14.25 \times 10^3$$
$$\cong 18.8 \text{ kilohms}$$

Therefore X_{CC2} should equal 18.8 kilohms at the lowest frequency, which is 2 kc. Solving for C_{C2},

$$C_{C2} = \frac{0.159}{18.8 \times 10^3 \times 2 \times 10^3}$$
$$\cong .042 \ \mu\text{f or larger}$$

Let $C_{C2} = 0.05 \ \mu\text{f}$.

Step 8: Construction and Check-out of the Circuit. The final step is to assemble the complete circuit and subject it to an analysis. The minimum and maximum frequencies can be checked with the oscilloscope, as can the output voltage waveform. Vary the 50-kilohm potentiometer being used for R_1 until the required amount of feedback is obtained, and adjust the 5-kilohm potentiometer used for R_{k2} for improvement of the output waveform.

11.17   The Phase-shift Oscillator.   A phase-shift oscillator is illustrated in Fig. 11.41.   The phase-shift network consists of C_1, C_2, C_3, R_1, R_2, and R_3.   They are selected so that a 180° phase shift is obtained across the network at the frequency of oscillations.   The vacuum tube also provides

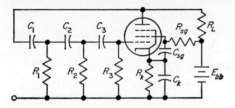

Fig. 11.41   A phase-shift oscillator.

a 180° phase shift, which results in a total phase shift of 360°, thereby establishing the possibility of positive feedback.   The feedback loop is shown from the plate of the tube to the beginning of the RC network C_1.

The frequency of oscillations is determined by

$$f_o = \frac{0.159}{2.45RC}$$

where
$$R = R_1 = R_2 = R_3$$
$$C = C_1 = C_2 = C_3$$

The output voltage of this phase-shift network is $\frac{1}{29}$ of the input, thereby requiring the tube to have a minimum A_v of 29 to ensure sufficient positive feedback to maintain oscillations.

The tube is operated in class A and may be designed in accordance with the conventional class A techniques.   Triodes are most often used, but a pentode can be utilized.   When a pentode-type tube is used, the screen voltage may be obtained by use of the screen-dropping resistor and bypass capacitor, or by connection to the correct voltage-divider tap when one is provided in the power supply.   The operating bias is obtained by use of R_k in most applications and is often bypassed by C_k so as to reduce degeneration.

11.18   Design of a Phase-shift Oscillator.   Let us design the phase-shift oscillator illustrated in Fig. 11.42.   Let $E_{bb} = 280$ volts, with a 100-volt tap for the screen supply; the 6SJ7 will be used; $f_o = 1$ kc.

Step 1: Determination of the RC-network Components
Recall that
$$f_o = \frac{0.159}{2.45RC}$$

and by transposing,

$$R = \frac{0.159}{2.45Cf_o}$$

Notice that R_3 serves as the grid resistor of the tube; therefore, it must be sufficiently large to prevent the input circuit of the tube from being severely shunted. In order to make it possible to use reasonably large

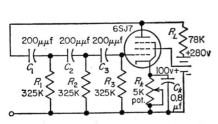

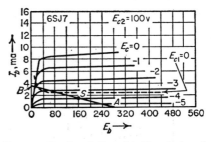

FIG. 11.42 A designed phase-shift oscillator.

FIG. 11.43 Load line for the designed phase-shift oscillator. (*Characteristics from RCA*)

values of R_1, R_2, and R_3, a relatively small value of C should be selected. Let $C = 200 \ \mu\mu\text{f} = C_1 = C_2 = C_3$.

Substituting values and solving,

$$R = \frac{0.159}{2.45 \times 200 \times 10^{-12} \times 1 \times 10^3}$$
$$\cong 325 \text{ kilohms}$$

Step 2: Determination of the Tube Parameters. Select a convenient operating point, such as S in Fig. 11.43, and construct the load line ASB. The zero signal parameters are:

$$E_{b,s} = 160 \text{ volts}$$
$$E_{c1,s} = -4 \text{ volts}$$
$$I_{b,s} = 1.5 \text{ ma}$$
$$I_{c2,s} \cong 0.5 \text{ ma}$$

and

$$R_L = \frac{E_{bb}}{I_{b(y)}} = \frac{280}{3.6 \times 10^{-3}} \cong 78 \text{ kilohms}$$

R_k can now be found:

$$R_k = \frac{E_{c1,s}}{I_{b,s} + I_{c2,s}} = \frac{4}{(1.5 + 0.5) \times 10^{-3}}$$
$$\cong 2 \text{ kilohms}$$

NOTE: A 5-kilohm potentiometer may be used for R_k when the circuit is first assembled. R_k is then adjusted for the most desirable output waveform.

$X_{Ck} = 0.1R_k$ at 1 kc = 200 ohms at 1 kc; and

$$C_k = \frac{0.159}{2 \times 10^2 \times 1 \times 10^3} \cong 0.8 \ \mu f \ (\text{minimum})$$

Step 3: Circuit Construction and Analysis. Adjust the R_k potentiometer for the desired output voltage waveform. The frequency of the oscillator can be checked with the oscilloscope. Adjustment of the frequency is generally most conveniently made by changes in R_1, R_2, and R_3. Increasing their values reduces the frequency, and vice versa.

11.19 Fundamentals of Vacuum-tube Multivibrator Action. A two-stage RC-coupled amplifier, where the output of stage 2 is coupled back to

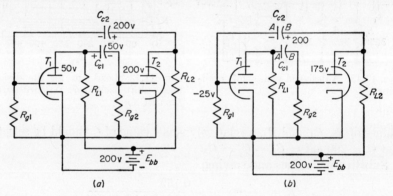

FIG. 11.44 The two multivibrator states: (a) T_1 ON, T_2 OFF, (b) T_1 OFF, T_2 ON.

the grid of stage 1, can be designed to oscillate at a predetermined frequency. This principle is utilized in the basic multivibrator circuit.

Let us consider the two states of the basic multivibrator with the arbitrarily selected potentials in Fig. 11.44. Assume T_1 is at such a conduction level that $E_{b1} = 50$ volts (R_{L1} drops the remainder of E_{bb}). Notice that C_{C1}, which blocks the unidirectional component of E_{b1}, is equal to 50 volts at this instant with the polarity shown in diagram a. Assume that T_2 is cut off at the same instant. Since T_2 has no plate current, R_{L2} drops zero volts and $E_{b2} = E_{bb} = 200$ volts at this instant, as shown in diagram a. C_{C2}, which blocks the unidirectional component of T_2, has a potential of 200 volts with the polarity as indicated.

Now consider the condition shown in diagram b. Assume the plate current is made to flow in T_2 by an abrupt pulse to its grid, thereby reducing E_{b2} by 25 volts. Recall that C_{C2} is 200 volts from condition A, but begins to decrease to the new value of E_{b2}, which is 175 volts. The time required for C_{C2} to change is determined by C_{C2}, R_{g1}, and r_{p2}. The electron discharge path of C_{C2} is from its plate A through R_{g1}, r_{p2}, and back to plate B of C_{C2}.   At the instant T_2 is abruptly driven into conduction, plate B of C_{C2} is $+200$ volts with respect to its plate A, but the plate of T_2 is only $+175$ volts with respect to common. C_{C2} and R_{g1} form a series circuit from the plate of T_2 and common and must have the same potential ($+175$ volts).   Since $e_{CC2} = +200$ volts, then e_{Rg1} must be -25 volts to provide the total of $+175$ volts.   Assume this -25 volts applied to the grid of T_1 is sufficiently negative to cut off T_1.   As T_1 begins to cut off, E_{b1} increases from 50 toward 200 volts.   Plate A of C_{C1} discharges electrons through R_{L1}, E_{bb}, R_{g2} on to plate B of C_{C1}.   This discharge action of C_{C1} drives the grid of T_2 positive, which develops grid current from T_2.   The T_2 grid current speeds up the discharge action of C_{C1}, thereby reducing the time required for C_{C1} to build up to a potential equal to E_{bb}.   This decreases the time required for T_2 to reach its maximum conduction state.

Assume T_2 is at its maximum conduction state when $E_{b2} = 50$ volts. At this time, e_{CC2} is still about 200 volts and e_{Rg1} must be -150 volts, so that

$$e_{CC2} + e_{Rg1} = E_{b2} = +50 \text{ volts}$$

While T_1 is at cutoff, C_{C2} discharges through R_{g1} with a current which exponentially decreases.   When the current is sufficiently small, e_{Rg1} will be less than the grid cutoff value of T_1, and T_1 begins to conduct.   As T_1 begins to conduct, E_{b1} decreases from $+200$ volts to, let us say, $+175$ volts. $e_{CC1} = +200$ volts at this instant, and since

$$e_{CC1} = e_{Rg2} + E_{b1},$$

then $e_{Rg2} = E_{b1} - e_{CC1} = 175 - 200 = -25$ volts

This cuts off T_2, causing E_{b2} to rise to the value of E_{bb}. $e_{CC2} = +50$ volts at this instant and since

$$e_{CC2} = e_{Rg1} + E_{b2}$$

then $e_{Rg1} = E_{b2} - e_{CC2} = 200 - 50 = +150$ volts

which permits the flow of T_1 grid current, reducing the time for the action to be completed.

$$T_1 = R_{g1} + \left(\frac{R_{L2} r_p}{R_{L2} + r_p} \right) C_{C2} \log_\epsilon \left(\frac{E_{b,\max} - E_{b,\min}}{E_{c,CO}} \right)$$

where $T_1 = T_p/2$

$T_p = 1/f_o =$ seconds

$E_{c,CO} =$ grid-cutoff voltage

$E_{b,\max} = E_{bb} =$ volts

$E_{b,\min} = E_b$ at the intersection of the load line and $E_c = 0$-volt characteristic

$R_{g1} = R_{g2} =$ ohms

$C_{C1} = C_{C2} =$ farads

When R_{g1} is many times larger (10 or more) than

$$\frac{R_{L2}r_p}{R_{L2} + r_p}$$

then
$$T_1 \cong R_g C \log_\epsilon \left(\frac{E_{b,\max} - E_{b,\min}}{E_{c,CO}} \right)$$

The time required for the output pulse to build up to 98 per cent of its peak value may be found by

$$t_b = 4C(R_L + r_p)$$

From the preceding equations, it is seen that the pulse time is directly proportional to the maximum and minimum plate voltages, R_g, R_L, r_p, and C, and inversely proportional to the required cutoff-grid potential. The conventional load-line techniques may be used for the determination of R_L and r_p. R_g of a large value is suggested for ease of design (but not above the recommended limits of the tube manufacturer), and C values must be selected to meet the frequency requirements as a first approximation. Final frequency adjustments can be made by varying R_g.

Figure 11.45 illustrates the waveforms of the free-running symmetrical multivibrator. As discussed in the preceding paragraphs, one tube

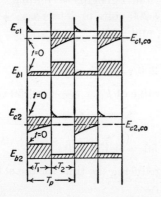

FIG. 11.45   Waveforms of the free-running symmetrical multivibrator.

conducts heavily at the time the second tube is at cutoff. The tubes exchange the conditions of ON and OFF at the rate determined by the preceding relationship. It will be noticed that the plate waveforms are essentially square wave in nature, while the control-grid potentials are exponential because of the RC time constant.

11.20 Design of a Free-running Symmetrical Multivibrator.
Let us design the free-running symmetrical multivibrator shown in Fig. 11.46. Refer to Fig. 11.40 for the average plate characteristics of the 12AU7.

Since the multivibrator is to be symmetrical, then

$$R_{L1} = R_{L2} \qquad R_{g1} = R_{g2} \qquad \text{and} \qquad C_{C1} = C_{C2}$$

Let $E_{bb} = 250$ volts; $f_o = 1$ kc.

Step 1: Load-line Construction and Determination of R_{L1} and R_{L2}.
Let $I_{b,SC} = 20$ ma; then the load line is drawn (see Fig. 11.40).

$$R_{L1} = R_{L2} = \frac{E_{bb}}{I_{b,SC}} = \frac{250}{20 \times 10^{-3}}$$
$$= R_{L2} = 12.5 \text{ kilohms}$$

Step 2: Determination of r_p and $E_{c,co}$

$$r_p = \frac{\Delta e_b}{\Delta i_b} \qquad \text{with } E_c \text{ fixed}$$

Since r_p is to be determined in the active region of the characteristics, let

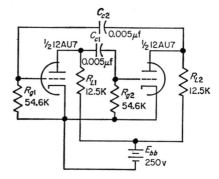

Fig. 11.46 A designed free-running symmetrical multivibrator.

us compute r_p with E_c constant at -2 volts and E_b varying from 100 to 150 volts;

$$r_p = \frac{150 - 100}{(13 - 6) \times 10^{-3}} = \frac{50}{7 \times 10^{-3}} \cong 7.14 \text{ kilohms}$$

$E_{c,co}$ is the grid-cutoff potential and is about -24 volts, as found in the characteristics.

Step 3: Determination of C_{C1}, C_{C2}, R_{g1}, and R_{g2}. In order to permit the use of the simplified design equation, let us select an R_g value which is 10 or more times the equivalent value of r_p and R_L.

$$R_{o,eq} = \frac{R_L r_p}{R_L + r_p} = \frac{12.5 \times 10^3 \times 7.14 \times 10^3}{(12.5 \times 10^3) + (7.14 \times 10^3)}$$
$$\cong 4.54 \text{ kilohms}$$

Therefore any R_g value greater than 45 kilohms will permit us to utilize the simplified equation for the determination of T_1.

C is selected so that X_{CC} is preferably less than the sum of $R_{o,eq}$ and R_g at the frequency of oscillation. The minimum acceptable value of R_g can be a means to determine the capacitor to be used.

$$X_{CC} \cong R_{o,\text{eq}} + R_g \text{ at 1 kc}$$
$$\cong 4.54\text{K} + 45\text{K} \cong 50 \text{ kilohms at 1 kc}$$

and
$$C = \frac{0.159}{1 \times 10^3 \times 5 \times 10^4} \cong 0.0032 \ \mu\text{f (minimum)}$$

Let us select $C = C_{C1} = C_{C2} = 0.005 \ \mu\text{f.}$
Since $f_o = 1$ kc, then
$$T_p = \frac{1}{f_o} = \frac{1}{1 \times 10^3} = 1 \times 10^{-3} \text{ sec}$$

and
$$T_1 \ (\text{half of } T_p) = 0.5 \times 10^{-3} \text{ sec}$$

The approximation design equation for T_1 is

$$T_1 = R_g C \log_\epsilon \frac{E_{b,\max} - E_{b,\min}}{E_{c,CO}}$$

Transposing,

$$R_g = \frac{T_1}{C \log_\epsilon \left[(E_{b,\max} - E_{b,\min})/E_{c,co} \right]}$$

where $E_{b,\min} = E_b$ at the intersection of the load line and $E_c = 0$ characteristic

$E_{b,\max} = E_{bb}$

Substituting values and solving,

$$R_g = \frac{5 \times 10^{-4}}{5 \times 10^{-9} \log_\epsilon \left[(250 - 100)/-24 \right]} \cong 54.6 \text{ kilohms}$$

NOTE: The use of a larger capacitor would result in a smaller R_g, and vice versa.

Step 4: Circuit Construction and Test. Upon constructing the circuit, the frequency of oscillation can be checked with an oscilloscope. The frequency can be reduced by increasing R_g and/or C_C.

11.21 Design of a Free-running Unsymmetrical Multivibrator.

Figure 11.47 illustrates the free-running unsymmetrical multivibrator and its waveforms, which is to be designed in this section. This multivibrator functions in the same way as its symmetrical counterpart, except the ON and OFF period of each tube is not the same. Following are the equations which determine the time of T_1 and T_2:

$$T_1 = \left(R_{g1} + \frac{R_{L2}r_p}{R_{L2} + r_p}\right) C_{C2} \log_\epsilon \frac{E_{bb2} - E_{b2,min}}{E_{c1,CO}}$$

and
$$T_2 = \left(R_{g2} + \frac{R_{L1}r_p}{R_{L1} + r_p}\right) C_{C1} \log_\epsilon \frac{E_{bb1} - E_{b1,min}}{E_{c2,CO}}$$

where
$$T_p = T_1 + T_2 = \frac{1}{f}$$

In some cases, it is desirable to make R_{L1} and R_{L2} different values, since T_1 and T_2 are not equal. This is not necessary in all cases, however,

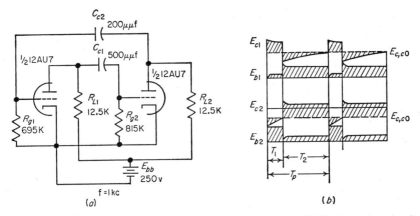

Fig. 11.47 The designed free-running unsymmetrical multivibrator: (a) circuit, (b) waveforms.

since the grid-load resistor R_g and coupling capacitor C_C have a greater effect upon the times of T_1 and T_2. Let us use the same R_L values found for the symmetrical circuit of Fig. 11.46; i.e.,

$$R_{L1} = R_{L2} = 12.5 \text{ kilohms}$$

Let $T_1 = 0.25T_p$; $T_2 = 0.75T_p$.
Since the frequency of oscillation is 1 kc, then

$$T_p = \frac{1}{1 \times 10^3} = 1 \times 10^{-3} \text{ sec}$$

and
$$T_1 = 0.25 \times 10^{-3} \text{ sec}$$
$$T_2 = 0.75 \times 10^{-3} \text{ sec}$$

C_{C1} and C_{C2} may be the same value of capacitance, provided R_{g1} and R_{g2} are not the same. In order to ensure a sufficiently large R_{g1} not to severely shunt the input of T_1, C_{C2} can be made smaller than C_{C1} (when

T_1 is shorter than T_2). Therefore,

$$C_{C2} = 200 \ \mu\mu f$$

and
$$C_{C1} = 500 \ \mu\mu f$$

Since each tube has the same R_L and load line (Fig. 11.40), the r_p value computed in the preceding section may be used (7.14 kilohms), and $E_{c,co}$ of each tube is also the same (-24 volts). $E_{bb} = 250$ volts, $E_{b,min}$ (determined by intersection of the load line and the $E_c = 0$-volt characteristic) $\cong 100$ volts.

Transposing the T_1 equation and substituting values, we may solve for R_{g1}:

$$
\begin{aligned}
R_{g1} &= \frac{T_1}{C_{C2} \log_\epsilon \left[(E_{bb2} - E_{b2,min})/E_{c1,CO} \right]} - \frac{R_{L2} r_p}{R_{L2} + r_p} \\
&= \frac{25 \times 10^{-5}}{2 \times 10^{-10} \log_\epsilon \left[(250 - 100)/-24 \right]} - \frac{12.5 \times 7.14 \times 10^6}{(12.5 \times 10^3) + (7.14 \times 10^3)} \\
&\cong 695 \text{ kilohms}
\end{aligned}
$$

Transposing the T_2 equation and substituting values, we may solve for R_{g2}:

$$
\begin{aligned}
R_{g2} &= \frac{T_2}{C_{C1} \log_\epsilon \left[(E_{bb1} - E_{b1,min})/E_{c2,CO} \right]} - \frac{R_{L1} r_p}{R_{L2} + r_p} \\
&= \frac{75 \times 10^{-5}}{5 \times 10^{-10} \log_\epsilon \left[(250 - 100)/-24 \right]} - \frac{12.5 \times 7.14 \times 10^6}{(12.5 \times 10^3) + (7.14 \times 10^3)} \\
&\cong 830 \text{ kilohms}
\end{aligned}
$$

NOTE: When R_g is many times larger than r_p and R_L, the preceding equations may be simplified to the bracketed quantity for a good first approximation.

The final adjustment of R_{g1} and R_{g2} may be made after assembling the circuit.

11.22 Design of a Free-running Positive-bias Multivibrator.

Let us design the free-running positive-bias multivibrator of Fig. 11.48. The positive bias for the grids of both tubes is obtained from the top of R_{B2}, although the desired positive potential for the grids could be obtained by other techniques.

R_{L1} and R_{L2} can be determined by use of the conventional load-line techniques used in the preceding multivibrators. For simplicity purposes, let $R_{L1} = R_{L2}$, although this is not a requirement. Let

$$R_{L1} = R_{L2} = 12.5 \text{ kilohms}$$

$E_{bb} = 250$ volts; desired positive bias $= 2$ volts. The load line is shown in Fig. 11.40. Symmetrical output is desired. Let us proceed to determine the remaining parameters of Fig. 11.48.

Step 1: *Determination of* R_{B1} *and* R_{B2}. Notice that R_{B1} and R_{B2} are directly across the power supply. Select the total R_B such that a low current drain is imposed on the supply. Let $R_{B1} + R_{B2} = 500$ kilohms; then

$$I_{RB} = \frac{E_{bb}}{R_{B1} + R_{B2}} = \frac{250}{500 \times 10^3} = 0.5 \text{ ma}$$

R_{B2} should be such a value that the potential across it is 2 volts.

$$R_{B2} = \frac{E_c}{I_{RB}} = \frac{2}{5 \times 10^{-4}} = 4 \text{ kilohms}$$

and
$$R_{B1} = \frac{E_{bb} - E_c}{I_{RB}} = \frac{248}{5 \times 10^{-4}} = 496 \text{ kilohms}$$

Let $f_o = 1$ kc, and $C_{C1} = C_{C2} = 0.005$ μf. The selection of R_{g1} and R_{g2} can be made the determining factor in ensuring the correct oscillation

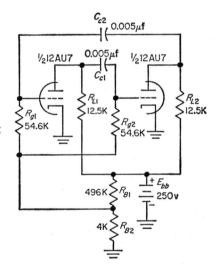

FIG. 11.48   The designed free-running positive-bias multivibrator.

frequency. When R_{g1} is many times larger than the parallel equivalent value of r_{p1} and R_{L1},

$$R_{g1} = \frac{T_1}{C_{C2} \log_\epsilon \left[(E_{bb} - E_{b,\min} + E_{c1})/(E_{c1} + E_{c1,co}) \right]}$$

where T_1 = time of first fractional period
$\quad T_p = T_1 + T_2 = 1/f_o$
$\quad E_{c1}$ = positive bias voltage for grid of T_1
$\quad E_{c1,co}$ = cutoff-grid potential of T_1

When R_{g2} is many times larger than the parallel equivalent value of r_{p1} and R_{L1},

$$R_{g2} = \frac{T_2}{C_{C2} \log_\epsilon \left[(E_{bb} - E_{b,min} + E_{c2})/(E_{c2} + E_{c2,CO})\right]}$$

where T_2 is the time of second fractional period.

Step 2: Determination of R_{g1} and R_{g2}. The $E_c = +2$-volt characteristic is drawn on the original family in Fig. 11.40. Since symmetrical operation is desired, then let $R_{g1} = R_{g2}$. The parameters obtained from the load line are:

$E_{b,min} = 75$ volts (intersection of $E_c = +2$-volt characteristic and the
        load line)
$E_{c,CO} = -24$ volts
   $E_c = +2$ volts
and $T_1 = 0.5T_p = 0.5(1/f_o)$
                 $T_1 = 0.5(1/1 \times 10^3) = 5 \times 10^{-4}$ sec
Solving for R_{g1},

$$R_{g1} \cong \frac{5 \times 10^{-4}}{5 \times 10^{-9} \log_\epsilon \left[(250 - 75 + 2)/(2 + (-24))\right]}$$
$$\cong 48 \text{ kilohms} \cong R_{g2}$$

NOTE: The use of a smaller C_{C1} and C_{C2} would result in the use of larger values of R_{g1} and R_{g2}.

11.23 Design of a One-shot Multivibrator. The synchronizing frequency is applied to T_1 (see Fig. 11.49). If a negative pulse is to be

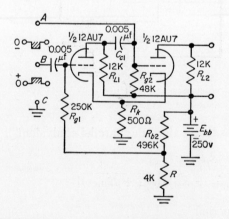

FIG. 11.49 The designed one-shot multivibrator.

used, it should be applied between points A and B, while a positive pulse is applied between points B and C. The synchronizing pulse upsets the initial circuit condition causing T_1 to conduct and cuts off T_2.

The free-running period of T_2 can be determined from

$$T_{1(2)} = R_{g2}C_{C2} \log_\epsilon \frac{E_{bb} - E_{b,\min} + E_{c2}}{E_{c2} + E_{c,co}}$$

where E_{bb}, $E_{b,\min}$, E_{c2} and $E_{c,co}$ are determined from the load line. Knowing $T_{1(2)}$, C_{C2} may be selected and then the value of R_{g2} is calculated:

$$R_{g2} = \frac{T_{1(2)}}{C_{C2} \log_\epsilon \left[(E_{bb} - E_{b,\min} + E_{c2})/(E_{c2} + E_{c,co}) \right]}$$

Notice that R_k is common to both cathodes. The grid of T_2 is very positive because it is connected to $+E_{bb}$. The cathode current of T_2 is relatively large, making e_{Rk} large enough to cut off T_1, since the positive bias for the grid of T_1 is relatively small e_{RB1}.

The pulse required to bring T_1 into conduction must be of sufficient magnitude to make the grid of T_1 less than its cutoff value. When such a pulse is applied to T_1, the grid of T_2 is driven beyond cutoff for a time determined by the preceding equation.

Using the load-line parameters of the preceding circuit (Fig. 11.48), the required R_{g2} would be the same value. R_{g1} may be selected to meet the input impedance requirements of stage 1.

PROBLEMS

11.61 Refer to the Armstrong oscillator of Fig. 11.35. $C_{g1} = 100$ $\mu\mu f$; $f_o = 5$ kc. In order to utilize grid-leak bias, R_g must be such a value that the C_{g1} discharge time constant is long. Find (a) T_p; (b) R_g (minimum).

11.62 Assume the Armstrong oscillator of Fig. 11.35 oscillates at 100 kc. $C_{g1} = 200$ $\mu\mu f$. Grid-leak bias is to be used. Find (a) T_p; (b) R_g (minimum).

11.63 An Armstrong oscillator has the following values: $L_g = 20$ mh; $C_{g2} = 50$ $\mu\mu f$; R_g is very large. Find f_o.

11.64 An Armstrong oscillator has the following parameters: $L_p = 30$ mh; $L_g = 10$ mh; $R_g = 20$ kilohms; $r_p = 10$ kilohms; $C_{g2} = 50$ $\mu\mu f$. Find f_o.

11.65 In Prob. 11.64, L_g is changed to 30 mh. Find f_o.

11.66 In Probs. 11.64 and 11.65, what effect does a change in L_g have upon f_o?

11.67 In Prob. 11.64, R_g is changed to 10 kilohms. Find f_o.

11.68 In Probs. 11.64 and 11.67, what effect does a change in R_g have upon f_o?

11.69 Refer to the Hartley oscillator of Fig. 11.36. $f_o = 10$ kc; $L_p = 40$ mh; $C_{g2} = 100$ $\mu\mu f$. Find beta.

11.70 Refer to the Hartley oscillator of Fig. 11.36. $L_p = 30$ mh; $L_g = 10$ mh; $C_{g2} = 50$ $\mu\mu f$; tube $\mu = 40$; the resistance of L_p is negligible when compared to r_p; k (coefficient of coupling) of L_p and $L_g = 1$. Find:
  (a) M of L_p and L_g (b) f_o
  (c) beta   (d) A_v
  (e) Z_L

11.71 In Prob. 70, C_{g2} is changed to 100 $\mu\mu f$. Find:
  (a) f_o (b) beta
  (c) A_v (d) Z_L

11.72 Refer to the Hartley oscillator of Fig. 11.36. $L_g = 20$ mh; $L_p = 60$ mh; $f_o = 5$ kc; k between L_g and $L_p = 1$; let $C_{g1} = C_{g2}$. Find:
(a) C_{g2}
(b) R_g (minimum) for good grid-leak bias action
(c) beta   (d) βA_v
(e) minimum tube mu required

11.73 Refer to the Hartley oscillator of Fig. 11.36. $L_g = 10$ mh; $L_p = 30$ mh; $f_o = 10$ kc; k between L_g and $L_p = 1$; let $C_{g1} = C_{g2}$. Find:
(a) C_{g2} (b) R_g (minimum)
(c) beta   (d) A_v
(e) minimum tube mu required

11.74 Refer to the Colpitts oscillator of Fig. 11.37. $C_1 = 200$ μμf; $C_2 = 100$ μμf; $L_1 = 50$ mh; tube $\mu = 50$; R_{L1} is negligible. Find:
(a) C_t (b) f_o
(c) beta   (d) A_v (with feedback)

11.75 Refer to the Colpitts oscillator of Fig. 11.37. $C_1 = 500$ μμf; $C_2 = 100$ μμf; $L_1 = 50$ mh; tube $\mu = 50$; R_{L1} is negligible. Find:
(a) C_t (b) f_o
(c) beta   (d) A_v (with feedback)

11.76 Refer to Probs. 11.74 and 11.75. What effect does an increase in C_1, with the remaining given parameters constant, have on (a) C_t; (b) f_o; (c) beta; (d) A_v (with feedback).

11.77 Refer to the Colpitts oscillator of Fig. 11.37. $L_1 = 30$ mh; $f_o = 10$ kc; R_{L1} is negligible. Calculate the required C_t.

11.78 In Prob. 11.77, assume a tube with a mu of 20 is used. What ratio of X_{C2}/X_{C1} must be used so that $\beta A_v = 1$?

11.79 C_t is that found in Prob. 11.77, and the ratio of X_{C2}/X_{C1} is that found in Prob. 11.78. Find (a) C_1/C_2 ratio; (b) C_1; (c) C_2.
Hint: Let $C_1 = x$; $C_2 = y$; then two equations can be established:

$$\frac{(x)(y)}{x + y} = C_t$$

and

$$\frac{x}{y} = \frac{X_{C2}}{X_{C1}}$$

Then solve by simultaneous equations.

11.80 In the Colpitts oscillator of Probs. 11.77 to 11.79, let $C_g = 500$ μμf. Find (a) T_p; (b) R_g (minimum).

11.81 Refer to the Colpitts oscillator of Fig. 11.37. $L_1 = 40$ mh; $f_o = 15$ kc; R_{L1} is negligible. Calculate the required C_t.

11.82 In Prob. 11.82, assume a tube with a mu of 10 is used. What ratio of X_{C2}/X_{C1} must be used so that $\beta A_v = 1$?

11.83 C_t is that found in Prob. 11.81, and the ratio of X_{C2}/X_{C1} is that determined in Prob. 11.82. Find (a) C_1/C_2 ratio; (b) C_1; (c) C_2.

11.84 In the Colpitts oscillator of Probs. 11.81 to 11.83, Let $C_g = 500$ μμf. Find (a) T_p; (b) R_g (minimum).

11.85 Refer to the Wien bridge oscillator of Fig. 11.38. Assume $A_{vt} = 50$; $C_a = C_b = 200$ μμf; $R_2 = 4$ kilohms; $f_o = 20$ kc. Find (a) R_1; (b) R_3 and R_4.

11.86 A Wien bridge oscillator has a frequency range of 50 to 500 cps. $R_3 = R_4 = 10$ megohms. Find (a) $C_{a,min}$; (b) $C_{a,max}$.

11.87 Refer to the Wien bridge oscillator of Fig. 11.38. Assume $E_{c1,s} = -6$ volts; $I_{b1,s} = 5$ ma. Compute the value of R_2.

11.88 Assume the Wien bridge oscillator of Fig. 11.38 has an over-all two-stage gain of 3. Find the minimum value of R_1.

11.89 In the Wien bridge oscillator of Probs. 11.87 and 11.88, assume $f_o = 10$ kc; $C_a = C_b = 100$ $\mu\mu$f. Find R_3 and R_4.

11.90 In Prob. 11.88, find R_3 and R_4 if $C_a = C_b = 50$ $\mu\mu$f.

11.91 Refer to the designed phase-shift oscillator of Fig. 11.42. $C_1 = C_2 = C_3 = 500$ $\mu\mu$f; f_o is to be 10 kc. Find (a) R_1; (b) R_2; and (c) R_3.

11.92 In Prob. 11.91, $C_1 = C_2 = C_3 = 200$ $\mu\mu$f. Find (a) R_1; (b) R_2; and (c) R_3.

11.93 Refer to the designed phase-shift oscillator of Fig. 11.42. $R_1 = R_2 = R_3 = 20$ kilohms; f_o is to be 5 kc. Find (a) C_1; (b) C_2; and (c) C_3.

11.94 In Prob. 11.93, $R_1 = R_2 = R_3 = 15$ kilohms. Find (a) C_1; (b) C_2; and (c) C_3.

11.95 Refer to the designed free-running symmetrical multivibrator circuit of Fig. 11.46. $f_o = 10$ kc; $E_{b,max} = 200$ volts; $E_{b,min} = 50$ volts; $E_{c,co} = -30$ volts; $C = 500$ $\mu\mu$f. Find (a) T_p; (b) T_1; (c) R_g.

11.96 In Prob. 11.95, $f_o = 15$ kc; all other parameters are the same. Find (a) T_p; (b) T_1; (c) R_g.

11.97 Refer to the designed free-running unsymmetrical multivibrator of Fig. 11.47. $R_{L1} = R_{L2} = 10$ kilohms; $r_p = 5$ kilohms (both tubes); $E_{bb1} = E_{bb2} = 200$ volts; $E_{b1,min} = E_{b2,min} = 50$ volts; $E_{c1,co} = E_{c2,co} = -20$ volts; $f_o = 15$ kc; $C_{c1} = 100$ $\mu\mu$f; $C_{c2} = 500$ $\mu\mu$f; $T_1 = 0.25T_1$. Find:

(a) T_p (b) T_1
(c) R_{g1} (d) T_2
(e) R_{g2}

11.98 In Prob. 11.97, $C_{c1} = 200$ $\mu\mu$f; $C_{c2} = 400$ $\mu\mu$f; all other parameters are as given. Find:

(a) T_p (b) T_1
(c) R_{g1} (d) T_2
(e) R_{g2}

11.99 In Probs. 11.95 and 11.96, with all other parameters the same, what is the relationship between the value of R_g and f_o?

11.100 In Probs. 11.95 and 11.96, assume $R_L = 20$ kilohms and $r_p = 10$ kilohms. Find the time required for the pulse to build up to 98 per cent of its peak value.

CHAPTER 12

SEMICONDUCTOR AND VACUUM-TUBE
LOGIC CIRCUITS

The fundamentals of switching circuits are analyzed first in this chapter. The basic concepts of ON and OFF are developed. This analysis is then followed by a study of actual switching circuits and techniques which may be employed for successful design of these circuits.

The basic gate circuits, which have one or more inputs for obtaining a single output, are examined. The transistor gates are treated first, and consideration is also given to the diode and vacuum-tube variety of these basic gate circuits. These fundamental gate circuits are of the switching variety, since the control device (transistor, diode, or tube) is in either its saturation or its cutoff region. In these circuits, the active region of the control device is generally used solely in changing from the saturation to the cutoff region, or vice versa. The output waveform is most often a pulse, and no attempt is made to faithfully reproduce the input signal. The duty of the input signal is to trigger the control device from one state to the other (saturation to cutoff, or vice versa). Only the most fundamental gate circuits used in pulse circuitry are analyzed in this chapter.

12.1 Fundamentals of Transistor Switching. A transistor switch has two conditions, namely ON and OFF. During the ON condition, the switch (transistor) is to display a very low resistance, whereas high resistance should be present during the OFF condition.

Figure 12.1 illustrates the output characteristics (common-emitter configuration) of a transistor and a load line. Recall that the transistor displays its minimum resistance value in the saturation region (between points A and B of the load line):

$$r_o = \frac{\Delta e_C}{\Delta i_C}$$

A glance at the AB region reveals that a relatively small change of e_C results in a relatively large change in i_C, which indicates a low dynamic resistance value. The OFF region lies between C and D. Using the preceding equation, we find that a relatively large change of e_C is accompanied by a very small change in i_c, resulting in a high dynamic resistance value in this region.

Depending upon the specific application, the transistor may be driven to any point between A and B (or just to the right of point B) during its ON time and anywhere between C and D during its OFF time. The dynamic resistance at OFF time can be maintained at a high value by the application of reverse bias to the base-emitter junction during the OFF

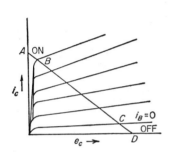

FIG. 12.1 ON-OFF conditions of a transistor switch.

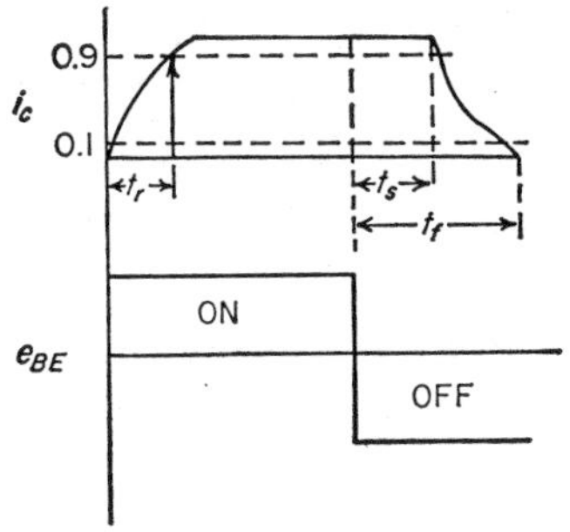

FIG. 12.2 Relationship of i_C and $e_{B\text{-}E}$ (PNP).

time. Note that point C of the load line is at the intersection of the zero base current characteristic, at which time zero bias is applied at the base. Reverse base bias results in a small reverse emitter current, which subtracts from I_{CO}, resulting in a lower I_{CO} value and a higher OFF time dynamic resistance.

Both ON and OFF conditions, as well as the region between them (called the active region), are stable states. The voltages and/or currents supplied to the transistor must be changed in order for it to switch from OFF to ON or vice versa.

The *rise time* t_r is the time required of i_C to arrive at 90 per cent of its maximum ON time value when a square-wave pulse is applied at the base of the transistor. This relationship is shown in Fig. 12.2.

The *storage time* t_s is the time required for the transistor to begin its turn OFF after the forward base bias is removed and a reverse base bias is applied. The storage time is caused by the minority-carrier storage in the base region. Recall that the carriers which passed from the emitter to the base must rely on the process of diffusion to cross the base region into the collector. When the transistor is driven far into its saturation region for an extended period of time, many of the emitter carriers become temporarily stored in the base. When the base-to-emitter voltage is reversed, the emitter is cut off. But many of the carriers that were previously passed by the emitter into the base have yet to clear the base region and migrate across the collector-base junction. Because of the base diffusion action, the transit time of the carriers in the base region is slower than in the emitter and collector regions. These carriers will continue to flow from the base to the collector after the emitter current is cut off. The lapse between i_E cutoff and the time when i_C actually begins to decrease is the storage time. This is shown in Fig. 12.2. The most successful technique for reducing the storage time to the point where it is no longer a serious problem is by not allowing the ON condition of the transistor to go all the way into its saturation region.

The *fall time t_f* is the time required for i_C to decrease to 10 per cent of its maximum value after the emitter current has been cut off, which is illustrated in Fig. 12.2. Notice the fall time includes the storage time plus the time during which i_C exponentially decays to its 10 per cent value.

12.2 The Common-emitter (Inverter) Switching Circuit.

Figure 12.3 illustrates the basic inverter switching circuit with an NPN and a PNP transistor. Refer to Fig. 12.3*a*, for the following discussion on the theory of operation of the NPN inverter. The base circuit of the transistor is to be under reverse bias so as to maintain the OFF collector current at a value slightly less than I_{CO}. The reverse base current required for maintaining this reverse bias for the OFF condition is about the same value as I_{CO}. Recall that I_{CO} is directly dependent upon temperature. Therefore the reverse base current to be selected, $i_{B,Q}$, should be approximately equal to the I_{CO} of the transistor at its maximum operating temperature. V_{bb} and R_{B2} are so selected that the base is under reverse bias with no signal applied. Upon examination of the NPN inverter, we see the following zero signal relationship:

$$e_{B\text{-}E,Q} = V_{bb} - e_{RB2}$$

where $e_{B\text{-}E,Q}$ is the desired reverse base-emitter bias at zero signal condition,

and
$$R_{B2} = \frac{V_{bb} - e_{B\text{-}E,Q}}{i_{B,Q} + i_{\text{bleeder}}}$$

where $i_{B,Q}$ = desired reverse base current at zero signal

    i_{bleeder} = arbitrarily selected by the designer

Once R_{B2} has been specified in the preceding manner, R_{B1} can be determined:

$$R_{B1} = \frac{e_{B\text{-}E,Q}}{i_{\text{bleeder}}}$$

The input pulse must be sufficiently large to overwhelm the reverse bias and bring the base current up to its ON value.   The rise time is increased by the length of time required to bring the base current from its reverse value to zero.   The magnitude of this pulse is most conveniently determined by the joint use of the $e_{C\text{-}E}$ versus i_C and $e_{B\text{-}E}$ versus i_C characteristics.   The ON base current value is found at the ON point along the load line of the output characteristics.   The maximum collector current is also found at the same point.   The maximum collector current value is then located on the $e_{B\text{-}E}$ versus i_C characteristic, from which $e_{B\text{-}E}$ for that condition is determined.   The input signal pulse amplitude must be equal

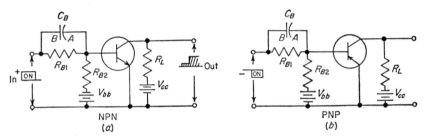

Fig. 12.3   Basic inverter switching circuits: (a) NPN, (b) PNP.

to the arithmetic sum of the OFF reverse $e_{B\text{-}E}$ and the ON forward $e_{B\text{-}E}$. In equation form,

$$E_i = e_{B\text{-}E,Q} + e_{B\text{-}E}\ \text{ON}$$

Recall, however, that $e_{B\text{-}E}$ can be determined by other methods.   When the $e_{B\text{-}E}$ versus i_C or $e_{B\text{-}E}$ versus i_B characteristics are not available, the following relationship enables us to approximate $e_{B\text{-}E}$:

$$e_{B\text{-}E,Q} \cong R_i i_{B,Q}$$

and
$$e_{B\text{-}E}\ \text{ON} \cong R_i i_B\ \text{ON}$$

where
$$R_i = \frac{h_{ie} + (h_{oe}h_{ie} - h_{fe}h_{re})R_L}{1 + h_{oe}R_L}$$

The purpose of the base capacitor C_B is to overdrive the transistor for a brief instant at the beginning of the ON pulse, thereby reducing the rise time of the collector current.   Let us examine the behavior of the circuit illustrated in Fig. 12.3a.

OFF *Condition.*   With no input signal pulse, the transistor is at cutoff, and a bleeder current flows through R_{B2}, R_{B1}, the input generator, and V_{bb}. i_{RB2} consists of the desired reverse base current and an arbitrarily selected bleeder current. The reverse base current flows into the base, and the bleeder current flows through R_{B1}. A negative voltage is present at the base, as determined by the preceding equation, and the remainder of V_{bb} appears across R_{B1} (assuming a negligible input generator resistance).

OFF *to* ON *Condition.* When the input generator develops a positive pulse, plate B of C_B discharges electrons and plate A takes on electrons, giving rise to a sharp increase in forward base current. This enables the rise time of the transistor to be considerably reduced, and $i_{C,M}$ is quickly developed. The charging time of the capacitor should be short, so as not to keep the transistor in the saturation region for too long a time.

ON *to* OFF *Condition.* When the input pulse is reduced to zero, the initial condition of reverse base bias is in effect. The minority carriers still in the base region cross the base-collector junction, and then i_C begins to decrease to its cutoff value.

The rise time can be approximated by

$$t_r \cong \frac{1}{f\alpha_{co}(1-\alpha)} \log_\epsilon \left(\frac{1}{1 - 0.9 i_{C,\max}/\beta i_{B,Q}} \right)$$

where $f\alpha_{co}$ = alpha-cutoff frequency
β = low-frequency beta
$i_{B,Q}$ = zero signal base current
The fall time can be approximated by

$$t_f \cong \frac{1}{f\alpha_{co}(1-\alpha)} \log_\epsilon (i_{C,\max} - i_{B,\max})$$

where $i_{B,\max}$ is the base current at the end of the ON period.

12.3 Design of an Inverter Switching Circuit.

Let us design an inverter switching circuit similar to Fig. 12.3a with the 2N333. $R_L = 1.2$ kilohms; $V_{bb} = 12$ volts; $V_{cc} = 12$ volts. The load line is shown in Fig. 12.4a; point $A (i_C = 9.2$ ma, $e_C = 1$ volt) is ON; point B is OFF; operating temperature = 50°C; pulse rate = 1 kc.

Step 1: *Determination of* $i_{B,Q}$ *and Bleeder Current.* At 50°C, assume $I_{co} \cong 0.015$ μa. In order to compensate for transistor aging and possible higher ambient temperature, let us select $i_{B,Q} = 1$ μa and set the bleeder current at 100 μa.

Step 2: *Determination of* $e_{B-E,Q}$. $e_{B-E,Q}$ can be quickly approximated from the i_C versus e_{B-E} characteristics of Fig. 12.4b. Since $i_{C,Q}$ is about zero, then $e_{B-E,Q}$ is about 0.4 volt (using the 25° curve).

Step 3: Determination of R_{B2} and R_{B1}. Recall that $e_{B\text{-}E,Q}$ is negative for the NPN for reverse bias. We may now find R_{B2}:

$$R_{B2} = \frac{V_{bb} - e_{B\text{-}E,Q}}{i_{B,Q} + i_{\text{bleeder}}}$$

$$= \frac{12 - 0.4}{101 \times 10^{-6}} = 115 \text{ kilohms}$$

and

$$R_{B1} = \frac{e_{B\text{-}E,Q}}{i_{\text{bleeder}}}$$

$$= \frac{4 \times 10^{-1}}{100 \times 10^{-6}} = 4 \text{ kilohms}$$

Step 4: Determination of C_B. The charging time constant of plate A of C_B should be very short, requiring the use of a small capacitor. Let

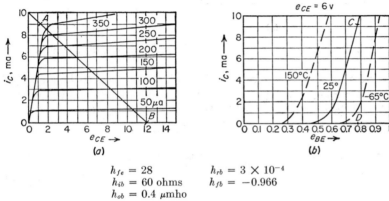

$$h_{fe} = 28 \qquad h_{rb} = 3 \times 10^{-4}$$
$$h_{ib} = 60 \text{ ohms} \qquad h_{fb} = -0.966$$
$$h_{ob} = 0.4 \text{ } \mu\text{mho}$$

FIG. 12.4   Characteristics of 2N333.   (*Transitron.*)

us select 200 $\mu\mu$f. Upon construction of the circuit, smaller and larger values can be tried. The final selection hinges on that value which produces the most desirable output waveform.

Step 5: Determination of the Positive Input Pulse Magnitude. Recall that $i_{C,M} = 9.2$ ma. Referring to Fig. 12.4b (using the 25°C characteristic), $e_{B\text{-}E}$ at that condition is about 0.77 volt. The input pulse must be $0.4 + 0.77 \cong 1.17$ volts.

Step 6: Circuit Construction. The components required are:

$$R_L = 1.2 \text{ kilohms}$$
$$R_{B2} = 115 \text{ kilohms}$$
$$R_{B1} = 4 \text{ kilohms}$$
$$C_B = 200 \text{ } \mu\mu\text{f}$$

12.4 The Inverter with Provisions for a Variable R_L. When used in a composite computer circuit, a given inverter may be called upon to drive one, two, or three inverters. R_L, in effect, will vary from that value of the original inverter to possibly one-third or less of that value. The collector potential of the inverter during its OFF time hinges on its load resistance and I_{CO}. With a larger R_L, this potential is higher than is the case with a smaller R_L. Connecting a diode to the collector, as shown in Fig. 12.5, results in the same collector potential at the OFF condition, regardless of the R_L value.

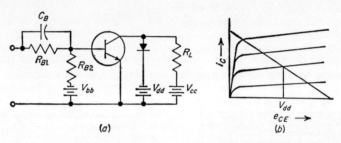

FIG. 12.5   An inverter with a diode clamp: (a) circuit, (b) load line.

e_{C-E} is a reverse voltage for D_1. The diode conducts as long as V_{dd} is greater than e_{C-E}. When e_{C-E} exceeds V_{dd}, the diode clamp is reverse-biased and it opens. $e_{D1} = V_{dd}$ in this condition, thereby clamping e_{C-E} to this value. Figure 12.5b illustrates this effect upon the load line.

12.5 The Nonsaturating Inverter. Figure 12.6 illustrates an NPN non-saturating inverter. Notice that D_1 and V_{dd} have an opposite effect in

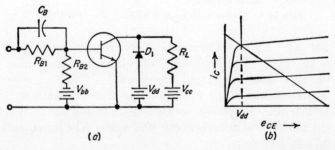

FIG. 12.6   NPN nonsaturating inverter: (a) circuit, (b) load line.

Fig. 12.6 to that illustrated in Fig. 12.5. Setting V_{dd} to the lowest desirable collector voltage will prevent e_{C-E} from decreasing beyond that value. This effectively keeps the transistor out of the saturation region, as shown in the load line of Fig. 12.6b.

PROBLEMS

12.1 Refer to Fig. 12.1. In what region is the output resistance of the transistor at a minimum? Explain.

12.2 In Fig. 12.1, what magnitude of output resistance does the transistor display in the (*a*) ON region; (*b*) OFF region; (*c*) active region.

12.3 How can the output resistance of a transistor switch be increased when it is in its OFF region? Explain.

12.4 In Fig. 12.1, why are all the regions "stable states"?

12.5 What is rise time t_r?

12.6 What is storage time t_s?

12.7 What is the primary cause of storage time?

12.8 State the relationship between the transit time of the carriers in the base region and storage time.

12.9 What is the best method for reducing the storage time of a given transistor?

12.10 What is fall time t_f?

12.11 Refer to the inverter-switching circuit of Fig. 12.3*a*. The 2N465 is to be used (see Fig. 8.5); $V_{cc} = 15$ volts; $V_{bb} = 15$ volts; $R_L = 2$ kilohms. Let condition ON be at the intersection of the load line and the 150-μa characteristic; and condition OFF is located at the intersection of the load line and the 0-μa characteristic; pulse rate = 10 kc; $C_B = 250$ $\mu\mu$f.

(*a*) Construct the load line.

(*b*) Locate the ON and OFF points on the load line.

(*c*) Determine i_C, e_C, and i_B for condition ON.

(*d*) Determine i_C and e_C for condition OFF.

NOTE: Assume i_B OFF is 1μa.

12.12 In the inverter switching circuit of Prob. 12.11, find (*a*) R_i of the transistor; (*b*) $e_{B-E,Q}$ (or OFF condition); (*c*) e_{B-E} ON.

12.13 Refer to Probs. 12.11 and 12.12. Let $i_{\text{bleeder}} = 100$ μa. Find (*a*) R_{B2}; (*b*) R_{B1}.

12.14 Determine the required magnitude of the input pulse for the inverter switching circuit of Probs. 12.11 to 12.13.

12.15 Refer to the inverter switching circuit of Fig. 12.3*a*. The 2N337 is to be used (see Fig. 8.26); $V_{cc} = 20$ volts; $V_{bb} = 20$ volts; $R_L = 1.5$ kilohms. Let condition ON be at the intersection of the load line and the 350-μa characteristic; and condition OFF is located at the intersection of the load line and the 0-μa characteristic; pulse rate = 20 kc; $C_B = 200$ $\mu\mu$f.

(*a*) Construct the load line.

(*b*) Locate the ON and OFF points on the load line.

(*c*) Determine i_C, e_C, and i_B for condition ON.

(*d*) Determine i_C and e_C for condition OFF.

NOTE: Assume i_B OFF is 1 μa.

12.16 In Prob. 12.15, using the i_C versus e_{B-E} characteristics, find (*a*) e_{B-E} OFF; (*b*) e_{B-E} ON.

12.17 Refer to Probs. 12.15 and 12.16. Let $i_{\text{bleeder}} = 100$ μa. Find (*a*) R_{B2}; (*b*) R_{B1}.

12.18 Determine the required magnitude of the input pulse for the inverter switching circuit of Probs. 12.15 to 12.17.

12.6 The Common-collector (Emitter-follower) Switching Circuit.

The basic emitter follower with an NPN transistor is shown in Fig. 12.7a. Unlike the inverter circuits discussed in the preceding sections, the emitter follower is designed so that the transistor operates in its active region (that region between saturation and cutoff). When the input signal is zero, the base is at ground potential. The emitter is negative with respect to common and is determined by

$$e_{E-B,Q} = V_{ee} - e_{RE,Q}$$

Therefore, the base-to-emitter junction is forward-biased by this potential at the zero signal operating point.

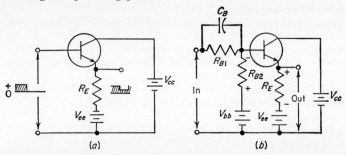

FIG. 12.7   The emitter follower: (a) basic circuit, (b) provisions for level restoration.

The base-to-collector junction is placed under reverse bias by V_{cc}. Notice that the input signal is actually applied across the base-collector junction. The base side of V_{cc} is negative, but the base side of the input pulse is positive, creating a series-opposing effect between these two potentials. The result is a reduction in e_{C-B}, and the transistor conducts more heavily with the application of a positive pulse at the base. There is a corresponding increase in the magnitude of i_E and e_{RE}. Taking e_{RE} as the output voltage of the circuit results in no low-frequency signal inversion and a voltage ratio slightly less than unity. The output level undergoes a negligible shift when the input and output of one stage is considered, but a substantial increase in output level is experienced when a cascade of several emitter followers is used.

Figure 12.7b illustrates a circuit that incorporates a voltage-divider network across the input circuit, which establishes the input signal level for each emitter follower. In this fashion, the original level is restored at the input of each emitter follower, thereby preventing the change in zero signal level. The placement of R_{B1} in series with the base tends to create overshoots in the waveform of e_{RE}, which is reduced by the use of C_B. The zero signal input level is established by

$$e_{B-C,Q} = e_{RB2} - V_{bb} + V_{cc} = e_{RB1} + V_{cc}$$

Notice that V_{bb} and V_{cc} are series-opposing, with their sum plus e_{RB2} appearing as the base-to-collector voltage. Since e_{RB1} equals the difference between e_{RB2} and V_{bb}, then e_{RB1} plus V_{cc} is also the base-to-collector input voltage. If the input signal of this emitter follower is the output of a preceding emitter-follower stage, then the base-to-common potential of this stage would be slightly less positive than the potential between the same points of the preceding stage. Therefore e_{RB1} should be that positive value required for the proper level restoration. e_{RB2} will drop the remainder of V_{bb}. R_{B1} is generally small as compared to R_{B2}, since the level restoration voltage is usually much lower than V_{bb}.

The value of R_{B1} may be arbitrarily selected (1 to 4 kilohms). With the desired level restoration potential e_{RB1} known, i_{RB1} can be readily calculated:

$$i_{RB1} = \frac{e_{RB1}}{R_{B1}}$$

The current flowing through R_{B2} is the sum of the bleeder current of R_{B1} and i_B. e_{RB2} is equal to the difference between V_{bb} and e_{RB1}; therefore

$$R_{B2} = \frac{V_{bb} - e_{RB1}}{i_{RB1} + i_{B,Q}}$$

The values of $i_{B,Q}$ and $i_{E,Q}$ can be determined with a fair degree of accuracy by use of the common-emitter output characteristics of the transistor. *Instead of using V_{cc} alone for the voltage supply plotted along the x axis, V_{ee} plus V_{cc} is used.* Notice that these two voltages are series-aiding. The x-axis intercept of the R_E load line, therefore, is determined by V_{ee} plus V_{cc}, shown as point A in Fig. 12.8. R_E should be

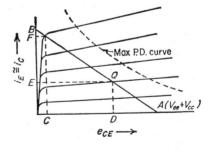

FIG. 12.8   Use of CE output characteristics for design of the emitter follower.

selected so that the maximum power dissipation curve of the transistor is not intercepted. The zero signal operating point Q is selected so that the maximum value of the positive input pulse does not drive the transistor into saturation.

In Fig. 12.8, the emitter-to-collector potential is at point D at rest condition and may be reduced to point C without being driven into saturation. $i_{B,Q}$ and $i_{C,Q}$ are determined at that point D in the conventional manner. Since $i_E \cong i_C$, this value may also be used as $i_{E,Q}$ in many practical cases. $i_{E,Q}$ can be more closely specified by determining the sum of $i_{C,Q}$ and $i_{B,Q}$. Recall that the output voltage is only slightly less than the input voltage in the common-collector configuration; therefore,

$$e_I \cong e_O$$

Since
$$e_I = e_{B\text{-}C}$$
and
$$e_O = e_{E\text{-}C}$$
then
$$e_{B\text{-}C} \cong e_{E\text{-}C}$$

As in the inverter switch, the storage time problem is best minimized by avoiding the saturation region for any portion of the input signal cycle. Since the input signal is series-opposing to V_{cc}, it serves to reduce the reverse bias to the collector. If the input signal amplitude is greater than V_{cc}, the base-to-collector junction will be under forward bias, and the transistor is driven deep into its saturation region. This is easily prevented by selecting a V_{cc} value which is 2 or more volts larger than the maximum input pulse value, thereby ensuring reverse bias to the base-collector junction for the complete input signal cycle.

Power dissipation is to be considered with the emitter follower, since its operating point is located in the active region. It is recommended that the maximum power dissipation of the transistor not be exceeded.

12.7 Design of an NPN Emitter Follower.

Let us use the Transitron 2N333 to design an emitter follower like that of Fig. 12.7b. $V_{cc} = 4$ volts; $V_{bb} = 8$ volts; $V_{ee} = 8$ volts; $R_E = 1.2$ kilohms; input pulse maximum value $= 2$ volts; CE output characteristics of the 2N333 are shown in Fig. 12.9; level restoration voltage $= 0.2$ volt.

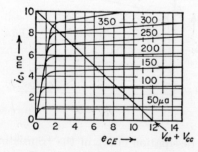

Fig. 12.9 Load line for the emitter follower. (*Characteristics from Transitron.*)

Step 1: Construction of the Load Line. The load line is to intercept the voltage axis at the $V_{cc} + V_{ee}$ value, which is 12 volts. The current axis-intercept point is determined in the conventional manner, resulting in the load line drawn in Fig. 12.9.

Step 2: Selection of Q and Zero Signal Parameters. Since the maximum input signal pulse = 2 volts, Q must be sufficiently down on the load line to allow a 2-volt excursion toward saturation but is not to actually reach that region. Let us select the intersection of the load line and the 200-μa characteristic for Q, which meets this requirement. The zero signal parameters are:

$$i_{B,Q} = 200 \ \mu\text{a}$$
$$i_{C,Q} \cong 6 \ \text{ma}$$
$$i_{E,Q} \cong 6.2 \ \text{ma} \ (i_{B,Q} + i_{C,Q})$$
$$e_{E\text{-}C,Q} \cong 5 \ \text{volts}$$

Step 3: Determination of Input Voltage Values. The level restoration potential is to be 0.2 volt, and $V_{bb} = 8$ volts; hence,

$$e_{RB1} = 0.2 \ \text{volt}$$
$$e_{RB2} = V_{bb} - e_{RB1} = 8 - 0.2 = 7.8 \ \text{volts}$$

Step 4: Determination of R_{B1}, R_{B2}, and C_B. Let $R_{B1} = 1$ kilohm; then

$$i_{RB1} = \frac{e_{RB1}}{R_{B1}} = \frac{2 \times 10^{-1}}{1 \times 10^3} = 200 \ \mu\text{a}$$

R_{B2} can now be calculated:

$$R_{B2} = \frac{V_{cc} - e_{RB1}}{i_{RB1} + i_{B,Q}}$$

Since both these currents flow through R_{B2},

$$R_{B2} = \frac{3.8}{4 \times 10^{-4}} = 9.5 \ \text{kilohms}$$

C_B is most conveniently determined empirically. Place an oscilloscope across the output, substituting several values of C_B. That capacitor which results in an output waveform most closely resembling the input pulse is selected.

PROBLEMS

12.19 In terms of the region in which they operate, what is the difference between the emitter-follower and the inverter switching circuits?

12.20 Refer to Fig. 12.7. How is forward bias made available to the base-emitter junction?

12.21 In Fig. 12.7, across what junction is the input potential applied?

12.22 What effect does the application of a positive pulse have on the circuit of Fig. 12.7? Explain.

12.23 What is the purpose of R_{B1} and R_{B2} in Fig. 12.7b? Explain.

12.24 State the function of C_B in Fig. 12.7b.

12.25 In Fig. 12.7b, what voltages establish the zero signal input level? Why is this sometimes necessary?

12.26 Refer to Fig. 12.7b. Assume $R_{B1} = 2$ kilohms; the desired level restoration potential = 4 volts. Find i_{RB1}.

12.27 In the circuit of Prob. 12.26, assume $i_{B,Q} = 100$ ma; $V_{bb} = 12$ volts. Find R_{B2}.

12.28 When designin an emitter-follower circuit with the use of the common-emitter output characteristics of the transistor, why is $V_{cc} + V_{ee}$ used on the x axis?

12.29 What is the best method for reducing storage time in the emitter-follower circuit?

12.30 Why should V_{cc} exceed the minimum input pulse in the emitter-follower circuit?

12.31 Refer to the emitter-follower circuit of Fig. 12.7b. The 2N465 is to be used (see Fig. 8.5); $V_{cc} = 10$ volts; $V_{bb} = 10$ volts; $V_{ee} = 10$ volts; $R_E = 2$ kilohms; input pulse maximum value = 5 volts; level restoration voltage = 0.3 volt; let $i_{B,Q} = 90$ μa.
(a) Construct the load line.
(b) Determine $i_{C,Q}$, $i_{E,Q}$, and $e_{E-C,Q}$.

12.32 In Prob. 12.31, find (a) e_{RB1}; (b) e_{RB2}.

12.33 In Probs. 12.31 and 12.32, let $R_{B1} = 2$ kilohms. Find (a) i_{RB1}; (b) R_{B2}.

12.34 Refer to the emitter-follower circuit of Fig. 12.7b. The 2N337 is to be used (see Fig. 8.26); $V_{cc} = 8$ volts; $V_{bb} = 0$ volts; $V_{ee} = 12$ volts; $R_E = 1.5$ kilohms; input pulse maximum value = 10 volts; level restoration voltage = 0.5 volt; let $i_{B,Q} = 100$ μa.
(a) Construct the load line.
(b) Determine $i_{C,Q}$, $i_{E,Q}$, and $e_{E-C,Q}$.

12.35 In Prob. 12.34, find (a) e_{RB1}; (b) e_{RB2}.

12.36 In Probs. 12.34 and 12.35, let $R_{B1} = 4$ kilohms. Find (a) i_{RB1}; (b) R_{B2}.

12.8 The Transistor AND Circuit $(A \cdot B)$.

The fundamental logic operations of the transistor are AND, OR, and INVERT, which are useful in the routing of information. There are two possible states in which the variables can be located:

0, which is nonexistence

1, which is existence

A basic AND circuit and a tabular description of its behavior with various input combinations (called a truth table) are illustrated in Fig. 12.10. There are two emitter followers with each transistor handling one input. T_1 has input A, T_2 has input B, and the common output is at point C. Each transistor is set at a zero signal level.

For example, in the circuit of Fig. 12.10, let -8 volts be "zero" and 0 volts be "one." Let us examine the four possible combinations of T_1 and T_2 inputs and the resultant output with the use of the truth table (Fig. 12.10b).

$A = -8$ $Volts$, $B = -8$ $Volts$. Since T_1 and T_2 are PNP types, both transistors will conduct heavily, resulting in $C = -8$ volts, which is "zero" (see horizontal line 1 of the truth table).

$A = 0$ *Volts*, $B = -8$ *Volts*. T_1 is at minimum conduction, T_2 is conducting heavily, and $C = -8$ volts or "zero" (line 2).

$A = -8$ *Volts*, $B = 0$ *Volts*. T_1 conducts heavily, T_2 is at minimum conduction, and C is again -8 volts or "zero" (line 3).

$A = 0$ *Volts*, $B = 0$ *Volts*.   Both T_1 and T_2 are at minimum conduction and $C = 0$ volts, which is "one" (line 4).

The AND circuit output is "one" only when the input to both transistors is "one."   The output is zero for the remaining three possible input combinations.

It should be noticed that the two transistors in Fig. 12.10 are the PNP type, and both are the emitter-follower circuit described in the preceding

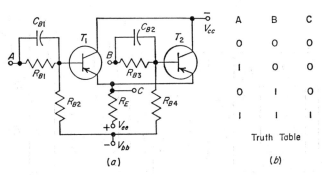

FIG. 12.10   PNP AND circuit: (*a*) circuit, (*b*) truth table.

section.   The design principles previously examined are applicable for the basic AND circuit.   Both emitter followers utilize a common load resistor R_E.

12.9   The Diode and Vacuum-tube AND Circuits.

The AND circuit illustrated in Fig. 12.11 utilizes semiconductor diodes.   Two input voltages, E_{i1} and E_{i2}, are shown in this circuit, although a larger number could be used, since the AND circuit may have two or more inputs. D_1 and D_2 conduct when E_{i1} and E_{i2} are zero volts.   In order to ensure that D_1 and D_2 undergo conduction at the zero signal condition, a clamping diode D_A with its clamping potential V_{aa} is incorporated.   In actual practice, V_{aa} is adjusted to that potential which enables the zero input signal value of the D_1 and D_2 currents to be at the desired value.   For simplicity purposes, the input circuit diodes are usually identical, and V_{aa} is set at such a value that each of these diode currents, which are about equal to each other, is greater than I_o.   Notice that $I_{D,A}$, the forward current of the diode clamp, will flow as long as one of the input diodes is conducting.   As long as D_A conducts, the output voltage is clamped to a value slightly less than V_{aa}, since D_A has a small forward resistance.   In such cases, the output voltage does not change. When all the input diodes are cut off, which occurs when all positive input pulses are equal or greater than V_{bb}, $I_{D,A}$ is zero, and the output voltage changes from V_{aa} to V_{bb}, thereby producing an output waveform.

It should be noticed that an output voltage waveform is obtained only when all the input diodes are cut off. In a diode AND circuit, such as Fig. 12.11, an output response is obtained only when a positive pulse of the previously stated magnitude is applied at the same instant to every input diode. In the event that positive pulses of different ON times are applied to the input diodes, the output waveform is developed only during the time the various positive input pulses overlap.

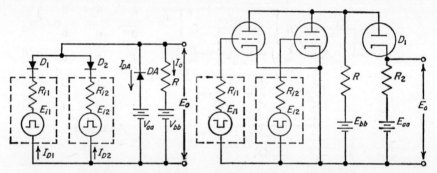

FIG. 12.11   A diode AND circuit.   FIG. 12.12   A vacuum-tube AND circuit.

Figure 12.12 illustrates a vacuum-tube AND circuit. A triode is used for each input circuit, and more than two can be used. The input triodes are conducting with the application of no input pulse, D_1 is cut off, and E_o is constant at the value of E_{aa}. When each input receives a negative pulse sufficient to cut off each triode, the triode plate voltage increases to E_{bb}. This is sufficient to make the plate of D_1 positive with respect to its cathode; D_1 conducts and E_o increases toward the magnitude of E_{bb}.

12.10   The Transistor OR Circuit $(A + B)$.

NPN emitter followers sharing a common R_E can be used for the basic OR circuit, as shown in Fig. 12.13. The OR circuit performs logical addition, which is best described by examining the truth table of Fig. 12.13b for each of the four possible conditions. The output C is "one" when it is similar to the most positive input (A or B). In the following discussion, let "one" be plus 4 volts for A, B, and C.

$A = 0$ *Volts*, $B = 0$ *Volts*. When the input of both T_1 and T_2 are zero, both transistors are operating at the zero signal operating point, and the output is zero (see line 1 of the truth table).

$A = 4 \ Volts, \ B = 0 \ Volts.$ Transistor 1 is conducting heavily while T_2 is conducting at its zero signal value, which means T_1 is "one" and T_2 is "zero." The output is positive; hence C is "one" (see line 2).

$A = 0 \ Volts, \ B = 4 \ Volts.$ Transistor 1 is conducting at its zero signal value while T_2 is conducting heavily (T_1 is "zero," T_2 is "one"), and the output C is "one." See line 3 of the truth table.

$A = 4 \ Volts, \ B = 4 \ Volts.$ Both transistors are conducting heavily (T_1 and T_2 are both "one"), and the output is positive ("one"). Line 4 of the truth table states this relationship.

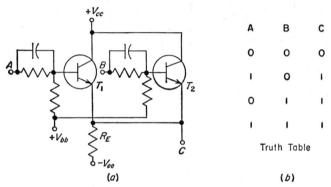

A	B	C
0	0	0
1	0	1
0	1	1
1	1	1

Truth Table

(a)   (b)

FIG. 12.13   NPN OR circuit: (a) circuit, (b) truth table.

Notice that the output is "one" when either one or both transistors are conducting heavily and "zero" only when both transistors are at their Q point. In other words, the output is "one" for three of the four possible combinations, whereas the output is "one" for only one of the four possibilities in the AND circuit.

12.11   The Diode and Vacuum-tube OR Circuits.

The OR circuit, like the AND circuit, has two or more inputs and a single output. The OR circuit is designed so that an output waveform is developed when a pulse is applied to any one of the input circuits. Diodes, triodes, or multigrid tubes may be used.

Refer to Fig. 12.14. The application of a negative pulse to any one of the semiconductor diodes will result in that diode conducting, and a negative pulse will appear across the output terminals. The selection of R_L such that it is many times greater than the forward resistance of the diodes results in the output pulse magnitude approaching the value of the input pulse. In the event that both diodes receive signals of unequal amplitudes, the output pulse will be that of the largest input pulse. This circuit can also be used for positive input pulses by merely reversing the input diode connections.

Refer to Fig. 12.15 for the triode OR circuit. The input signal triodes are biased beyond cutoff by E_{cc}. R_L is to be as large as possible in most cases where there are many inputs, so that the output pulse will not be too sensitive to the number of inputs that are activated. When one or

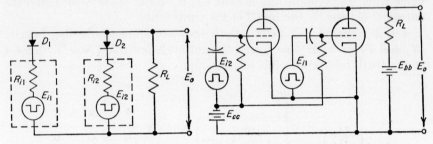

FIG. 12.14 A diode OR circuit. FIG. 12.15 A triode OR circuit.

more triodes are brought into conduction by a positive input pulse of sufficient magnitude, a negative pulse is obtained across the output.

12.12 The Transistor NOT-AND Circuit $(\bar{A} \cdot \bar{B})$. Recall that an inverter circuit, as indicated by its name, inverts the signal applied to its input. For example, the application of a "one" signal to its input results in a "zero" output. Using the symbol A for the input signal, the "zero" output can be designated as $\bar{A}$, or *not A*. The use of a second inverter circuit for the signal B results in a *not B* $(\bar{B})$ circuit. Combining the two inverters such that they have a common load resistance results in the $\bar{A} \cdot \bar{B}$ or NOT-AND circuit. Figure 12.16 illustrates this circuit with NPN

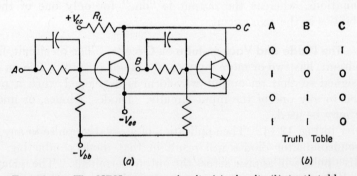

A	B	C
0	0	I
I	0	0
0	I	0
I	I	0

Truth Table

(a) (b)

FIG. 12.16 The NPN NOT-AND circuit: (a) circuit, (b) truth table.

transistors, which means it is used only when both inputs are negative and the output is positive.

As shown in the truth table, a "one" output is obtained only when a "zero" input is applied to both the inverters. Each inverter circuit can be designed with the techniques studied previously; the reader will recall that they share a common R_L.

12.13   The Vacuum-tube NOT **Circuit.**   The NOT circuit is also called an *inverter* circuit, since the output pulse polarity is inverted with respect to the input pulse.   A basic NOT circuit is illustrated in Fig. 12.17.   The

FIG. 12.17   A diode NOT circuit.

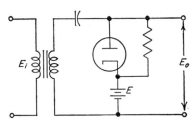

actual inversion is done by the input transformer.   This circuit, in conjunction with other gate circuits, has many applications.   A 1/1 transformation ratio is commonly used, and the transformer is the pulse type.

12.14   The Transistor AND-NOT **Circuit** $(A \cdot \bar{B})$.   The AND-NOT operation can be performed by use of an NPN inverter and a PNP emitter follower, as shown in Fig. 12.18.   Notice that the AND-NOT circuit is actually a

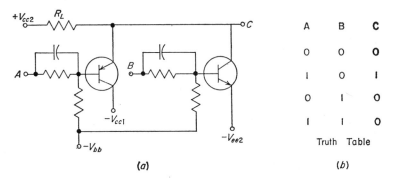

A	B	C
0	0	0
1	0	1
0	1	0
1	1	0

Truth Table

(a)   (b)

FIG. 12.18   The AND-NOT circuit: (a) circuit, (b) truth table.

complementary-symmetry arrangement.   The truth table shows that the output is "one" only at the condition when the PNP emitter follower has a "one" input and the NPN inverter has a "zero" input.   This circuit may be designed by use of the previously investigated techniques.

12.15   The Transistor Blocking Oscillator.   The blocking oscillator incorporates the technique of positive feedback from the collector back to the base like many of the oscillators previously analyzed. A transformer which possesses a high coefficient of coupling is used in the feedback loop so that a large amount of feedback is made available.

Figure 12.19 illustrates a blocking oscillator. Recall that the base emitter must be under forward bias in order to conduct when the circuit is closed. The conventional voltage-divider bias technique may be utilized for this purpose; $e_{RB2,Q}$ should be slightly greater than $e_{RE,Q}$. The self-biasing network is R_E and C_E. C_B restricts the unidirectional component of the base current to the voltage-divider circuit.

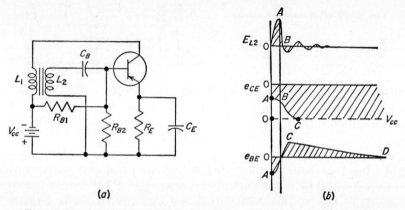

Fig. 12.19 A transistor blocking oscillator: (a) circuit, (b) waveforms.

When the circuit is first closed, the transistor conducts. The initial surge of collector current creates an expanding flux field about the primary which induces a voltage across the secondary. The secondary-induced voltage is abrupt and makes the base negative. During this interval, the top plate of C_E is charged, thereby placing the emitter at a relatively large negative potential, which cuts off the collector current (O to C of the e_{B-E} waveform); i.e., the transistor is "blocked." The top plate of C_E then begins to discharge through R_E to the lower plate of C_E. The OFF time of the period is determined by the time constant of R_E and C_E. When the top plate of C_E has discharged sufficiently so that e_{RE} is equal to e_{RB2}, the transistor again conducts. The ON time of the period is greatly determined by the resonant frequency of the transformer. A practical means of selecting the exact OFF time and period time is by use of a variable R_E.

12.16 Design of a Transistor Blocking Oscillator.
The blocking oscillator of Fig. 12.20 is designed in this section. A ½ iron-core transformer with a coefficient of coupling approaching unity is used. $V_{cc} = 12$ volts, and the transistor is the 2N465. The period OFF time is to be 0.001 sec.

Step 1: *Determination of R_E and C_E.*   As a first approximation, R_E and C_E are selected so the time constant is equal to 0.001 sec.   Recall from the preceding section that the time constant of these two components determines the period OFF time.   The period ON time will not

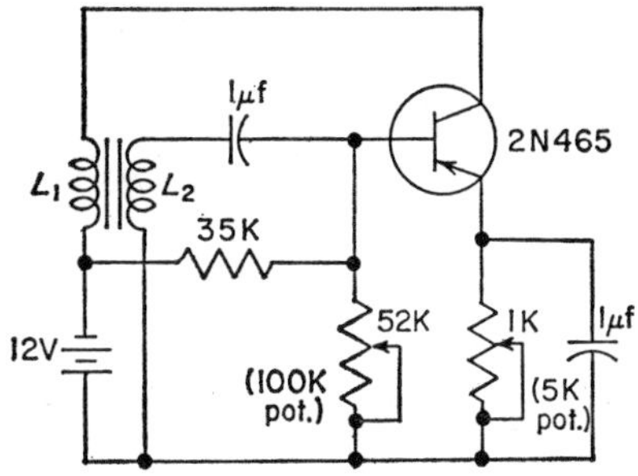

FIG. 12.20 A designed transistor blocking oscillator.

exceed 0.25 T_p; therefore the OFF time will be at least 0.75 T_p.   Using this relationship,

$$T_p = 1.33 \text{ OFF time} = 0.00133 \text{ sec}$$

and the period repetition rate is

$$T_p = \frac{1}{0.00133} \cong 750 \text{ cps}$$

C_E should be so selected that its reactance at the frequency of the transformer pulse is considerably less than R_E.   The transformer pulse frequency, as indicated by the waveform in Fig. 12.19, is higher than the period repetition rate.   By selecting X_{CE} at 750 cps to be less than R_E, sufficient bypassing of R_E is assured.

$$R_E C_E = 0.001 \text{ sec}$$

Transposing,

$$C_E = \frac{1 \times 10^{-3}}{R_E}$$

Let $R_E = 1$ kilohm; then

$$C_E = \frac{1 \times 10^{-3}}{1 \times 10^3} = 1 \text{ μf}$$

It is suggested that a 5-kilohm potentiometer be used for R_E, which can be adjusted for the exact period OFF time.

Step 2: *Determination of R_{B1}, R_{B2}, and C_B.* R_{B2} and R_{B1} form the conventional voltage-divider bias network.   When the circuit is first connected, e_{RB2} should be sufficiently large to permit the transistor to conduct.   The d-c load line is in great part determined by R_E, which can be used for a first approximation.   The load line for 1 kilohm with the 2N465 is shown in Fig. 7.43. Let us select the initial base current as 90 μa; and $i_C = 5$ ma. R_i of the 2N465, with a 1-kilohm load resistor, was determined in step 3 of Sec. 7.22 as 1,380 ohms; the required $e_{B\text{-}E}$ is then

$$e_{B\text{-}E} = i_B R_i$$
$$= 90 \times 10^{-6} \times 1.38 \times 10^3 = 0.124 \text{ volt}$$

e_{RE} at this condition is

$$e_{RE} = (i_C + i_B)R_E = 5.09 \times 10^{-3} \times 1 \times 10^3$$
$$= 5.09 \text{ volts}$$

and $\qquad e_{RB2} = e_{RE} + e_{B\text{-}E} = 5.09 + 0.124$
$$\cong 5.214 \text{ volts}$$

Let i_{RB2}, the bleeder current, $= 100$ μa, then

$$R_{B2} = \frac{e_{RB2}}{i_{RB2}} = \frac{5.21}{1 \times 10^{-4}} \cong 52 \text{ kilohms}$$

and $\qquad R_{B1} = \dfrac{V_{cc} - e_{RB2}}{i_B + i_{\text{bleeder}}} = \dfrac{12 - 5.21}{190 \times 10^{-6}}$

$$= \frac{6.79}{1.9 \times 10^{-4}} \cong 35.7 \text{ kilohms}$$

C_B shunts the voltage-divider network and may be made equal to C_E, which is 1 μf.

It is suggested that R_{B2} be a 100-kilohm potentiometer and can be adjusted at the value required to fire the transistor into conduction.

Step 3: Circuit Construction and Analysis. R_E is adjusted for the correct period OFF time, and R_{B2} is adjusted for firing the transistor into conduction. The waveforms can be analyzed with the oscilloscope.

12.17 The Vacuum-tube Blocking Oscillator. Figure 12.21 illustrates a blocking oscillator and its associated waveforms. Positive feedback from the plate to grid is achieved by use of a transformer with a high coefficient of coupling. Notice that grid-leak bias is utilized (R_g and C_g). Let us consider the operation of this oscillator for one complete period. When the circuit is first closed, the increase in plate current creates an expanding flux from within L_1, which induces a voltage across L_2. This drives the control grid positive (AB in the E_c waveform) to such an extent that electron flow within the tube is from the cathode to the control grid, and the plate current abruptly drops to zero (C to D of the I_b waveform). The sharp increase in grid current quickly charges plate A of C_g (B to C of the E_c waveform) and electron flow is then from the grid to the cathode via R_g. Since this current rise is abrupt, e_{Rg} is likewise abrupt, and E_c is almost instantaneously driven far into its cutoff region (C to D of the E_c waveform). Therefore the tube is cut off or "blocked." Plate A of C_g discharges through R_g, then through L_2 back to plate B of C_g. The discharge action is exponential in nature (D to E of the E_c waveform), and the plate current remains blocked until E_c returns to its cutoff value, at which time the cycle of events is repeated.

The ON time of the period is determined in great part by transformer design (the resonant frequency of the transformer). The OFF time is determined by the discharging time constant of C_g and R_g. Larger values of C_g and/or R_g lengthen the OFF time and the period time. The

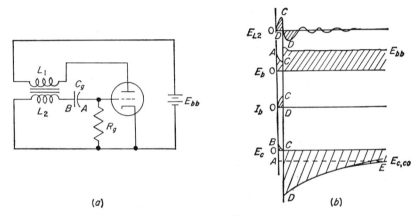

(a)   (b)

FIG. 12.21   The vacuum-tube blocking oscillator: (a) circuit, (b) waveforms.

plate voltage of the tube is at cutoff during the OFF time and decreases when the tube is fired. Since the plate current is abruptly blocked by the grid action, the plate potential increases almost instantaneously to E_{bb}. Notice the waveform of the transformer secondary. The peak voltage at C abruptly returns to zero (point D) when the tube is blocked, and the stored energy is dissipated within the transformer, as indicated by the damped waveform. When the damped waveform extends over a substantial portion of the OFF time, the E_b and E_c waveforms tend to become more rounded. Loading down the secondary with a damping resistor reduces the transient response of the secondary, and the E_b and E_c waveforms are improved.

In practical applications, the transformation ratio is at least $1/1$. R_g is often made variable so the exact OFF time and period time desired by the designer can be achieved.

PROBLEMS

12.37   What type(s) of switching circuits are used in the AND circuit?

12.38   In Fig. 12.16, under what input conditions will the output be "one"?

12.39 What type(s) of switching circuits are used in the NOT-AND circuit?

12.40 What type(s) of switching circuits are used in the AND-NOT circuit?

12.41 In Fig. 12.18, under what input conditions will the output be "one"?

12.42 In the designed blocking oscillator of Fig. 12.20, what is the major determining factor of the ON time?

12.43 How can the OFF time of the circuit in Fig. 12.20 be controlled?

12.44 What purpose does R_{B1} and R_{B2} serve in Fig. 12.20? How are their values determined?

12.45 State the purpose of C_B in Fig. 12.20 and how its value may be determined.

12.46 What provisions are made for positive feedback in the blocking oscillator of Fig. 12.21?

12.47 Explain the action of the blocking oscillator of Fig. 12.21 which leads to cutoff.

12.48 What type of transformer is required for the blocking oscillator? Explain.

12.49 What determines the ON time of the blocking oscillator?

12.50 What determines the OFF time of the blocking oscillator?

12.51 What is a practical method for setting the desired OFF time of a blocking oscillator?

CHAPTER 13

SEMICONDUCTOR AND VACUUM-TUBE MODULATION AND DETECTION CIRCUITS

The first section of this chapter briefly discusses the nature and classification of radio waves, which play an important part in modulation and detection as related to the field of communications. The majority of the most significant types of modulation are considered in the second and third sections. Amplitude modulation, perhaps one of the most simple varieties, is considered in some detail. No attempt is made in this text to consider those circuits which are suitable for other types of modulation, with the exception of a simple f-m modulator. The other types of modulations deal with specific applications which will be considered in great detail in another text.

The fundamental principles of modulation and detection can be utilized in transistor and vacuum-tube circuits in a similar manner. A-m modulation circuits of each of the three transistor terminals are paralleled with similar circuits for each of the three basic vacuum-tube terminals. In the case of the transistor-modulation circuits, actual circuits have been designed. As in the other chapters in this text, all designed circuits have been constructed and carefully evaluated, thereby minimizing possible troublesome conditions.

The principles of mixing are considered after modulation, with several transistor and vacuum-tube types analyzed in detail. Detection, or demodulation, is the last topic of the chapter. Both transistor and vacuum-tube types are considered.

13.1 The Nature and Classification of Radio Waves. Radio waves are actually electromagnetic waves, which consist of electric energy in electrostatic and magnetic states. It can be assumed that an electromagnetic wave would appear as shown in Fig. 13.1.

Let us first emphasize the fact that an electromagnetic wave is invisible. Secondly, it travels at the speed of light. The wave is conveniently

analyzed in terms of its major characteristics, which are:
1. Wavelength
2. Plane of polarization
3. Direction
4. Intensity
5. Frequency

Radio waves can be produced by alternating current. The intensity of the radio wave, with all other factors equal, is a direct function of the frequency of the alternating current.

Referring to Fig. 13.1, the wave travels at right angles to the direction of the electrostatic and magnetic flux, which are at right angles to each

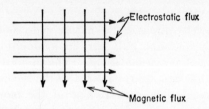

FIG. 13.1 Electromagnetic wave traveling toward the reader.

other. If the electrostatic and magnetic flux are in the directions shown, the wave would travel toward the reader. If either the electrostatic or the magnetic flux direction were reversed, the wave would travel away from the reader. If both flux were reversed, their directions with respect to each other would be the same, and the electromagnetic wave would travel in its original direction.

The strength or intensity of an electromagnetic wave is generally measured in *microvolts stress per meter*. This voltage stress is that produced in space by the electrostatic flux of the wave and is nearly identical to the voltage induced by the magnetic flux of the same wave in a conductor which is 1 m long.

The frequency of an electromagnetic wave is the same as that of the alternating current which produced it. Electromagnetic waves are classified in accordance with their frequency. The wavelength in space is a direct function of its velocity (which is that of light) and an inverse function of the alternating current frequency; i.e.,

$$\lambda = \frac{c}{f} = \frac{300,000,000}{f}$$

where λ = lambda = wavelength in meters
c = velocity of light = 300,000,000 m/sec
f = frequency of the alternating current in cps

In analyzing the preceding equation, it is noted that an alternating current of low frequency generates an electromagnetic wave of relatively long wavelength.

The plane of polarization is determined by the direction of the electrostatic flux. In Fig. 13.1, where the electrostatic flux is shown to be horizontal, the wave is said to be horizontally polarized. On the other hand, if the electrostatic flux is on the vertical plane, the wave is vertically polarized. Table 13.1 lists the chief classifications of radio waves:

Table 13.1   Classification of Electromagnetic Waves

Class	Frequency Range	
Very low frequency (vlf).........	10–30	kc
Low frequency (l-f)..............	30–300	kc
Medium frequency (m-f).........	300–3,000	kc
High frequency (h-f).............	3–30	Mc
Very high frequency (vhf)........	30–300	Mc
Ultrahigh frequency (uhf)........	300–3,000	Mc
Super-high frequency (shf).......	3,000–30,000	Mc
Microwave frequency...........	2,000 Mc and above	

13.2   The Fundamentals of Modulation.   The prime function of electromagnetic waves is to carry information from a source (called the transmitter) to some other point (called the receiver). In order to accomplish this primary objective, some portion of the radio wave must be made to vary in accordance with the signal variations. This objective can be achieved in a number of ways, which are listed and described in the following section.

The electromagnetic wave is called the *carrier*, since it carries the desired information, while the signal itself is called the *intelligence*. An electromagnetic wave which contains intelligence is called a *modulated wave*. The process of blending the carrier and signal is called *modulation*.

The function of the transmitting station is to first modulate the carrier wave and then radiate it through space with sufficient energy to enable it to reach the intended receivers. At the reception end, the modulated carrier induces voltages in the receiver antenna. This induced voltage contains all the characteristics of the electromagnetic wave itself, and it is made to drive a current which also retains these properties. The received modulated carrier is then amplified in many cases. Eventually, the carrier and intelligence are separated, which is called *detection* or *demodulation*. The recovered intelligence is then free of its carrier and available for whatever use it was intended.

13.3   Types of Modulation.   There are several types of modulation. *Amplitude modulation* is obtained when the amplitude of the carrier voltage is made to change as the desired signal information. *Frequency* *modulation* is obtained if the frequency of the carrier is made to vary as the desired signal information, with the carrier voltage amplitude held constant.   *Phase modulation* occurs when the desired signal information varies the phase angle of the carrier voltage.   *Pulse-width modulation* occurs when the signal voltage varies the width of the carrier, which consists of short duration pulses in such cases.   Short-width pulses are also used for the carrier in *pulse-amplitude modulation;* only in this case, the signal voltage is made to vary the amplitude of the carrier pulses. Another type of modulation with the short width pulse carrier is *pulse-position modulation*, in which case the signal changes the starting-position of each carrier pulse.

This text concerns itself only with amplitude modulation, the principles of which are analyzed in the following section.

13.4 The Principles of Amplitude Modulation. Let us consider the nature of a modulated sine-wave carrier, which is illustrated in Fig. 13.2.

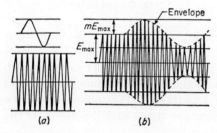

Fig. 13.2   Amplitude modulation: (a) before, (b) after.

For simplicity purposes, both the carrier and signal are shown to be simple sine waves in Fig. 13.2a.   When the carrier is amplitude-modulated, it appears as shown in Fig. 13.2b.   The modulated wave actually consists of three waves, whose frequencies are:

  1. f_c
  2. $f_c + f_s$
  3. $f_c - f_s$

where f_c = carrier frequency

     f_s = modulation frequency

Waves 2 and 3 are called *sideband* frequencies and actually carry the intelligence of the composite wave.   It can be seen that the sideband frequencies are determined by the modulation or intelligence frequency.

The voltage of a single frequency carrier may be determined by

$$e_c = E_m \cos (2\pi f_c t + \theta)$$

Since the phase angle does not act as a part of amplitude modulation, it may be omitted, simplifying the preceding relationship to

$$e_c = E_m \cos 2\pi f_c t$$

where e_c = instantaneous carrier voltage

E_m = maximum carrier voltage (unmodulated)

f_c = carrier frequency

t = time at which e_c is being determined

As shown in Fig. 13.2b, the envelope of the modulated waveform is identical to the waveform of the original signal. The ratio of the envelope amplitude to the carrier amplitude is termed the *modulation m.* This ratio is often given as a per cent, and is called *per cent modulation.*

The voltage of the intelligence may be determined by

$$e_{\text{mod}} = mE_m \cos 2\pi f_s t$$

When the carrier is amplitude-modulated, with both the carrier and intelligence being simple sine waves, its instantaneous voltage may be determined by

$$e = (1 + m \cos 2\pi f_s t)E_m \cos 2\pi f_c t$$

and expanding,

$$e = E_m \cos 2\pi f_c t + mE_m \cos 2\pi f_c t \cos 2\pi f_s t$$

The trigonometric expansion formula is

$$(\cos a)(\cos b) = \tfrac{1}{2} \cos (a + b) + \tfrac{1}{2} \cos (a - b)$$

Substituting, we obtain the equation of an amplitude-modulated wave:

$$e = E_m \cos 2\pi f_c t + \frac{mE_m}{2} \cos 2\pi(f_c + f_s)t + \frac{mE_m}{2} \cos 2\pi(f_c - f_s)t$$

Notice that the preceding equation reveals that three waves are present in the composite waveform. The carrier is denoted by the term $(E_m \cos 2\pi f_c t)$, the upper sideband by $\left[\left(\dfrac{mE_m}{2} \cos 2\pi f_c + f_s\right)t\right]$, and the lower sideband by $\left[\dfrac{mE_m}{2} \cos 2\pi(f_c - f_s)t\right]$.

Amplitude modulation results in two additional waves, called the upper and lower sidebands, being added to the original carrier. It should be noticed that the upper-sideband frequency is above the carrier frequency by the same amount that the lower-sideband frequency is below the carrier frequency. For example, if the carrier frequency is 600 kc and the modulation frequency is 10 kc, then

$$\text{Upper sideband } (f_c + f_s) = 600 \text{ kc} + 10 \text{ kc} = 610 \text{ kc}$$

and

$$\text{Lower sideband } (f_c - f_s) = 600 \text{ kc} - 10 \text{ kc} = 590 \text{ kc}$$

and

$$\text{Bandwidth} = (f_c + f_s) - (f_c - f_s)$$
$$= 610 \text{ kc} - 590 \text{ kc} = 20 \text{ kc}$$

Notice that the bandwidth is equal to twice the modulation frequency.

Referring to the equation of an amplitude-modulated wave, it is seen that the per cent of modulation (indicated by m) determines the magnitude of the sidebands. When $m = 0$, there are no sidebands, since there is no amplitude modulation. When $m = 1$ (100 per cent modulation), the carrier amplitude varies from a maximum of twice the unmodulated carrier amplitude to a minimum of zero. When m is greater than 1, overmodulation occurs, resulting in a portion of the wave being zero. Fig. 13.3 illustrates these three conditions.

Up to this point in our analysis, the amplitude-modulated wave has been represented as voltage versus time. Several other types of representation are possible, as shown in Fig. 13.4. Refer to Fig. 13.4a, which

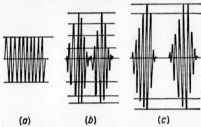

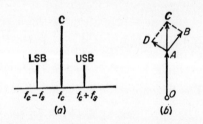

FIG. 13.3 The effect of m upon the a-m waveform: (a) $m = 0$, (b) $m = 1$, (c) $m > 1$.

FIG. 13.4 Additional ways of representing an a-m wave: (a) e versus f, (b) phase relationships.

illustrates the relationship of voltage versus frequency. Voltage appears only at the carrier, lower- and upper-sideband frequencies. When the modulation frequency increases, lower sideband and upper sideband move farther apart, whereas they come closer together when the modulation frequency decreases. Their amplitudes increase when the per cent of modulation is increased, and vice versa.

Refer to Fig. 13.4b, which is a phasor diagram of the carrier and two sidebands. E_m, the amplitude of the unmodulated carrier, is represented by OA. AB represents the upper sideband and moves clockwise with increases of modulation frequency. The lower sideband is designated by AD, which moves counterclockwise with increases of modulation frequency. AC is the resultant of AD and AB. The resultant of AC and OA is the composite voltage of the entire waveform. The resultant of AC and OA, if traced out, would result in the e versus t waveform of Fig. 13.2b.

In conclusion, it should be noted that while energy is present in both the carrier and sidebands, signal energy exists only in the sidebands. The energy of the carrier is that energy required to transmit the signal to its destination.

The frequency of modulation is different for various types of applications. Table 13.2 lists some of the chief applications of amplitude modulation and their modulation frequencies:

Table 13.2   Modulation Frequencies of Several A-M Applications

Application	Frequency, cps
Telegraph	0–120
Speech:	
High fidelity	40–15,000
Commercial broadcast	100–5,000
Long-distance telephone	250–3,500

The carrier frequency is generally assigned to a station by the Federal Communications Commission (FCC), which is the law-enforcement body in such matters. Commercial a-m radio lies within the carrier frequency range of 550 to 1,600 kc. A given a-m radio station is assigned a carrier frequency by the FCC, with a permissible bandwidth of ±5 kc. For example, a station assigned a carrier frequency of 1,360 kc actually has permissible sidebands extending from 1,355 kc (1,360 kc − 5 kc) to 1,365 kc (1,360 kc + 5 kc.) Notice that this restricts the maximum modulation frequency to 5 kc.

13.5   A-M Collector Modulation. As pointed out in the preceding sections, amplitude modulation is achieved by the process of mixing the intelligence frequency with a carrier frequency. In Fig. 13.5, the mixing action occurs in the collector circuit of $T2$. $T1$ is the modulator amplifier, and its purpose is to deliver a modulation power output to the secondary of its output transformer, which is equal to 50 per cent of the power output of $T2$. When this power relationship is achieved, 100 per cent modulation occurs.

In those cases where T_1 and T_2 are dissimilar transistors, the output transformer of T_1 is selected to ensure the proper impedance matching between the two transistors.

T_1 is operated in the class A mode. R_{E1} provides stabilization while C_{E1} is designed to reduce the signal degenerative feedback. R_{B3} and R_{B4} form the conventional forward bias network for T_1. A push-pull or complementary stage operated in class A, AB, or B may be used in place of T_1 when additional modulation power is desired.

T_2 is operated in class C for maximum efficiency. R_{B1} and R_{B2} form the conventional forward bias for T_2 starting. R_{E2} and C_{E2} serve as the class C self-bias network. When the transistor conducts, the top plate of C_{E2} charges via T_2, which is a short time-constant circuit. When this plate is charged, T_2 is cut off, and the top plate discharges through R_{E2}

on to the bottom plate. R_{E2} should be sufficiently large to maintain T_2 at cutoff for more than half the carrier signal cycle. Increasing R_{E2} decreases the angle of collector current, which increases the efficiency but also reduces the output power of T_2. C_3 is such a value that it places the tank circuit at common potential for the carrier frequency, but offers sub-

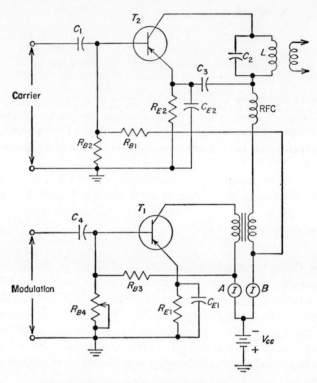

Fig. 13.5 A-m collector-modulation circuit.

stantial reactance to the modulation frequency. The tank circuit, which consists of C_2 and L, resonates at the carrier frequency. When the circuit is properly balanced, the unidirectional collector current of both transistors remains steady. In many cases, the collector potential of T_2 is kept at a lower value than the collector voltage of T_1, so as to facilitate the achievement of 100 per cent modulation.

13.6 Design of an A-M Collector-modulation Circuit. The 2N497 (NPN silicon) is to be used for both transistors. $V_{cc} = 10$ volts. The collector and i_C versus e_{B-E} characteristics are illustrated in Fig. 13.7. Carrier frequency = 10 kc; modulation frequency = 100 cps.

Step 1: *Determination of the Tank-circuit Components* (C_2 *and* L). L is arbitrarily selected as 50 mh.   Since

$$f_o = \frac{159 \times 10^{-3}}{(LC)^{1/2}}$$

and

$$C = \frac{1}{4(\pi)^2(f_o)^2 L}$$

where

$$f_o = 10 \text{ kc}$$
$$L = 50 \times 10^{-3} \text{ henry}$$

Substituting values and solving,

$$C = \frac{1}{(4)(3.14)^2(1 \times 10^4)^2(50 \times 10^{-3})}$$
$$= 0.005 \ \mu\text{f}$$

Step 2: *Determination of* R_{E2}, C_{E2}, *and* T_2 *Starting Parameters.*   As a first approximation, R_{E2} may be considered as the load resistance of T_2 at

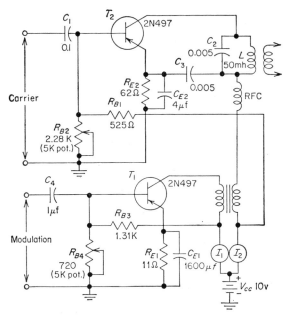

Fig. 13.6   The designed a-m collector-modulation circuit.

the starting condition.   Select R_{E2}, such that the maximum dissipation ratings of the transistor are not exceeded.   Let the load line be drawn from 10 volts on the x axis and 160 ma on the y axis, which fulfills this requirement.

Then

$$R_{E2} = \frac{10}{160 \times 10^{-3}} = 62.5 \text{ ohms}$$

C_{E2}, in order to have a short charging time constant as compared to its discharge time constant, should have a reactance at the carrier frequency which is 10 per cent or less of R_{E2}; i.e.,

$$X_{CE2} \cong 6 \text{ ohms at 10 kc}$$

Solving for C_{E2},

$$C_{E2} = \frac{159 \times 10^{-3}}{6 \times 1 \times 10^4} \cong 2.65 \ \mu\text{f (minimum)}$$

Let $C_{E2} = 4 \ \mu\text{f}$.

Returning to the load line of R_{E2}, let the starting operating point be at the intersection of the load line and the $i_B = 3$ ma characteristic. The T_2 starting parameters are:

$$i_{B2,Q} = 3 \text{ ma}$$
$$i_{C\text{-}E2,Q} = 90 \text{ ma}$$
$$e_{C\text{-}E2,Q} = 4.3 \text{ volts}$$

Step 3: Determination of R_{B1}, R_{B2}, C_1, and C_3. The required $e_{B\text{-}E,Q}$ value can be graphically determined since the $e_{B\text{-}E}$ versus i_C characteristic is available. Find $i_{C,Q}$ on the y axis of Fig. 13.7b. Draw a line from this

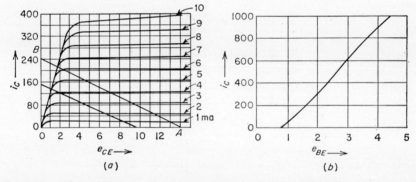

Fig. 13.7   Typical characteristics of the 2N497. (*Characteristics from Transitron.*)

point parallel to the x axis until the characteristic is intercepted. Now construct a perpendicular line from this point of intersection to the x axis. $e_{B\text{-}E,Q}$ is found to be about 1.15 volts. R_{E2} develops a reverse bias, which can be determined by

$$e_{RE2,Q} = V_{cc} - e_{C\text{-}E2,Q} = 10 - 4.3 = 5.7 \text{ volts}$$

e_{RB2} must be 1.15 volts greater than e_{RE2} so as to ensure that

$$e_{B\text{-}E} = 1.15 \text{ volts}$$

i.e.,

$$e_{RB2} = e_{RE2} + e_{B\text{-}E} = 5.7 + 1.15 = 6.85 \text{ volts}$$

The bleeder current through R_{B2} is arbitrarily selected, which in this case is to be 3 ma. R_{B2} can now be computed:

$$R_{B2} = \frac{6.85}{3 \times 10^{-3}} = 2.28 \text{ kilohms}$$

and

$$R_{B1} = \frac{V_{cc} - e_{RB2}}{i_{B,Q} + i_{\text{bleeder}}} = \frac{10 - 6.85}{6 \times 10^{-3}} \cong 525 \text{ ohms}$$

C_1 is to offer negligible impedance to the carrier frequency, and any value greater than 0.005 μf will suffice in this case. Let $C_1 = 0.1$ μf. C_3 is to offer low impedance to the carrier but high reactance to the modulation frequency. A value similar to that of C_2 is a good approximation; i.e., let $C_3 = 0.005$ μf.

Step 4: Determination of T_1 Load Line, T_1 Zero Signal Parameters, and Output Transformer Turns Ratio. T_1 is a class A transformer-coupled amplifier. If the reflected primary impedance of the transformer is allowed to be equal to the starting load resistance of T_2, a transformer with a unity ratio can be used. Therefore, the load line AB is R_L (reflected) of T_1. Using the zero signal operating point, the zero signal parameters are

$$i_{B,Q} = 3 \text{ ma}$$
$$i_{C-E1,Q} = 90 \text{ ma}$$
$$e_{C-E1,Q} = 9 \text{ volts}$$

Step 5: Determination of R_{E1}, C_{E1}, R_{B3}, R_{B4}, and C_4. Let

$$e_{RE1,Q} = 1 \text{ volt (reverse bias)}$$
$$i_{RE1,Q} = i_{C,Q} + i_{B,Q} = 93 \text{ ma}$$

and

$$R_{E1} = \frac{e_{RE1,Q}}{i_{RE1,Q}} = \frac{1}{93 \times 10^{-3}} \cong 11 \text{ ohms}$$

X_{CE1} should be 10 per cent of R_{E1} at the modulation frequency,

$$C_{E1} = \frac{159 \times 10^{-3}}{1 \times 1 \times 10^{2}} \cong 1600 \text{ } \mu\text{f (minimum)}$$

Since such a large value of C_{E1} may not be readily available, the largest obtainable value should be used.

We next compute the voltage-divider resistor values:

$$R_{B4} = \frac{e_{RE1} + e_{B-E,Q}}{i_{\text{bleeder}}}$$

From the computations of R_{B2}, $e_{B-E,Q}$ was found to be 1.15 volts, and $e_{RE1,Q}$ was determined to be 1 volt in the preceding paragraph. Let $i_{bleeder} = 3$ ma. Substituting values and solving,

$$R_{B4} = \frac{1 + 1.15}{3 \times 10^{-3}} \cong 720 \text{ ohms}$$

and
$$R_{B3} = \frac{V_{cc} - e_{RB4}}{i_{B1,Q} + i_{bleeder}} = \frac{10 - 2.15}{(3 + 3) \times 10^{-3}}$$
$$= 1.31 \text{ kilohms}$$

C_4 is to offer negligible impedance to the modulation frequency. One microfarad or larger will suffice in this case.

Step 6: *Construction and Testing of the Circuit.* In order to check the angle of collector current in the modulator amplifier, a small resistor may be placed in series with the tank circuit. The waveform across this resistor will be that developed by the T_2 collector current. A lower angle of collector current increases circuit efficiency and decreases the power output. It is suggested that the T_2 circuit be completely checked before connecting it to the output of T_1. The modulation amplifier should also be checked separately, to ensure an undistorted modulation output. When both stages are known to be operating as required, then they may be coupled by means of the modulation amplifier output transformer.

The tank circuit of T_2 will have the modulated carrier frequency across it and can be checked with an oscilloscope. For 100 per cent modulation, the carrier amplitude should vary from a maximum to zero. The envelope is most clearly seen by setting the oscilloscope horizontal frequency at the modulation frequency. The use of 5-kilohm potentiometers for R_{B4} and R_{B2} will enable the designer to set the bias levels of the two transistors for proper circuit adjustments. If the maximum undistorted modulation is about 50 to 70 per cent, the difficulty may exist in the use of the same collector potential for both the modulation and modulator stages. In such instances, provisions may be made for the reduction of the T_2 collector potential. Reducing V_{cc} of T_2 to 80 per cent of the T_1 V_{cc} is a good first approximation. Improvement of the modulation waveform can often be made by placing a damping resistor across the secondary of the modulation amplifier output transformer.

A relatively large value of R should be used in such cases, to present a constant but light load across the modulation amplifier output. Increasing the value of R_{B3} is sometimes helpful in improving the modulation waveform.

The designed a-m collector modulation circuit of Fig. 13.6 with a carrier input of 10 mv rms and modulation input of 10 mv rms possessed approximately 80 per cent undistorted modulation.

13.7   A-M Plate Modulation.   Prior to delving into the analysis of several basic a-m modulation techniques, it is suggested that the reader review the fundamental principles revealed in Secs. 13.1 to 13.5.   It was shown that amplitude modulation results from the process of mixing an intelligence signal (such as audio) with a carrier signal (such as a radio frequency).   The manner in which the mixing is performed determines the type of modulation circuit.   In Fig. 13.8, the mixing is per-

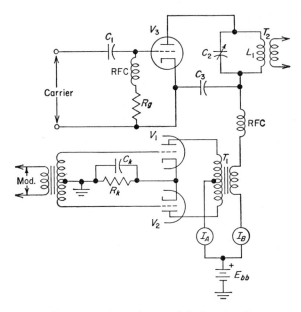

FIG. 13.8   A-m plate-modulation circuit.

formed in the plate of V_3 and is, therefore, called a plate-modulation circuit.   This is the counterpart of the transistor collector modulation circuit.

V_1 and V_2 are the modulator tubes, which can be operated in class A, AB, or B.   The modulation is delivered to the input of V_1 and V_2.   By selection of the proper tubes, the modulation power delivered to the secondary of T_1 should be 50 per cent of the V_3 d-c plate signal power. V_3 develops the desired power of the carrier applied to its grid.

$$V_3 \text{ d-c plate input power} = E_{b3,s}I_{b3,s}$$

where E_b and I_b are measured without the presence of modulation.

The modulation circuit must be capable of furnishing a power output equal to half of the V_3 plate input power for 100 per cent modulation; i.e.,

$$P_{o,\text{mod}} = 0.5(E_{b3,s}I_{b3,s})$$

The turns ratio of T_1 is selected so that proper impedance matching between the class C carrier amplifier and modulator amplifier is achieved. Recall that,

$$a = \left(\frac{Z_p}{Z_s}\right)^{\frac{1}{2}}$$

where, in this case,

  a = primary/secondary turns ratio
  Z_p = load impedance of the modulator amplifier
  Z_s = load impedance of the class C carrier amplifier
and from Fig. 13.8,

$$Z_s = \frac{E_{b3,s}}{I_{b3,s}}$$

$$Z_p = \frac{4E_{bb}}{I_{2m,y}}$$

where Z_p = plate-to-plate impedance of V_1 and V_2
  $I_{m,y}$ = the intersection of the load line and the y axis

For 100 per cent modulation, E_{b3} must be capable of varying from zero to $2E_{b3,s}$.

C_3 places the tank circuit of V_3 at common potential for the carrier frequency but should present relatively high impedance to the modulation frequency. C_2 and L_1 form a tank circuit which resonates at the carrier frequency.

The d-c plate current of V_3 should remain at the same level with and without modulation. If the d-c plate current does change, nonlinearity is present. Furthermore, if the modulation circuit is functioning properly, these d-c plate currents should remain constant. The plate current relationships can be checked by observing the behavior of the d-c ammeters (A and B) placed in the plate circuits.

13.8 A-M Base Modulation. A second method for producing amplitude modulation is the base-modulation circuit, one of which is illustrated in Fig. 13.9 (for PNP transistors). T_2 is the modulation amplifier and is operated in class A to ensure a faithful reproduction of the intelligence variations. T_1 is the carrier-frequency amplifier, which is operated in the class C mode. The periodic conduction of T_1 need only replenish the tank circuit (C_2 and L_1) to maintain oscillations at the resonant frequency of the tank. The reverse bias required for class C operation is obtained by use of V_{bb}.

The modulation output of T_2 is transformer-coupled to the input circuit of T_1. Notice that the secondary of the output transformer is in series with the fixed reverse bias V_{bb}. Therefore the reverse bias varies in accordance with the modulation variations, causing T_1 to follow these variations during its conduction time.   In this way, the carrier is modulated in the base circuit of the transistor, which is the reason it may be called a base-modulation circuit.   The turns ratio of the T_2 output transformer is selected for matching R_o of T_2 to R_i of T_1 during the T_1 conduction time.

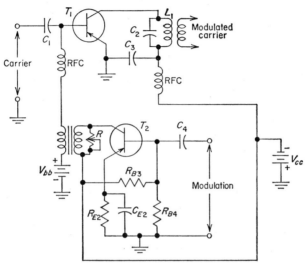

FIG. 13.9   The a-m base-modulation circuit.

13.9 Design of an A-M Base-modulation Circuit.   The a-m base-modulation circuit of Fig. 13.10 is designed in this section.   The 2N465 (PNP) transistors will be used. $V_{cc} = 12$ volts; carrier frequency = 10 kc; modulation frequency is to vary from 100 cps to 1 kc.

Step 1: *Determination of* T_1 *Parameters.*   Let the reflected primary impedance of the output transformer be assumed to be 1 kilohm.   The signal load line AC for this R_L is illustrated in Fig. 7.43 (AB is the d-c load line determined by R_{E2}).   Select $i_{B,Q} = 90$ μa; then

$$i_{C,Q} \cong 5.2 \text{ ma}$$
$$e_{C-E,Q} \cong 10 \text{ volts}$$

Let $e_{RE2} = 2$ volts.   Solving for R_{E2},

$$R_{E2} = \frac{e_{RE2}}{i_{E2,Q}} = \frac{2}{5.09 \times 10^{-3}} = 363 \text{ ohms}$$

X_{CE2} should be about 40 ohms at 100 cps.   Solving for C_{E2},

$$C_{E2} = \frac{159 \times 10^{-3}}{4 \times 10^1 \times 1 \times 10^2} \cong 40 \ \mu\text{f}$$

$R_i \cong h_{ie} \cong 1{,}400$ ohms, and

$$e_{B\text{-}E,Q} = i_{B,Q}R_i = 90 \times 10^{-6} \times 1.4 \times 10^{3}$$
$$\cong 0.126 \text{ volt}$$

Let $i_{\text{bleeder}} = 100$ μa, then

$$R_{B4} = \frac{e_{RB4}}{i_{RB4}} = \frac{e_{RE,Q} + e_{B\text{-}E,Q}}{i_{\text{bleeder}}}$$
$$= \frac{2.126}{100 \times 10^{-6}} \cong 21.26 \text{ kilohms}$$

and
$$R_{B3} = \frac{V_{cc} - e_{RB4}}{i_{B,Q} + i_{\text{bleeder}}} = \frac{12 - 2.126}{(90 + 100) \times 10^{-6}}$$
$$\cong 52 \text{ kilohms}$$

C_4 should offer negligible impedance to all modulation frequencies; therefore, let $C_4 = 1$ μf or larger.

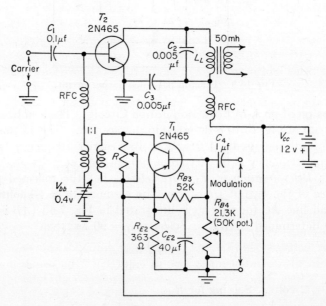

Fig. 13.10   The designed a-m base-modulation circuit.

Step 2: *Determination of T_2 Parameters.*   Recall that T_2 is to operate in class C; therefore, the reverse bias voltage V_{bb} should be large enough to prevent the transistor from conducting for over half of the carrier-frequency amplitude.   This means the selection of V_{bb} determines the

level of the input carrier signal. Also, as revealed in step 1, 0.1 volt
(forward bias) applied to the base results in about 5 ma of collector current
with R_L = 1 kilohm. Considering these factors, the carrier should drive
the base to about 0.2 volt as a maximum.

Over half of the carrier cycle should take place between $e_{B\text{-}E}$ = 0
(cutoff) and V_{bb} (see Fig. 13.11) for class C operation. Let ⅔ of the
carrier cycle take place in the cutoff region (causing the angle of collector

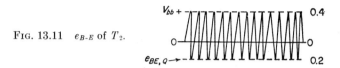

FIG. 13.11 $e_{B\text{-}E}$ of T_2.

current to be about 120°), thereby requiring V_{bb} = 0.4 volt and the
peak-to-peak carrier potential = 0.6 volt (0.21 volt rms).

The input resistance of T_2, when it is conducting, is approximately
equal to h_{ie}, as stated in step 1, which is 1,400 ohms. Since R_i of T_2 is not
too different from R_o of T_1, a 1/1 transformer ratio can be used.

The tank circuit is identical to that designed in the collector-modulation
circuit; i.e., with L_1 = 50 mh, C_2 = 0.005 μf. Let C_3 = 0.005 μf; and
C_1 = 0.1 μf. The r-f chokes should have very low resistance.

Step 3: Circuit Construction and Test. After the circuit is assembled,
the waveforms and percentage of modulation can be easily analyzed with
the oscilloscope. The bias level for the base of T_2 can be varied by
altering the magnitude of V_{bb}, and the bias level of T_1 can be adjusted
by R_{B4} (which should be a 50-kilohm potentiometer).

13.10 A-M Grid Modulation. Figure 13.12 illustrates a control-grid
modulation circuit. As the title of the circuit implies, the modulation
and the carrier are mixed in the control-grid circuit of the carrier amplifier.

V_1 is the modulation amplifier and is operated in the class A mode.
Cathode bias with some degeneration is utilized; i.e., C_k is selected so that
it serves only as a partial bypass to the modulation variations. It
should be pointed out that the modulation amplifier can be a push-pull
type of circuit, operated in class A, AB, or B. A push-pull modulation
amplifier is incorporated in those cases where larger modulation power is
required.

Fixed bias for the carrier amplifier is preferred over the grid-leak type of
bias with control grid modulation. This fixed bias is denoted as E_{cc} in
Fig. 13.12. C_2 is to serve as a low impedance path for the carrier frequencies but should offer substantial opposition to the modulation
frequencies. E_{cc} is so selected that class C operation of V_2 is assured.
The modulation voltage variations are delivered to the grid circuit of

V_2, which is thereby made to vary at the modulation rate; and the plate current of V_2 is amplitude-modulated by this action. The tank circuit, which consists of C_3 and L_1, is made to resonate at the carrier frequency. The carrier-frequency variations of the tank circuit will be amplitude-modulated. R_1 is placed across the primary of V_1 in order to minimize the variations in the load felt by the modulator amplifier. R_1 should be of such a value that it presents a lighter load to the modulator than would be the case if V_1 were functioning with a normal output rather than

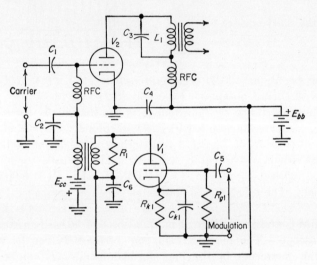

FIG. 13.12   A-m grid-modulation circuit.

into the grid circuit of the class C amplifier. The placement of R_1 in this location tends to reduce distortion of the modulation signal.

C_4 serves as a low carrier impedance path between the bottom of the tank circuit and common, but a high impedance path for the modulation variations. The two radio-frequency chokes (in the grid and plate circuits of V_2) restrict the carrier variations to the V_2 circuits. The transformation ratio of T_1 is usually in the order of 1/1.

13.11   Design of an A-M Emitter-modulation Circuit.

The a-m emitter-modulation circuit of Fig. 13.13 is designed in the same manner as the a-m base-modulation circuit with several changes. The cutoff bias for T_2 is obtained by the self-bias technique (R_{E1} and C_{E1}), and the secondary of the T_1 output transformer is connected to the emitter of T_2. The turns ratio of this transformer can be approximately the same as that used in the base-modulation circuit. The primary of the output transformer is shunted with R, which is adjusted for the best modulation waveform.

R_{B2}, R_{B1}, and R_{E1} can be the same values as R_{B4}, R_{B3}, and R_{E2}, respectively, thereby placing T_2 at the same starting operating point as T_1. All other parameters are determined in the manner stated in the preceding section concerned with the base-modulation circuit.

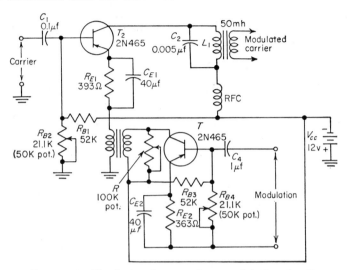

FIG. 13.13   The designed a-m emitter-modulation circuit.

13.12 A-M Cathode Modulation.   A basic a-m cathode-modulation circuit is illustrated in Fig. 13.14. V_2, the carrier amplifier, is operated in the class C mode, as in the previous amplitude-modulation circuits. Fixed grid bias is shown for V_2 in Fig. 13.14, but the use of grid-leak bias

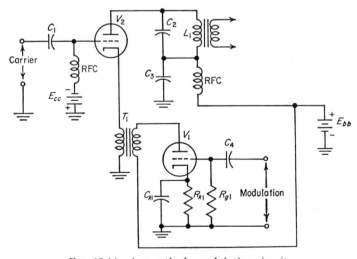

FIG. 13.14   A-m cathode-modulation circuit.

is acceptable for higher percentages of modulation. C_2 and L_1 form the tank circuit which resonates at the carrier frequency. C_3 is to place the tank circuit at common potential for the carrier frequency but offers relatively high impedance to the modulation frequency.

V_1 is the modulation amplifier and may be the push-pull circuit of the class A, AB, or B variety, depending upon the modulation power requirements. Cathode bias is commonly utilized in this stage, when operated in class A.

It should be noted that cathode modulation is actually a combination of grid and plate modulation. The application of the modulation frequency to the cathode circuit via T_1 results in modulating both the grid bias and plate potential of the carrier amplifier.

The output impedance of the modulation amplifier V_1 may be approximated by

$$Z_{o,V1} = m \frac{E_b}{I_b}$$

where m = percentage of plate modulation as a decimal (for example, 50 per cent modulation has m = 0.5).

$E_b = E_{b1,s}$

$I_b = I_{b1,s}$

$Z_{o,V1}$ is the impedance to be seen by the modulation amplifier. Selection of the correct transformation ratio of T_1 will ensure this condition.

A-m modulation can also be achieved by use of pentode tubes. In one case, the modulation and carrier are mixed in the screen grid circuit and is called *screen modulation*. *Suppressor modulation* is also possible by mixing the carrier and modulation in the suppressor grid circuit.

13.13 The Transistor Automatic Frequency Control Circuit.

Figure 13.15 illustrates a transistor automatic frequency control circuit, which can also be used as an *f-m reactance modulator*. The over-all purpose of this circuit is to cause the carrier frequency to vary in direct proportion to the amplitude of the base-emitter voltage. The tank circuit, C_2, C_3, and the 50 mh coil, is made to resonate at the rest tank frequency. The basic oscillator, because of the capacitor voltage-divider arrangement for feedback, is of the Colpitts variety.

Notice that the top of the tank circuit is connected to the base of the transistor by the blocking capacitor C_c and R_1. R_1 is so selected that its ohmic value is many times the reactance of the input capacitance of the transistor X_{Ci}. This enables the rest frequency current in the C_cR_1 branch to be very nearly in phase with the tank voltage. But i_{Ci} leads e_{Ci} by 90°. Since e_{Ci} appears in parallel with $e_{B\text{-}E}$, these two voltages are equal and in phase. Therefore $e_{B\text{-}E}$ lags i_{Ci} by 90°. The collector current is in phase with e_{B-E}.   Therefore, the collector current must lag i_{ci} by 90°, which means the collector current lags the tank voltage by 90°. At this condition, since the tank-circuit current possesses a lagging characteristic, the tank appears inductive.   A change in e_{B-E} creates a corresponding change in the lagging collector current, which effectively varies the inductive characteristic "seen" by the tank circuit.   In the circuit of Fig. 13.15, where a PNP transistor is utilized, a positive swing of the e_{B-E} signal reduces the forward bias of the transistor base circuit,

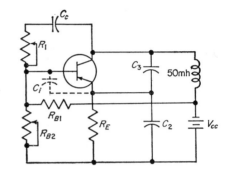

FIG. 13.15   A transistor afc circuit.

resulting in a reduction of collector current.   A reduction in the lagging collector current is "seen" as a reduction in the inductance across the tank circuit, and the frequency of the tank circuit is increased.   A larger positive pulse at the base would decrease the collector current even more, further reducing the inductance "seen" across the tank, which would increase the tank circuit resonant frequency even more.   When the e_{B-E} signal undergoes a negative pulse, the lagging collector current increases, which appears as more inductance across the tank circuit. This change results in a reduced resonant frequency of the tank circuit.

In conclusion, it is seen that the collector current is made to lag the tank voltage by the introduction of C_c and R_1 into the circuit.   The lagging collector current appears as an inductance to the tank circuit, thereby altering the resonant frequency of the tank.   The modulation-amplitude variations that are introduced in the base circuit can be made to vary the lagging collector current, which effectively varies the inductance "seen" across the tank circuit.   In this way, the e_{B-E} variations alter the carrier frequency of the tank circuit and automatic frequency control or frequency modulation is achieved.

The e_{B-E} variations have a more pronounced effect upon the carrier-frequency change by increasing the ratio of L_1/C_T and by decreasing the ohmic value of R_1.   A well-regulated V_{cc} is required to ensure that no collector-current variations are produced by a fluctuating supply voltage. R_{B2} and R_{B1} form the conventional voltage-divider bias network for class

A starting.   Refer to Sec. 7.15 for a detailed analysis of the Colpitts oscillator.

13.14 Design of a Transistor Automatic Frequency Control Circuit.
The transistor afc circuit of Fig. 13.16 is designed in this section.   The

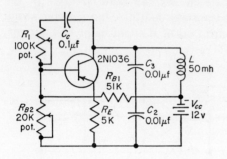

FIG. 13.16   A designed transistor afc circuit.

2N1036 transistor is used. $V_{cc} = 12$ volts; the rest carrier frequency = 10 kc.

Step 1: *Construction of the Basic Oscillator.*   The Colpitts type of oscillator is here used and is designed in accordance with the criteria set down in steps 1 to 8 of Sec. 7.15.   The basic oscillator is the designed Colpitts circuit of Fig. 7.33.

Step 2: *Determination of C_c and R_1.* R_1 should be large as compared to X_C of the transistor input capacitance (which is unknown in this case). R_1 is therefore determined empirically by using a 100-kilohm potentiometer. C_c should present negligible reactance to the carrier frequency. Let $C_c = 0.1$ μf.

Step 3: *Circuit Construction, Test, and Evaluation.*   Upon construction of the designed circuit, R_{B2} and R_1 are adjusted for optimum operation.

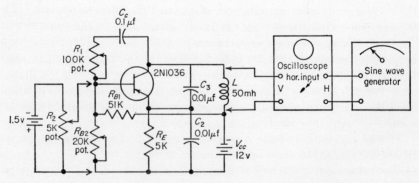

FIG. 13.17   Circuit for evaluation of the effect of e_{B-E} changes upon the tank-circuit carrier frequency.

With the oscilloscope connected across the tank circuit, the amplitude and waveforms of the circuit can be brought to the most desirable condition by the variations of R_{B2} and R_1.

To evaluate the effect of $e_{B\text{-}E}$ changes on the tank-circuit frequency, the circuit of Fig. 13.17 can be connected. Change $e_{B\text{-}E}$ by one-half volt or so; check the change in the tank-circuit frequency by use of the oscilloscope and a generator connected as shown in Fig. 13.17. The use of Lissajous patterns is a convenient method of frequency determination. In the designed circuit, a 200-cps frequency variation for a $0.5e_{B\text{-}E}$ change was obtained.

13.15 The Vacuum-tube F-M Reactance Modulator. A pentode f-m reactance modulator is illustrated in Fig. 13.18. C_1 and L_1 form a tank

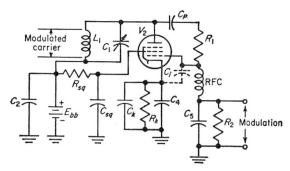

Fig. 13.18 A pentode f-m reactance modulator.

circuit which is made to resonate at the rest carrier frequency. The top of this tank circuit is coupled to the control grid of V_2, the modulator tube, by way of the d-c blocking capacitor C_p and R_1. The bottom of the tank is at common potential in terms of the carrier frequency, as is the cathode of V_2. Therefore, the control grid of V_2 is effectively across the tank circuit. The value of R_1 is made many times larger than X_{Ci} (the reactance of the input capacitance of the tube). In this way, the carrier-frequency current in the C_pR_1 branch i_{ci} is just about in phase with the tank voltage. Recalling the relationship between current and voltage of a capacitor, i_{ci} leads e_{ci} by 90°. The carrier plate current is in step with the control-grid voltage. But the control-grid voltage, which is e_{ci}, lags i_{ci} by 90°; therefore, i_p lags i_{ci} by 90°. This results in i_p lagging 90° behind the tank voltage. At this condition, then, a lagging plate current flows, and the tank appears inductive.

The modulation frequency is also applied to the control grid but is isolated from the carrier frequency by the radio-frequency choke. The control-grid voltage variations produced by the modulation cause the tube transconductance to change with the same variations.   The changes in tube transconductance produce corresponding variations in the lagging plate current of the modulator tube.   A reduction in the amplitude of the plate current increases the tank frequency, and an increase in plate current magnitude decreases the tank frequency.   This varying plate current, which is always lagging the tank voltage by 90°, appears as a variable inductance across the tank circuit, thereby varying the resonant frequency of the tank circuit as a function of the modulation signal amplitude.   Therefore, frequency modulation is achieved by the reactance modulator circuit of Fig. 13.18.

The transconductance of the modulator tube has much to do with the sensitivity of the modulator (the effect of a given grid voltage change on the variation of plate-current amplitude).   The circuit is most effective with the incorporation of high transconductance pentodes.   An increase in sensitivity can also be brought about by increasing the ratio of L_1/C_1 and by reducing the value of R_1.   Since a change in plate potential also causes a change in the plate current, which will create a frequency variation, a very well-regulated E_{bb} is required.   The screen grid voltage may be obtained by use of a screen-dropping resistor R_{sg} and a bypass capacitor C_{sg} or by use of a separate voltage-divider tap from the power supply. C_2 places the bottom of the tank circuit at common potential for the signal frequencies.   Cathode bias is used for establishing the zero signal conditions for the modulator tube.

The modulator tube can be made to appear as a capacitor across the tank circuit by making X_{Ci} many times larger than R_1.

13.16 Transistor Mixers and Converters.   Since each radio and television station has its own carrier frequency, a receiver must be tunable over the entire broadcasting range if it is to be capable of receiving any transmitted signal which reaches its antenna.   The received signal is usually small, in the micro- or millivolt range.   Therefore, it must be magnified many times prior to its delivery to the speaker, which means the use of a cascaded-amplifier arrangement.   If each of the amplifiers had to be tuned each time a new station was selected, serious alignment problems would arise.   This problem is greatly minimized by incorporating the principle of mixing.

Let us describe the principle of mixing with the use of Fig. 13.19.   The incoming signal is transformer-coupled to the base of the transistor when C_B and L_B are made to resonate at that frequency, say 1,500 kc. C_E is "ganged" with C_B, so that both capacitors are varied at the same time. C_E and L_E, called the local oscillator, are so selected that its resonant frequency is higher than the incoming signal by a constant difference, say 455 kc.   Hence, with an incoming signal of 1,500 kc, the local oscillator frequency is 1,500 kc + 455 kc = 1,955 kc. The local oscillator delivers this frequency, 1,955 kc, to the emitter while the 1,500-kc signal is being applied to the base.

The application of two signals of unlike frequencies to the base and emitter of the transistor results in a collector current which has four chief frequency components:

1. Input signal frequency
2. Local oscillator frequency
3. Sum of 1 and 2
4. Difference of 1 and 2 (always constant)

Recall that the oscillator frequency is usually higher than the input signal by the same frequency (such as 455 kc in commercial a-m receivers).

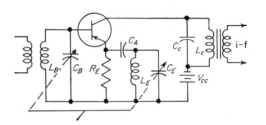

FIG. 13.19   A transistor mixer.

Therefore, the collector current will have one of its components as the constant-difference frequency (with the input signal intelligence) regardless of the station selected! In our earlier example, i_C will have a 455 kc component with the input signal intelligence of any selected station within that broadcast band. The collector tank circuit (C_C and L_C) is designed to resonate at the intermediate frequency i-f.

The local oscillator and the collector tank circuit should appear as low impedance to the signal frequency in order to minimize the problem of degeneration. Furthermore, the local oscillator and the base-tuned circuit should appear as low impedance to the i-f for the same reason. Selection of the transistor is greatly determined by its alpha-cutoff frequency, which must be above the maximum input frequency to be handled by the transistor. High beta transistors are also recommended for efficient operation. The impedance of the collector tank circuit should be high for the i-f.

In order to enable the circuit of Fig. 13.19 to function, the local oscillator is driven with a signal at the i-f. Since the collector tank circuit resonates at this frequency, a portion of this energy can be taken back to the local oscillator. In this way, a self-oscillating transistor mixer is obtained. The collector-to-local oscillator feedback is most commonly achieved by use of transformer coupling, as shown in Fig. 13.20 (L_o is the primary, and L_E is the secondary).

The term *converter* is closely related to the principle of the *mixer*. The mixer of Fig. 13.20 is also a converter, since the modulated signal frequency was converted from that of the incoming signal to the i-f. Figure 13.21 illustrates a typical mixer section of a transistor radio.

C_2 and the 435-μh coil are tunable to all frequencies in the commercial a-m broadcasting range. C_2 and C_1 are "ganged" so that the local

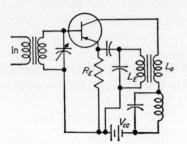

Fig. 13.20 A transistor self-oscillating mixer.

oscillator (C_1 and its coil) is tuned to a frequency which is equal to the incoming signal frequency plus 455 kc. The incoming signal is applied to the base, and the oscillator signal is delivered to the emitter. The 30-turn coil in the collector serves as a "tickler," since it is the primary of the oscillator return transformer (L of the local oscillator is the secondary). The collector tank circuit is a tuned primary circuit, which resonates to 455 kc, the intermediate frequency. The i-f signal is transformer-coupled to the base of the first i-f amplifier.

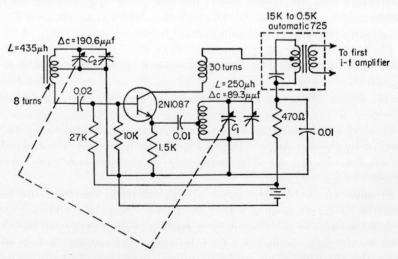

FIG. 13.21 A transistor converter stage. [*Taken with permission from "Transistor Manual" (4th Edition), published by General Electric Co., Charles Building, Liverpool, N.Y.*]

The 1.5-kilohm emitter resistor provides the stability features analyzed in previous circuits. The 470-ohm resistor and 0.01 capacitor in the collector circuit serve as a filter network for the undesirable collector currents. The 10- and 27-kilohm resistors form the popular voltage-divider bias network studied previously. The transistor is commonly biased such that the zero signal collector current is at a low value.

Notice that the oscillator is the common-base type, which is analyzed in Chap. 7. The mixer section, on the other hand, is in the common-emitter configuration, since the incoming signal is applied to the base.

13.17 Vacuum-tube Mixer Circuits.   Figure 13.22 illustrates a mixer or converter circuit. The modulated carrier is delivered to G_3 from a pre-

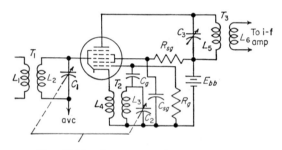

Fig. 13.22   A vacuum-tube mixer circuit.

ceding circuit by the tuned transformer T_1. Therefore G_3 functions as a control grid, and the variations of the modulated carrier will be felt by this electrode. G_1 is also a control grid, and it is energized by the variations of the tank circuit composed of L_3 and C_2. This tank circuit is an oscillator and resonates at a frequency that is always a fixed value above or below the carrier frequency delivered to the second control grid G_3. In other words, the secondary of T_1 and the oscillator tank are gang-tuned, which ensures a constant-difference frequency between the two signals.

The mixer or converter tube performs several functions at the same time. The cathode, G_1, and G_2 serve as a triode for the oscillator tank circuit (L_3 and C_2). The required positive feedback is achieved by transformer T_2, and grid-leak bias is developed by R_g and C_g. G_2 serves as the plate of the oscillator triode. But G_2 is of grid construction and therefore does not capture all the electrons which are streaming toward it with the oscillator frequency variations. Therefore, a substantial portion of this electron flow passes through the G_2 region and is then under the influence of the second control grid G_3. Recall that the G_3 potential is varying at the modulated carrier frequency. The already varying electron stream is subjected to these new voltage variations, which brings out the mixing action. The electrons flow toward the plate under the influence of one of four frequency variations:

1. Modulated carrier frequency
2. Oscillator frequency
3. The sum frequency of 1 and 2
4. The difference frequency of 1 and 2

Number 4, the modulated frequency equal to the difference between the original carrier frequency and the oscillator frequency, is customarily the frequency selected, since it is a fixed frequency with the modulation of the G_3 signal imposed upon it. The significant advantage of acquiring a fixed frequency for all signal carrier frequencies is the simplification of tuning to this output. This fixed-difference frequency, called the *intermediate frequency* i-f, is developed in the plate tank circuit (C_3 and L_5), which is usually the primary of a tuned transformer T_3. These variations are then coupled into a tuned amplifier, called the *intermediate-frequency amplifier*, for amplification. In some instances, more than one i-f stage is incorporated, thereby magnifying the original modulation to the desired amplitude. Immediately following the i-f stage is the detector stage, which may be a diode or one of the triode varieties analyzed in the following sections. The modulation without the carrier (which is the detected signal) is then amplified in one or more stages before delivery to its ultimate load (such as a loud-speaker in the case of a conventional radio receiver).

Automatic Volume Control. Notice that the second control grid G_3 is connected to an automatic volume control (avc) network. Avc is a bias which varies directly with the strength of the incoming signal. When the received signal is strong, this bias increases, thereby reducing the transconductance of the mixer tube. When the incoming signal is comparatively small, the bias is correspondingly reduced, and the transconductance of the mixer tube is not affected as much. Since this type of bias automatically adjusts itself to the incoming signal strength, it tends to make all signals, regardless of original strength, appear at the same average magnitude in the output tank circuit of the mixer stage. This ultimately maintains the volume output of the modulation variations relatively constant and is therefore called automatic volume control.

A typical method for obtaining avc bias potential is from the output of the detector stage, as shown in Fig. 13.23. Since this is the modulation of the original signal, its average magnitude is directly related to the strength of the original signal. These modulation variations are smoothed out by an *RC* filter and delivered to the second control grid of the mixer. In order to increase the effectiveness of avc action, this bias is often applied to the control grid of the i-f amplifier as well.

In Fig. 13.23, the modulation variations are smoothed out by the C_4-R_1 filter network. This results in an average unidirectional voltage appearing across C_4 which is directly proportional to the original modulation signal strength. The left side of R_1 is connected to the grid circuits of those amplifiers which are to receive the bias.

Since this type of bias varies with the incoming signal strength, it can only be effectively used with tubes which display remote cutoff characteristics. The incorporation of avc with sharp cutoff tubes could result in

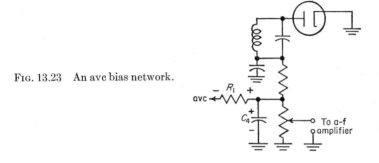

FIG. 13.23   An avc bias network.

"squelched" operation or cutoff when the incoming signal strength is strong.

13.18   Transistor A-M Detector.

Recall that detection is the process by which the intelligence is recovered from the carrier and passed on to the succeeding stage without the carrier (see Sec. 9.12). Detection, in both semiconductor and vacuum-tube circuits, is commonly accomplished by means of the diode in conjunction with an appropriate filter circuit. The leading disadvantage of this technique is that the detector stage does not provide signal gain. By use of a transistor with the proper parameters, the signal is amplified as well as detected.

Figure 13.24 illustrates a possible transistor a-m detector followed by an amplifier stage. L_2 and C_2 resonate at the carrier frequency. C_B and C_L are of such value that they offer very low impedance to the carrier frequency. In this manner, the carrier-frequency variations are bypassed while the intelligence-frequency variations are retained. It is in this manner that the carrier frequency is removed and the modulation signal is retained, thereby resulting in detection.

In observing Fig. 13.2b, it is seen that the intelligence is present in both the positive and negative regions of the carrier. Both are the same intelligence, and only one need be retained in the detection process. This means that only the positive or negative portion of the signal must be retained.

In a class B common-emitter amplifier using a PNP transistor, the negative signal excursions drive the transistor into its active region, while the positive signal swings bring the transistor into its cutoff region. It would therefore seem that biasing the detector amplifier at cutoff in the absence of an input signal would adequately serve the purpose. But it should be recalled that nonlinear distortion is apt to occur if true class B operation is to be utilized; therefore class AB operation is suggested. The base-emitter circuit should have sufficient forward bias to keep the negative signal swings (with the PNP) out of the nonlinear region of its i_C versus $e_{B\text{-}E}$ characteristic. The conventional voltage-divider technique

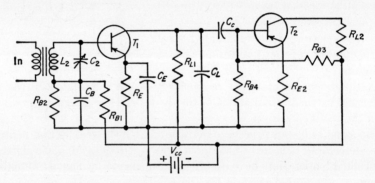

FIG. 13.24   A transistor detector and one amplifier stage.

(R_{B1} and R_{B2} of Fig. 13.24) may be utilized to provide the required forward bias.

R_E may be used in the emitter lead to provide greater linearity and temperature stabilization. The negative feedback aspect of R_E with respect to the desired intelligence frequency may be minimized by the use of a bypass capacitor C_E which offers low impedance to both the carrier and intelligence frequencies. The value of R_E should be kept small (if used at all) since it results in a reduction of gain.

For optimum results, the impedance of the source should be high as compared to R_i of the transistor when the transistor is conducting and low as compared to R_i of the transistor when it is at cutoff. These input resistance values can be approximated by use of

$$R_{i\text{ ON}} \cong \frac{e_{B\text{-}E\text{ ON}}}{i_{B\text{ ON}}}$$

and

$$R_{i\text{ OFF}} \cong \frac{e_{B\text{-}E\text{ OFF}}}{i_{B\text{ OFF}}}$$

In order to ensure high gain, a transistor whose alpha is close to unity should be used. Also, the cutoff frequency of the transistor should be higher than that of the carrier frequency to be handled by the detector amplifier.   The emitter and collector circuits should offer low impedance to the carrier frequency.   The base and emitter circuits should offer low impedance to the intelligence frequency.   The value of R_L should be as high as permissible.

13.19   A-M Plate Detection.   Prior to the study of the detector circuits analyzed in this and the following section, it is recommended that the student review Sec. 9.12, which deals with the basic concepts of a-m detection and the a-m diode detector circuits.

It will be recalled from Sec. 9.12 that detection, also called demodulation, is a two-step process:

1. Rectification of the modulated carrier
2. Retention of the amplitude variations and bypassing the carrier variations

With the diode detector, the rectification of the modulated carrier was performed by the normal action of the diode, and the filter network acted upon the rectified output.

The same two-step process is performed with the triode- and pentode-type detectors.   Figure 13.25 illustrates a triode a-m plate detector. The tube is biased just above cutoff, i.e., in class AB.   The use of cathode bias R_k is a common method for the achievement of the required bias. The zero signal plate current $I_{b,s}$ is usually very small in magnitude (less than 0.5 ma in most cases). C_k is to serve as a bypass for both the carrier and modulation frequencies.   These values are determined in the conventional manner; i.e.,

$$R_k = \frac{E_{c,s}}{I_{b,s}}$$

where $E_{c,s}$ and $I_{b,s}$ are determined by the load-line analysis.   In order to serve as a good bypass capacitor, C_k should be 10 per cent or less than the value of R_k at the lowest modulation frequency, and

$$C_k = \frac{159 \times 10^{-3}}{(X_{C,k} \text{ at } f_{\text{mod,min}})(f_{\text{mod,min}})}$$

By establishing the zero signal operating point at a near cutoff value, the triode will function as a rectifier for all practical purposes, in that only the positive portion of the modulated carrier delivered to the grid will be reproduced in the plate circuit.   By proper biasing, the portion of the negative part of the modulated carrier that is reproduced in the plate can be made negligible.   The filter circuit, designed to bypass the carrier variations to common but retain the amplitude variations (which is the modulation), consists of C_2, C_3, and RFC in the plate circuit. C_2 and C_3 should offer negligible reactance to the carrier frequency, thereby shunting these variations away from the load resistor R_L. The RFC, which offers high reactance to the carrier frequency, encourages the carrier variations to select the C_2 to common path. C_3 supplements this filtering action and increases its effectiveness. It should be pointed out that the carrier-frequency filter circuit is sometimes not as elaborate as the pi arrangement shown in Fig. 13.25. In many circuits, only C_2 is utilized, with RFC and C_3 not included. In such cases, C_2 serves as a carrier-frequency shunt across R_L and E_{bb}, and the use of the proper capacitor value still results in reasonably effective filtering.

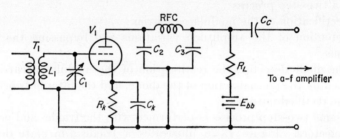

FIG. 13.25   A triode a-m plate detector.

Because of the filtering action, only the modulation variations of the positive portion of the original modulated carrier signal appear across R_L. These variations are coupled by C_c to the input of an amplifier, which magnifies the modulation variations. In Fig. 13.25, this is shown as the input of an audio amplifier.

Since the process of rectification and separation of carrier and modulation is performed in the plate circuit of the tube, this circuit is called a *plate detector*. The plate detector circuit is also known as a *grid-bias detector*.

Design Criteria for the Plate Detector. In determining the parameters of the circuit, the load line is determined, for all practical purposes, by R_L. The load line is selected in the conventional manner, taking the appropriate measures to ensure linear operation during the conduction portion of the input cycle. When such a load line is determined, the required R_L value is computed by

$$R_L = \frac{E_{bb}}{I_{b,sc}}$$

where $I_{b,sc}$ = the intersection of the load line and the plate-current axis.

The zero signal operating point is then selected (near cutoff), from which $E_{c,s}$ and $I_{b,s}$ can be determined. With this information, R_k and

C_k are computed in accordance with the relationship stated in the beginning of this section.

The maximum permissible input signal is determined by, and is equal to $E_{c,s}$, since the grid should not be driven positive in the interest of keeping the distortion down to a minimum.

The filter network can then be designed by setting X_{C2} and X_{C3} equal to 10 per cent or less of R_L at the carrier frequency.   RFC is selected so that its reactance at the carrier frequency is very high as compared to X_{C2}.

One of the chief advantages of the plate detector is its ability to handle a large input signal.   The chief disadvantages are its high distortion and its relative insensitiveness to small input signals.

13.20   A-M Grid-leak Detection.   Figure 13.26 shows one possible a-m grid-leak detector circuit.   Notice that there is no zero signal grid bias,

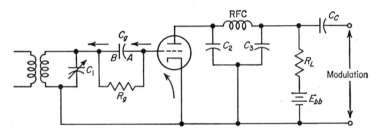

FIG. 13.26   An a-m grid-leak detector circuit.

since grid-leak bias is used.   The grid-leak action of R_g and C_g causes the cathode-grid portion of the tube to function as a diode, thereby rectifying the input modulated carrier. C_g is to offer a low reactance to the carrier variations (X_{C_g} is often no larger than $0.1R_g$ at the carrier frequency). This permits C_g to have a short charging time constant (see arrows for charging path) and a long discharge time constant (through R_g).   The carrier frequency is filtered in the plate circuit (C_2, RFC, and C_3) as in the plate detector circuit. R_L and the filter circuit are determined in the same manner described in the preceding section.

The chief disadvantage of the grid-leak detector is its relatively high distortion, but it has the significant advantage of being sensitive to small input signals.

PROBLEMS

13.1   What are the two states of electric energy in a radio wave?

13.2   Refer to Fig. 13.1.   State two methods in which the direction of the radio wave can be reversed.

13.3   What is the relationship between the frequency and wavelength of a radio wave?

13.4 The carrier frequency of a certain radio station is 910 kc. Determine its wavelength.

13.5 The wavelength of a certain transmission is 200 m. Find the frequency.

13.6 What determines the frequency of an electromagnetic wave?

13.7 What is the plane of polarization of a radio wave?

13.8 State the relationship between microvolts stress per meter and the magnetic flux intensity of a radio wave.

13.9 In what frequency class is commercial a-m broadcasting?

13.10 What is modulation?

13.11 What is detection?

13.12 State the relationship between Probs. 13.6 and 13.7.

13.13 Refer to Fig. 13.2. f_c = 1,280 kc. Find the sideband frequencies when the modulation frequency is:

(*a*) 100 cps   (*b*) 500 cps

(*c*) 1,000 cps   (*d*) 5,000 cps

13.14 In Prob. 13.13, find the bandwidth for *a* to *d*.

13.15 Refer to Fig. 13.2. f_c = 910 kc; m = 0.5; modulation frequency = 1,500 cps; E_{max} = 4 volts. Find:

(*a*) time of 1 cycle T_p

(*b*) carrier voltage at $0.2T_p$

(*c*) upper- and lower-sideband voltages at $0.2T_p$

13.16 Repeat part *c* of Prob. 13.15 with m = 0.9.

13.17 State the relationship between sideband amplitude and m.

13.18 In Prob. 13.13, state the relationship between sideband frequency and bandwidth.

13.19 What happens to an amplitude-modulated wave when m exceeds 1? Explain.

13.20 Refer to the designed a-m collector-modulation circuit of Fig. 13.6. L = 20 mh; f_o = 20 kc. Find C.

13.21 In the circuit of Fig. 13.6, assume R_{E2} = 50 ohms; carrier frequency = 20 kc. Find C_{E2}.

13.22 In the a-m collector-modulation circuit considered in Probs. 13.20 and 13.21, let V_{cc} = 15 volts; $e_{C\text{-}E}$ = 7 volts; $e_{B\text{-}E,Q}$ = 0.9 volt; $i_{C,Q}$ = 6 ma; $i_{B,Q}$ = 500 μa; $i_{bleeder}$ = 200 μa. Find:

(*a*) $e_{RE2,Q}$ (*b*) e_{RB2}

(*c*) R_{B2} (*d*) R_{B1}

13.23 In Fig. 13.6, what class of operation is used for T_2? Explain.

13.24 Refer to the a-m collector-modulation circuit considered in Probs. 13.20 to 13.23. Let $e_{B\text{-}E,Q}$ = 1 volt; $f_{S,min}$ = 100 cps; $i_{C1,Q}$ = 50 ma; $i_{B1,Q}$ = 1 ma; $i_{bleeder}$ = 1 ma; $e_{RE1,Q}$ = 2 volts. Find:

(*a*) R_{E1} (*b*) C_{E1}

(*c*) R_{B4} (*d*) R_{B3}

13.25 In Fig. 13.8, why is the carrier amplifier operated in class *C*?

13.26 At what frequency is the tank circuit of V_3 tuned in the a-m plate modulator of Fig. 13.8?

13.27 In Fig. 13.8, state the function of V_1 and V_2.

13.28 In Fig. 13.8, what power relationship must exist for 100 per cent modulation?

13.29 What determines the turns ratio of T_1 in Fig. 13.8?

13.30 I_A and I_B in Fig. 13.8 are d-c ammeters. What relationship exists between these readings when the circuit is properly adjusted? Explain.

13.31 Refer to the designed a-m base-modulation circuit of Fig. 13.10. Let V_{cc} = 15 volts; modulation frequency = 100 cps to 20 kc; $i_{C1,Q}$ = 7 ma; $e_{C\text{-}E,Q}$ = 5 volts; $e_{RE2,Q}$ = 1 volt. Find (*a*) R_{E2}; (*b*) C_{E2}.

13.32 In the circuit considered in Prob. 13.31, let $i_{B,Q} = i_{bleeder} = 100$ μa; $s_{B\text{-}E,Q} = 0.8$ volts. Find (a) R_{B4}; (b) R_{B3}.

13.33 State the similarities and differences of the three a-m modulation circuits.

13.34 Refer to the a-m grid modulation circuit of Fig. 13.12. What type of bias is used for the carrier amplifier?

13.35 What determines the turns ratio of the V_1 output transformer in Fig. 13.12?

13.36 What is the purpose of R_1 in Fig. 13.12? Explain.

13.37 Why are the modulation power requirements of Fig. 13.12 less severe than those in Fig. 13.8?

13.38 In the a-m cathode-modulation circuit of Fig. 13.14, where is the carrier actually modulated? Explain.

13.39 What is the relationship between Figs. 13.8, 13.12, and 13.14?

13.40 Refer to the automatic frequency control circuit of Fig. 13.16. State the design criteria for the Colpitts oscillator used in this circuit.

13.41 Explain the purposes of C_c and R_1 in Fig. 13.16.

13.42 State the complete theory of operation of the f-m reactance modulator of Fig. 13.18.

13.43 Explain the role of C_i in the operation of Fig. 13.18.

13.44 Why does the tube appear as a variable inductance across the tank circuit in Fig. 13.18?

13.45 Under what condition could the tube in Fig. 13.18 be made to appear as a variable capacitance across the tank circuit?

13.46 Refer to Fig. 13.19. Assume the incoming signal frequency is 710 kc. Find (a) local oscillator frequency; (b) collector-current frequencies; (c) selected collector-current frequency.

13.47 In Fig. 13.19, assume the incoming signal frequency is 1,360 kc. Find (a) local oscillator frequency; (b) collector-current frequencies; (c) selected collector-current frequency.

13.48 Refer to part c of Probs. 13.46 and 13.47. Explain the relationship.

13.49 Refer to Fig. 13.21. The input signal circuit is tuned to 1,080 kc. Find the value of C_2 at this setting.

13.50 In Prob. 13.49, find the value of C_2 when the input signal circuit is tuned to 600 kc.

13.51 State the two-step process of demodulation.

13.52 State the complete theory of operation of the triode a-m plate detector circuit of Fig. 13.25.

13.53 State the complete theory of operation of the a-m grid-leak detector of Fig. 13.26.

13.54 Make a comparison of triode versus diode detectors, stating the advantages and drawbacks of each type.

13.55 State the complete theory of operation of the mixer circuit illustrated in Fig. 13.22.

13.56 In Fig. 13.22, assume the oscillator tank circuit is to oscillate at frequencies that are 455 kc higher than the incoming modulated signal. What relationship (in size) must exist between C_2 and C_1?

13.57 To what frequency is the C_3,L_5 tank circuit in Fig. 13.22 usually tuned? Why is this so?

13.58 Explain the theory of operation of the avc network illustrated in Fig. 13.23.

CHAPTER 14

PRINCIPLES OF GAS TUBES

This chapter is concerned with the basic principles of gaseous conduction, which includes the volt-ampere characteristics of these devices. Several gas diodes are analyzed in terms of their differences and similarities with regard to each other. The manner in which ionization is developed, with the several techniques used for its development, is treated in some detail. The use of the gas diode for voltage regulation, with several variations, is discussed. The gas triode, called the thyratron, is analyzed along with several of its common applications. The cold-cathode triode is studied, and a typical application is examined. A discussion on the use of gaseous discharge lamps as sources of light is included at the end of the chapter.

14.1 Fundamentals of Gaseous Tube Conduction. In the analysis of the vacuum-tube diode in Chap. 10, it was pointed out that a negative space charge, due to the liberated electrons, exists between the cathode and the plate. The introduction of a controlled amount of certain gases can result in a reduction of this interelectrode space charge. Figure 14.1 reveals the general relationship of this space charge in a typical vacuum diode, free space, and a gas-filled diode.

From Fig. 14.1, it is seen that the space charge in the gas-filled tube is most concentrated in the cathode region, resulting in the development of a high potential between the cathode and a nearby point (such as A in Fig. 14.1) and a low potential across the remainder of the tube. Let us examine the basic operation of a gas tube, which helps to clarify the relationship depicted in Fig. 14.1. Assume a controlled amount of gas (such as mercury vapor, argon, or neon) is introduced into the tube. As in the vacuum diode, the electrons are liberated from the cathode and speed toward the plate. As these electrons travel toward the plate, many of them collide with the gas molecules which are floating about within the cathode-plate space. If the electron-molecule collision occurs with sufficient force, one or more electrons will be liberated from the gas molecule, which then becomes a positive ion. Since the gas ion has an

over-all positive charge, it is attracted toward the cathode.   Because of
its relatively large mass (about 1,850 times that of an electron), the
positive ion moves toward the plate at a relatively low velocity.   The
positive ion nullifies the charge of one electron (if it lost one electron)
in the area, thereby reducing the space-charge effect in that vicinity.

Cathode electron emission in a gas tube may be accomplished in several
ways.   The *cold-cathode* type utilizes the principle that a sufficiently

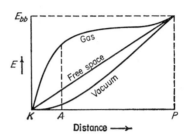

Fig. 14.1 E versus distance
relationships in diodes.

positive anode can actually "pull" electrons out of the cathode material;
the electrons then collide with the gas molecules and trigger the ionization
process.   The *ionic-heated* cathode is constructed of a material which
will emit electrons when it is bombarded by positive ions.   The *heated
cathode* is commonly used, of which there are two types.   The filament
type is preferred for heavy duty gas tubes because of its relative ease in
obtaining a copious supply of electrons and its ability to stand up under
the ion bombardment.   The indirectly heated cathode is generally
restricted to some of the low-current-carrying gas diodes.

The voltage distribution within the gas-filled tube is shown with more
detail in Fig. 14.2.   The *cathode sheath* consists of a congregation of posi-

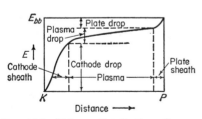

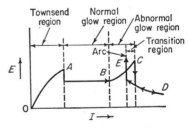

Fig. 14.2   Voltage distribution of a gas-
filled tube as a function of distance from
the cathode.

Fig. 14.3   Volt-ampere characteristics of
a cold-cathode gas-filled tube.

tive ions which surround the cathode.   The *plate sheath* is made up
of a large number of free electrons which surround the plate.   Most
of the plate-supply voltage is dropped across the cathode sheath.   The
plasma is "seen" by the tube as a conductor which extends from the edge

of the cathode sheath to the edge of the plate sheath.   The voltage drop across the plasma and across the plate sheath is relatively small, as shown in Fig. 14.2.

Figure 14.3 illustrates the generalized volt-ampere characteristics of a cold-cathode gas-filled tube.   The operation of the cold-cathode type can be best described in conjunction with the volt-ampere characteristics. As the potential is increased from zero, the current builds up, as shown by region OA.   Since a limited number of ionized atoms are present in the tube before the breakdown (ionization) voltage is reached, this current does develop.   This region is called the *Townsend region*.   At point A, the breakdown voltage has been achieved, and ionization then occurs at a more rapid rate.   Visible electromagnetic radiation occurs at this condition, which is produced by free electrons recombining with the ions and occupying positions in the outer shells.   This is called the *normal-glow* region since only a portion of the cathode area is covered by the glow discharge. It should be noticed that the voltage across the tube remains relatively constant in this region (A to B of Fig. 14.3) for relatively large changes in the current.   The normal-glow region is utilized in gas-filled voltage-regulator tubes.   It should be pointed out that the current flowing in the tube must be limited by a series resistor; otherwise the current will continue to increase and the tube will destroy itself.   Current limiting resistance must be introduced in the gas tube circuit for this reason.   If the series-limiting resistor is decreased sufficiently, the tube current will increase, and the tube will then be in its *abnormal-glow* region (B to C of Fig. 14.3) of its volt-ampere characteristics.   Notice that the constant-voltage characteristic is gone, since the voltage gently increases as a direct function of the current.   The abnormal-glow region is identified by the fact that the cathode is completely covered by the glow of the gaseous discharge.

If the series-limiting resistor is decreased still further, the current will continue to increase.   At a certain magnitude of current (point C), the voltage decreases and the entire discharge is concentrated on a small portion of the cathode.   The tube is now in the *arc region* (points C, D, E). Notice that this is a region of *negative resistance* (increase of current with a decrease of voltage).   The control of the arc relies heavily on the series-limiting resistor.   All of the regions of the volt-ampere characteristic are used in one application or another.

14.2 Types of Gas Diodes.   Vacuum-tube diodes, because of their relatively high internal resistance, develop plate voltages of excessive values when large amounts of current are drawn.   As a result of this limitation, practical vacuum-tube diodes are designed for plate currents of only a few amperes.   It will be recalled from the preceding section that the introduction of a controlled amount of gas within the tube results in a substantial reduction in the internal resistance of the tube because the ionization process effectively neutralizes the bulk of the space charge and its associated effects. The reduction of tube resistance is accompanied by a corresponding reduction in tube power loss, which becomes significant with high values of plate current. For this reason, a number of gas diodes are preferred over vacuum diodes for many high-current applications. Several of these gas diodes are considered in the following sections.

14.3 The Rectigon or Tungar Diode. The rectigon or tungar diode is of the high-current low-voltage type. The gas used is either argon alone, or argon and mercury vapor. A filament-type cathode of heavy tungsten construction is utilized. The gas serves to neutralize the space charge within the tube and reduces the problem of filament vaporization. As the melting point of the tungsten filament is approached, it tends to vaporize, but the pressurized gas retards this tendency to a significant degree. In this way, the cathode current can be allowed to reach much higher values then would be the case if no gas were present.

The maximum peak inverse voltage of this type of tube is relatively low (several hundred volts). Conduction in the reverse direction, called *flashback* or *arcback*, will occur if the maximum peak inverse voltage is exceeded. Conduction can occur in the reverse direction, since ionization can be achieved with the application of potential regardless of its polarity. The gas-diode temperature must be maintained in accordance with the manufacturer's specifications. If the tube is allowed to overheat, the internal gas pressure will be increased, and the maximum peak inverse voltage is reduced. Typical rectigon or tungar gas diodes possess current ratings which range from less than 1 amp up to 20 amp with plate potentials from under 10 volts up to 250 volts.

14.4 Mercury-vapor Rectifiers. The mercury-vapor type of tube is a low-pressure gas diode. Because of reduced gas pressure, this type of gas diode features high voltage and medium current capabilities. The indirectly heated cathode is used. Liquid mercury and mercury vapor at relatively low pressure are utilized. The maximum peak inverse voltage is higher than the previous type, and the voltage drop of the tube is higher by a small amount.

The purpose of the liquid mercury is to maintain the desired vapor pressure within the tube. The liquid mercury evaporates when the tube becomes hot and then condenses in the coolest region of the tube, which is usually the base. The vapor pressure of the tube is regulated by the temperature of this relatively cool area.

The cathode of this type of tube requires preheating. Common mercury-vapor rectifier cathodes must be heated by passing their pre-rated current for periods of 5 sec to 2 min prior to the application of any plate potential. The plate voltage of the tube has much to do with the life of the cathode. With plate potentials of less than 20 volts, the bombardment of the cathode by the ions results in no significant cathode damage. When the plate voltage exceeds this value, the cathode may be damaged, and the tube may require replacement. Parameters for typical mercury vapor tubes are plate current, from less than 1 amp up to 25 amp; maximum peak inverse voltage, 1,500 volts to 20 kv; cathode current, 2 to 40 amp; cathode potential, 2.5 to 5 volts. Twelve volts is the plate potential for typical tubes.

14.5 Mercury-arc Rectifier Tubes. In this type of tube, there is a mercury pool and several electrodes. The fact that the plate is positive does result in a significant flow of current until ionization occurs. The ionization process can be mechanically initiated in several ways:

1. A starting electrode with the appropriate charge is mechanically placed into the pool of mercury. It is then removed, thereby breaking the circuit, which initiates the arc and ionization begins.

2. The unit can be mechanically rocked, which places the pool of mercury in physical contact with the charged electrode. When the mercury pool withdraws from contact with the electrode during its rocking motion, the circuit is broken and ionization is initiated.

When ionization occurs, a specific spot on the surface of the mercury pool becomes the source of electrons. This *cathode spot* can be recognized because it is a little bright region on the mercury pool surface. It is believed that the chief cause of electron emission at the cathode spot is electrostatic attraction; i.e., the electrons are "pulled" out of the pool at that point. The required potential for the establishment of this electrostatic attraction is developed by the cathode sheath of positive ions directly on the surface of the pool. The cathode spot tends to move about the mercury pool surface at a rapid rate. When larger magnitudes of current are made to flow, several cathode spots may be present. Each cathode is capable of delivering at least 10 amp.

It should be noted that the action will stop if the plate voltage drops below some minimum positive value, unless other provisions are made. In applications where the plate potential is varying (such as in the rectification process), it is impractical to resort to one of the mechanical procedures described in a preceding paragraph. Therefore "keep-alive" auxiliary electrodes are used. These auxiliary electrodes enable the ionization process to continue while the plate voltage decreases below the minimum ionization value.

The mercury-arc rectifier is noted for its ability to handle large amounts of power (megawatts) with high efficiency.   The larger units are sometimes water-cooled.

14.6   The Ignitrons.   The introduction of an additional electrode to regulate tube conduction results in the development of the ignitron tube.   As in the mercury-arc rectifier, the mercury pool is the cathode, and the plate is of a graphite material.   A "keep-alive" electrode is used, and there are several ignitor electrodes.   The ignitor possesses high resistance and is in contact with the mercury.   With the passage of current through the ignitor, a substantial amount of heat is developed.   When the heat is sufficiently high, there is a temporary break in the circuit between the tip of the ignitor and the pool of mercury.   When the break is closed, a cathode spot will be formed.   The purpose of the ignitor is only to develop the cathode spot, and this is the instant at which the keep-alive electrode takes over; then the ignitor current falls off.   The keep-alive electrode then maintains the cathode spot until the plate voltage is sufficiently positive to initiate the main arc.   At that point, the keep-alive electrode ceases to pass current.   During the next cycle, the sequence of events is again repeated at the right instant; i.e.:

1. The ignitor develops a cathode spot.

2. The keep-alive electrode maintains the cathode spot until the plate takes over.

3. The plate then develops the main arc.

Ignitrons are temperature sensitive, as are most gas tubes.   Operating temperatures which are too low result in erratic changes in plate current. High temperature, on the other hand, increases the vapor pressure, which reduces the arcback voltage of the tube.   This problem is commonly solved by use of water jackets with regulated temperature values.

14.7   Gas Tube Voltage-regulator Circuits.   The cold-cathode type of gas diode has often been used as a voltage-regulator tube.   A fundamental voltage-regulator circuit is illustrated in Fig. 14.4.

Recall that the normal-glow region of the volt-ampere characteristic of the gas diode (A to B of Fig. 14.3) is a constant-voltage region.   For

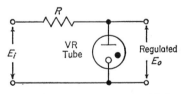

Fig. 14.4   Basic gas-diode voltage-regulator circuit.

relatively large changes in tube current, the tube voltage will be maintained at a constant value.   The voltage-regulator tube is, therefore, rated in terms of this constant voltage and the plate current values within that range. For example, there are vr tubes rated at 150 volts, with a current range of about 5 to 40 ma.

The input voltage to the circuit should preferably be at least 25 per cent greater than the desired regulated input voltage, as this would ensure a sufficiently high value of initial potential to fire the tube into ionization.

The purpose of the series resistance is two-fold:

1. It limits the current to values within the normal-glow region.

2. It provides a voltage drop which varies with changes in the input voltage and gas-tube current, thereby maintaining a constant output potential across the vr tube.

14.8 Design of a Basic Voltage-regulator Circuit. Let us design a circuit similar to that shown in Fig. 14.4. Assume $E_i = 270$ volts, and the desired E_o (regulated) = 150 volts.

Step 1: Selection of a VR Tube. Since the desired regulated output potential is 150 volts, we shall attempt to locate a vr tube with this voltage rating. Assume such a vr tube is found to be commercially available, with a current rating of 10 to 50 ma.

Step 2: Determination of R. Let us arbitrarily select the tube current to be somewhere near the center of its normal-glow range, such as 30 ma in this case. Since R is in series with the vr tube, its current is the same. The voltage drop of the resistor must always be equal to the difference between the input voltage and the vr tube voltage drop. Therefore

$$R = \frac{E_i - E_b \text{ (vr)}}{I_b \text{ (vr)}}$$

Substituting values,

$$R = \frac{270 - 150}{30 \times 10^{-3}} = 4 \text{ kilohms}$$

Step 3: Determination of Allowable Input Voltage Change. If the tube current should decrease to less than 10 ma, the vr tube will be in its Townsend region and will no longer display its constant-voltage characteristic. When the current exceeds 40 ma, the vr tube will be in its abnormal-glow region, and the output voltage will also change as a function of the voltage. Therefore, it is useful to know the minimum and maximum input voltage values that can be permitted while still maintaining a regulated output voltage.

Examining the voltage distribution of the circuit in Fig. 14.4, we find that

$$E_i = E_R + E_b \text{ (vr)}$$

Since R has been determined to be 4 kilohms in the preceding step, $E_{R,\max}$ and $E_{R,\min}$ can be computed by Ohm's law:

$$E_{R,\max} = I_{b,\max}R = 40 \times 10^{-3} \times 4 \times 10^3 = 160 \text{ volts}$$
and $$E_{R,\min} = I_{b,\min}R = 10 \times 10^{-3} \times 4 \times 10^3 = 40 \text{ volts}$$

Restating the preceding voltage relationship in terms of minimum and maximum values,

$$E_{i,\min} = E_{R,\min} + E_b \text{ (vr)}$$
and $$E_{i,\max} = E_{R,\max} + E_b \text{ (vr)}$$

Substituting, these values can be found:

$$E_{i,\min} = 40 + 150 = 190 \text{ volts}$$
and $$E_{i,\max} = 160 + 150 = 310 \text{ volts}$$

In the designed circuit, the input voltage can vary from 190 to 310 volts and still maintain a regulated output potential of 150 volts.

Figure 14.5 illustrates a voltage-regulator circuit which provides a constant current (load) as well as a constant output voltage.

Notice that the load resistance R_L is connected in parallel with the vr tube, which indicates that

$$E_b \text{ (vr)} = E_{RL}$$

As in the previous circuit, the vr tube is operated in its normal-glow region, thereby maintaining a constant voltage across the tube for the

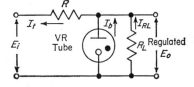

Fig. 14.5 Gas-diode voltage regulator with provision for regulated output voltage and constant current.

current values of that region of its volt-ampere characteristics. Since R_L is a fixed resistance and its voltage drop is held constant by the vr tube action, then the current through R_L is constant.

14.9 Design of the Constant-current Regulator.

Let us design a circuit like that of Fig. 14.5. Assume a fixed load of 50 ma is required;

$$E_i = 270 \text{ volts}$$

Step 1: *Selection of the VR Tube.* Since no load voltage is specified, let us select a vr tube that has the following parameters:

$$\text{Normal-glow region current} = 5 \text{ to } 40 \text{ ma}$$
$$\text{Normal-glow region } E_b = 150 \text{ volts}$$

Step 2: *Determination of R.* The current flowing through the series resistance R is the total current of the vr tube and the load; i.e.,

$$I_R = I_b \text{ (vr)} + I_{RL}$$

Select a value of vr current near the center of its normal-glow region, such as 25 ma. Therefore,

$$I_R = 25 \text{ ma} + 50 \text{ ma} = 75 \text{ ma}$$

The voltage drop of the series resistance can next be found by

$$E_R = E_i - E_b \text{ (vr)}$$

Substituting values,

$$E_R = 270 - 150 = 120 \text{ volts}$$

and

$$R = \frac{E_R}{I_R} = \frac{120}{75 \times 10^{-3}} = 1.6 \text{ kilohms}$$

Step 3: *Determination of Allowable Input Voltage Variations.* Variations of the input voltage which do not reduce the vr tube current below 5 ma or increase it above 40 ma can be permitted. This is so since these values are the normal-glow limits of the vr tube being used, thereby maintaining a fixed potential across R_L, which enables the load I_{RL} to remain constant.

We should first determine the allowable changes in I_R:

$$I_{R,\min} = I_{b,\min} \text{ (vr)} + I_{RL}$$

and
$$I_{R,\max} = I_{b,\max} \text{ (vr)} + I_{RL}$$

Substituting values,

$$I_{R,\min} = 5 \text{ ma} + 50 \text{ ma} = 55 \text{ ma}$$

and
$$I_{R,\max} = 40 \text{ ma} + 50 \text{ ma} = 90 \text{ ma}$$

Since R was determined to be 1.6 kilohms in the preceding steps, the permissible E_R variations can be found:

$$E_{R,\min} = I_{R,\min} R$$

and
$$E_{R,\max} = I_{R,\max} R$$

Substituting values,

$$E_{R,\min} = 55 \times 10^{-3} \times 1.6 \times 10^3 = 88 \text{ volts}$$

and
$$E_{R,\max} = 90 \times 10^{-3} \times 1.6 \times 10^3 = 144 \text{ volts}$$

The corresponding variations of the input voltage can then be determined:

$$E_{i,\max} = E_{R,\max} + E_b \text{ (vr)}$$

and
$$E_{i,\min} = E_{R,\min} + E_b \text{ (vr)}$$

Substituting values,

$$E_{i,\max} = 144 + 150 = 294 \text{ volts}$$

and
$$E_{i,\min} = 88 + 150 = 238 \text{ volts}$$

Therefore the input voltage can vary from 238 to 294 volts and still maintain a fixed load of 50 ma.

14.10   A High-voltage-regulator Circuit.   Figure 14.6 illustrates a technique which can be used when the required regulated output voltage is

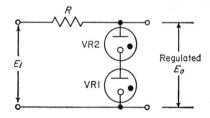

Fig. 14.6   A high-voltage-regulator circuit.

larger than obtainable by use of one vr tube.   Since the vr tubes are in series, the regulated output potential is the sum of the two fixed voltages: E_b (vr1) and E_b (vr2).   Three or more vr tubes in series can also be used for even larger values of regulated output voltages.   The vr tubes utilized in such an arrangement must be similar in current characteristics, since the same current flows (series circuit), and the initial condition should be near the center of the normal-glow region of the volt-ampere characteristics.

The series resistor R can be determined by

$$R = \frac{E_R}{I_R}$$

Expanding,

$$R = \frac{E_i - [E_b \text{ (vr1)} + E_b \text{ (vr2)}]}{I_b \text{ (vr)}}$$

where I_b (vr1) $= I_b$ (vr2), and is near the center of the normal-glow region current values.

14.11 A High-current-regulator Circuit. Figure 14.7 illustrates a circuit which can be used when the desirable current magnitude exceeds values

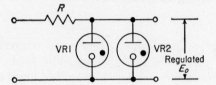

Fig. 14.7 A high-current-regulator circuit.

within the normal-glow region of a single vr tube. By use of two or more vr tubes, the desired current is divided accordingly.

The required series resistor can be computed by

$$R = \frac{E_R}{I_R}$$

Expanding,

$$R = \frac{E_i - E_b \text{ (vr1)}}{I_b \text{ (vr1)} + I_b \text{ (vr2)}}$$

where $E_b \text{ (vr1)} = E_b \text{ (vr2)}$

and $I_b \text{ (vr1)} = I_b \text{ (vr2)}$

14.12 External Current Control of a Gas Diode. The firing of a gas tube may be accomplished by internal or external methods. The ignitor type of electrode discussed in conjunction with the analysis of the ignitron is one technique by which the ionization process is triggered internally. The use of a control grid between the plate and cathode is a second technique for the establishment of ionization, and is discussed in Sec. 14.14. The mechanical rocking of the tube, or mechanically inserting and removing a charged electrode, is a third technique (as in the mercury arc type) which has been used for the ionization trigger.

Figure 14.8 illustrates a gas-filled stroboscope tube. The ionization of this tube is triggered by use of the external electrode, which is shown

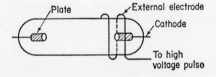

Fig. 14.8 A gas diode with external current control.

wound around the cathode end of the tube. An electrostatic field is established between the external electrode and cathode. A very high voltage pulse (often in excess of 1,000 volts) is applied between the external electrode and cathode. With the use of sufficiently high voltage, the impulse is enough to trigger the ionization process within the tube. During the interval of the brief high voltage, a pulse of intense light is produced. The cathode consists of a pool of mercury in this type of gas diode.

14.13 Magnetic Field Control of a Gas Diode. Ionization triggering can also be performed with the use of an external magnetic field. A brief description of this action is best understood with the aid of Fig. 14.9,

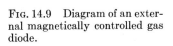
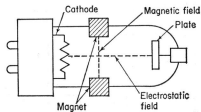

FIG. 14.9 Diagram of an external magnetically controlled gas diode.

which illustrates the general layout of a gas diode whose ionization is triggered by an external magnetic field. The electromagnet is located between the cathode and plate such that its magnetic field is at right angles to the electrostatic field which exists between the positive plate and negative cathode.

Recall that an electrostatic field (as exists between the plate and cathode) urges electrons to move in the direction of that field. The magnetic field, on the other hand, tends to deflect the electrons before they acquire the velocity required to produce ionization. A reduction of the magnetic field strength, then, allows these electrons to again move toward the plate with increasing acceleration. The velocity of a substantial number of once diverted electrons will reach sufficient values to produce the ionization process by collisions with the gas molecules. Therefore, ionization is controlled by the strength of the external magnetic field. Notice that ionization can occur only if the magnetic field strength is reduced to a certain critical value, which is related to the plate potential. With high plate voltages, the critical magnetic field strength value is also higher, and vice versa.

14.14 The Thyratron. The thyratron, as a gas triode may be called, has a control grid as well as the usual plate and cathode. The purpose of the control grid is different from that in vacuum tubes. Ionization triggering is produced by the potential applied between the control grid and cathode. Once ionization occurs, however, the control grid has no influence over the cathode-to-plate current. The magnitude of the current is limited by a series resistor, as was the case with the gas diode. In order to stop the current flow, the plate potential must be reduced to near zero volts, which is called the extinction potential.

The thyratron may be one of two types, which are identified by the polarity of their grid potential:

1. Negative-grid thyratrons
2. Positive-grid thyratrons

In either type, the control grid loses control soon after ionization begins. Let us examine the starting action of the negative-grid type. When the grid is made less negative by the application of a sufficiently positive pulse, cathode-emitted electrons are permitted to pass through the grid

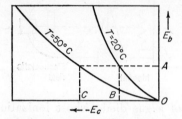

FIG. 14.10 Starting E_c versus E_b characteristics for a negative-grid thyratron.

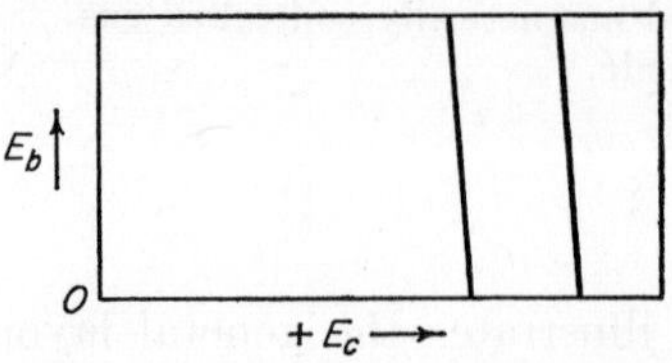

FIG. 14.11 Starting E_c versus E_b characteristics for the positive-grid thyratron.

network and on toward the plate with increasing velocity. As these electrons are accelerated, they will acquire sufficient velocity to initiate ionization by electron-gas molecule collisions. Once the ionization process begins, the positive ions drift toward the negative control grid, thereby forming a control-grid sheath of positive charges. This sheath is of sufficient positive charge to cancel the negative charge of the control grid, thereby prohibiting this electrode from exercising any control upon the current passing through its meshwork.

In the case of most negative control-grid thyratrons, a negative potential must be maintained on the control grid during its periods of desired nonconduction. The magnitude of the negative potential at which ionization occurs is determined in great part by the existing plate potential. The relationship between starting control-grid voltage and plate potential is illustrated in Fig. 14.10.

From Fig. 14.10, it is seen that firing is achieved at a more negative grid voltage for more positive plate voltages, and vice versa. Also illustrated in Fig. 14.10 is the fact that the starting characteristics of the thyratron are affected by temperature. For a given plate voltage (point A), the tube will fire at a larger negative grid voltage (point C) when the tube is at a higher temperature (as compared to point B). In other words, at higher temperatures, a smaller positive pulse at the control grid will fire the tube into conduction.

The starting characteristics of the positive control-grid thyratron are shown in Fig. 14.11. From the characteristics, it is seen that a positive pulse of sufficient magnitude must be applied to the control grid in order to trigger the ionization process. Since this type of thyratron will not conduct until this positive grid potential is achieved, the grid may be placed at zero volts during its nonconducting periods.

14.15 A Thyratron Saw-tooth Oscillator.   A thyratron saw-tooth oscillator circuit is shown in Fig. 14.12a. The operation of this circuit is

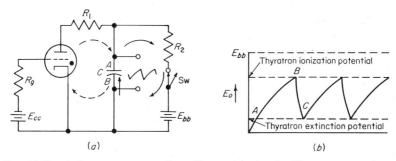

FIG. 14.12   A thyratron saw-tooth oscillator: (a) circuit, (b) output waveform.

most easily described in conjunction with the starting E_c versus E_b characteristics in Fig. 14.10 and the output waveform of Fig. 14.12b.

Upon closing the switch, C will exponentially charge to the magnitude of E_{bb} if the tube does not fire. Assume the grid voltage is at value B of Fig. 14.10. At this value of grid potential, the tube will ionize at E_b of A volts. Therefore plate A takes on electrons and plate B discharges electrons (see solid arrows in the circuit diagram), until e_c is equal to the ionization potential of the thyratron (points A to B in the output waveform diagram). When e_c achieves the potential designated by point B, the thyratron will fire into conduction. When the tube conducts, positive plate A of the capacitor will take on electrons and plate B discharges electrons. The discharge path of the plate B electrons is shown by the broken line arrows in Fig. 14.12a. Since the conducting tube presents very low resistance, the discharging time constant is relatively short, and e_c rapidly decreases. At some low plate potential called the *extinction potential*, the thyratron deionizes (stops conducting). This is shown as point C of the waveform diagram. The discharge path for plate B electrons is no longer open, thereby concluding the discharge action.   Plate A immediately begins to again take on electrons, and the exponential rise of the next saw-tooth pulse begins.

The period time of the saw-tooth voltage can be varied by adjustments in E_{bb}, E_{cc}, R_2, and C. Increasing R_2 or C will increase the charging time constant of plate B, thereby causing the leading edge of the saw-tooth waveform to be less linear. R_2 cannot be made too small, however, for another reason. R_2 should be selected so that the plate B charging current is 1 ma or less. If R_2 is smaller than this critical value, the plate B charging current will be large enough to keep the thyratron in its conduction state. C will remain charged at the voltage drop value of the thyratron, and a d-c output is the result. As a first approximation,

$$R_2 = \frac{E_{bb}}{I_{M,\text{charge}}}$$

where $I_{M,\text{charge}}$ = less than 1 ma.

R_g is the protective grid resistor, and R_1 serves as a protective resistor for the plate. Since plate A of the capacitor charges through R_1 and the tube, R_1 should be kept to the minimum value allowed.

14.16 Design Criteria for the Thyratron Saw-tooth Oscillator. This design technique is developed in conjunction with Figs. 14.10 and 14.12.

Step 1: Determination of E_{bb} and E_{cc}. E_{bb} and E_{cc} are selected from the starting E_c versus E_b characteristic of the thyratron. The values selected are determined by the desired magnitude of the output waveform, since e_c cannot exceed the starting E_b of the tube. Also, selecting a small e_c as compared to E_{bb} results in a more linear output saw-tooth waveform. Keeping these two considerations in mind, the use of the starting characteristics is sufficient for a wise selection of E_{bb} and E_{cc}. Note that the use of a larger E_{cc} will increase the amplitude of e_c.

Step 2: Selection of R_2 and C. R_2 and C are selected such that one period T_p of the output waveform is a fraction of one time constant of the plate B charge. The smaller the fraction of one time constant used, the more linear the leading edge of the output waveform. And yet R_2 cannot be made too small because the capacitor charging current must be kept to less than 1 ma. Therefore,

$$R_2 = \frac{E_{bb}}{I_{M,\text{charge}}}$$

where $I_{M,\text{charge}}$ is less than 1 ma.

C can be determined once the desired fraction of one time constant has been selected. For example, assume the period time is to be 0.1 tc; i.e.,

$$T_p = 0.1 \text{ tc} = 0.1 R_2 C$$

and

$$C = \frac{T_p}{0.1 R_2}$$

where T_p is the desired period time of the output voltage.

Step 3: Determination of R_1 and R_g. R_1, for all practical purposes, is a series-limiting resistor. Its value is selected such that the maximum current of the tube will be within its safe limits; i.e.,

$$R_1 = \frac{E_{b,\text{ON}}}{I_{b,\text{max}}}$$

where $E_{b,\text{ON}}$ = ionization potential of the thyratron

$I_{b,\text{max}}$ = maximum plate current

R_g is selected in the same general manner. It should be sufficiently large to prohibit the grid current from exceeding a specified safe value, usually indicated by the manufacturer.

$$R_g = \frac{E_{cc}}{I_{c,\text{max}}}$$

where E_{cc} = grid voltage supply

$I_{c,\text{max}}$ = maximum grid current

14.17 A Thyratron Trigger Circuit. When accurate precision firing of a thyratron is desired, the circuit illustrated in Fig. 14.13 warrants con-

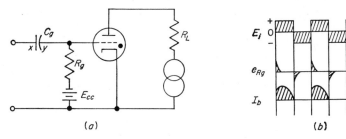

Fig. 14.13   A thyratron trigger circuit: (a) circuit, (b) waveforms.

sideration. The input voltage E_i applied between the input terminals can be the square-wave output of a multivibrator. The plate supply is a sinusoidal source which is in phase with E_i. The square wave is then differentiated by the C_g-R_g network. In order to differentiate effectively, the time constant of this network should be short as compared to the pulse time of the square wave. A convenient rule of thumb is to make this time constant about one-tenth the square-wave pulse time. With such a time constant relationship, the waveform across R_g is as shown in Fig. 14.13b.

When a positive pulse is applied to terminal A, plate y of the capacitor takes on electrons, and plate x discharges electrons. This current instantaneously rises to a maximum value which is limited by the size of R_g. By proper design, this positive potential is sufficient to fire the tube into conduction, since the plate voltage is also positive at this instant. Since the time constant of C_g is short, its current quickly decreases to zero by the time a fraction of the positive pulse time has passed. This places the control grid at zero potential for the remainder of the positive pulse. The thyratron will conduct for most of the positive pulse.

When the negative input pulse is applied, plate x of C_g takes on electrons and plate y discharges, which develops a voltage across R_g and places the control grid at a negative potential. The tube cannot be fired at this time because the plate potential is also negative.

The waveforms of this circuit are shown in Fig. 14.13b. Notice that the differentiated waveform e_{Rg} has a sharp leading edge, making it an ideal trigger.

14.18 Cold-cathode Gas Triodes. As implied by its title, the cold-cathode gas triode utilizes an unheated cathode. The control element, which may be called the starter anode, is physically closer to the cathode than to the plate. Therefore, ionization occurs between the cathode and starter anode at a much lower potential than that required for plate-to-cathode ionization. The gap between the cathode and control anode is first ionized; then the plate-to-control anode gap becomes ionized if the plate is at the correct potential. The required plate-to-cathode voltage for this transfer of ionization is determined by the current in the cathode-control anode circuit (called the transfer current).

Figure 14.14 illustrates the relationship between the transfer current and the plate-to-cathode ionization voltage. When there is no transfer

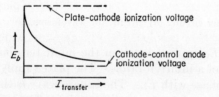

Fig. 14.14 Cold-cathode gas triode transfer characteristic.

current, there is no ionization action in the gap between the control anode and the cathode, thereby requiring the maximum plate-to-cathode voltage for tube conduction. With increasing values of transfer current, lower plate-to-cathode voltages are required for firing the tube. The exact contour of this characteristic is determined by structure of the tube elements, type of gas utilized, and the pressure at which the gas is maintained in the tube. Notice that the minimum plate-to-cathode voltage cannot be less than the sustaining voltage of the cathode-to-control anode gap.

14.19 A Cold-cathode Gas Triode Saw-tooth Oscillator. Figure 14.15a illustrates a cold-cathode gas triode saw-tooth oscillator. The theory of operation is somewhat similar to that of the thyratron saw-tooth oscillator. E_{cc} and R_g are selected to establish the desired magnitude of transfer current (usually in microamperes). Once the magnitude of the transfer current is selected, the required plate-to-cathode firing voltage is also established (see the transfer characteristics of Fig. 14.14).

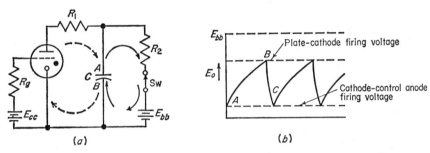

(a)   (b)

FIG. 14.15   A cold-cathode gas triode saw-tooth oscillator: (a) circuit, (b) output waveform.

This also determines the maximum output voltage (point A in Fig. 14.15b).

Assume that C is initially uncharged. Upon closing the switch, plate B of the capacitor takes on electrons in the path shown by the solid line arrows. Therefore e_c exponentially increases until the voltage value of point B is achieved, which is the plate-cathode ionization potential. At this instant, the ionization is transferred to the control anode to plate gap, and the tube conducts. Once the tube conducts, plate A takes on electrons and e_C decreases to point C of the waveform diagram, at which time the ionization between the control anode and plate ceases. But the ionization between the control anode and cathode is sustained by E_{cc}. The charging path of plate A is designated by the broken line arrows. When the plate current ceases, plate B of the capacitor begins to take on electrons again, thereby repeating the action.

14.20 Gaseous Discharge Lamps. Lamps of the gaseous discharge type are more efficient than the type which utilize incandescent filaments.

The so-called "neon" signs are glow lamps which contain an inert gas under pressure. The electrodes are not heated in this type of glow lamp, thereby requiring a high ionization potential. Once ionization is achieved, however, the potential needed to maintain conduction is relatively low (about 150 volts per ft). The majority of the light made available by this type of glow tube originates from the plasma of the tube. The color of the light is determined by the type of gas; helium glow is yellow; mercury blended with neon and argon has a blue glow. The use of colored glass tubing makes it possible to obtain other colors.

Mercury-vapor lamps under high pressure are an efficient source of light. The arc is started by a second gas (such as argon or neon). The establishment of the arc causes the temperature of the electrodes to increase sufficiently so that they will emit electrons. The high temperature causes the mercury to vaporize. The chief drawback of the mercury-vapor lamp is the required high pressure.

Fluorescent lamps, when used with alternating voltage, have an emitter at both ends of the unit. With one voltage polarity, the emitter on the left side serves as the cathode, and the emitter on the right side functions as the plate. Upon reversal of the line voltage, the two emitters exchange roles. A thin layer of fluorescent material is placed on the inside walls of the tube which emits light when bombarded. A high starting voltage is required for operation of the lamp, which may be obtained by utilizing the inductive surge of a starting device or a step-up transformer.

Sodium-vapor lamps operate with gas at relatively low pressure. The filament is oxide-coated. The metallic sodium vaporizes as the lamp temperature increases. Because of the low pressure of the sodium vapor, neon gas is added at a higher pressure. In actual operation, the filament is heated first. Since the neon is at a higher pressure, most of the filament-emitted electrons strike the neon atoms, giving off an orange-red color. Once ionization is established, the emitted electrons succeed in ionizing more of the sodium atoms. As an increased number of sodium atoms become ionized, the glow color slowly changes to yellow.

PROBLEMS

14.1 Describe the manner in which ionization can be developed in the gas-filled diode.

14.2 Why does the ion move toward the cathode in a gas-filled diode?

14.3 Describe the manner in which ionization reduces the space charge in the gas-filled diode.

14.4 Name and describe three ways in which cathode electron emission can be achieved.

14.5 What does the cathode sheath consist of, and where is it located in a gas tube?

14.6 What does the plate sheath consist of, and where is it located in a gas tube?

14.7 What is the plasma in a gas tube?

14.8 Describe the Townsend region of the volt-ampere characteristics of a gas tube.

14.9 Describe the normal-glow region of the volt-ampere characteristics of a gas tube.

14.10 In what region does the tube voltage remain relatively constant? Explain.

14.11 Describe the abnormal-glow region of the volt-ampere characteristics of a gas tube.

14.12 How can the abnormal-glow region be identified?

14.13 What region of the volt-ampere characteristics displays negative resistance? Explain.

14.14 State two advantages of the gas diode over the vacuum-tube diode.

14.15 What is flashback?

14.16 What precautions must be taken to avoid flashback?

14.17 Describe the effect of temperature on the behavior of a gas diode.

14.18 What are the voltage and current capabilities of low-pressure gas diodes?

14.19 State the purpose of liquid mercury in a mercury-vapor rectifier.

14.20 Why must the cathode of a mercury-vapor rectifier be preheated?

14.21 What is the preheating time required for mercury-vapor rectifiers?

14.22 State two ways in which the ionization process can be started in a mercury-arc rectifier tube.

14.23 In the mercury-arc rectifier, what causes electron emission at the cathode spot?

14.24 Why are "keep-alive" auxiliary electrodes sometimes used in mercury-arc rectifiers?

14.25 Describe the role played by the ignitor in the ignitron.

14.26 State the series of events that lead to conduction in the ignitron.

14.27 What type of gas tube is most commonly used in voltage-regulator circuits?

14.28 What region of the volt-ampere characteristics of the gas tube is utilized in voltage-regulator circuits?

14.29 Refer to Fig. 14.4. What is the purpose of R?

14.30 Refer to Fig. 14.4. $E_i = 250$ volts; desired E_o (regulated) $= 150$ volts; vr tube has a voltage rating of 150 volts and a current rating of 5 to 45 ma; center of the normal-glow region is at 25 ma. Find:

    (a) R (b) wattage dissipation of R

    (c) $E_{R,max}$ (d) $E_{R,min}$

    (e) $E_{i,max}$ (f) $E_{i,min}$

14.31 Refer to Fig. 14.5. A fixed load of 40 ma is required; $E_i = 250$ volts; a vr tube is selected which has a normal-glow region current of 5 to 35 ma and a normal-glow region E_b of 150 volts. Assume 20 ma is the vr current at the center of its normal-glow region. Find:

    (a) I_R (b) E_R

    (c) R (d) $I_{R,min}$

    (e) $I_{R,max}$ (f) $E_{R,min}$

    (g) $E_{R,max}$ (h) $E_{i,max}$

14.32 Refer to Fig. 14.6. $E_i = 400$ volts; E_b of each vr tube is 150 volts; I_b of each vr at the center of the normal-glow region is 30 ma. Find R.

14.33 Refer to Fig. 14.6. $E_i = 350$ volts. The vr tubes specified in Prob. 14.32 are to be used. Find R.

14.34 Refer to Fig. 14.7. Two identical vr tubes are to be used; $E_i = 250$ volts; $E_b = 150$ volts; I_b in the center of the normal-glow region $= 30$ ma. Find R.

14.35 Describe the manner in which the gas-filled stroboscope tube of Fig. 14.8 is fired.

14.36 Describe how a gas tube can be fired by use of an external magnetic field.

14.37 State the purpose of the control grid in the thyratron.

14.38 State the two types of thyratrons.

14.39 Why does the control grid of the thyratron lose control after the tube is fired?

14.40 State the relationship between E_c and E_b of a negative-grid thyratron.

14.41 Refer to Fig. 14.12. Describe the action of C when the tube is not conducting.

14.42 Refer to Fig. 14.12. Describe the action of C when the tube is conducting.

14.43 Refer to Fig. 14.12. What portion of the output waveform is developed during the OFF time of the tube and what portion during the ON time of the tube?

14.44 Refer to Fig. 14.13. Assume the input voltage and plate supply are at a frequency of 200 cps. The time constant of R_g and C_g is to be one-tenth the pulse time of E_i; $R_g = 10$ kilohms. Find:

(a) T_p of E_i (b) pulse time of E_i

(c) tc of R_g and C_g (d) C_g

14.45 In Prob. 14.44, assume $R_g = 33$ kilohms. Find C_g.

14.46 In the circuit considered in Prob. 14.44, assume $C_g = 20$ μf. Find R_g.

14.47 Refer to Fig. 14.13. Why is a short time constant used for C_g?

14.48 State the relationship between the transfer current and plate voltage of a cold-cathode gas triode.

14.49 What is the minimum plate potential of a cold-cathode gas triode? Explain.

14.50 Refer to Fig. 14.15. Describe the action of C at the time the tube is not conducting.

14.51 Refer to Fig. 14.15. Describe the action of C during the time the tube is conducting.

14.52 Refer to Fig. 14.15. What portion of the output waveform is developed (a) when the tube is not conducting; (b) when the tube is conducting?

CHAPTER 15

PHOTOELECTRIC DEVICES

The utilization of light as a source of energy to develop and/or to facilitate current flow is considered in this chapter. The characteristics and behavior of photoconductance cells are the first devices to be analyzed, followed by a similar treatment of the photovoltaic cells. Phototubes and gas phototubes are examined as to their behavior. The operation of photosemiconductor diodes and phototransistors, the latest additions to the photoelectric device family, is considered in the last sections of this chapter.

15.1 Fundamental Classification of Photoelectric Devices. The fundamental principle behind all photoelectric devices is that the device converts light energy into electrical energy. The manner in which the electrical energy is made available upon conversion determines the category in which the device is classified. There are three general categories of photoelectric devices:

1. Photoemissive cells
2. Photovoltaic cells
3. Photoconductive cells

The name of each category implies the manner in which the electrical energy is made available. The photoemissive cell is a device in which the impingement of a cathodelike material by light rays produces an emission of electrons. In such cases, the electrical energy is in the current which results from the emission action. The photoemissive cell is a high internal-resistance device.

The photovoltaic cell is a device in which the impingement of certain types of surfaces by light rays produces an electromotive force. This emf, in turn, may be utilized for driving current which is a direct function of the intensity of the impinging light. The photovoltaic cell is often called a barrier-layer cell and is a low impedance device.

Photoconductance cells are devices which undergo a change in resistance as a direct function of the impinging light intensity. This category of photoelectric devices has been substantially enlarged with the advent of the semiconductor photodiodes and phototransistors.

15.2   Photoelectric Terminology.   A number of the more commonly used photometric terms are defined in this section.

International Candle. The basic unit of light intensity (luminous intensity) is the candle. The unit is the light intensity emitted by a certain type of lamp when it is operated under standard conditions.

$$1 \text{ candle } = 4\pi \text{ lumens}$$

Luminous Flux. This term refers to the visible sensation produced by radiant energy.

Lumen. The unit of luminous flux is the lumen.

$$1 \text{ lumen } = \text{ flux within a unit angle from a source of 1 candle}$$

Foot-candle

$$\text{Foot-candles} = \frac{\text{lumens}}{\text{square feet}} \quad \text{for a uniformly illuminated surface}$$

15.3   Cathodes of Photoemissive Cells.   In Sec. 15.1 it was pointed out that photoemissive devices emit electrons when their cathode surfaces are exposed to an impinging light. Recall that light may be considered as either a wave or as consisting of electromagnetic particles called photons. Using the photon concept, the manner in which electrons are emitted from the cathodelike surface is readily explained. Certain metals are such that a significant number of their free electrons are at sufficiently high energy levels in the conduction energy band so that the increment of energy acquired by collisions with the bombarding photons will enable them to break through the surface barrier of the metal. This action is electron emission. Recall that the same outcome was achieved in the conventional vacuum tube by the phonon bombardment of the cathode (application of heat to the cathode). Not all metals lend themselves to this action conveniently at desirable light frequencies. The majority of photoemissive cathodes are made of alkali metals because of their frequency response. Typical photoemissive cathode materials are cesium, lithium, potassium, rubidium, and sodium. The selected metal is often combined in a carefully prepared manner with hydrogen or oxygen.

15.4   Vacuum Phototubes.   Figure 15.1 illustrates the conventional symbol for the vacuum phototube. The cathode is coated with a photoemissive material, as described in the preceding section. The second

electrode in the envelope is the plate, which is placed at a positive potential when in a circuit.   In this way, the cathode-emitted electrons are
drawn to the positive plate.   The cathode is mounted in such a way that
it is easily exposed to the light source from which the photoemissive

FIG. 15.1   Symbol of the vacuum phototube.

action is initiated.   The area between the plate and cathode is evacuated
in the same manner as the conventional vacuum tube.   Since the resistance between cathode and plate is high, the plate current is small in
magnitude.   Increasing the plate voltage from an initial low magnitude
to a higher value results in a steep climb in plate current, and then it
levels off at further increases in plate potential.   The typical volt-ampere
characteristics of the vacuum phototube are shown in Fig. 15.2.

From Fig. 15.2 it is seen that the current is directly affected by the
intensity of the impinging light.   Near point O, only a small amount of

FIG. 15.2   Volt-ampere characteristics
of a vacuum phototube.

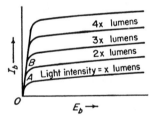

current flows because of the space charge which exists at relatively low
values of plate potential.   As the plate potential is increased between
points O and A, the current rises sharply because the space-charge effect
is being overcome.   Beyond point A, the current levels off, since the
plate is now receiving most of the electrons which are being emitted by
the cathode.   If the impinging light intensity is increased (such as from
x lumens to $2x$ lumens in Fig. 15.2), the quantity of emitted electrons is
increased, and the saturation current (point B) is higher than that found
at the lower light intensity.

Figure 15.2 indicates that the volt-ampere characteristics of the
vacuum phototube are somewhat similar to the conventional vacuum-
tube pentode.   In the region to the right of the knee, the slope of the
characteristics is very small, indicating a high cathode-to-plate resistance.
Since I_b is usually in microamperes, whereas E_b is in volts, the dynamic

plate resistance of the vacuum phototube is generally in the order of megohms.

15.5 Design of Vacuum Phototube Circuits. The vacuum phototube circuit may be designed by use of the techniques utilized in pentode circuits. If R_L is the ultimate load resistance, then

$$R_L = \frac{E_{bb}}{I_b(SC)}$$

where a suitable load is selected so that the variations in light intensity will produce a linear rate of change in I_b. If the output of the phototube

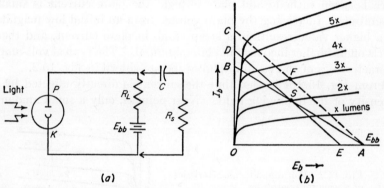

Fig. 15.3 Vacuum phototube circuit: (a) circuit, (b) load line.

is to be RC-coupled, then the considerations of RC coupling must be incorporated. C must be selected such that X_C at $f_{co,\min}$ is low, so that most of the signal variations will appear across R_s. The signal load line must be considered in such cases, where

$$R_{L1,eq} = \frac{R_{L1}R_s}{R_{L1} + R_s}$$

Using E_{bb} as the x-axis intercept, the y-axis intercept (point C) can be calculated from

$$I_b = \frac{E_{bb}}{R_{L1,eq}}$$

Point C and A are then connected with a broken line (see Fig. 15.3b).

The line parallel to CA and which passes through the operating point S is the signal load line.

15.6 A Direct-coupled Phototube Amplifier. Figure 15.4a illustrates a phototube amplifier which is capable of responding to gradual changes in illumination. The power supply E_{bb} is the d-c type. VT1 is a vacuum-tube triode and is biased at or near cutoff when the phototube PT1 is passing only its dark current. This initial bias condition is established by R_1 and R_k. When light is made to impinge upon the cathode of PT1, photocurrent will flow through R_g in the direction indicated by the arrow. Notice that the top of R_g is directly connected to the control grid of VT1. Therefore the flow of photocurrent places the positive voltage drop of R_g across the grid to common terminals. The grid of VT1 is made to change in the positive direction by the initiation of photocurrent. As a result of this action, the plate current of VT1 increases. When the illumination is of sufficient magnitude, the control grid of VT1 voltage

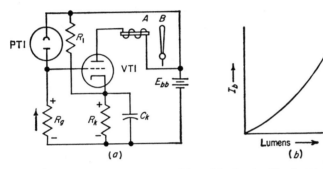

FIG. 15.4 A direct-coupled phototube amplifier: (a) circuit, (b) phototube amplifier characteristics.

excursion will be large enough to cause I_b of VT1 to increase to the value required to activate the relay.

Notice that PT1 and R_g are in series, and

$$e_{PT1} + e_{Rg} = E_{bb} \qquad \text{at all times}$$

When PT1 is not exposed to illumination, only a small value of dark current will flow. At this time,

$$e_{PT1} \cong E_{bb}$$

Therefore, PT1 appears very close to an open circuit in the absence of illumination. When subjected to light, photocurrent is made to flow. The development of this photocurrent causes e_{PT1} to decrease and e_{Rg} to increase. Remembering the simple series relationship, then

$$\Delta e_{PT1} = -\Delta e_{Rg}$$

Since R_g is a resistance of fixed ohmic value, a change in its voltage drop must indicate a change in the resistance in series with it. This series resistance is that of the phototube. Therefore, the resistance of the phototube must decrease as a direct function of illumination. The phototube, in other words, functions as a light-controlled resistance. The OFF condition of the amplifier is the condition of no light. The ON condition occurs when the illumination is sufficient to allow the required relay energizing current to flow in the vacuum-tube plate circuit.

The vacuum tube is used in conjunction with the phototube because the photocurrent is of insufficient magnitude to energize the relay. By using the photocurrent variations for changing the grid potential of the vacuum tube, the plate current of the vacuum tube will be a reproduction of the photocurrent changes. By utilizing the vacuum tube, this current will be large enough to activate the relay. Since the vacuum-tube grid is directly coupled to the phototube, this amplifier is particularly useful in those applications where the changes in illumination are gradual or slow.

15.7 An RC-coupled Phototube Amplifier.

An RC-coupled phototube amplifier is illustrated in Fig. 15.5. The theory of operation of this

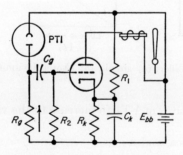

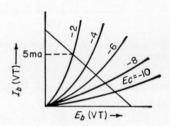

FIG. 15.5 An RC-coupled phototube amplifier.

FIG. 15.6 Load line and E-I characteristics of VT1.

amplifier is fundamentally the same as the direct-coupled version in the preceding section. One of the chief differences is the manner in which the variations of R_g are coupled to the control grid of the vacuum tube. RC coupling is utilized in this case. C_g is selected so that its reactance at the lowest desirable frequency is sufficiently low to allow the variations to be felt by the grid. As in previous RC coupling circuits,

$$X_{C_g} \text{ at } f_{co,\min} \cong R_g + R_2$$

R_g is determined by the magnitude of the photocurrent at which the relay should be energized. R_2 serves as a d-c path to common for the control grid. The photocurrent is often expressed in microamperes because of its low magnitude. Knowing this desired value of photocurrent, along with the required energizing current of the relay, the value of R_g can be determined.

Example: Assume PT1 is to activate the relay when the photocurrent $= 20\ \mu a$; required relay energizing current $= 5$ ma; the vacuum-tube load line is shown in Fig. 15.6; minimum light frequency $= 10$ cps; $e_{Rk} = 10$ volts; $R_2 = 500$ kilohms. Find (a) R_g; (b) C_g.

Solution: When $i_{PT1} = 20\ \mu a$, i_{Rg} also equals $20\ \mu a$, since they are in series. The grid-to-cathode potential is equal to the sum of the R_g and R_k voltages; i.e.,

$$e_{g\text{-}k} = e_{Rg} + e_{Rk}$$

Reading from the load line and characteristics of Fig. 15.6, $e_{g\text{-}k}$ must be -2 volts when I_b is 5 ma (the required energizing relay current). Transposing the preceding equation and substituting values,

$$\begin{aligned} e_{Rg} &= e_{g\text{-}k} - e_{Rk} \\ &= -2 + 10 = 8 \text{ volts} \end{aligned}$$

NOTE: The polarity of each voltage is obtained by the Kirchhoff loop method. Knowing e_{Rg} and i_{Rg}, the required R_g can be computed:

$$\begin{aligned} R_g &= \frac{e_{Rg}}{i_{Rg}} = \frac{8}{20 \times 10^{-6}} \\ &= 400 \text{ kilohms} \end{aligned}$$

Using the preceding relationship,

$$\begin{aligned} X_{Cg} \text{ at 10 cps} &= R_g + R_2 \\ &= 400\text{K} + 500\text{K} = 900 \text{ kilohms} \end{aligned}$$

Now solving for C_g,

$$C_g = \frac{159 \times 10^{-3}}{10 \times 9 \times 10^5} \cong 0.018\ \mu\text{f (minimum)}$$

NOTE: If $f_{co,\text{min}}$ were a higher value, a smaller capacitor would be permissible.

15.8   A Circuit for Limiting Phototube Amplifier Operation between Two Illumination Levels.   In many applications, the phototube amplifier must operate between two levels of illumination, rather than the extreme conditions of darkness and light.   The circuit is illustrated in Fig. 15.7c. In the design of this circuit, it is suggested that a load-line analysis of both the phototube and vacuum tube be utilized.   Following are suggested steps in the design:

Step 1.   The two illumination levels are located on the load line of the phototube characteristics, as illustrated in Fig. 15.7a.   The minimum and maximum phototube current values (points A and B) can then be read off of the y axis. Assume point A is 1 μa and point B is 2.5 μa.

Step 2. Inasmuch as the vacuum-tube plate current is to be near zero until the illumination level increases to $2x$ lumens, I_b of the vacuum tube should be near zero at this condition (shown as point C in Fig. 15.7b).

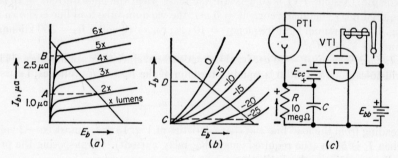

Fig. 15.7 A circuit for limiting phototube operation between two illumination levels: (*a*) phototube characteristics, (*b*) vacuum-tube characteristics, (*c*) circuit.

Therefore a bias of -25 volts (in this case) is required for the vacuum-tube amplifier when the phototube is operating at point A (1 μa). By referring to the circuit of Fig. 15.7c, it is seen that

$$e_{g\text{-}k} = E_{cc} + e_R$$

and

$$E_{cc} = e_{g\text{-}k} - e_R$$

Assume $R = 10$ megohms, then

$$e_R = i_{\text{PT1}}R$$
$$= 1 \times 10^{-6} \times 10 \times 10^6 = 10 \text{ volts}$$

Now the value of E_{cc} can be determined:

$$E_{cc} = e_{g\text{-}k} - e_R = -25 - 10 = -35 \text{ volts}$$

Step 3. The peak vacuum tube current can be determined next:

$$e_{g\text{-}k} = E_{cc} + e_R$$

where

$$e_R = i_{\text{PT1}}R = 2.5 \times 10^{-6} \times 10 \times 10^6$$
$$= 25 \text{ volts}$$

Therefore,

$$e_{g\text{-}k} = -25 + 25 = 0 \text{ volts}$$

This is shown as point D in Fig. 15.7b.

Notice that a relatively small change of the phototube current can create a large change in the vacuum tube I_b. For example, if the vacuum tube I_b changed from 1 to 20 ma for the 1- to 2.5-μa phototube I_b change, then the current change is multiplied by

$$\frac{19 \times 10^{-3}}{1.5 \times 10^{-6}} = 12{,}667$$

In other words, the change in the phototube I_b (in this example) is multiplied by 12,667.

15.9 Sound Reproduction with a Phototube. Figure 15.8 illustrates a phototube sound reproduction circuit. This type of circuit can be used

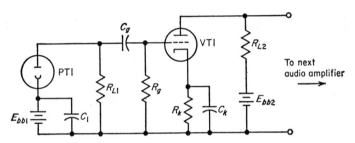

Fig. 15.8 A phototube sound reproduction circuit.

in developing the sound of a motion picture film. The light rays passing through the sound track of the motion picture film are directed to the cathode of the phototube. The variations in light intensity develop similar variations in the photocurrent of PT1. Notice that RC coupling is used between the output of PT1 and the input of VT1. C_g serves as a block for the unidirectional component of the photocurrent but offers negligible reactance to the signal variations. The phototube stage is a transducer because variations in light intensity are converted into identical variations of current and voltage. These signal voltage variations are developed across R_g, thereby causing the grid-to-cathode potential of the first vacuum-tube voltage amplifier to vary in the same manner. In this way, the plate current of VT1 possesses the same variations, and reproduction has been achieved. Succeeding stages of amplification may be used to bring the audio signal up to a desired level.

15.10 Design Techniques for a Phototube Sound Reproduction Circuit.
The design of the circuit illustrated in Fig. 15.8 utilizes the same techniques used for vacuum-tube and transistor RC-coupled audio amplifiers. The signal load line of the phototube is shown in Fig. 15.9.

Following is a simple graphical technique for the construction of the signal load line. Refer to Figs. 15.8 and 15.9.

Step 1: *Construct the D-C Load Line*

Given E_{bb1} and R_{L1},

$$I_{b(SC)} = \frac{E_{bb1}}{R_{L1}} = \text{point } A$$

Select a suitable operating point, such as S in Fig. 15.9.

Step 2: *Construction of the Signal Load Line.* The load resistance "seen" by PT1 is determined by

$$R_{L1,eq} = \frac{R_{L1}R_g}{R_{L1} + R_g}$$

Using the value of $R_{L1,eq}$ and E_{bb1}, determine I_b for this condition:

$$I_b = \frac{E_{bb1}}{R_{L1,eq}} = \text{point } B$$

Join point B and E_{bb1} by a broken line (see Fig. 15.9). The signal load line is parallel to line BE_{bb1} and passes through the operating point S.

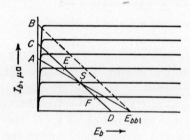

FIG. 15.9 Signal load line of a phototube.

With the use of a straight edge, the signal load line is now constructed. Line CSD is the signal load line.

Step 3: *Determination of VT1 Parameters.* The load line for VT1 is determined in a similar manner, as shown in both transistor and vacuum-tube designs in preceding chapters. The operating point of VT1 is selected so that the plate voltage excursions of PT1 will be in the active region of the VT1 characteristics. Once the operating point is determined, the cathode-bias resistor can be computed:

$$R_k = \frac{E_{c,s} \,(\text{VT1})}{I_{b,s} \,(\text{VT1})}$$

When VT1 is a pentode,

$$R_k = \frac{E_{c,s}}{I_{b,s} + I_{c2,s}}$$

C_k serves the usual function as a signal bypass for R_k. Using the established criterion,

$$X_{Ck} \text{ at } f_{\min} \cong 0.1 R_k$$

and

$$C_k = \frac{159 \times 10^{-3}}{X_{Ck(f,\min)} f_{\min}}$$

15.11 Electron-multiplier Tubes. Figure 15.10 illustrates the theory of electron multiplication as occurs in an electron-multiplier tube. There

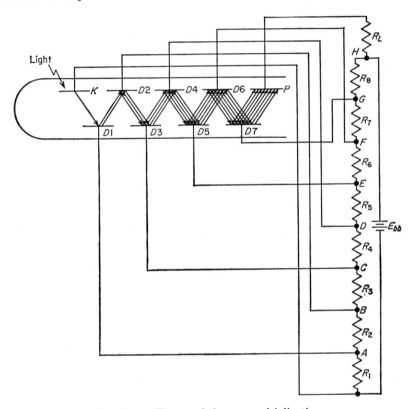

Fig. 15.10 Theory of electron multiplication.

are eight plate structures in Fig. 15.10, and the potential of each is progressively higher. Therefore eight different voltages are required. e_{D1} is obtained from point A of the voltage divider across E_{bb}, e_{D2} from point B, etc. These plate structures are called dynodes. The cathode is coated with a photoemissive material and will emit electrons when illuminated. The cathode is at the common point of the circuit. The principle of secondary emission is utilized in this type of tube.

When the light rays impinge upon the cathode, electrons are emitted.

The emitted electrons are drawn to dynode 1 (D1 in Fig. 15.10), since it is at a positive potential. When these free electrons strike D_1, they bounce off, and other electrons are knocked off of D_1 as well. These electrons are drawn to D_2, where the secondary emission action is repeated. In this way, more and more free electrons are involved in the process. After $D7$, the free electrons are drawn to the plate, which is fed into the load resistor R_L. The number of electrons arriving at the plate is considerably greater than the number emitted by the cathode. This process is called electron multiplication. Up to nine or ten dynodes can be used in such tubes. The gain of an electron multiplier is substantially greater than the conventional phototube.

15.12 Gas Phototubes.

The gas phototube incorporates the principles of gaseous conduction with photoemission of electrons. A small amount

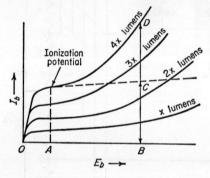

FIG. 15.11 Volt-ampere characteristics of the gas phototube.

of inert gas (such as argon) is contained in the tube envelope. The cathode is exposed to the light radiation, which produces electron emission. The emitted electrons are attracted toward the positive plate, increasing their velocity as they approach the plate. Many of these electrons acquire sufficient acceleration so that they ionize a gas molecule in moving toward the plate. In the meantime, the positive ion wanders toward the cathode, thereby encouraging more electrons to leave the cathode. In this way, the plate current can be made 10 times larger than possible in the vacuum phototube. The *gas-amplification factor* is the ratio of plate current in the gas tube to the plate current that would flow if the gas were not present. In order to safeguard the cathode from destruction by the ions, plate potentials of less than 100 volts are generally used.

Figure 15.11 illustrates the generalized volt-ampere characteristics of the gas phototube. The region between points O and A are somewhat similar to the vacuum phototube, since this is the region of the characteristics at which the ionization potential of the gas has not been achieved, and conduction is solely determined by the photoemissivity of the cathode. At and above the ionization plate potential, the current rises sharply because of the ionization action. This rise in plate current is more pronounced at higher plate potentials and at higher input light intensities. The gas amplification factor can be graphically determined·

$$\text{Gas } \mu = \frac{D - B}{C - B}$$

15.13 Design of Gas Phototube Circuits.
Figure 15.12 illustrates a gas phototube circuit and a load line drawn upon the *E-I* characteristics.

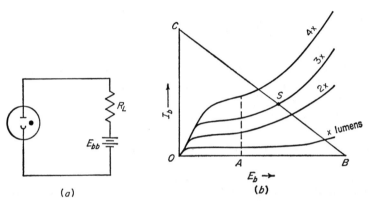

FIG. 15.12   Gas phototube analysis: (*a*) circuit, (*b*) load line.

Since the characteristics are less linear than found in the vacuum phototubes, precautions are usually taken so that the load line intersects the active region of the characteristics to the right of the ionization potential point (*A* in Fig. 15.12*b*). This results in R_L values smaller than used for the vacuum phototube circuits. The nonlinear relationship between input light intensity and I_b is one of the major disadvantages of the gas phototube. The frequency response of the tube is poorer than the vacuum phototube because of the relatively long transit time required by the ions. R_L is determined in the conventional manner:

$$R_L = \frac{E_{bb}}{I_b(SC)}$$

15.14 Photovoltaic Cells.
Figure 15.13 illustrates a basic photovoltaic cell circuit. As pointed out in Chap. 3, semiconductors are photosensitive. When the semiconductor is exposed to light radiation, photons bombard the crystal structure. If the light is of sufficient intensity, the

photons will be large enough to enable a number of valence electrons to break their bonds and become liberated. If the semiconductor is in contact with a metal, the liberated electrons will flow out of the semiconductor into the metal.

The over-all charge of the semiconductor prior to being excited by the light energy is zero, as is also the over-all charge of the metal. When

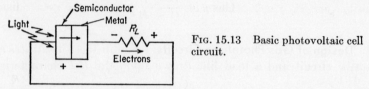

Fig. 15.13   Basic photovoltaic cell circuit.

the mobile electrons leave the semiconductor, the semiconductor side of the junction becomes positive, and the metal side is negative. This is an emf across this junction.

If a complete circuit is provided, as in Fig. 15.13, electron current will flow in the indicated direction. A single cell of this type will generate less than 0.5 volt in bright light.

One drawback of this type of cell is that it can lose its sensitivity to light intensity variations if the cell temperature goes above 60°C or so. At these higher temperatures, the phonons (heat energy) are liberating more electrons than the photons are, and the effect of the light intensity variations is masked by the greatly increased current flow.

In the circuit of Fig. 15.13, R_L is generally very low, thereby allowing currents as high as 1 ma to flow. These cells may be connected in parallel for increased current capacity and in series for increased voltages. The high capacitance of the junction makes it virtually impossible to incorporate this device in modulated light circuits.

15.15 Semiconductor Photodiodes. The generation of free carriers across an unbiased PN junction as a function of light intensity is not significant as compared to the thermal generation of mobile carriers. In order to make the PN junction more sensitive to variations in light intensity, the thermal generation of mobile carriers must be kept to a negligible value. This can be easily accomplished by applying reverse bias to the PN junction.

A basic PN junction photodiode circuit is illustrated in Fig. 15.14. Notice that the polarity of V_{cc} is such that the junction is under reverse-bias condition. In this condition, no thermal generated forward current will flow across the junction. A small magnitude of reverse current flows, which is the typical reverse current of a reverse-biased semiconductor diode. Since this current flows in the absence of light, it is called the *dark current.*

Let us consider the action of the circuit in Fig. 15.14 when the junction is exposed to light radiation.   As the photons bombard the junction area, a number of minority carriers acquire sufficient increments of energy to permit them to break their covalent bonds and migrate across the junction.   Only those liberated minority carriers in the immediate vicinity of the junction contribute to this flow of *reverse photocurrent*.   Those minority carriers which achieve liberation at relatively great distances

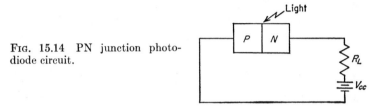

FIG. 15.14 PN junction photo-diode circuit.

from the junction undergo recombination before they have the opportunity to reach the edge of the junction.

It should be emphasized that photocurrent in the PN junction photodiode is a reverse current.   In this way, the thermal effect of carrier generation is minimized, and the bulk of the liberated carriers are a result of the impinging light energy.   The effect of thermal energy upon current flow is restricted to the original dark current, which can be kept down to less than 10 μa under proper reverse biasing.   Because the PN junction is under a reverse-bias condition, the diode impedance is high, generally in the order of 1 to 10 megohms.   The light aimed at the junction may be modulated, and the diode photocurrent will follow the variations in light intensity up to frequencies of 500 kc.

15.16 Phototransistors.   Figure 15.15 illustrates a basic phototransistor circuit.   Let us consider the manner in which photocurrent is generated within the transistor.   The radiant energy is focused at the collector-base junction, which is reverse-biased by V_{cc}.   The emitter-base junction is under the influence of forward bias, which is achieved by the voltage-divider network (R_{B2} and R_{B1}) in Fig. 15.15.   Recall that this type of bias network is incorporated in many of the circuits in Chaps. 5 to 9.

FIG.   15.15   Phototransistor circuit.

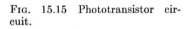
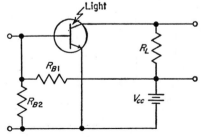

Prior to the introduction of the radiant energy, a certain amount of current flows in the circuit, as determined by the magnitude of the forward emitter-base bias and V_{cc}. When the radiant energy is focused on the collector-base junction, electrons and holes are generated in the collector region. The liberated electrons are pulled out of the collector by the positive terminal of V_{cc}, and the holes migrate to the base region. The additional holes in the base region permit a larger number of electrons to move from the emitter across the emitter-base junction to the base, and on to the collector. This action is best understood by analyzing what the additional holes do to the base side of the base-emitter junction. The base side of the base-emitter junction is negative. The introduction of additional holes (which are positive) reduces this junction potential (makes the base side less negative). Reducing the base-emitter junction potential is equivalent to increasing the forward bias of this junction, thereby increasing the current. In the conventional transistor, the emitter-base junction is modulated by the input signal. In the phototransistor, this same junction potential is modulated by the radiant energy via the process just described. Therefore the additional current, which is the photocurrent, follows the variations of the radiant energy applied to the collector-base junction.

The phototransistor is much more sensitive than the photodiode described in the preceding section. The dark current, which is the forward current which flows without the application of radiant energy to the collector-base junction, is in the order of 10 to 100 μa. The transistor impedance can be in the range of 10 to 100 kilohms, and the device can handle modulation frequencies as high as 100 kc.

15.17 Sound Reproduction with a Phototransistor. A phototransistor circuit for the conversion of changing light intensity into electric variations (which are eventually converted to sound) is shown in Fig. 15.16. The beam of light must be carefully directed onto the base of the photo-

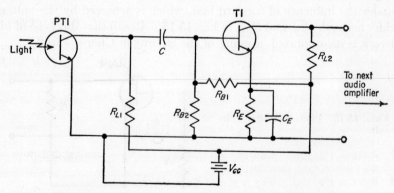

Fig. 15.16   A phototransistor sound reproduction circuit.

transistor. The developed photocurrent is directly proportional to the light intensity. As in the case of a motion film sound strip, these variations of light intensity are the sound variations in a converted form. Therefore, the collector current of PT1 is a reproduction of the changes in light intensity. The variations are RC-coupled to the base of T1, which faithfully amplifies the signal.

The d-c and signal load lines for PT1 are constructed in the manner described in Sec. 15.10. R_E serves to enhance thermal stability, and C_E is a signal frequency bypass. R_{B2} and R_{B1} form the conventional voltage-divider network for the establishment of forward bias for the base of $T1$. Recall that

$$R_{B2} = \frac{e_{B\text{-}E,Q}(T1) + e_{RE,Q}}{i_{\text{bleeder}}}$$

where i_{bleeder} and $e_{RE,Q}$ are selected by the designer.

and

$$R_{B1} = \frac{V_{cc} - e_{RB2,Q}}{i_{B,Q} + i_{\text{bleeder}}}$$

The coupling capacitor C is determined by

$$X_C \text{ at } f_{\min} \cong R_{o,\text{eq}} \,(PT1) + R_{i,\text{eq}}(T2)$$

PROBLEMS

15.1 State the fundamental principle of all photoelectric devices.

15.2 State the three general categories of photoelectric devices.

15.3 State the manner in which electrons are emitted in the photoemissive cell.

15.4 What effect does the impingement of light rays have on a photovoltaic cell?

15.5 What is another name for the photovoltaic cell?

15.6 What effect does the impingement of light rays have on photoconductance cells?

15.7 What is a candle?

15.8 What is a lumen?

15.9 What is a foot-candle?

15.10 Explain electron emission that results from photon bombardment.

15.11 Why is the plate current of a vacuum phototube small in magnitude?

15.12 State the relationship between plate current and the intensity of the impinging light in the vacuum phototube.

15.13 What is the general magnitude of the dynamic plate resistance of the vacuum phototube?

15.14 Refer to Fig. 15.3b. Between what points is the (a) d-c load line; (b) signal load line.

15.15 Refer to Fig. 15.3a and b. State the equation for determination of the (a) d-c load line; (b) signal load line.

15.16 Refer to Prob. 15.15. State the graphical technique for final replacement of the signal load line.

15.17 Refer to Fig. 15.4. State one advantage of directly coupling a phototube to the vacuum-tube amplifier.

15.18 Refer to Fig. 15.4. Why is $e_{PT1} \cong E_{bb}$ in the absence of light?

15.19 Refer to Fig. 15.4. Why does e_{PT1} decrease with the introduction of light?

15.20 In Fig. 15.4a, let R_g = 10 megohms; PT1 dark current = 1 μa; PT1 current when illuminated = 4 μa. Find e_{R_g} at condition of (a) dark current; (b) illumination.

15.21 Refer to Prob. 15.20. Assume $e_{g-k,s}$ of VT1 = -5 volts at dark current condition and I_b of VT1 = 1 ma at that time. Find the following at the dark current condition:
(a) e_{Rk}; (b) R_k.

15.22 Refer to Fig. 15.5 for the RC-coupled phototube amplifier. Assume f_{min} = 5 cps; R_g = 10 megohms; R_2 = 500 kilohms. Find the required C_g (minimum).

15.23 Refer to Fig. 15.5. Assume PT1 is to activate the relay when the photocurrent = 25 μa. The required energizing current for the relay = 10 ma; minimum light frequency = 5 cps; e_{Rk} = 8 volts; R_2 = 470 kilohms; e_{g-k} (when relay is energized) = -1 volt. Find:
(a) e_{R_g} (b) R_g
(c) X_{C_g} at f_{min} (d) C_g (minimum)

15.24 Refer to Fig. 15.7. What determines points A and B in diagram A?

15.25 Refer to Fig. 15.7. State how a small change in phototube current can create a large change in the vacuum tube I_b.

15.26 Refer to Fig. 15.7. Assume a photocurrent change from 2 to 10 μa causes the vacuum tube I_b to change from 0.5 to 12.5 ma. Find the over-all current gain.

15.27 Refer to Fig. 15.8. Describe how the phototube current can be made to vary in accordance with the original sound variations of the film.

15.28 Why is a phototube called a transducer?

15.29 Refer to Fig. 15.8. Let R_{L1} = 5 megohms; R_g = 1 megohm. What is the load resistance seen by PT1 under (a) zero signal condition (direct current); (b) signal condition?

15.30 In Prob. 15.29, explain why the load resistance seen by PT1 with a signal is not the same as that seen by PT1 in the absence of a signal.

15.31 Refer to Fig. 15.8. Let $E_{c,s}$(VT1) = -4 volts; $I_{b,s}$(VT1) = 2 ma. Find R_k.

15.32 In Prob. 15.32, find C_k for f_{min} = 10 cps.

15.33 Refer to Fig. 15.10. Explain why more electrons leave D_1 than arrive at D_1 from the cathode.

15.34 What is electron multiplication?

15.35 State the relationship between the gain of an electron-multiplier tube as compared to a conventional phototube. Explain.

15.36 Describe the operation of a gas phototube.

15.37 What is the gas-amplification factor?

15.38 Refer to Fig. 15.11. Explain the operation of the tube between points (a) O and A; (b) A and B.

15.39 Why is the frequency response of the gas phototube poorer than the vacuum phototube?

15.40 Refer to Fig. 15.13. Upon being excited, the semiconductor side of the junction becomes positive and the metal side becomes negative. Explain.

15.41 State the reason a photovoltaic cell decreases in light sensitivity with increases in temperature.

15.42 What must be done to a PN junction to make it more sensitive to light intensity variations? Explain.

15.43 What kind of current is the photocurrent of the PN junction photodiode?

15.44 Refer to Fig. 15.15. At what junction is the light applied?

15.45 Refer to Fig. 15.15. Describe the development of photocurrent in the phototransistor.

INDEX